Stability and Stabilization of Time-Delay Systems

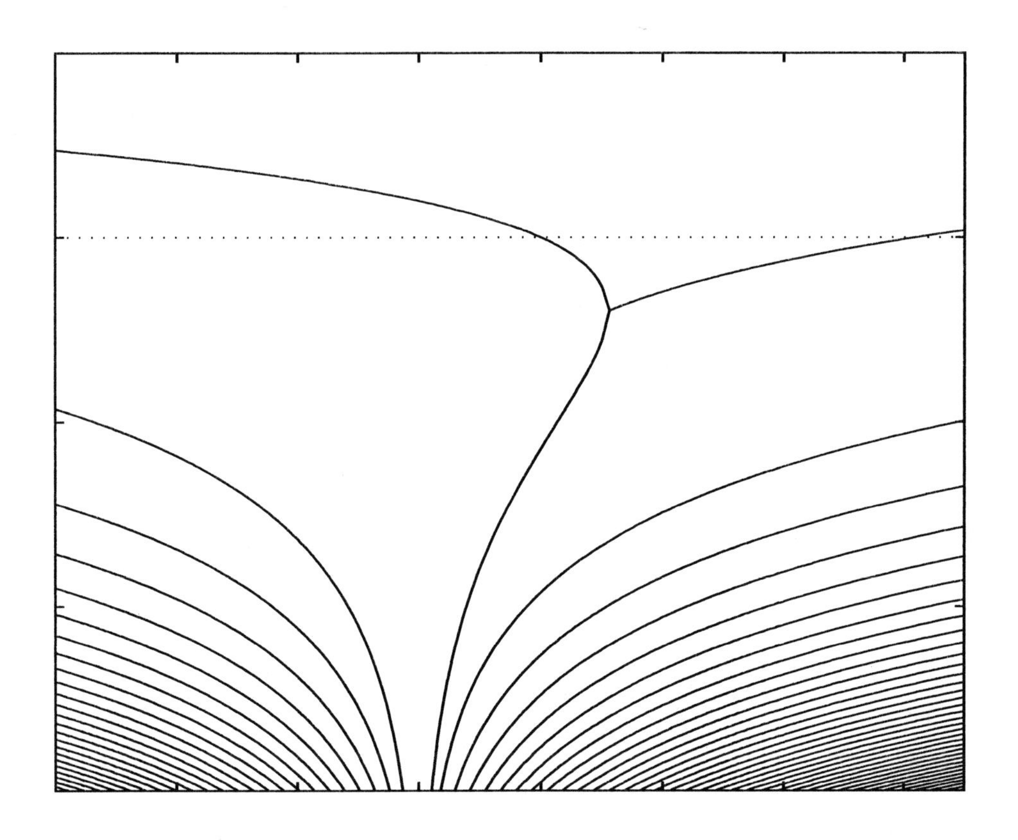

Advances in Design and Control

SIAM's Advances in Design and Control series consists of texts and monographs dealing with all areas of design and control and their applications. Topics of interest include shape optimization, multidisciplinary design, trajectory optimization, feedback, and optimal control. The series focuses on the mathematical and computational aspects of engineering design and control that are usable in a wide variety of scientific and engineering disciplines.

Series Volumes

Michiels, Wim and Niculescu, Silviu-Iulian, *Stability and Stabilization of Time-Delay Systems: An Eigenvalue-Based Approach*
Ioannou, Petros, and Fidan, Baris, *Adaptive Control Tutorial*
Bhaya, Amit, and Kaszkurewicz, Eugenius, *Control Perspectives on Numerical Algorithms and Matrix Problems*
Robinett III, Rush D., Wilson, David G., Eisler, G. Richard, and Hurtado, John E., *Applied Dynamic Programming for Optimization of Dynamical Systems*
Huang, J., *Nonlinear Output Regulation: Theory and Applications*
Haslinger, J. and Mäkinen, R. A. E., *Introduction to Shape Optimization: Theory, Approximation, and Computation*
Antoulas, Athanasios C., *Approximation of Large-Scale Dynamical Systems*
Gunzburger, Max D., *Perspectives in Flow Control and Optimization*
Delfour, M. C. and Zolésio, J.-P., *Shapes and Geometries: Analysis, Differential Calculus, and Optimization*
Betts, John T., *Practical Methods for Optimal Control Using Nonlinear Programming*
El Ghaoui, Laurent and Niculescu, Silviu-Iulian, eds., *Advances in Linear Matrix Inequality Methods in Control*
Helton, J. William and James, Matthew R., *Extending H^∞ Control to Nonlinear Systems: Control of Nonlinear Systems to Achieve Performance Objectives*

Stability and Stabilization of Time-Delay Systems

An Eigenvalue-Based Approach

Wim Michiels
Katholieke Universiteit Leuven
Leuven, Belgium

Silviu-Iulian Niculescu
Laboratoire des Signaux et Systèmes
Gif-sur-Yvette, France

Society for Industrial and Applied Mathematics
Philadelphia

10 9 8 7 6 5 4 3 2 1

Library of Congress Cataloging-in-Publication Data

Michiels, W., (Wim)
Stability and stabilization of time-delay systems : an Eigenvalue-based approach / Wim Michiels, K.U. Leuven, Silviu-Iulian Niculescu.
p. cm. -- (Advances in design and control ; 12)
Includes bibliographical references and index.
ISBN 978-0-898716-32-0 (alk. paper)
1. Automatic control. 2. Time delay systems. 3. Stability. 4. Eigenvalues. I. Leuven, K. U. II. Niculescu, Silviu-Iulian. III. Title.

TJ213.M485 2007
629.8'3–dc22

2007061746

Contents

Preface

The interconnection between two (or more) physical systems is always accompanied by *transfer* phenomena (material, energy, information), such as *transport* and *propagation*. Mathematically speaking, transport and propagation phenomena can be represented by *delay elements*. In this way the corresponding overall systems are governed by a special type of differential equations, namely *delay differential equations* (DDEs).

DDEs are also used in modeling various other phenomena coming from biosciences (heredity in population dynamics [144, 168]), chemistry (behaviors in chemical kinetics [290, 262]), or economics (dynamics of business cycles [291]). Further examples in engineering can be found in [286, 223, 96].

As mentioned by El'sgol'ts and Norkin [71] or Răsvan [255], time-delay systems have a long history and, to the best of our knowledge, the first DDEs are encountered in the work of Bernoulli and Condorcet. However, the theory started to be developed in the second half of the 20th century with the work of the East European mathematical school—Myshkis [219], Krasovskii [142], and Halanay [100] (to cite only a few)—who devoted most attention to the extension of the Lyapunov theory to such class of differential equations. In the 1960s, an increasing interest in the topic appeared also in North America as confirmed by the monographs of Pinney [251] and Bellman and Cooke [13], the first one almost forgotten, with a particular interest in the complex-domain approach and related frequency-domain techniques and methods. Next, the theory arrived to some degree of maturity in the 1970s as proven by the publications and the monographs devoted to the field in that period. Among them, we mention the pioneering work of Hale [101] (the second edition of the monograph published in 1971), which is one of the most cited references in the field for its fundamental results and approaches, but also for its quality and clarity of the presentation. For further references and a deeper historical perspective, we refer to [223, 258, 235].

It is important to point out that various references devoted to time-delay systems in engineering existed even before the 1950s (for example, the papers co-authored by Callender [39, 40] and the editors of the journal *Engineer* [285]), with some contradictory conclusions concerning the effects induced by the delay presence in dynamical systems: sometimes *destabilizing* (mainly by using "huge" gains), and sometimes *stabilizing* (mainly in controlling some oscillatory modes). The explanation of such "dichotomic" behaviors was done case by case, without any attempt for a comprehensive explanation of the situations where stabilizing/destabilizing effects may occur.

Although by now the fundamental results in the theory of functional differential equations (FDEs) are well known and well understood (see, for instance, [13, 101, 106] to cite only a few), the increasing number of applications involving large-scale systems with corresponding complex decision making strategies in which the *delay* (transport, propagation, communication, decision) becomes a "critical" parameter made the development of efficient numerical algorithms and methods for evaluating critical delays and related stability/instability properties necessary. This monograph presents some approaches and techniques in this sense.

Recent approaches in *robust control* opened interesting perspectives and issues in dealing with delays in dynamical systems, where delays are eventually treated as *uncertainty* [96, 223, 20]. Some of them (frequency-sweeping tests, matrix pencil approaches) will be largely discussed in the monograph. Such interpretations of delays as uncertainty are at the origin of an abundant literature in the control area. The corresponding results are expressed in terms of solutions of appropriate Riccati equations [170], linear matrix inequalities [20] in connection or not with the μ-formalism. An exhaustive overview concerning these approaches in the context of stability analysis can be found in [223].

At the same time the increasing number of efficient algorithms for dealing with *nonlinear eigenvalue problems* [179] represented another important issue in treating delay systems. As in the finite-dimensional case, essential properties of time-delay systems (asymptotic behavior, stability, instability, oscillations) are connected with the spectrum location of the corresponding linearized systems. As we shall explain in the next chapters, time-delay systems are infinite-dimensional systems, but with particular spectral properties. Such properties will be explicitly exploited in deriving the main (stability and stabilization) results and related algorithms. In this context, particular attention will be paid to the distinction between *retarded* and *neutral* systems because, although both belong to the class of time-delay systems, their spectral properties are distinct. Most of the approaches presented in this book concern *retarded* delay systems, yet they can be easily extended to the *neutral* case.

It is important to point out that, excepting the FDEs based representation, there are several ways to *represent* time-delay systems—as evolution equations over abstract spaces [15] (infinite-dimensional setting), 2-D systems [158], systems over rings of operators [130], and behavioral based representations [90]. Throughout the volume, we have adopted the FDE based representation. We further assume that the nominal models are completely known. In other words, we do not focus on delay modeling, identification, or identifiability.

Book outline and content

Our intention is to present the stability analysis and synthesis by delayed (state and output) feedback in the linear case by using a *unitary* methodology: the *eigenvalue based approach*. Without any loss of generality we mainly concentrate on the following aspects that, to our best knowledge, have not received a full treatment in the literature:

(a) sensitivity analysis with respect to delays and to other systems' parameters (continuity of the spectrum with respect to the parameters based on Rouché type theorems and variants, pseudospectra, and related properties);

(b) pole placement strategies in stabilization and (nonlinear) optimization of the spectral abscissa function or robustness indicators. Although such approaches are rather classical in the finite-dimensional case, the extensions to delay systems need some special treatment due to the infinite-dimensional nature of the system. However, the particular spectral properties will be helpful to perform such control strategies in both the retarded and the neutral cases, with some precautions in the latter case.

Many examples complete the presentation and illustrate the main results proposed in the monograph. Most of the major ideas are explained by using (several) extremely simple, easy-to-follow (low-order) examples. Finally, the last part of the monograph is devoted to several applications spanning various fields from engineering to biology. All the applications considered start from some generic remarks on the way in which the models are derived, but without any deep discussions on the model derivation and its limitations. The choice of the applications was mainly explained by their impact in engineering, biosciences, and related fields, but also by our own interest in the corresponding topics.

How to read the book?

The book is organized in three parts:

(a) Stability and robust stability. This part deals with the *analysis* of linear time-delay systems from a stability point of view. It starts with an overview of spectral properties of both retarded and neutral systems. To make the fundamental results apparent, eigenvalue plots are used extensively throughout the text. Then the robustness of stability and related problems are studied using pseudospectra and related quantities such as stability radii. The next three chapters deal with the characterization of stability regions in parameter spaces, both qualitatively (shape of regions, etc.) and quantitatively (explicit computational algorithms). Finally, extensions of the presented results for systems with constant parameters to systems with periodically varying parameters are briefly discussed.

(b) Stabilization and robust stabilization. The second part is devoted to the *synthesis* problems that correspond to the analysis problems treated in the first part, with the focus on stabilization. The first chapter is devoted to an eigenvalue based stabilization approach that is inspired by the classical pole placement method for systems without delay. Next, a numerical case study is presented to illustrate how delays in the control loop affect the stabilizability with state feedback. The following chapter is devoted to the robust stabilization problem, and corresponds to the chapter on pseudospectra presented in the first part. Finally, a new stabilization approach is presented which is based on recently developed methods for nonsmooth optimization.

(c) Applications. A wide class of applications is presented, from congestion analysis in high-performance networks to output feedback stabilization and the analysis of predictor-type controllers, from consensus problems in traffic flows to the stability analysis of various delay models in biosciences. We tried to achieve the right correlation between the theory presented in the first two parts of the monograph and the applications which we consider. In some cases, we present several alternative approaches handling the same stability analysis or control problem.

It is important to point out that we have made the parts independent of each other as much as possible. However, a number of fundamental results are needed for the whole theoretical development; these are presented in the first chapter of the monograph. Since such results can be found in excellent references devoted to the theory of FDEs, we decided to only mention them here, and to pay more attention to some particular approaches and related methodologies that have not received full attention in the literature, such as the *sensitivity analysis* (to cite only one approach).

Acknowledgments

The idea of writing this monograph appeared a couple of years ago, but, as usual, the *delay factor* interfered with the whole process. However, we believe that, finally, the delay had a *positive* impact.

We greatly acknowledge the financial support of the Belgian research projects IAP P5, Dyamical Systems and Control: Computation, Identification and Modeling and IAP P6, DYSCO (Dyanamical Systems, Control and Optimization), funded by the program on Interuniversity Poles of Attraction, initiated by the Belgian State, Prime Minister's Office for Science, Technology and Culture, of the Center of Excellence on Optimization in Engineering of the K.U. Leuven, of the Fund for Scientific Research—Flanders, from which WM received a postdoctoral fellowship in the period of writing of the book, and of the French CNRS (National Center for Scientific Research), which covered in part the traveling between Belgium and France in the period 2002–2006. Starting with 2006, a bilateral French–Flemish collaboration project (Tournesol in 2006, and H. Curien in 2007), entitled *Distributed delays in dynamical systems: Analysis and applications*, helped us to continue and to finish the monograph.

Parts of the book have been presented at the European graduate school FAP (Formation d'Automatique à Paris) in the framework of the CTS (Control Training Site) in Automatic Control in 2005 and 2006. The students' feedback was constructive, and helped us to reorganize some material and to present results from a different point of view.

We are grateful to ELIZABETH GREENSPAN from SIAM, Philadelphia, for her help and patience during the preparation of the manuscript.

We would like to thank our friends and past and present collaborators, who implicitly or explicitly made a significant contribution to the research results presented in the book. Among them, we mention DIRK ROOSE, TATYANA LUZYANINA, STEFAN VANDEWALLE, JORIS VANBIERVLIET, KOEN VERHEIDEN (K.U. Leuven, Belgium), KOEN ENGELBORGHS (Materialise, Belgium), PATRICK VANSEVENANT (TVH, Belgium), VINCENT VAN ASSCHE (Université de Picardie, Soissons, France), KURT LUST (Rijksuniversiteit Groningen, the Netherlands), THOMAS PLOMTEUX (BASF, Belgium), LUC MOREAU (SIDMAR, Belgium), DIRK AEYELS (Universiteit Gent, Belgium), RODOLPHE SEPULCHRE (Université de Liège, Belgium), DENIS DOCHAIN (Université Catholique de Louvain, Belgium), ABHIJIT GANGULI (Université Libre de Bruxelles, Belgium), FABIEN CHATTÉ (NEOPOST, France), SABINE MONDIÉ, VLADIMIR KHARITONOV (CINVESTAV-IPN, Mexico), JAIME MORENO (UNAM, Mexico), TOMÁŠ VYHLÍDAL (Czech Technical University at Prague, the Czech Republic), ERIK I. VERRIEST (Georgia Institute of Technology, USA), JEAN PIERRE RICHARD (Ecole Centrale de Lille, France), MICHEL DAMBRINE (Université de Valenciennes, France), FRÉDÉRIC MAZENC (INRA Montpellier, France), DANIEL MELCHOR AGUILAR (IPICYT at San Luis Potosí, Mexico), CHAOUKI T. ABDALLAH (University of New Mexico at Albuquerque, USA),

KIRK GREEN (Vrije Universiteit Amsterdam, the Netherlands), THOMAS WAGENKNECHT (University of Manchester, UK), SONDIPON ADHIKARI (University of Bristol, UK), HENK NIJMEIJER (T.U. Eindhoven), HENRI HUIJBERTS (Queen Mary University of London, UK), VLADMIR RĂSVAN (University of Craiova, Romania), KEQIN GU (Southern Illinois University at Edwardsville, USA), JIE CHEN, PEILIN FU (University of California at Riverside, USA), RIFAT SIPAHI (Northeastern University at Boston, USA), CONSTANTIN-IRINEL MORĂRESCU (University "Politehnica" at Bucharest, Romania), QING-CHANG ZHONG (University of Liverpool, UK), DORON LEVY, PETER S. KIM (Stanford University, USA), and ROGELIO LOZANO (CNRS, Compiègne, France).

Last but not least WM would like to thank BARBARA for the extremely nice time they had together and for her patience in the busy period of finishing the book (*mpf, ik zie je graag!*). He is also thankful to his parents, family, and friends for creating a nice atmosphere and for all the support. Concerning SIN, there is a special person in his life, LAURA, to whom he owes the exceptional support that she gave to Silviu to overcome all the difficulties both professional and extra-professional in the last fifteen years. We dedicate this monograph to all of them, in love and gratitude.

Leuven, Belgium, January 2007	WIM MICHIELS
Gif-sur-Yvette, France, January 2007	SILVIU-IULIAN NICULESCU.

Symbols

A^{-1}	inverse of the matrix A
$A^{\dagger}$	Moore–Penrose inverse of the matrix A
A^*	complex conjugate transpose of the matrix A
$\mathbb{C}$	set of complex numbers
$\mathbb{C}_-$, $\mathbb{C}_+$	open left half-plane, open right half-plane
$\mathrm{clos}(Z)$	closure of set Z
$\mathcal{C}([-\tau,\ 0],\ \mathbb{F})$	space of continuous functions from $[-\tau,\ 0]$ to $\mathbb{F}$
$d(\rho, E), \quad \rho \in \mathbb{C},\ E \subseteq \mathbb{C}$	$\inf_{t \in E} \lvert \rho - t \rvert$
$D(E, F), \quad E, F \subseteq \mathbb{C}$	$\sup_{\rho \in E} d(\rho, F)$
$D_h(E, F), \quad E, F \subseteq \mathbb{C}$	$\max\{D(E, F), D(F, E)\}$, Hausdorff distance
$\det(A)$	determinant of the matrix A
$\mathcal{D}(A)$	domain of the operator A
e	$e := \exp(1)$
ϕ	empty set
I	identity matrix (of appropriate dimension)
I_k	identity matrix of size k-by-k
j	imaginary unit, $j = \sqrt{-1}$
λ	characteristic root
$\Lambda(z)$	matrix pencil with indeterminate z
$\mu_\Delta(A)$	structured singular value of A with respect to the uncertainty set Δ
$\mathbb{N}$	set of natural numbers (zero is not included)
ω	frequency, imaginary part of eigenvalue
$\mathbb{Q}$	set of rational numbers
$\mathbb{R}$	set of real numbers
$j\mathbb{R}$	imaginary axis
$\mathbb{R}_+$	$\{r \in \mathbb{R} :\ r \geq 0\}$
$\mathbb{R}^*, \mathbb{R}^*_+$	$\mathbb{R} \setminus \{0\}$, $\mathbb{R}_+ \setminus \{0\}$
$\Re(\lambda), \Im(\lambda), \quad \lambda \in \mathbb{C}$	real and imaginary part of λ
$\vec{r} \in \mathbb{R}^m, \vec{n} \in \mathbb{N}^m, \ldots$	short notation for $(r_1, \ldots, r_m)$, $(n_1, \ldots, n_m)$,...
$\sigma_i(A)$	ith singular value of the matrix $A \in \mathbb{R}^{n \times n}$, $\sigma_1(A) \geq \sigma_2(A) \geq \cdots \geq \sigma_n(A)$
$\sigma(A)$	spectrum of the matrix or operator A

$r_\sigma(A)$	spectral radius of A
$r_e(A)$	radius of the essential spectrum of A
$\sigma(\Lambda(z)),\ \Lambda = zM + N$	spectrum of matrix pencil Λ, $\sigma(\Lambda(z)) = \{z \in \mathbb{C} :\ \det(zM + N) = 0\}$
$\mathbb{Z}$	set of integer numbers
$\bar{E}$	closure of the set E
$\mathcal{P}(E)$	power set of E, set of all subsets
$\Vert x\Vert,\ x \in \mathbb{C}^n$	Euclidean norm of x
$\Vert x\Vert_p,\ x \in \mathbb{C}^n$	Hölder p-norm of x
$\Vert A\Vert_p,\ A \in \mathbb{C}^{n\times m}$	induced matrix p-norm
$\Vert\phi\Vert_s$	supremum norm of $\phi \in \mathcal{C}([-\tau,\ 0], \mathbb{R}^n)$, $\Vert\phi\Vert_s = \sup_{\theta\in[-\tau,\ 0]} \Vert\phi(\theta)\Vert_2$
$x \geq 0,\ x \in \mathbb{R}^m$	$x_i \geq 0, i = 1, \ldots, m$, where $x = (x_1, \ldots, x_m)$
$\mathcal{B}^m_+$	$\{\vec{r} \in \mathbb{R}_{+^m} :\ \Vert\vec{r}\Vert = 1\}$
$\Vert M(j\omega)\Vert_\infty$	$\mathcal{H}_\infty$ norm of the stable transfer matrix $M(j\omega)$, $\Vert M(j\omega)\Vert_\infty = \sup_{\omega\geq 0} \sigma_1(M(j\omega))$

Acronyms

AQM	active queue management
ARE	algebraic Riccati equation
DDE	delay differential equation
FDE	functional differential equation
FSA	finite spectrum assignment
IQC	integral quadratic constraints
LHP	left half-plane
LMI	linear matrix inequality
MIMO	multiple input, multiple output
NFDE	neutral functional differential equation
ODE	ordinary differential equation
PDE	partial differential equation
RHP	right half-plane
SISO	single input, single output
ssv	structured singular value
TCP	transmission control protocol

Part I

Stability analysis of linear time-delay systems

Chapter 1

Spectral properties of linear time-delay systems

As specified in the Preface, most reactions of real (physical) systems to external actions and signals never take place instantaneously, mainly due to transport and propagation phenomena, and one of the ways to overcome such problems is to include some information on the past in the corresponding model of the systems' dynamics. Such systems are generically called *time-delay systems*.

Roughly speaking, a *time-delay system* is a dynamical system represented by differential equations in some unknown function (and certain of its derivatives) evaluated at arguments which are distributed over some intervals in the past. Differential equations where the right-hand side does not only depend on the state variable at the present time but is a functional evaluated at a solution segment are generally called *functional differential equations* (FDEs) (see, for instance, [101, 106, 52] for the exact definition, and related justifications).

Among the general problems of interest in the theory of time-delay systems, we mention the correct formulation of the initial value problem together with the representation of solutions, and the asymptotic behavior of solutions correlated to the concept of stability.

In this introductory chapter we present several fundamental definitions, properties, and results concerning linear delay differential equations of retarded and neutral type, with the emphasis on their relation with the spectra of appropriately defined operators. Such results, needed in the forthcoming chapters, are presented in a tutorial way. However, some notions and properties essential for the *eigenvalue based approach* considered in this monograph will receive particular attention, and are completed by elementary or sketched proofs. The presentation closely follows standard references in the field of FDEs, such as [13, 100, 101, 106, 67, 52]. Several relatively simple examples and some discussions concerning the computation of the characteristic roots complete the presentation.

The chapter is organized as follows: Section 1.1 is devoted to linear systems of *retarded* type. The initial value problem, the asymptotic growth rate of solutions, asymptotic stability, and some quantitative and qualitative properties of the spectrum are considered. The section ends with some aspects concerning the computation of the characteristic roots. Section 1.2 introduces the same notions in the context of linear systems of *neutral* type, including a complete discussion of the delay sensitivity problem. Some notes together with a list of references end the chapter.

1.1 Time-delay systems of retarded type

We discuss spectral properties of linear time-delay systems of retarded type described by the following delay differential equations (DDEs):

$$\dot{x}(t) = A_0 x(t) + \sum_{i=1}^{m} A_i x(t - \tau_i), \tag{1.1}$$

where $x(t) \in \mathbb{R}^n$ is the state variable at time t, $A_i \in \mathbb{R}^{n\times n}$, $i = 0, 1, \ldots, m$, are real matrices, and $0 < \tau_1 < \tau_2 \cdots < \tau_m$ represent the *time-delays*. The time-delays are *pointwise* or *discrete*, and, hence, they describe the situation where the memory effect is "selective." Such differential equations are sometimes also called differential-difference equations (see [13] for some arguments).

Although in this chapter we will not discuss the case of *distributed delays* over some delay intervals, most of the results presented below can be extended to this class of time-delay systems.

1.1.1 Initial value problem

The initial condition for the time-delay system (1.1) is the function segment

$$\phi \in \mathcal{C}([-\tau_m,\ 0], \mathbb{R}^n),$$

where $\mathcal{C}([-\tau_m,\ 0],\ \mathbb{R}^n)$ is the Banach space of continuous functions mapping the interval $[-\tau_m,\ 0]$ into $\mathbb{R}^n$ and equipped with the supremum norm, $\|\cdot\|_s$.

Due to the linearity of the mapping $f : \mathcal{C}([-\tau_m,\ 0], \mathbb{R}^n) \to \mathbb{R}^n$, defined by

$$f(\phi) := A_0\phi(0) - \sum_{i=1}^{m} A_i\phi(-\tau_i), \tag{1.2}$$

the existence and uniqueness of solutions are guaranteed for all initial conditions.

Furthermore, for a given initial condition the corresponding solution can be explicitly constructed by using the *method of steps*. More precisely, the function ϕ defined on $[-\tau_m,\ 0]$ allows us to define the evolution of the delay system (1.1) on the interval $[0,\ \tau_m]$ as the solution of a standard ordinary differential equation (ODE) of the form

$$\dot{\xi}_1(t) = A_0\xi_1(t) + \sum_{i=1}^{m} A_i\phi(t - \tau_i),$$

with the initial condition $\xi_1(0) = x(0) = \phi(0)$. The next step consists of computing the solution on the interval $[\tau_m,\ 2\tau_m]$ by using the solution ξ_1 defined on $[0,\ \tau_m]$. More precisely, the evolution of (1.1) on the interval $[\tau_m,\ 2\tau_m]$ is given by the solution of the ODE:

$$\dot{\xi}_2(t) = A_0\xi_2(t) + \sum_{i=1}^{m} A_i\xi_1(t - \tau_i),$$

with the initial condition $\xi_2(\tau_m) = x(\tau_m) = \xi_1(\tau_m)$. The solution of the original delay system (1.1) is obtained by considering the collection of all these *pieces of trajectories*

ξ_k defined on $[(k-1)\tau_m, k\tau_m]$, for all positive integers k. Hence, the computation of the solution of (1.1) at some given time-instant $T > 0$ for a given initial condition ϕ reduces to solving q ODEs, where q satisfies the condition $(q-1)\tau_m \leq T \leq q\tau_m$. This process yields a *unique, globally defined forward solution* of (1.1). It is important to note that this solution becomes *smoother* as the time t is increased: if we start with a continuous initial condition on $[-\tau_m,\ 0]$, then the corresponding solution of (1.1) becomes continuously differentiable on $(0,\ \tau_m]$, twice-continuously differentiable on $(\tau_m, 2\ \tau_m]$, and so on. This also implies that a backward continuation for negative time $t \leq -\tau_m$ requires additional smoothness properties of the initial function ϕ defined on $[-\tau_m, 0]$ (see, e.g., [101] and the references therein).

Although in our case the method of steps guarantees the existence and uniqueness of solutions directly by construction, the existence results for general time-delay systems of retarded type are typically proved using fixed point theorems (see [13, 100, 223]).

Example 1.1. *Consider the scalar equation*

$$\dot{x}(t) = x(t) - \alpha x(t-\tau), \tag{1.3}$$

where α and τ are real parameters with $\tau \geq 0$.

If the initial condition ϕ on $[-\tau,\ 0]$ is given by the constant function $\phi \equiv 1$, then the corresponding solution on the interval $[k\tau,\ (k+1)\tau]$ satisfies

$$x(t) = \alpha^{k+1} + (1-\alpha)\sum_{i=0}^{k} \alpha^i f_i(t) e^{t-i\tau},$$

where the functions f_i are given by

$$f_i(t) := \sum_{h=0}^{i} (-1)^h c_{i-h} \frac{t^i}{i!},\quad i = 0, \ldots, k,$$

with the coefficients c_i defined as follows: $c_0 = 1$ and

$$c_i := 1 + i \sum_{h=0}^{i-1} \frac{(i-h)^{h-1}}{h!} \tau^h,\quad i = 1, \ldots, k.$$

Indeed, it is easy to see that the solution on the first delay interval $[0,\ \tau]$ corresponds to the solution of the initial value problem:

$$\dot{\xi}_1(t) = \xi_1(t) - \alpha, \qquad \xi_1(0) = 1.$$

The remaining steps can be performed by recursion.

A particular case of interest corresponds to $\alpha = 1$, for which the solution $x(t)$ reduces to $x(t) = 1$ for all $t \in [-\tau, \infty)$.

In what follows, let us define

$$x(\phi) : t \in [-\tau_m,\ \infty) \to x(\phi)(t) \in \mathbb{R}^n$$

as the unique forward solution of (1.1) with initial condition $\phi \in \mathcal{C}([-\tau_m,\ 0], \mathbb{R}^n)$; that is,

$$x(\phi)(\theta) = \phi(\theta) \quad \forall \theta \in [-\tau_m,\ 0].$$

Then the state at time t is given by the function segment $x_t(\phi) \in \mathcal{C}([-\tau_m,\ 0],\ \mathbb{R}^n)$, defined as

$$x_t(\phi)(\theta) = x(\phi)(t+\theta),\ \theta \in [-\tau_m,\ 0].$$

Denote with $\mathcal{T}(t),\ t \geq 0$ the solution operator, mapping the initial data onto the state at time t,

$$(\mathcal{T}(t)\phi)(\theta) = x_t(\phi)(\theta) = x(\phi)(t+\theta),\quad \theta \in [-\tau_m,\ 0]. \tag{1.4}$$

Sometimes, this solution operator is also called the operator of translation along trajectories [52], since

$$x_{t+t_0}(\phi) = \mathcal{T}(t_0)x_t(\phi), \tag{1.5}$$

for all $t \geq 0,\ t_0 \geq 0$, and $\phi \in \mathcal{C}([-\tau_m,\ 0],\ \mathbb{R}^n)$. The solution operator is a strongly continuous semigroup. The semigroup property ($\mathcal{T}(0) = I, \mathcal{T}(t+s) = \mathcal{T}(t)\mathcal{T}(s)$, for all $t \geq 0, s \geq 0$) follows from the definition and the uniqueness of solutions. A simple proof of the property of the semigroup to be strongly continuous, that is,

$$\forall t \geq 0\ \forall \phi \in \mathcal{C}[-\tau_m,\ 0], \mathbb{R}^n)\ \lim_{s\to t} \|\mathcal{T}(t)\phi - \mathcal{T}(s)\phi\|_s = 0,$$

can be found in [52].

Let $\mathcal{A}$ be the *infinitesimal generator* of $\mathcal{T}(t)$. This operator satisfies

$$\begin{aligned} \mathcal{D}(\mathcal{A}) = \Big\{\phi \in \mathcal{C}([-\tau_m,\ 0],\ \mathbb{R}^n) :\ & \tfrac{d\phi}{d\theta} \in \mathcal{C}([-\tau_m,\ 0],\ \mathbb{R}^n), \\ & \phi(0) = A_0\phi(0) + \textstyle\sum_{k=1}^m A_k\phi(-\tau_k)\Big\}, \\ \mathcal{A}\,\phi = \tfrac{d\phi}{d\theta}, & \end{aligned} \tag{1.6}$$

and allows us to rewrite equation (1.1) as an abstract ODE:

$$\dot{x}_t = \mathcal{A}x_t. \tag{1.7}$$

Summarizing, the same time-delay system can be described in three different ways: by means of a delay (or, more general, functional) differential equation like (1.1), which is of the form

$$\dot{x}(t) = f(x_t),$$

with f defined in (1.2); by means of a mapping like (1.5); or as an abstract ODE over an infinite-dimensional function space like (1.7). In what follows, the choice of the delay system representation will be dictated by the problem under consideration. However, in most of the cases and examples throughout the monograph, we will use the DDE based representation.

1.1.2 Spectrum: definitions

The substitution of a sample solution of the form $e^{\lambda t}v$, with $v \in \mathbb{C}^{n\times 1} \setminus \{0\}$, leads us to the *characteristic equation*, of (1.1),

$$\det \Delta(\lambda) = 0, \tag{1.8}$$

where

$$\Delta(\lambda) := \lambda I - A_0 - \sum_{i=1}^{m} A_i e^{-\lambda \dot{\tau}_i}$$

is the *characteristic matrix*. The left-hand side of (1.8) is called the characteristic function. Similar to the finite-dimensional case, the roots of (1.8) are called the *characteristic roots* of (1.1).

The spectra of the linear operators $\mathcal{A}$, $\mathcal{T}(t)$, and the characteristic roots are related in the following way. The characteristic roots are the eigenvalues of the operator $\mathcal{A}$, which only features a *point spectrum*, that is, $\sigma(\mathcal{A}) = P\sigma(\mathcal{A})$. Furthermore, the algebraic multiplicity of a complex number λ as an eigenvalue of $\mathcal{A}$ is equal to its multiplicity as a zero of the characteristic matrix $\Delta(\lambda)$, while its geometric multiplicity is equal to the dimension of the null space of $\Delta(\lambda)$. An eigenvalue of $\mathcal{A}$ is called *simple* if its algebraic multiplicity is equal to *one*, and *multiple* otherwise. In the multiple eigenvalue case, we will make a clear distinction between *semisimple* and *nonsemisimple* eigenvalues, depending on the relation between the algebraic and geometric multiplicity: a semisimple eigenvalue corresponds to the case where both multiplicities are equal. For $\lambda \in \sigma(\mathcal{A})$, the corresponding eigenfunctions take the form

$$ve^{\lambda\theta}, \ \theta \in [-\tau_m, \ 0], \tag{1.9}$$

where $v \in \mathbb{C}^n \setminus \{0\}$ satisfies

$$\Delta(\lambda)v = 0. \tag{1.10}$$

Sometimes, the vector v is called a *right eigenvector* associated to the characteristic root λ. By similarity, we can construct a *left eigenvector* w, which satisfies $w \neq 0$, $w^*\Delta(\lambda) = 0$, and is related to the eigenfunctions of the adjoint of $\mathcal{A}$ (see [101] for more about the construction of the adjoint space).

The spectra of $\mathcal{A}$ and $\mathcal{T}(t)$ are related by

$$\sigma(\mathcal{T}(t)) = \exp\left(t\sigma(\mathcal{A})\right) \ \text{ plus possibly } \{0\}\,. \tag{1.11}$$

Furthermore, if $\lambda \in \sigma(\mathcal{A})$, then $e^{\lambda t} \in P\sigma(\mathcal{T}(t))$, with corresponding eigenfunctions given by (1.9). Conversely, if $z(t) \in \sigma(\mathcal{T}(t))$ and $z(t) \neq 0$, then there exists some $\lambda \in \sigma(\mathcal{A})$ such that $z(t) = e^{\lambda t}$.

Example 1.2. *Recall the scalar system (1.3), with $\alpha = 2$, and $\tau = 1/3$. In Figure 1.1, we plot the characteristic roots λ which are the eigenvalues of the operator $\mathcal{A}$, as well as the eigenvalues z of the operator $\mathcal{T}(1)$. These eigenvalues are connected by the relation $z = e^{\lambda}$.*

Example 1.3. *For $\alpha = 1$, the characteristic matrix of (1.3) becomes*

$$\Delta(\lambda) = \lambda - 1 + e^{-\lambda\tau}.$$

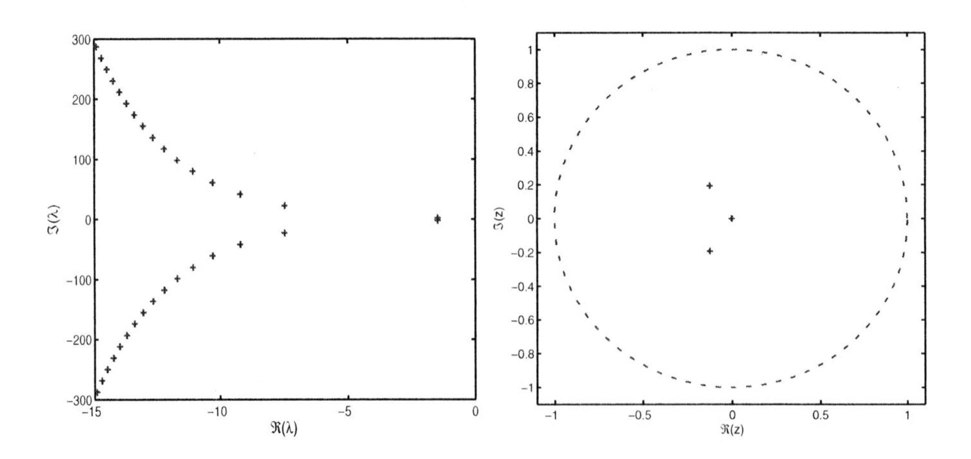

Figure 1.1. *(left) Characteristic roots of the system (1.3) with* $\alpha = 2$ *and* $\tau = 1/3$. *(right) Eigenvalues of the corresponding solution operator* $\mathcal{T}(1)$.

If $\tau = 1$, *then the characteristic root* $\lambda = 0$ *has multiplicity two since* $\Delta(0) = \Delta'(0) = 0$ *and* $\Delta''(0) \neq 0$. *If* $\tau \neq 1$, *then the root* $\lambda = 0$ *is* simple, *since* $\Delta(0) = 0$, *but* $\Delta'(0) \neq 0$. *Furthermore, the characteristic root at zero is* invariant *w.r.t. delay changes; that is,* $\Delta(0) = 0$ *for all positive delays. Such an invariance property will prove its interest in the forthcoming chapters.*

1.1.3 Asymptotic growth rate of solutions and stability

We have the following definitions.

Definition 1.4. *The null solution of (1.1) is asymptotically stable*[1] *if and only if*

$$\forall \epsilon > 0\ \exists \delta > 0\ \forall \phi \in \mathcal{C}([-\tau_m,\ 0], \mathbb{R}^n)\ (\|\phi\|_s < \delta) \Rightarrow (\forall t \geq 0\ \|x_t(\phi)\|_s < \epsilon),$$
$$\forall \phi \in \mathcal{C}([-\tau_m,\ 0], \mathbb{R}^n)\ \lim_{t \to +\infty} x(\phi)(t) = 0.$$

Definition 1.5. *The null solution of (1.1) is exponentially stable if and only if there exist constants* $C > 0$ *and* $\gamma > 0$ *such that*

$$\forall \phi \in \mathcal{C}([-\tau_m,\ 0], \mathbb{R}^n)\ \|x_t(\phi)\|_s \leq Ce^{-\gamma t}\|\phi\|_s.$$

For the case of linear DDEs of retarded type under consideration, exponential stability and asymptotic stability are equivalent. For the relationship between these notions in a more general setting, we refer to the monographs of Halanay [100] and Stépán [286].

[1] For reasons of conciseness we will often use the less precise formulation, "The system (1.1) is asymptotically stable."

The asymptotic behavior of the solutions of (1.1) and, thus, their stability properties are determined by the spectral radius $r_\sigma(\mathcal{T}(1))$. We have, for instance, the following.

Proposition 1.6. *The null solution of (1.1) is exponentially stable if and only if*

$$r_\sigma(\mathcal{T}(1)) < 1,$$

or, equivalently, all characteristic roots of (1.1) are located in the open left half-plane.

Example 1.7. *The system (1.3) with $\alpha = 2$ and $\tau = 1/3$ is asymptotically stable. As shown in Figure 1.1, all characteristic roots are in the open left half-plane. Equivalently, all eigenvalues of the operator $\mathcal{T}(1)$ are located inside the unit circle.*

Throughout this monograph, we will mostly restrict ourselves to the asymptotic stability notion. However, in some of the applications we will need other types of stability notions. These will be defined only in the corresponding parts and related with the stability notions presented above.

1.1.4 Spectrum: qualitative properties

Although the characteristic function of the time-delay system (1.1) is *transcendental* and has an infinite number of zeros, it has some nice, simple, and interesting properties. Some of these properties, which are useful in the forthcoming chapters, are presented below.

Proposition 1.8. *If there exists a sequence $\{\lambda_k\}_{k\geq 1}$ of characteristic roots of (1.1) such that*

$$\lim_{k\to\infty} \mid \lambda_k \mid \to +\infty,$$

then

$$\lim_{k\to\infty} \Re(\lambda_k) \to -\infty.$$

Corollary 1.9. *The following assertions hold:*

(i) There are only a finite[2] *number of characteristic roots in any vertical strip of the complex plane, given by*

$$\{\lambda \in \mathbb{C} : \alpha < \Re(\lambda) < \beta\},$$

with $\alpha, \beta \in \mathbb{R}$, and $\alpha < \beta$.

(ii) There exists a number $\gamma \in \mathbb{R}$ such that all characteristic roots are confined to the half-plane

$$\{\lambda \in \mathbb{C} : \Re(\lambda) < \gamma\}.$$

This result can be strengthened. The following proposition, which plays an important role in the study of continuity properties of the spectrum, allows to construct an *envelope*

[2] multiplicity taken into account

curve around the characteristic roots of (1.1) (see also [13] for further discussions and properties).

Proposition 1.10. *If λ is a characteristic root of the system (1.1), then it satisfies*

$$|\lambda| \leq \|A_0\|_2 + \sum_{i=1}^{m} \|A_i\|_2 e^{-\Re(\lambda)\tau_i}. \tag{1.12}$$

Proof. The expression $\Delta(\lambda) = 0$ is equivalent to

$$\lambda \in \sigma\left(A_0 + \sum_{i=1}^{m} A_i e^{-\lambda\tau_i}\right).$$

Interpreting the argument of $\sigma(\cdot)$ as a matrix leads to

$$|\lambda| \leq \left\| A_0 + \sum_{i=1}^{m} A_i e^{-\lambda\tau_i} \right\|_2,$$

from which (1.12) follows straightforwardly. $\square$

Example 1.11. *For the system (1.3), the estimate (1.12) becomes*

$$|\lambda| \leq 1 + |\alpha| e^{-\tau\Re(\lambda)}.$$

For $\alpha = 2$ and $\tau = 1/3$, the resulting envelope curve, $|\lambda| = 1 + 2e^{-\Re(\lambda)\tau}$, is depicted in Figure 1.2.

To conclude this section, we address a modal expansion of the solutions of (1.1).

Proposition 1.12. *[67] Consider the solution $x(\phi)$ of system (1.1) corresponding to some initial function $\phi \in \mathcal{C}([-\tau_m, 0], \mathbb{R}^n)$. For any $\zeta \in \mathbb{R}$ such that*

$$\det \Delta(\lambda) \neq 0$$

for all $\lambda \in \mathbb{C}$ on the line $\Re(\lambda) = \zeta$, the following asymptotic expansion holds:

$$x(\phi)(t) = \sum_{k=1}^{\ell} p_k(t) e^{\lambda_k t} + o(e^{\zeta t}), \text{ for } t \to \infty,$$

where $\lambda_1, \ldots, \lambda_\ell$ are the (finitely many) characteristic roots with real part exceeding ζ, and $p_k(t)$, $k = 1, \ldots, \ell$, are $\mathbb{C}^n$-valued polynomial of degree less than or equal to $m_k - 1$, with m_k the multiplicity of λ_k as a root of the characteristic equation.

It follows that if all the characteristic roots are in the open left half-plane, then all the solutions of (1.1) converge to zero exponentially as $t \to +\infty$. This is in accordance with Proposition 1.6.

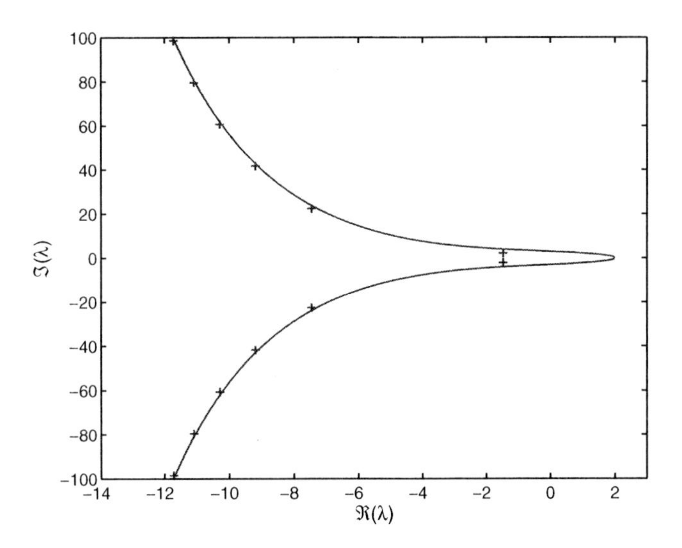

Figure 1.2. *Envelope curve on the characteristic roots for the system (1.3) with* $\alpha = 2$ *and* $\tau = 1/3$ *(solid line). The characteristic roots are indicated with "+."*

1.1.5 Spectrum: continuity properties

We discuss continuity properties of the characteristic roots and related quantities such as spectral abscissa. As we explicitly address the dependence of the characteristic roots on parameters, we will sometimes write, for instance,

$$\Delta(\lambda;\ \vec{\tau}, A_0, \ldots, A_m)$$

instead of $\Delta(\lambda)$, where the two types of arguments, separated by a dot-comma, correspond to variables and parameters, respectively. The short notation $\vec{\tau}$ stands for $(\tau_1, \ldots, \tau_m)$.

Based on Corollary A.1 of the appendix, a corollary of Rouché's theorem, we have the following statement on the continuity of the individual characteristic roots, which says, roughly speaking, that the characteristic roots behave continuously w.r.t. variations of the system matrices and delays.

Proposition 1.13. *Let* λ_0 *be a characteristic root of (1.1) with multiplicity k. There exists a constant* $\bar{\epsilon} > 0$ *such that for all* $\epsilon > 0$ *satisfying* $\epsilon < \bar{\epsilon}$*, there is a number* $\delta > 0$ *such that*

$$\Delta(\lambda;\ \vec{\tau} + \delta\vec{\tau}, A_0 + \delta A_0, \ldots, A_m + \delta A_m),$$

where

$$\begin{array}{ll} \delta\vec{\tau} \in \mathbb{R}^m, & \|\delta\vec{\tau}\| < \delta,\ \vec{\tau} + \delta\vec{\tau} \geq 0, \\ \delta A_k \in \mathbb{R}^{n\times n}, & \|\delta A_k\|_2 < \delta,\ k = 0, \ldots, m, \end{array}$$

has exactly k zeros[3] *in the disk* $\{\lambda \in \mathbb{C} :\ |\lambda - \lambda_0| < \epsilon\}$.

[3] multiplicity taken into account

Next, let the *spectral abscissa function* corresponding to the system (1.1) be defined as follows:

$$\alpha(\vec{\tau}, A_0, \ldots, A_m) := \sup\left\{\Re(\lambda) : \det \Delta(\lambda;\ A_0, \ldots, A_m) = 0\right\}. \tag{1.13}$$

Note from Corollary 1.9 that the spectral abscissa always exists and is finite. Furthermore, there always exist (rightmost) characteristic roots λ such that $\Re(\lambda) = \alpha$. Hence, the supremum operator in (1.13) can be replaced with a maximum and the exponential stability conditions from Proposition 1.6 can be rephrased as

$$\alpha(\vec{\tau}, A_0, \ldots, A_m) < 0.$$

Regarding the continuity of the spectral abscissa we have the following result.

Theorem 1.14. *The function* $\alpha : \mathbb{R}_+^m \times \mathbb{R}^{n\times n\times(m+1)} \to \mathbb{R}$,

$$(\vec{\tau}, A_0, \ldots, A_m) \mapsto \alpha(\vec{\tau}, A_0, \ldots, A_m),$$

is continuous.

Proof. The assertion follows from the combination of Proposition 1.13, Corollary 1.9, and the bound (1.12) on the characteristic roots. For the case where only the delays are varied a detailed proof is given in [59]. □

It is important to point out that the spectral abscissa function is also continuous at these points where the time-delay system reduces to a system without delays. This happens for instance at $\vec{\tau} = (0, \ldots, 0)$ or at parameter values where the system matrices corresponding to the delayed terms vanish.

The following corollary justifies the methods for computing stability regions in parameter spaces, discussed in the subsequent chapters.

Theorem 1.15. *If the matrices* $A_0, \ldots, A_m$ *and the delays* $\tau_1, \ldots, \tau_m$ *are varied, then a loss or acquisition of exponential stability of the null solution of (1.1) is associated with characteristic roots on the imaginary axis.*

Example 1.16. *Figure 1.3 depicts the rightmost characteristic roots of the system (1.3) as a function of the parameter* α *for a fixed value of* τ *and vice versa. Notice that the spectral abscissa is continuous, yet not everywhere differentiable. A more elaborate study of continuity properties of the spectral abscissa will be performed in Chapter 10.*

For $\alpha = 0$*, the system reduces to*

$$\dot{x}(t) = x(t). \tag{1.14}$$

It follows from Proposition 1.13 and Theorem 1.14 that one characteristic root converges to the characteristic root $\lambda = 1$ *of (1.14) as* $\alpha \to 0$*, while the real parts of the other characteristic roots move off to* $-\infty$*. A similar situation occurs as* $\tau \to 0+$*, where the system reduces to*

$$\dot{x}(t) = (1 - \alpha)x(t). \tag{1.15}$$

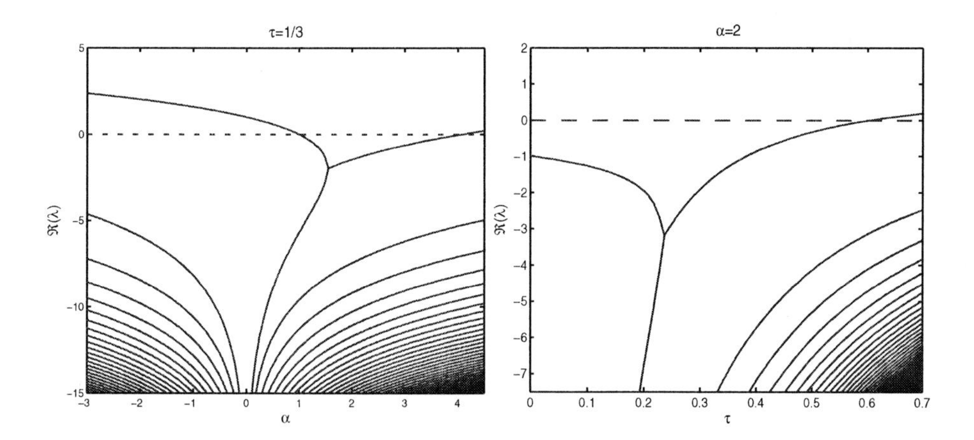

Figure 1.3. *(left) The rightmost characteristic roots of the system (1.3) as a function of the parameter α for $\tau = 1/3$. (right) The rightmost characteristic roots as a function of the delay-parameter τ for $\alpha = 2$.*

Note in particular that asymptotic stability for $\tau = 0$ is preserved for small positive values of the delay.

Let us now exploit the continuity results in order to prove the asymptotic stability of (1.3) for $\tau = 1/3$ and $\alpha = 2$. We fix α and consider τ as a free parameter. The system is obviously asymptotically stable for $\tau = 0$. Next, by Theorem 1.15 a loss of stability is associated with characteristic roots on the imaginary axis. Substituting $\lambda = j\omega_0$, $\omega_0 > 0$, in the characteristic equation yields

$$j\omega_0 - 1 + 2e^{-j\omega_0\tau} = 0. \tag{1.16}$$

Solving this equation results in $\omega_0 = \sqrt{3}$ and

$$\tau = \frac{\angle(1 + j\sqrt{3}) + 2\pi l}{\sqrt{3}}, \quad l \geq 0.$$

The continuity w.r.t. the delay value implies that the system, stable for $\tau = 0$, remains stable for all delays $\tau \in [0,\ \tau_m)$, where the delay margin *τ_m is given by $\tau_m = (\angle(1 + j\sqrt{3}))/\sqrt{3}$. The remaining step is to compare τ_m with the nominal delay value. In our case, we have $\tau_m > 1/3$, which allows us to conclude that the system is asymptotically stable for $\tau = 1/3$.*

1.1.6 Computation of characteristic roots

There are several numerical techniques for computing the rightmost characteristic roots. Since the characteristic equation is an analytic function, methods for computing zeros of analytic functions can be used directly, such as the contour integration based method of [143] and the quasipolynomial root finder approach described in [318]. The latter has proved its effectiveness for linear time-delay systems. However, such methods require that the system's dimension be small. Otherwise, working directly with the determinant in formula (1.8) is to be avoided for reasons of numerical stability [91]. Instead, one can search for pairs (λ, v) appearing in the nonlinear eigenvalue problem (1.10). Systems of equations such as (1.10) can be solved efficiently and accurately by using *Newton's method* because (1.10) is finite-dimensional, albeit nonlinear. However, Newton's method only works if good initial guesses for the solutions (λ, v) are available. These can be found by solving an approximate linear eigenvalue problem, obtained by discretizing the infinite-dimensional linear operators $\mathcal{A}$ and $\mathcal{T}(t)$, where t is fixed. Next, these approximations are corrected using Newton's method. We now describe these two steps in more detail. In the first step, one considers alternatives to (1.10), induced by the formulations (1.5)–(1.7). The discretization of the infinite-dimensional operators $\mathcal{A}$ and $\mathcal{T}(t)$ yields approximate *linear* eigenvalue problems. Both approaches have been studied in the literature. In [25, 26] and the references therein (see also [314]) the infinitesimal generator $\mathcal{A}$ is discretized into a matrix. While [26] is based on a spectral discretization, in [25] the derivatives appearing in (1.6) are approximated using a Runge–Kutta method. In [77, 78, 313] the solution operator $\mathcal{T}(h)$, which performs an integration over a time-step h, is discretized into a matrix using a k-step *linear multistep* method combined with Lagrange interpolation to evaluate the delayed terms (see the book [12], which contains a detailed analysis of methods for time-integration of DDEs). Computing the eigenvalues of this matrix yields approximate eigenvalues z of $\mathcal{T}(h)$, from which the approximate characteristic roots can be obtained (note that the equation $z = e^{\lambda h}$ is not one to one, yet for a given μ the characteristic root λ can be selected using the form of the corresponding eigenfunction). This method is implemented in the steady state stability routine of the software package DDE-BIFTOOL [74], which we have used throughout the book for the computation of characteristic roots. In the routine a heuristic described in [77] is implemented for an automatic choice of the steplength h which aims at a large h—and thus a low computational cost—for which the approximations to the characteristic roots in a given right half-plane are still excellent. This steplength heuristic is further improved in [313] for a class of special-purpose linear multistep methods.

In the second step, Newton's method is used to correct the computed characteristic roots up to a desired accuracy. Hence, the final accuracy is *not* affected by the discretization error in the first step. The discretization method should indeed only return good starting values for Newton's method. Whether a starting value is good enough to allow for Newton's method to converge is, in general, problem dependent.

To conclude we mention some other approaches for the computation of characteristic roots. The so-called semidiscretization method (see, e.g., [121]) is strongly related to the method described above for approximating $\mathcal{T}(h)$. Here, the underlying idea consists of approximating the delays with a sawtooth function, in such a way that along a solution the delayed arguments of the differential equation become piecewise constant functions of time, and the system, sampled on discrete time-instants, becomes equivalent to a discrete-time system, of which the stability information is obtained by solving a finite-dimensional linear eigenvalue problem. So the difference lies in the fact that the approximation is not induced by a numerical time-integration step, but by approximating the delays in the original system, after which the integration step is performed exactly. Finally, in [76] a method is described for the computation of the dominant eigenvalues of $\mathcal{T}(t)$, with t fixed, which is based on subspace iteration. In that case one iteration step requires the computation of m solutions of the delay equation over a time-interval t, with m the dimension of the subspace. This can again be done using time-integration methods as described in [12]. Note here that a large value of t typically leads to fewer iterations required, yet to a large computational cost per iteration.

1.2 Time-delay systems of neutral type

We discuss spectral properties of the neutral system

$$\frac{d}{dt}\left(x(t)+\sum_{k=1}^{m} H_k x(t-\tau_k)\right) = A_0 x(t) + \sum_{k=1}^{m} A_k x(t-\tau_k), \tag{1.17}$$

where $x(t) \in \mathbb{R}^n$ is the state variable at time t and $0 < \tau_1 < \tau_2 \cdots < \tau_m$ represent the time-delays. In what follows we use the short notation $\vec{\tau} = (\tau_1, \ldots, \tau_m)$.

1.2.1 Initial value problem

The initial condition for the neutral system (1.17) is the function segment

$$\phi \in \mathcal{C}([-\tau_m,\ 0], \mathbb{R}^n),$$

where $\mathcal{C}([-\tau_m,\ 0],\ \mathbb{R}^n)$ is the Banach space of continuous functions mapping the interval $[-\tau_m,\ 0]$ into $\mathbb{R}^n$ and equipped with the supremum norm, $\|\cdot\|_s$.

Due to the fact that the map $\mathcal{N} : \mathcal{C}([-\tau_m,\ 0], \mathbb{R}^n) \to \mathbb{R}^n$, defined by

$$\mathcal{N}(\phi) := \phi(0) - \sum_{k=1}^{m} H_k \phi(-\tau_k),$$

is atomic at zero; the existence and unicity of solutions of (1.17) are guaranteed. Let

$$x(\phi) : t \in [-\tau_m,\ \infty) \to x(\phi)(t) \in \mathbb{R}^n$$

be the unique forward solution with initial condition $\phi \in \mathcal{C}([-\tau_m,\ 0], \mathbb{R}^n)$, that is,

$$x(\phi)(\theta) = \phi(\theta) \quad \forall \theta \in [-\tau_m,\ 0].$$

Then the state at time t is given by the function segment $x_t(\phi) \in \mathcal{C}([-\tau_m,\ 0],\ \mathbb{R}^n)$ defined as

$$x_t(\phi)(\theta) = x(\phi)(t+\theta),\ \theta \in [-\tau_m,\ 0].$$

Denote with $\mathcal{T}_N(t),\ t \geq 0$, the solution operator, mapping initial data onto the state at time t,

$$(\mathcal{T}_N(t)\phi)(\theta) = x_t(\phi)(\theta) = x(\phi)(t+\theta),\ \ \theta \in [-\tau_m,\ 0]. \tag{1.18}$$

This is a strongly continuous semigroup. Let $\mathcal{A}_N$ be its infinitesimal generator, that is,

$$\begin{aligned}
\mathcal{D}(\mathcal{A}_N) = \Big\{&\phi \in \mathcal{C}([-\tau_m,\ 0],\ \mathbb{R}^n) : \tfrac{d\phi}{d\theta} \in \mathcal{C}([-\tau_m,\ 0],\ \mathbb{R}^n),\\
&\phi(0) + \textstyle\sum_{k=1}^{m} H_k \tfrac{d\phi}{d\theta}(-\tau_k) = A_0\phi(0) + \textstyle\sum_{k=1}^{m} A_k \phi(-\tau_k)\Big\},\\
\mathcal{A}_N\, \phi = \tfrac{d\phi}{d\theta}&.
\end{aligned}$$

The associated delay-difference equation of (1.17) is given by $\mathcal{N}(x_t) = 0$, or

$$x(t) - \sum_{k=1}^{m} H_k x(t - \tau_k) = 0. \tag{1.19}$$

For any initial condition $\phi \in \mathcal{C}_D([-\tau,\ 0], \mathbb{R}^n)$, where

$$\mathcal{C}_D([-\tau_m,\ 0], \mathbb{R}^n) = \{\phi \in \mathcal{C}([-\tau_m,\ 0], \mathbb{R}^n) : \mathcal{N}(\phi) = 0\}$$

is a closed subspace of $\mathcal{C}([-\tau,\ 0], \mathbb{R}^n)$, a forward solution of (1.19) is uniquely defined. Let

$$y(\phi) : t \in [-\tau_m,\ \infty) \to y(\phi)(t) \in \mathbb{R}^n$$

be the unique forward solution of (1.19) with initial condition $\phi \in \mathcal{C}_D([-\tau_m,\ 0], \mathbb{R}^n)$. Then the state at time t is given by the function segment $y_t(\phi) \in \mathcal{C}_D([-\tau_m,\ 0],\ \mathbb{R}^n)$,

$$y_t(\phi)(\theta) = y(\phi)(t + \theta),\ \theta \in [-\tau_m,\ 0].$$

We denote by $\mathcal{T}_D(t)$ the corresponding solution operator.

1.2.2 Spectrum: definitions

The substitution of a sample solution of the form $e^{\lambda t} v$, with $v \in \mathbb{C}^{n \times 1} \setminus \{0\}$, leads us to the *characteristic equation* of (1.17)

$$\det\left(\Delta_N(\lambda)\right) = 0,$$

where

$$\Delta_N(\lambda) := \lambda \left(I - \sum_{k=1}^{m} H_k e^{-\lambda \tau_k} \right) - A_0 - \sum_{k=1}^{m} A_k e^{-\lambda \tau_k} \tag{1.20}$$

is the *characteristic matrix*.

The zeros of (1.20) are called the *characteristic roots* of (1.17). The spectra of the linear operators $\mathcal{A}_N$, $\mathcal{T}_N(t)$ and the characteristic roots are related in the following way. The characteristic roots are the eigenvalues of the operator $\mathcal{A}$, which only features a *point spectrum*, that is, $\sigma(\mathcal{A}_N) = P\sigma(\mathcal{A}_N)$. Furthermore, the algebraic multiplicity of a complex number λ as an eigenvalue of $\mathcal{A}_N$ is equal to its multiplicity as a zero of $\Delta_N(\lambda)$, while its geometric multiplicity is equal to the dimension of the null space of $\Delta_N(\lambda)$. For $\lambda \in \sigma(\mathcal{A})$, the corresponding eigenfunctions take the form

$$v e^{\lambda \theta},\ \theta \in [-\tau_m,\ 0], \tag{1.21}$$

where the eigenvector $v \in \mathbb{C}^n \setminus \{0\}$ satisfies

$$\Delta_N(\lambda) v = 0. \tag{1.22}$$

The spectra of $\mathcal{A}_N$ and $\mathcal{T}_N$ are related by

$$\sigma(\mathcal{T}_N(t)) = \operatorname{clos}\left(\exp\left(t\sigma(\mathcal{A}_N)\right)\right) \text{ plus possibly } \{0\}. \tag{1.23}$$

Furthermore, if $\lambda \in \sigma(\mathcal{A}_N)$, then $e^{\lambda t} \in P\sigma(\mathcal{T}_N(t))$, with corresponding eigenfunctions given by (1.21). It is important to mention that $\mathcal{T}_N(t)$ in general features an essential spectrum. This is the part of the spectrum that cannot be removed with a compact perturbation,

$$\sigma_e(\mathcal{T}_N(t)) := \bigcap_{K \text{ compact}} \sigma(\mathcal{T}_N(t) + K).$$

For the system (1.17) the essential spectrum of $\mathcal{T}(t)$ appears as the collection complex numbers λ for which there are infinitely many eigenvalues in any open disk centered at λ (multiplicity taken into account).

The characteristic equation of the associated delay-difference equation (1.19) is given by $\Delta_D(\lambda) = 0$, where

$$\Delta_D(\lambda) := \left(I - \sum_{k=1}^{m} H_k e^{-\lambda \tau_k} \right). \tag{1.24}$$

The zeros of (1.24) are called the characteristic roots of the delay-difference equation (1.19). These characteristic roots are related with the spectrum of $\mathcal{T}_D(t)$ as follows:

$$\sigma(\mathcal{T}_D(t)) = \text{clos}\left\{ e^{\lambda t} : \lambda \in \mathbb{C} \text{ and } \Delta_D(\lambda) = 0 \right\} \text{ plus possibly } \{0\}. \tag{1.25}$$

The spectrum of $\mathcal{T}_D(t)$ only features an essential spectrum, that is

$$\sigma_e(\mathcal{T}_D(t)) = \sigma(\mathcal{T}_D(t)). \tag{1.26}$$

A very important result in what follows, which connects the spectra of $\mathcal{T}_N(t)$ and $\mathcal{T}_D(t)$, is the following:

$$\sigma_e(\mathcal{T}_N(t)) = \sigma(\mathcal{T}_D(t)). \tag{1.27}$$

Example 1.17. *We consider the neutral system*

$$\frac{d}{dt}\left(x(t) - \frac{3}{4}x(t-\tau_1) + \frac{1}{2}x(t-\tau_2) \right) = \frac{1}{4}x(t) + \frac{3}{4}x(t-\tau_1), \tag{1.28}$$

where

$$\vec{\tau} = (1, 2). \tag{1.29}$$

In Figure 1.4 we plot the characteristic roots λ, which are the eigenvalues of the operator $\mathcal{A}_N$, as well as the eigenvalues z of the operator $\mathcal{T}_N(1)$. These are connected via the relation $z = e^{\lambda}$. The operator $\mathcal{A}_N$ only features a point spectrum; in particular, the characteristic roots are all isolated and of finite multiplicity. The operator $\mathcal{T}_N(1)$ features an essential spectrum that corresponds to the accumulation points of the eigenvalues given by

$$z_e^{\pm} = \frac{3 \pm \sqrt{23}j}{8}. \tag{1.30}$$

In Figure 1.5 we plot the characteristic roots of the associated delay-difference equation

$$x(t) = \frac{3}{4}x(t-\tau_1) - \frac{1}{2}x(t-\tau_2). \tag{1.31}$$

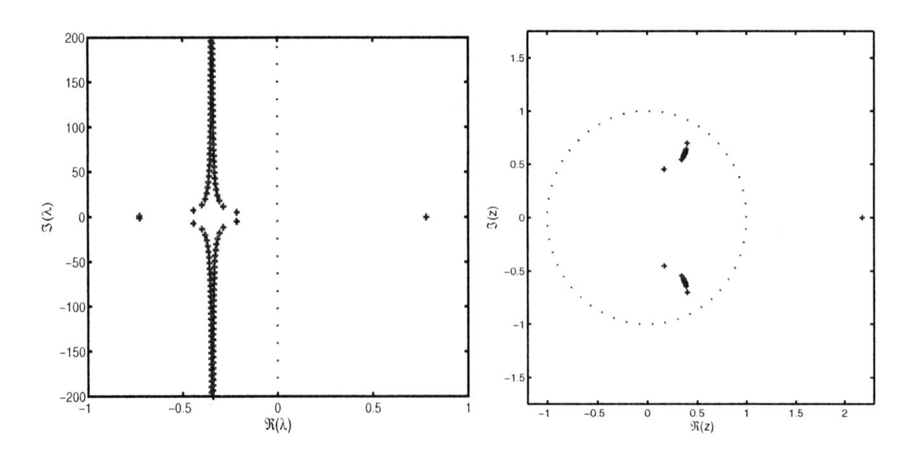

Figure 1.4. *(left) Characteristic roots of the neutral system (1.28)–(1.29). (right) Eigenvalues of the corresponding operator $\mathcal{T}_N(1)$.*

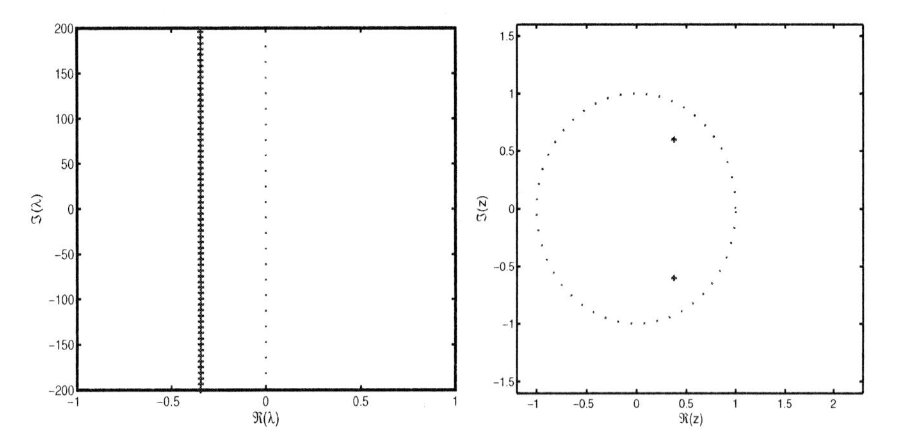

Figure 1.5. *(left) Characteristic roots of the delay-difference equation (1.31)–(1.29). (right) Eigenvalues of $\mathcal{T}_D(1)$.*

The characteristic roots can be computed analytically as follows:

$$1 - \frac{3}{4}e^{-\lambda} + \frac{1}{2}e^{-2\lambda} = 0 \tag{1.32}$$

$$\Leftrightarrow e^{\lambda} = \frac{3 \pm \sqrt{23}j}{8} \tag{1.33}$$

$$\Leftrightarrow \lambda = -\log 2 \pm j\left(\operatorname{atan}\frac{\sqrt{23}}{3} + 2\pi l\right), \; l \in \mathbb{Z}. \tag{1.34}$$

The fact that the right-hand sides of (1.30) and (1.33) are equal is a consequence of (1.26)–(1.27).

1.2.3 Asymptotic growth rate of solutions and stability

The definition of stability notions is similar for ODEs and DDEs of retarded type.

Definition 1.18. *The null solution of (1.17) is asymptotically stable*[4] *if and only if*

$$\forall \epsilon > 0\ \exists \delta > 0\ \forall \phi \in \mathcal{C}([-\tau,\ 0], \mathbb{R}^n)\ (\|\phi\|_s < \delta) \Rightarrow (\forall t \geq 0\ \|x_t(\phi)\|_s < \epsilon),$$
$$\forall \phi \in \mathcal{C}([-\tau,\ 0], \mathbb{R}^n)\ \lim_{t \to +\infty} x(\phi)(t) = 0.$$

Definition 1.19. *The null solution of (1.17) is exponentially stable if and only if there exist constants $C > 0$ and $\gamma > 0$ such that*

$$\forall \phi \in \mathcal{C}([-\tau,\ 0], \mathbb{R}^n)\ \|x_t(\phi)\|_s \leq Ce^{-\gamma t}\|\phi\|_s.$$

It is clear that exponential stability implies asymptotic stability. Contrary to the case of linear delay equations of retarded type, the converse does not hold in general, as illustrated with an example in [315] (see also [30]).

The asymptotic behavior of the solutions of (1.17) and thus their stability properties are determined by the spectral radius $r_\sigma(\mathcal{T}_N(1))$. In particular, we have the following result.

Proposition 1.20. *The null solution of (1.17) is exponentially stable if and only if*

$$r_\sigma(\mathcal{T}_N(1)) < 1$$

or, equivalently, all characteristic roots are located in the open left half-plane and bounded away from the imaginary axis.

For the delay-difference equation (1.19) associated with (1.17), stability definitions and their relation with spectral properties are similar. We have, for instance, the following definition.

Definition 1.21. *The null solution of (1.19) is exponentially stable if and only if there exist constants $C > 0$ and $\gamma > 0$ such that*

$$\forall \phi \in \mathcal{C}_D([-\tau,\ 0], \mathbb{R}^n)\ \|y_t(\phi)\|_s \leq Ce^{-\gamma t}\|\phi\|_s.$$

Proposition 1.22. *The null solution of (1.19) is exponentially stable if and only if*

$$r_\sigma(\mathcal{T}_D(1)) < 1$$

or, equivalently, all characteristic roots of (1.19) are located in the open left half-plane and bounded away from the imaginary axis.

---

[4]For reasons of conciseness we will often use the less precise formulation, "The system (1.17) is asymptotically stable."

Let us now relate the exponential stability of the null solution of (1.19) with the exponential stability of null solution of (1.17). The following result is a direct corollary of (1.27) and Propositions 1.20 and 1.22.

Proposition 1.23. *A necessary condition for the exponential stability of the null solution of the neutral equation (1.17) is the exponential stability of the null solution of the delay-difference equation (1.19).*

1.2.4 Spectrum: qualitative properties

Instrumental to the study of qualitative features of the spectra related to the neutral equation (1.17) we first study some qualitative features of the spectra related to the delay-difference equation (1.19).

Delay-difference equation. Define the collection of the real parts of all the characteristic roots of (1.19) as

$$Z_D := \{\Re(\lambda) : \det \Delta_D(\lambda) = 0\} \tag{1.35}$$

and let the spectral abscissa c_D be its supremum:

$$c_D := \sup \{\Re(\lambda) : \det \Delta_D(\lambda) = 0\}. \tag{1.36}$$

It is clear that $c_D < \infty$. The following result characterizes the set Z_D.

Proposition 1.24. *Define*

$$\mathcal{I}_D = \left\{ \zeta \in \mathbb{R} : \det \left(I - \sum_{k=1}^{m} A_k e^{-\zeta \tau_k} e^{i\theta_k} \right) = 0 \text{ for some } \vec{\theta} \in [0,\ 2\pi]^m \right\}.$$

Then the following holds:

- $\bar{Z}_D \subseteq \mathcal{I}_D$;
- *If the delays $\vec{\tau}$ are rationally independent,*[5] *then $\bar{Z}_D = \mathcal{I}_D$; hence, $\bar{Z}_D$ consists of a finite number of intervals. If the delays $\vec{\tau}$ are commensurate, then Z_D consists of a finite number of points.*

The next proposition from [5] describes the typical behavior of the spectrum, characterized by chains of characteristic roots whose real parts are bounded, yet whose imaginary parts tend to infinity.

Proposition 1.25. *If $\zeta \in \bar{Z}_D$, then there exists a sequence of characteristic roots of (1.19), $\{\lambda_n\}_{n\geq 1}$, satisfying*

$$\lim_{n\to\infty} \Re(\lambda_n) = \zeta, \quad \lim_{n\to\infty} = \Im(\lambda_n) = \infty.$$

[5]The m components of $\vec{\tau} = (\tau_1, \ldots, \tau_m)$ are rationally independent if and only if $\sum_{k=1}^{m} n_k \tau_k = 0,\ n_k \in \mathbb{Z}$, implies $n_k = 0$, for all $k = 1, \ldots, m$. For instance, two delays τ_1 and τ_2 are rationally independent if their ratio is an irrational number. See Section A.4 of the appendix for more information on the interdependency of numbers.

If the delays are commensurate, then the chains of roots can be computed analytically. For $\vec{\tau} = \vec{n}\tau_0$ with $\tau_0 \in \mathbb{R}_+$ and $\vec{n} \in \mathbb{N}^m$, the characteristic equation is given by

$$p(z) := \det\left(I - \sum_{k=1}^{m} H_k z^{-n_k}\right) = 0,$$

where $z = e^{-\lambda\tau_0}$. Note that p becomes a polynomial in z after a multiplication with z^{n_m}. Hence, the characteristic roots are given by

$$\left\{\lambda \in \mathbb{C} : \ \lambda = \frac{-\mathrm{Log}(z_k) + l2\pi j}{\tau_0}, \ p(z_k) = 0, \ l \in \mathbb{Z}\right\}.$$

Note that Z_D consists of at most n points, namely

$$\left\{r \in \mathbb{R} : \ r = \frac{\mathrm{Log}(z_k|}{\tau_0}, \ p(z_k) = 0\right\},$$

a property in accordance with Proposition 1.24. We refer to Example 1.17 and, in particular, Figure 1.5 for a numerical example. For details about the spectrum of delay-difference equations in the noncommensurate case we refer to [181].

Neutral equation. Equation (1.17) also features chains of characteristic roots, whose position is determined by the associated *delay-difference equation.*

Proposition 1.26. *If $\zeta \in \bar{Z}_D$, with Z_D defined by (1.35), then there is a sequence of characteristic roots $\{\lambda_n\}_{n\geq 1}$ of (1.17) satisfying*

$$\lim_{n\to\infty} \Re(\lambda_n) = \zeta, \quad \lim_{n\to\infty} = \Im(\lambda_n) = \infty.$$

Proof. The proof follows from the relation (1.27) and Proposition 1.25. □

Intuitively this result is expected, since for $\lambda \neq 0$ the characteristic equation can be written in the form

$$\Delta_D(\lambda) = \frac{1}{\lambda}\left(A_0 + \sum_{k=1}^{m} A_k e^{-\lambda\tau_k}\right).$$

If $|\lambda| >> \Re(\lambda)$, then the right-hand side is very small. Hence, characteristic roots with a large modulus but small real part are expected to be approximate zeros of $\Delta_D(\lambda)$.

In the half-plane $\Re(\lambda) > c_D$ the set of characteristic roots of neutral systems has many properties similar to the retarded case. The next proposition is an example of this.

Proposition 1.27. *For any $\epsilon > 0$, the system (1.17) has only a finite number of characteristic roots in the right half- plane*

$$\Re(\lambda) \geq c_D + \epsilon,$$

where c_D is defined in (1.36).

Proof. The assertion follows from the fact that $\mathcal{T}_N(t)$ only has point spectrum in the set $|z| \geq r_e(\mathcal{T}_D(1))$, consisting of eigenvalues of finite multiplicity, and the relation between the spectrum of this operator and the spectrum of $\mathcal{A}_D$. □

Also here, one can derive some envelope curves containing all characteristic roots with $\Re(\lambda) > c_D$, yet as we shall see in the next section this is only useful when taking small perturbations explicitly into account.

To clarify the above results we again refer to Example 1.17. In particular, the correspondence between the characteristic roots with large moduli shown in Figures 1.4 and 1.5 is a consequence of Proposition 1.26. In the right half-plane there is only one characteristic root, in accordance with Proposition 1.27.

1.2.5 Spectrum: continuity properties

The exponential stability of the delay-difference equation (1.19) associated with (1.17) may be sensitive to infinitesimal delay perturbations, which strongly affects the continuity properties of the characteristic roots of (1.17). For this, we first give this phenomenon a closer look and discuss its relation with the spectral properties of the neutral equation. Next we discuss some continuity properties of the characteristic roots of the neutral equation (1.17) and related quantities such as spectral abscissa.

As we explicitly address the dependence of characteristic roots on parameters, we will write, for instance,

$$\Delta_N(\lambda;\ \vec{\tau}, H_1, \ldots, H_m, A_0, \ldots, A_m)$$

instead of $\Delta_N(\lambda)$, where the two types of arguments (variable and parameters) are separated with a dot-comma.

Delay sensitivity problem of the associated delay-difference equation. It is well known that the spectral radius $r_\sigma(\mathcal{T}_D)$, although continuous in the system matrices H_k, is *not* continuous in the *delays* $\vec{\tau}$ (see, e.g., [106, 107, 113]), which carries over to the spectral abscissa

$$c_D(\vec{\tau};\ H_1, \ldots, H_m) = \sup\left\{\Re(\lambda) : \det\left(I - \sum_{k=1}^{m} H_k e^{-\lambda\tau_k}\right) = 0\right\}. \tag{1.37}$$

One consequence is that arbitrarily small perturbations on the delays may destroy stability of the delay-difference equation. This has led to the introduction of the concept of *strong stability* in [107]: we say that the null solution of (1.19) is strongly exponentially stable if it remains exponentially stable when subjected to small variations in the delays. Theorem 2.2 and Corollary 2.2 of [107] provide the following conditions.

Proposition 1.28. *The null solution of the delay-difference equation (1.19) is strongly exponentially stable if and only if $\gamma_0 < 1$, where*

$$\gamma_0(H_1, \ldots, H_m) := \max_{\vec{\theta}\in[0,\ 2\pi]^m} r_\sigma\left(\sum_{k=1}^{m} H_k e^{j\theta_k}\right).$$

Furthermore, if $\gamma_0 > 1$ then Equation (1.19) is exponentially unstable for rationally independent delays.

Notice that the quantity γ_0 does not depend on the value of the delays; in other words, exponential stability locally in the delays is equivalent with exponential stability globally in the delays [107].

Even if the delay-difference equation is strongly exponentially stable, it is very useful to have *more precise* information about the position of the real parts of its characteristic roots, and in particular the upper bound (1.37). Due to *lack of continuity* of this quantity w.r.t. the delays, from a practical point of view we are once again led to the smallest upper bound, which is *insensitive* to small delay changes. More precisely, we define this "safe" upper bound $\bar{C}_D(\vec{\tau})$ as follows.

Definition 1.29. *Let $\bar{C}_D(\vec{\tau};\ H_1, \ldots, H_m) \in \mathbb{R}$ be defined as*

$$\bar{C}_D(\vec{\tau};\ H_1, \ldots, H_m) = \lim_{\epsilon \to 0+} c_\epsilon(\vec{\tau};\ H_1, \ldots, H_m),$$

where

$$c_\epsilon(\vec{\tau},\ H_1, \ldots, H_m) = \sup\left\{c_D(\vec{\tau} + \delta\vec{\tau};\ H_1, \ldots, H_m) :\ \delta\vec{\tau} \in \mathbb{R}^m \text{ and } \|\delta\vec{\tau}\| \leq \epsilon\right\}.$$

Clearly, we have $\bar{C}_D(\vec{\tau};\ H_1, \ldots, H_m) \geq c_D(\vec{\tau};\ H_1, \ldots, H_m)$ and, as we shall illustrate, the inequality can be *strict*.

In order to present a computational expression for $\bar{C}_D(\vec{\tau};\ H_1, \ldots, H_m)$, we define the function $f:\ \mathbb{R} \to \mathbb{R}_+$ in the following way:

$$f(c;\ \vec{\tau}, H_1, \ldots, H_m) = \max_{\vec{\theta} \in [0,\ 2\pi]^m} r_\sigma\left(\sum_{k=1}^{m} H_k e^{-c\tau_k} e^{j\theta_k}\right). \tag{1.38}$$

This function is *continuous* in both its *argument* c and *parameters* $\vec{\tau}$ and $H_k,\ k = 1, \ldots, m$. Notice that $\gamma_0(H_1, \ldots, H_m) = f(0;\ \vec{\tau}, H_1, \ldots, H_m)$. We have the following result [199, Theorem 6].

Theorem 1.30. *The quantity $\bar{C}_D(\vec{\tau};\ H_1, \ldots, H_m)$ is equal to the unique zero of the strictly decreasing function*

$$c \in \mathbb{R} \to f(c;\ \vec{\tau}, H_1, \ldots, H_m) - 1,$$

where f is defined in (1.38). Furthermore, $\bar{C}_D(\vec{\tau};\ H_1, \ldots, H_m)$ is continuous in both the delays $\vec{\tau} \in \mathbb{R}_+^m$ and the parameters $H_1, \ldots, H_m$.

Remark 1.31. *As $f(c,\ \vec{\tau}, H_1, \ldots, H_m)$ is strictly decreasing, a (robust) bisection algorithm is appropriate for the computation of $\bar{C}_D(\vec{\tau};\ H_1, \ldots, H_m)$.*

Remark 1.32. *Since $f(c;\ \vec{\tau}, H_1, \ldots, H_m) \leq \sum_{k=1}^{m} \|H_k\| e^{-c\tau_k}$, an upper bound on $\bar{C}_D(\vec{\tau})$ is given by the unique solution of*

$$\sum_{k=1}^{m} \|H_k\| e^{-c\tau_k} - 1 = 0. \tag{1.39}$$

Accordingly $\sum_{k=1}^{m} \|H_k\| < 1$ *is a* sufficient *condition for strong exponential stability.*

We mention some special cases where the above expressions become very simple. In the case of one delay ($m = 1$) we have $\gamma_0 = r_\sigma(H_1)$,

$$f(c;\ \tau_1, H_1) = r_\sigma(H_1)e^{-c\tau_1},\ \bar{C}_D(\tau_1) = \frac{1}{\tau_1}\log r_\sigma(H_1).$$

When the equation is scalar ($m = 1$) we have

$$\gamma_0(H_1, \ldots, H_m) = \sum_{k=1}^{m} |H_k|,\quad f(c;\ \vec{\tau}, H_1, \ldots, H_m) = \sum_{k=1}^{m} |H_k|e^{-c\tau_k}.$$

Example 1.33. *Consider the delay-difference equation (1.31). For* $\vec{\tau} = (1, 2)$ *its null solution is exponentially stable but not strongly exponentially stable because* $\gamma_0(-3/4, 1/2) = |3/4| + |1/2|$ *is larger than one. Furthermore, we have* $c_D((1, 2);\ -3/4, 1/2) \approx -0.3466$, *which is* strictly smaller *than* $\bar{C}_D((1, 2);\ -3/4, 1/2) \approx 0.1616$. *This illustrates the noncontinuity of* $c_D(\vec{\tau};\ -3/4, 1/2)$ *w.r.t.* $\vec{\tau}$. *In the left frame of Figure 1.6 we plot the characteristic roots of (1.31) for both the nominal delays* $\vec{\tau} = (1, 2)$ *(indicated with +), and the perturbed delays* $\vec{\tau} = (0.99, 2)$ *(indicated with o). In the right frame of Figure 1.6 we plot the eigenvalues of the operator* $\mathcal{T}_D(1)$. *The dashed curves are described by*

$$\Re(\lambda) = \bar{C}_D((1, 2);\ -3/4, 1/2)\ |z| = \exp(\bar{C}_D((1, 2);\ -3/4, 1/2)). \tag{1.40}$$

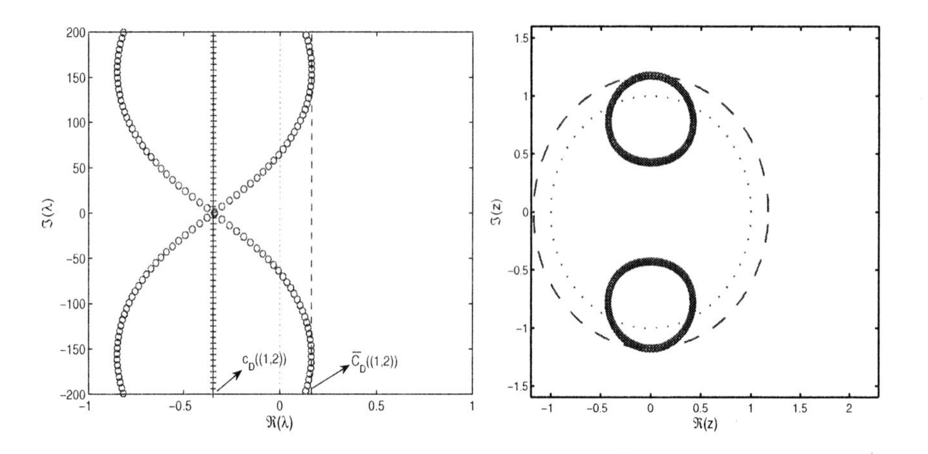

Figure 1.6. *(left) Characteristic roots of the delay-difference equation (1.31). (right) Corresponding eigenvalues of* $\mathcal{T}_D(1)$.

Relation with the spectrum of the neutral equation. From (1.26)–(1.27) it follows that not only the delay-difference equation (1.19), but *also* the neutral equation (1.17) has characteristic roots with real part *arbitrarily close* to $\bar{C}_D(\vec{\tau};\ H_1, \ldots, H_m)$ for certain (arbitrarily small) delay perturbations.

Example 1.34. *In the left frame of Figure 1.7 we plot the characteristic roots of (1.28) for both the nominal delays* $\vec{\tau} = (1, 2)$ *(indicated with +), and the perturbed delays* $\vec{\tau} = (0.99, 2)$ *(indicated with o). In the right frame of Figure 1.6 we plot the eigenvalues of the operator* $\mathcal{T}_N(1)$. *The dashed curves again correspond to (1.40).*

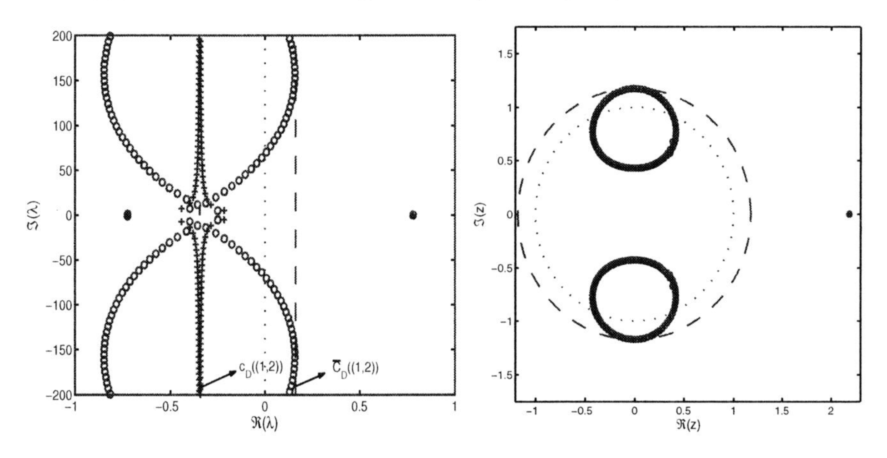

Figure 1.7. *(left) Characteristic roots of the neutral equation (1.28). (right) Corresponding eigenvalues of* $\mathcal{T}_N(1)$.

If $\Re(\lambda) > c_D(\vec{\tau};\ H_1, \ldots, H_m)$, then the matrix $\Delta_D(\lambda;\ \vec{\tau}, H_1, \ldots, H_m)$ is invertible. If, in addition, $\Re(\lambda) > \bar{C}_D(\vec{\tau};\ H_1, \ldots, H_m)$, then the following estimate holds:

$$\begin{aligned} \|(\Delta_D(\lambda;\ \vec{\tau}, H_1, \ldots, H_m)^{-1})\|_2 &= \left\| \left(I - \sum_{k=1}^{m} H_k e^{-\Re(\lambda)\tau_k} e^{-j\Im(\lambda)\tau_k} \right)^{-1} \right\|_2 \\ &\leq \max_{\vec{\theta}\in[0,\ 2\pi]^m} \left\| \left(I - \sum_{k=1}^{m} H_k e^{-\Re(\lambda)\tau_k} e^{j\theta_k} \right)^{-1} \right\|_2 \end{aligned} . \quad (1.41)$$

The right-hand side of (1.41) is well defined because $f(\Re(\lambda);\ \vec{\tau}, H_1, \ldots, H_m) < 1$ if $\Re(\lambda) > \bar{C}_D(\vec{\tau}; H_1, \ldots, H_m)$. This leads to a lemma, which will play a crucial role in the proof of some continuity properties of the spectrum discussed in the next section.

Lemma 1.35. *If* λ *is a characteristic root of the neutral system (1.17) with* $\Re(\lambda) > \bar{C}_D(\vec{\tau};\ H_1, \ldots, H_m)$, *then it satisfies*

$$|\lambda| \leq b(\Re(\lambda);\ \vec{\tau}, H_1, \ldots, H_m, A_0, \ldots, A_m), \quad (1.42)$$

where

$$\begin{aligned} &b(\Re(\lambda);\ \vec{\tau}, H_1, \ldots, H_m, A_0, \ldots, A_m) := \\ &\quad \max_{\vec{\theta}\in[0,\ 2\pi]^m} \left\| \left(I - \textstyle\sum_{k=1}^{m} H_k e^{-\Re(\lambda)\tau_k} e^{j\theta_k} \right)^{-1} \right\|_2 \\ &\quad \times \left(\|A_0\|_2 + \textstyle\sum_{i=1}^{m} \|A_k\|_2 e^{-\Re(\lambda)\tau_k} \right). \end{aligned} \quad (1.43)$$

Proof. Because $\Delta_D(\lambda;\ \vec{\tau}, H_1, \ldots, H_m)$ is invertible, we can write the characteristic equation in the form

$$\det\left(I - \Delta_D(\lambda;\ \vec{\tau}, H_1, \ldots, H_m)^{-1}\left(A_0 + \sum_{k=1}^{m} A_k e^{-\lambda\tau_k}\right)\right) = 0.$$

This equation can be interpreted as

$$\lambda \in \sigma\left(\Delta_D(\lambda;\ \vec{\tau}, H_1, \ldots, H_m)^{-1}\left(A_0 + \sum_{k=1}^{m} A_k e^{-\lambda\tau_k}\right)\right),$$

which implies

$$|\lambda| \le \left\|\Delta_D(\lambda;\ \vec{\tau}, H_1, \ldots, H_m)^{-1}\left(A_0 + \sum_{k=1}^{m} A_k e^{-\lambda\tau_k}\right)\right\|_2.$$

By further working out this estimate and using (1.41) one arrives at (1.42). □

It is important to mention that for a given $\Re(\lambda) > \bar{C}_D$, the quantity(1.43) is *continuous in the delays* at the nominal delay values and, hence, not sensitive to infinitesimal delay perturbations.

Example 1.36. *Let us revisit the example (1.31), for which the estimate (1.42)–(1.43) becomes*

$$|\lambda| \le \frac{\frac{1}{4} + \frac{3}{4}e^{-\Re(\lambda)\tau_1}}{1 - \frac{3}{4}e^{-\Re(\lambda)\tau_1} - \frac{1}{2}e^{-\Re(\lambda)\tau_2}}. \tag{1.44}$$

With $\vec{\tau} = (1, 2)$ the curve defined by (1.44) is shown in bold in Figure 1.8, as well as the characteristic roots for the nominal delays $\vec{\tau}(1, 2)$ and the perturbed delays $\vec{\tau} = (0.99, 2)$. The left and right frames correspond to a different scaling of the real and imaginary axes. Despite of the discontinuity of the function $\vec{\tau} \rightarrow c_D(\vec{\tau};\ -3/4, 1/2)$ at $\vec{\tau} = (1, 2)$, the envelope curve (1.44) is not affected by small delay perturbations.

Continuity properties and stability switches. Also in the neutral case the individual characteristic roots behave continuously w.r.t. the system's parameters (following from Corollary A.1).

Proposition 1.37. *Let λ_0 be a characteristic root of the neutral equation (1.17) with multiplicity k. There exists a constant $\bar{\epsilon} > 0$ such that for all $\epsilon > 0$ satisfying $\epsilon < \bar{\epsilon}$, there is a number $\delta > 0$ such that*

$$\Delta_N(\lambda;\ \vec{\tau} + \delta\vec{\tau}, H_1 + \delta H_1, \ldots, H_m + \delta H_m, A_0 + \delta A_0, \ldots, A_m + \delta A_m),$$

where

$$\begin{array}{ll} \delta\vec{\tau} \in \mathbb{R}^m, & \|\delta\vec{\tau}\| < \delta,\ \vec{\tau} + \delta\vec{\tau} \ge 0, \\ \delta H_k \in \mathbb{R}^{n\times n}, & \|\delta H_k\|_2 < \delta,\ k = 1, \ldots, m, \\ \delta A_k \in \mathbb{R}^{n\times n}, & \|\delta A_k\|_2 < \delta,\ k = 0, \ldots, m, \end{array}$$

has exactly k zeros[6] in the disk $\{\lambda \in \mathbb{C} :\ |\lambda - \lambda_0| < \epsilon\}$.

[6]multiplicity taken into account

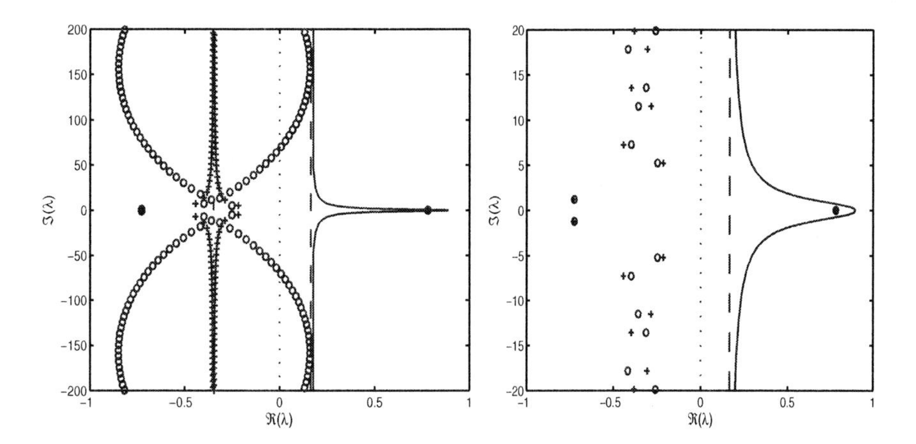

Figure 1.8. *Characteristic roots of the neutral equation (1.28), and the curve defined by (1.44).*

Let the *spectral abscissa function* corresponding to the neutral equation (1.17) be defined as follows:

$$\begin{aligned}&\alpha(\vec{\tau}, H_1, \dots, H_m, A_1, \dots, A_m) := \\ &\qquad \sup\left\{\Re(\lambda) : \ \det \Delta_N(\lambda;\ \vec{\tau}, H_1, \dots, H_m, A_0, \dots, A_m) = 0\right\}.\end{aligned}$$

In contrast to the retarded case, this function is in general not continuous, as we illustrate with the following example.

Example 1.38. *The characteristic matrix of the system*

$$\begin{aligned}&\tfrac{d}{dt}\left(x(t) - \tfrac{3}{4}x(t-\tau_1) + \tfrac{1}{2}x(t-\tau_2)\right) \\ &\qquad = -\left(x(t) - \tfrac{3}{4}x(t-\tau_1) + \tfrac{1}{2}x(t-\tau_2)\right)\end{aligned} \tag{1.45}$$

is given by

$$\Delta_N(\lambda) = (\lambda+1)\left(1 - \frac{3}{4}e^{-\lambda\tau_1} + \frac{1}{2}e^{-\lambda\tau_2}\right) = (\lambda+1)\Delta_D(\lambda).$$

Hence, the characteristic roots of (1.45) consist of the characteristic roots of the delay-difference equation (1.31), in addition to a characteristic root at $\lambda = -1$. The discontinuity of the function $\vec{\tau} \mapsto c_D(\vec{\tau}; -3/4, 1/2)$ at $\vec{\tau} = (1, 2)$, discussed in Example 1.33, carries over in this case to the function $\vec{\tau} \mapsto \alpha(\vec{\tau}, -3/4, 1/2, 3/4, -1/2)$.

Again, the problem can be solved by modifying the spectral abscissa function in such a way that small delay perturbations are taken explicitly into account. We have the following result.

Theorem 1.39. *The function $\beta : \mathbb{R}_+^m \times \mathbb{R}^{n\times n\times(2m+1)} \to \mathbb{R}$,*

$$\begin{aligned}&(\vec{\tau}, H_1, \dots, H_m, A_0, \dots, A_m) \mapsto \beta(\vec{\tau}, H_1, \dots, H_m, A_0, \dots, A_m) \\ &\qquad := \max\left\{\alpha(\vec{\tau}, H_1, \dots, H_m, A_0, \dots, A_m), \bar{C}_D(\vec{\tau};\ H_1, \dots, H_m)\right\}\end{aligned}$$

is continuous.

Proof. We first prove continuity at a point where

$$\alpha(\vec{\tau}, H_1, \dots, H_m, A_0, \dots, A_m) \geq \bar{C}_D(\vec{\tau};\ H_1, \dots, H_m).$$

Fix $\epsilon > 0$. Let $\gamma > 0$ be such that

$$\begin{aligned} b_\gamma(\Re(\lambda)) := \sup \Big\{ & b(\Re(\lambda);\ \vec{\tau} + \delta\vec{\tau},\ H_1 + \delta H_1, \dots, H_m + \delta H_m, \\ & A_0 + \delta A_0, \dots, A_m + \delta A_m) :\ \delta\vec{\tau} \in \mathbb{R}^m,\ \|\delta\vec{\tau}\|_2 < \gamma,\ H_k \in \mathbb{R}^{n\times n}, \\ & \|H_k\|_2 < \gamma,\ k = 1, \dots, m,\ A_k \in \mathbb{R}^{n\times n},\ \|A_k\|_2 < \gamma,\ k = 0, \dots, m \Big\} \end{aligned}$$

is defined and bounded if $\Re(\lambda) \geq \alpha(\vec{\tau}, H_1, \dots, H_m, A_0, \dots, A_m) + \epsilon$ (see Lemma 1.35 for the definition of b). Define the compact set

$$\Omega := \left\{ \lambda \in \mathbb{C} :\ \Re(\lambda) \geq \alpha(\vec{\tau}, H_1, \dots, H_m, A_0, \dots, A_m) + \epsilon,\ |\lambda| \leq b_\gamma(\Re(\lambda)) \right\}. \quad (1.46)$$

It is clear that on Ω the characteristic matrix

$$\Delta_N(\lambda;\ \vec{\tau}, H_1, \dots, H_m, A_0, \dots, A_m)$$

has no zeros. From Rouché's Theorem (see Section A.1 of the appendix) it follows that there exists a number γ_2 such that

$$\Delta_N(\lambda;\ \vec{\tau} + \delta\vec{\tau},\ H_1 + \delta H_1, \dots, H_m + \delta H_m,\ A_0 + \delta A_0, \dots, A_m + \delta A_m) \quad (1.47)$$

has no zeros in Ω whenever

$$\|\delta\vec{\tau}\|_2 < \gamma_2,\ \|\delta H_k\| < \gamma_2,\ k = 1, \dots, m,\ \ \|\delta A_k\| < \gamma_2,\ k = 0, \dots, m. \quad (1.48)$$

If γ_2 is taken smaller than γ, then, by Lemma 1.35, (1.48) also implies that (1.47) has no zeros satisfying

$$\Re(\lambda) \geq \alpha(\vec{\tau}, H_1, \dots, H_m, A_0, \dots, A_m) + \epsilon.$$

Since the above analysis can be repeated for any $\epsilon > 0$ we arrive at

$$\begin{aligned} & \forall\epsilon\ \exists\gamma_2\ \big(\|\delta\vec{\tau}\|_2 < \gamma_2\ \&\ \|\delta H_k\| < \gamma_2,\ k = 1, \dots, m\ \&\ \|\delta A_k\| < \gamma_2,\ k = 0, \dots, m\big) \Rightarrow \\ & \big(\forall\lambda \in \mathbb{C} \text{ with } \Re(\lambda) \geq \alpha(\vec{\tau},\ H_1, \dots, H_m, A_0, \dots, A_m) + \epsilon : \\ & \ \Delta_N(\lambda;\ \vec{\tau} + \delta\vec{\tau},\ H_1 + \delta H_1, \dots, H_m + \delta H_m,\ A_0 + \delta A_0, \dots, A_m + \delta A_m) \neq 0\big). \end{aligned}$$

Roughly speaking, this statement expresses that infinitesimal perturbations cannot lead to a (discontinuous) growth of the spectral abscissa function α. Combining this fact with the continuity of $\bar{C}_D(\vec{\tau};\ H_1, \dots, H_m)$ (Theorem 1.30) and the continuity of the individual characteristic roots (Proposition 1.37), we arrive at the continuity of $\beta(\vec{\tau}, H_1, \dots, H_m, A_0, \dots, A_m)$.

Next, we consider the case where

$$\alpha(\vec{\tau}, H_1, \dots, H_m, A_0, \dots, A_m) < \bar{C}_D(\vec{\tau};\ H_1, \dots, H_m).$$

The proof is by contradiction. Since the function $\bar{C}_D$ is continuous, a violation of the statement of the theorem implies the existence of a number $\epsilon > 0$ such that

$$\begin{aligned}
&\forall \beta > 0\ \exists \delta\vec{\tau} \text{ with } \|\delta\vec{\tau}\|_2 < \beta,\ \exists \delta H_k \text{ with } \|\delta H_k\|_2 < \beta, k = 1, \dots, m, \\
&\exists \delta A_k \text{ with } \|\delta A_k\|_2 < \beta, k = 0, \dots, m : \\
&\alpha(\vec{\tau} + \delta\vec{\tau},\ H_1 + \delta H_1, \dots,\ H_m + \delta H_m,\ A_0 + \delta A_0, \dots,\ A_m + \delta A_m) \\
&> \bar{C}_D(\vec{\tau};\ H_1, \dots, H_m) + \epsilon.
\end{aligned} \tag{1.49}$$

Let $\{\beta_n\}_{n\geq 1}$ be a sequence of strictly positive real numbers satisfying $\lim_{n\to\infty} \beta_n = 0$. Expression (1.49) implies the existence of a corresponding sequence $\{\lambda_n\}_{n\geq 1}$ of complex numbers satisfying $\Re(\lambda) \geq \bar{C}_D(\vec{\tau};\ H_1, \dots, H_m) + \epsilon$, and sequences of perturbations $\{\delta\vec{\tau}_n\}_{n\geq 1}$, $\{\delta H_{k,n}\}_{n\geq 1}$, $\{\delta A_{k,n}\}_{n\geq 1}$ with $\|\delta\vec{\tau}_n\|_2 < \beta_n$, $\|\delta H_{k,n}\|_2 < \beta_n$, $\|\delta A_{k,n}\|_2 < \beta_n$, such that

$$\Delta_N(\lambda;\ \vec{\tau} + \delta\vec{\tau}_n,\ H_1 + \delta H_{1,n}, \dots,\ H_m + \delta H_{m,n},\ A_0 + \delta A_{0,n}, \dots,\ A_m + \delta A_{m,n}) \tag{1.50}$$

has a zero for $\lambda = \lambda_n$. Since for sufficiently small β_n, all zeros of (1.50) satisfying $\Re(\lambda) \geq \bar{C}_D(\vec{\tau};\ H_1, \dots, H_m) + \epsilon$ can be constrained to a compact set (analogous to (1.46)), the sequence $\{\lambda_n\}_{n\geq 1}$ has a converging subsequence with limit $\bar{\lambda}$. It is easy to show, using Rouché-type arguments, that

$$\Delta_N(\bar{\lambda};\ \vec{\tau}, H_1, \dots, H_m, A_0, \dots, A_m) = 0.$$

It follows that

$$\begin{aligned}
&\alpha(\vec{\tau}, H_1, \dots, H_m, A_0, \dots, A_m) \\
&\qquad \geq \bar{C}_D(\vec{\tau},\ H_1, \dots, H_m)
\end{aligned}$$

and we arrive at a contradiction. □

The next result lays the theoretical basis for the methods for computing stability regions in parameter spaces, discussed in the next chapters.

Theorem 1.40. *Assume that the delay-difference equation associated with the neutral equation (1.17) is strongly exponentially stable, that is,* $\gamma_0(H_1, \dots, H_m) < 1$.

If the matrices $A_0, \dots, A_m$ *and the delays* $\tau_1, \dots, \tau_m$ *are varied, then a loss or acquisition of exponential stability of the null solution of (1.17) is associated with characteristic roots on the imaginary axis.*

Proof. Let U be an *arbitrary* compact subset of $\mathbb{R}_+^m$, to which the delays are constrained. As $\gamma_0(H_1, \dots H_m) = f(0;\ \vec{\tau}, H_1, \dots, H_m) < 1$, Theorem 1.30 implies that

$$\max_{\vec{\tau}\in U} \bar{C}_D(\vec{\tau};\ H_1, \dots, H_m) < 0. \tag{1.51}$$

By Theorem 1.39 it then follows that the spectral abscissa function

$$(\vec{\tau}, A_0, \dots, A_m) \mapsto \alpha(\vec{\tau}, H_1, \dots, H_m, A_0, \dots, A_m)$$

is *continuous* whenever

$$\alpha(\vec{\tau}, H_1, \dots, H_m, A_0, \dots, A_m) \geq \max_{\vec{\tau}\in U} \bar{C}_D(\vec{\tau}; H_1, \dots, H_m).$$

A loss of exponential stability is thus characterized by $\alpha = 0$. Since for any $\epsilon > 0$ there are only a finite number of characteristic roots with $\Re(\lambda) \geq \bar{C}_D(\vec{\tau})$, a situation where $\alpha = 0$ corresponds to the presence of characteristic roots on the imaginary axis. □

Remark 1.41. *If the matrices H_k, $k = 1, \ldots, m$, are also varied, then the situation is more complex, since γ_0 depends on these matrices and strong exponential stability of the delay-difference equation may be lost. According to Theorem 1.39, a transition from $\beta < 0$ to $\beta > 0$ then corresponds to either the case discussed above ($\alpha = 0$ for the critical parameters and characteristic roots on the imaginary axis) or $\bar{C}_D = 0$ for the critical parameters. Note that in the latter case the condition $\beta > 0$ does not necessarily imply instability, yet instability can always be achieved by applying infinitesimal perturbations to the delays.*

1.2.6 Computation of characteristic roots

Several methods are available for the computation of characteristic roots of linear time-delay systems of neutral type.

First, let us mention the methods based on a discretization of the solution operator $\mathcal{T}_N(t)$. As in the retarded case, the approach consists of computing the eigenvalues of the resulting discretized operator (matrix), transforming these eigenvalues to approximate characteristic roots, and performing Newton corrections of these roots using the characteristic equation. Even though more attention has been paid to developing and testing this method for retarded systems, it can also be used for computing a part of the spectrum of a neutral system. As shown in [75] for the linearized solution operator around a periodic solution, a discretization based approach may provide good approximations of the characteristic roots with $\Re(\lambda) > \bar{C}_D$ at least. Advantages of the discretization based methods are their numerical robustness and reliability. Recall that they can be implemented using matrix-vector operations only and do not require the generally ill-conditioned step of computing explicitly the characteristic equation. A disadvantage of this approach is that the accuracy of the computed characteristic roots (of the discretized system) decays when their moduli increase. For this it is important to compute $\bar{C}_D$ in a preliminary step, which can be done as outlined in Remark 1.31 and discussed in detail in [199].

From a stability analysis point of view, most important are the knowledge of the spectral bound $\bar{C}_D$ and the characteristic roots with real part larger than $\bar{C}_D$. According to Lemma 1.35 we can restrict ourselves to looking for characteristic roots in the *compact* set

$$\Re(\lambda) \geq \bar{C}_D(\vec{\tau};\ H_1, \ldots, H_m) + \epsilon,\ |\lambda| \leq b(\Re(\lambda);\ \vec{\tau}, H_1, \ldots, H_m, A_0, \ldots, A_m),$$

where $\epsilon > 0$ is a small number. Experience reveals that the number of such characteristic roots is typically very small [199, 75] (if ϵ is not chosen extremely small). This opens the possibility to apply methods and software for directly computing all zeros of analytic functions in a compact set to the characteristic function [149]. Here, we mention the methods described in [143], which are based on contour integration, and the quasipolynomial mapping based technique of [318]. The basic idea of the latter technique consists of mapping the contours $\Re(\det \Delta_N(\lambda)) = 0$ and $\Im(\det \Delta_N(\lambda)) = 0$ using a level curve tracing algorithm and locating the intersection points of the contours which are the root approximations. Using such approaches, unlike if a discretization based method is used, all the characteristic roots located in a defined region of the complex plane are approximated within the same accuracy. These methods works well if the quasipolynomial structure is not too complicated and the quasipolynomial is not ill-conditioned.

1.3 Notes and references

We introduced some classes of linear time-delay systems and outlined spectral properties, thereby taking a stability analysis point of view and using eigenvalue plots to illustrate the main results. Our goal was not to present a complete theory, but to focus on these properties which play an important role in the rest of the book. Key references for the general theory of FDEs, of which the delay equations presented form a special class, are [101, 106, 138, 139, 67]. For a general introduction to infinite-dimensional systems from a systems theory point of view we refer to [58].

The part on retarded systems is based on Cooke [52], with complements from [101, 106]. The terminology piece of trajectories in defining the state notion x_t of a time-delay system was suggested by Krasovskii in [142]. The construction of solutions using the step-by-step method follows closely the arguments in [223] (see also [100]). To the best of our knowledge, Bellman was the first to propose such a construction of the solutions in the context of delay-difference equations. For a more general discussion on the method of steps we refer to [71]. Further remarks and comments on the distribution of zeros of the characteristic function for linear systems of retarded type can be found in Bellman and Cooke [13] (see also some discussions in Kolmanovskii and Myshkis [138]).

The discussion on neutral equations presented in Sections 1.2.1 to 1.2.3 is based on results of [106, 107, 108, 199, 181, 249], which are applied to (1.17) and further developed taking into account the specific structure of this equation. The results on delay sensitivity in Section 1.2.4 are based on [199]. To the best of our knowledge, a detailed analysis of continuity properties of the spectrum, in particular of the spectral abscissa, as presented in Section 1.2.5, is not performed in the existing literature, although some of the ideas and results are implicity present in some papers (for instance, in [98]). This clarifies why the proofs in Section 1.2.5 are fully developed.

Throughout the chapter we have restricted ourselves to linear time-delay systems with *pointwise* (discrete) delays (although most of the results in this chapter can be generalized to more general classes of FDEs), because the remainder of the book is almost exclusively devoted to problems involving this type of delays. If other types of delays are considered in particular places (e.g., in Chapters 15 and 16), they concern systems which can be brought into a form with pointwise delays using particular model transformations in the sense of [162, 205] (see also [97]). At such places the necessary additions to the stability theory of this chapter will be provided.

Chapter 2

Pseudospectra and robust stability analysis

2.1 Introduction

Closeness to instability is a key issue in understanding the behavior of physical systems subject to perturbations. The computation of pseudospectra has become an established tool in analyzing and gaining insight into this phenomenon (see, for instance, [303, 301, 116] and the references therein). More explicitly, pseudospectra of a system are sets in the complex plane to which eigenvalues or characteristic roots can be shifted under a perturbation of a given size. In the simplest case of a matrix (or linear operator) A, the ϵ-pseudospectrum $\Lambda_\epsilon(A)$ is defined as

$$\Lambda_\epsilon(A) := \{\lambda \in \mathbb{C} : \lambda \in \Lambda(A + P), \text{ for some } P \text{ with } \|P\| < \epsilon\}, \tag{2.1}$$

where Λ denotes the spectrum and $\|\cdot\|$ denotes an arbitrary matrix (or operator) norm. Equation (2.1) is known to be equivalent to the following:

$$\Lambda_\epsilon(A) = \{\lambda \in \mathbb{C} : \|R(\lambda, A)\| > 1/\epsilon\}, \tag{2.2}$$

where $R(\lambda, A) = (\lambda I - A)^{-1}$ denotes the corresponding resolvent operator.

Although most systems can be written in a first-order form, it is often advantageous to exploit the underlying structure in their analysis; for example, one may wish to compute pseudospectra of higher-order or DDEs. In particular, this can be of importance in sensitivity investigations, where it is desirable to respect the structure of the governing system. For example, many physical problems involving vibration of structural systems and vibro-acoustics are modeled by second-order differential equations of the form

$$A_2\ddot{x}(t) + A_1\dot{x}(t) + A_0x(t) = 0,$$

where A_2, A_1, and A_0 represent mass, damping, and stiffness matrices, respectively. Stability is inferred from the characteristic roots, found as solutions of

$$\det(A_2\lambda^2 + A_1\lambda + A_0) = 0.$$

To understand the sensitivity of the characteristic roots with respect to complex perturbations with weights α_i applied to A_i, $i = 0, 1, 2$, the ϵ-pseudospectrum of the matrix polynomial $P(\lambda) = A_2\lambda^2 + A_1\lambda + A_0 \in \mathbb{C}^{n\times n}$ can be defined as

$$\begin{aligned}\Lambda_\epsilon(P) := \{\lambda \in \mathbb{C} \quad : \quad & (P(\lambda) + \Delta P(\lambda))x = 0 \text{ for some } x \neq 0 \\ & \text{and } \Delta P(\lambda) = \delta A_2\lambda^2 + \delta A_1\lambda + \delta A_0 \\ & \text{with } \delta A_i \in \mathbb{C}^{n\times n} \text{ and } \|\delta A_i\| < \epsilon\alpha_i,\ i = 0, 1, 2\}.\end{aligned} \tag{2.3}$$

We refer to [298] for a survey on the quadratic eigenvalue problem, including numerical solutions and applications, and to [297] for pseudospectra of polynomial matrices. More recently, pseudospectra for nonlinear eigenvalue problems that correspond to linear time-delay systems have been defined and analyzed, starting with [94]. In its simplest form of one fixed, discrete delay $\tau \in \mathbb{R}_+$, the characteristic matrix takes the form

$$Q(\lambda) := \lambda I - A_0 - A_1 e^{-\lambda\tau},$$

as we have seen in Chapter 1. Similar to (2.3) the associated pseudospectrum is defined in [94] as

$$\begin{aligned}\Lambda_\epsilon(Q) := \{\lambda \in \mathbb{C} \quad : \quad & (Q(\lambda) + \Delta Q(\lambda))x = 0 \text{ for some } x \neq 0 \\ & \text{and } \Delta Q(\lambda) = \delta A_0 + \delta A_1 \epsilon^{-\lambda\tau} \\ & \text{with } \delta A_i \in \mathbb{C}^{n\times n} \text{ and } \|\delta A_i\| < \epsilon\alpha_i,\ i = 0, 1\}.\end{aligned} \tag{2.4}$$

The aim of this chapter is two-fold. The first goal is to present a unified theory for the definition and computation of pseudospectra of general nonlinear eigenvalue problems of the form

$$\det(F(\lambda)) = 0, \tag{2.5}$$

where the *characteristic matrix* F is given by

$$F(\lambda) := \sum_{i=0}^{m} A_i p_i(\lambda), \tag{2.6}$$

with $A_i \in \mathbb{C}^{n\times n}$ and p_i entire. The spectrum of F is defined as

$$\Lambda(F) := \{\lambda \in \mathbb{C} : \det(F(\lambda)) = 0\}. \tag{2.7}$$

It is easy to see that all the cases described above are included in this class of nonlinear eigenvalue problem. The second goal is to rephrase and interpret the results for the case of time-delay systems.

The chapter is organized as follows. In Section 2.2 we define and analyze pseudospectra for the nonlinear eigenvalue problem (2.5) under the assumption of complex perturbations on the coefficient matrices A_i. Thereby, the structure of the matrix function (2.6) is fully exploited. Various perturbation measures are discussed and computable formulae are derived that are tractable from a numerical point of view. One of the practical applications concerns the associated complex *stability radius* of (2.6), that is, a measure of the distance to instability. See [116, 196, 250, 87] for the concept and various types of stability radii. Specifically, if we decompose $\mathbb{C}$ into two disjoint regions, a desired region $\mathbb{C}_d$ and an undesired region $\mathbb{C}_u$, the complex stability radius of (2.6) is defined as

$$\begin{aligned} r_{\mathbb{C}}(F;\ \mathbb{C}_d, \|\cdot\|_{\text{glob}}) := \inf_{\lambda\in\mathbb{C}_u} \inf_{\epsilon>0} \big\{\epsilon :\ \det\big(\textstyle\sum_{i=0}^{m}(A_i+\delta A_i)p_i(\lambda)\big) = 0 \\ \text{for some } \Delta = (\delta A_0,\dots,\delta A_m) \in \mathbb{C}^{n\times n\times(m+1)} \text{ with } \|\Delta\|_{\text{glob}} < \epsilon\big\}, \end{aligned} \tag{2.8}$$

where $\|\Delta\|_{\text{glob}}$ is a global measure of the perturbation Δ, a combination of the complex perturbations δA_i; this is discussed in detail in Section 2.2. In other words, $r_{\mathbb{C}}$ defines the norm of the smallest perturbation that destroys the $\mathbb{C}_d$-stability, that is, having all the roots confined to $\mathbb{C}_d$. Furthermore, it corresponds to the smallest ϵ value at which the ϵ-pseudospectrum has a nonempty intersection with $\mathbb{C}_u$. Note that for a system with continuous time, for example the time-delay systems discussed in this book, $\mathbb{C}_d = \mathbb{C}_-$; for discrete time systems, $\mathbb{C}_d = \{\lambda \in \mathbb{C} :\ |\lambda| < 1\}$.

Section 2.3 is devoted to the definition and derivation of computable expressions for so-called *structured* pseudospectra and stability radii of (2.5). The main difference with the (unstructured) pseudospectra and stability radii discussed in Section 2.2 lies in the fact that, *in addition* to exploiting the structure of the matric function (2.6), a particular structure can be imposed on the perturbations of the individual coefficient matrices A_i, at the price of a higher computational cost. The motivation stems from the observation that in many practical applications the coefficient matrices have a certain structure that should be respected in the sensitivity analysis [321]. In Sections 2.2 and 2.3 the results are applied explicitly to time-delay systems. Two numerical examples are presented in Section 2.4, and some concluding remarks end the chapter.

2.2 Pseudospectra for nonlinear eigenvalue problems

Definitions and expressions for pseudospectra of (2.6) are presented. The connection with stability radii is clarified. Next, computational issues are discussed. Finally, the main results are rephrased for a class of time-delay systems.

2.2.1 Definition and expressions

We study the zeros of the equation given by (2.6), where $A_i \in \mathbb{C}^{n\times n}$, $i = 0,\dots,m$, and the functions $p_i : \mathbb{C} \to \mathbb{C}$, $i = 0,\dots,m$, are entire. In particular, we are interested in the effect of bounded perturbations of the matrices A_i on the position of the roots. For this, we analyze the perturbed equation,

$$\det\left\{\sum_{i=0}^{m}(A_i+\delta A_i)p_i(\lambda)\right\} = 0. \tag{2.9}$$

The first step in the robustness analysis is to define the class of perturbations under consideration, as well as a measure of the combined perturbation

$$\Delta := (\delta A_0,\dots,\delta A_m). \tag{2.10}$$

In this section we assume that the allowable perturbations δA_i, $i = 0,\dots,m$, are *complex* matrices, that is,

$$\Delta \in \mathbb{C}^{n\times n\times(m+1)}. \tag{2.11}$$

Introducing weights $w_i \in \overline{\mathbb{R}}_+$, $i = 0, \ldots, m$, where $\overline{\mathbb{R}}_+ = \mathbb{R}_+ \cup \{\infty\}$, we define three global measures of the perturbations:

$$\|\Delta\|_{\text{glob}} := \|[w_0\, \delta A_0 \ldots w_m\, \delta A_m]\|_p, \tag{2.12}$$

or

$$\|\Delta\|_{\text{glob}} := \left\| \begin{bmatrix} w_0\, \delta A_0 \\ \vdots \\ w_m\, \delta A_m \end{bmatrix} \right\|_p, \tag{2.13}$$

where $\|M\|_p$ is the induced matrix norm given by $\|M\|_p = \sup_{\|x\|_p=1} \|Mx\|_p$, $p \in \overline{\mathbb{N}} := \mathbb{N} \cup \{+\infty\}$. Notice that $w_j = \infty$ for some j means that no perturbation on A_j is allowed when the combined perturbation Δ is required to be bounded; that is, $w_j = \infty \Rightarrow \delta A_j = 0$, for some j. Finally, we also consider a measure of *mixed* type:

$$\|\Delta\|_{\text{glob}} := \left\| \begin{bmatrix} w_0 \|\delta A_0\|_{p_1} \\ \vdots \\ w_m \|\delta A_m\|_{p_1} \end{bmatrix} \right\|_{p_2}, \quad p_1, p_2 \in \overline{\mathbb{N}}_0. \tag{2.14}$$

For instance, when $p_2 = \infty$ and all weights are equal to one, the condition $\|\Delta\|_{\text{glob}} < \epsilon$ corresponds to the natural assumptions of taking perturbations satisfying $\|\delta A_i\|_{p_1} < \epsilon$, $i = 0, \ldots, m$. In this special case, (2.14) is also equal to the p_1-norm of the *block-diagonal* perturbation matrix $\operatorname{diag}(\delta A_0, \ldots, \delta A_m)$, considered in [250, 87] for polynomial matrices. Notice that if all weights are finite, then the measures given by (2.12)–(2.14) are norms.

For any of the above definitions of $\|\Delta\|_{\text{glob}}$, we define the ϵ-pseudospectrum of (2.6) as follows.

Definition 2.1.

$$\begin{aligned} \Lambda_\epsilon(F;\ \|\cdot\|_{\text{glob}}) := \Big\{\lambda \in \mathbb{C} :\ & \det\Big(\textstyle\sum_{i=0}^m (A_i + \delta A_i) p_i(\lambda)\Big) = 0 \text{ for some} \\ & \Delta = (\delta A_0, \ldots, \delta A_m) \in \mathbb{C}^{n \times n \times (m+1)} \text{ with } \|\Delta\|_{\text{glob}} < \epsilon\Big\}. \end{aligned} \tag{2.15}$$

Defining the function $f : \mathbb{C} \to \overline{\mathbb{R}}^+$ as the inverse of the size of the smallest perturbation which shifts a root to λ if such perturbations exist, and zero otherwise, more precisely,

$$f(\lambda;\ \|\cdot\|_{\text{glob}}) = \begin{cases} 0, & \text{if } \det\Big(\sum_{i=0}^m (A_i + \delta A_i) p_i(\lambda)\Big) \neq 0,\ \forall \Delta \in \mathbb{C}^{n\times n\times(m+1)}, \\ +\infty, & \text{if } \lambda \in \Lambda(F), \\ \Big(\inf\Big\{\|\Delta\|_{\text{glob}} :\ \det\Big(\sum_{i=0}^m (A_i + \delta A_i) p_i(\lambda)\Big) = 0\Big\}\Big)^{-1}, & \text{otherwise,} \end{cases} \tag{2.16}$$

we can also define the ϵ-pseudospectra as

$$\Lambda_\epsilon(F;\ \|\cdot\|_{\text{glob}}) = \Big\{\lambda \in \mathbb{C} :\ f(\lambda;\ \|\cdot\|_{\text{glob}}) > \epsilon^{-1}\Big\}. \tag{2.17}$$

The boundary of pseudospectra is thus formed by the level sets of the function f, which can be written in a computational form as follows [183, Theorem 1].

Theorem 2.2. *For the perturbation measures (2.12)–(2.14) the function (2.16) satisfies*

$$f(\lambda;\ \|\cdot\|_{\text{glob}}) = \begin{cases} \left\| \left(\sum_{i=0}^{m} A_i p_i(\lambda) \right)^{-1} \right\|_{\alpha} \cdot \|w(\lambda)\|_{\beta}, & \lambda \notin \Lambda(F), \\ +\infty, & \lambda \in \Lambda(F), \end{cases}$$

where

$$w(\lambda) = \begin{bmatrix} \frac{p_0(\lambda)}{w_0} \\ \vdots \\ \frac{p_m(\lambda)}{w_m} \end{bmatrix} \tag{2.18}$$

and

$$\begin{array}{lc|l} \alpha = p,\ \beta = p, & & \text{perturbation measure (2.12)}, \\ \alpha = p,\ \beta = q, & \frac{1}{p} + \frac{1}{q} = 1, & \text{perturbation measure (2.13)}, \\ \alpha = p_1,\ \beta = q_2, & \frac{1}{p_2} + \frac{1}{q_2} = 1 & \text{perturbation measure (2.14)}. \end{array}$$

The proof is frequency domain based. It relies on a feedback interpretation of the perturbed system, and an explicit construction of worst case perturbations.

2.2.2 Connection with stability radii

As outlined in Section 2.1 the concept of stability radii given by (2.8) is closely related to pseudospectra. To clarify this relationship and to arrive at a computable formula, we need the following continuity property of the individual roots of (2.6) with respect to changes of matrices A_i, which can be once again shown using Rouché's Theorem.

Proposition 2.3. *For all $\mu > 0$ and $\lambda_0 \in \mathbb{C}$, there exists a $\nu > 0$ such that for all $\Delta = (\delta A_0, \ldots, \delta A_m) \in \mathbb{C}^{n \times n \times (m+1)}$ with $\|\Delta\|_{\text{glob}} < \nu$, (2.9) has the same number of roots*[7] *as (2.6) in the disk $\{\lambda \in \mathbb{C} :\ |\lambda - \lambda_0| < \mu\}$.*

Assume that all the roots of (2.6) are in $\mathbb{C}_d$. Let Δ_c be an arbitrary perturbation with $\|\Delta_c\|_{\text{glob}}$ finite, for which there is at least one root in $\mathbb{C}_u$ (such perturbations always exist, by Theorem 2.2). Next, apply the perturbation $\Delta := \epsilon\ \Delta_c$, where $\epsilon \geq 0$ is a parameter. Clearly, the function

$$\epsilon \in [0,\ 1] \rightarrow \epsilon\ \Delta_c$$

is continuous with respect to the measure $\|\cdot\|_{\text{glob}}$. Consequently, by Proposition 2.3 one of the following phenomena must happen to the roots of the perturbed system when ϵ is continuously varied from zero to one:

1. some roots move from $\mathbb{C}_d$ to $\mathbb{C}_u$;

2. roots coming from infinity appear in $\mathbb{C}_u$ (only for unbounded $\mathbb{C}_u$).

If the second case can be excluded, a loss of stability is always associated with roots on the boundary of $\mathbb{C}_d$ and it becomes sufficient to scan this boundary in the outer optimization

[7] multiplicity taken into account

of (2.8). In other words, the stability radius is the smallest value of ϵ for which an ϵ-pseudospectrum contour reaches the boundary of $\mathbb{C}_d$. Formally, using (2.16) one has the following corollary.

Corollary 2.4. *Assume that all the roots of (2.6) are in $\mathbb{C}_d$. Then*

$$r_{\mathbb{C}}(F;\ \mathbb{C}_d, \|\cdot\|_{\text{glob}}) = \inf_{\lambda\in\Gamma_{\mathbb{C}_d}} \frac{1}{f(\lambda;\ \|\cdot\|_{\text{glob}})} = \frac{1}{\sup_{\lambda\in\Gamma_{\mathbb{C}_d}} f(\lambda;\ \|\cdot\|_{\text{glob}})}, \tag{2.19}$$

where $\Gamma_{\mathbb{C}_d}$ is the boundary of the set $\mathbb{C}_d$.

The following example from [183] demonstrates that Corollary 2.4 does not hold if perturbations create roots coming from infinity in $\mathbb{C}_u$.

Example 2.5. *The equation $p(\lambda) = 0$, with*

$$p(\lambda) = \lambda + 1 + \delta a\, e^{\lambda},$$

is $\mathbb{C}_-$-stable for $\delta a = 0$. With $\|\Delta\|_{\text{glob}} = |\delta a|$, we have

$$\inf_{\lambda\in\Gamma_{\mathbb{C}_-}} \frac{1}{f(\lambda;\ \|\cdot\|_{\text{glob}})} = \inf_{\omega\geq 0} \frac{|1+j\omega|}{|e^{j\omega}|} = 1,$$

that is, shifting roots to the imaginary axis requires $|\delta a| \geq 1$. However, the stability radius is zero because for any real $\delta a \neq 0$, there are infinitely many roots in the open right half-plane, whose real parts move off to plus infinity as $|\delta a| \to 0+$. To see this, note that $p(-\lambda)$ can be interpreted as the characteristic function of the DDE $\dot{x}(t) = x(t) + \delta a\, x(t-1)$, which has infinitely many characteristic roots located in a logarithmic sector of the left half-plane, as we have seen in Chapter 1.

2.2.3 Computational issues

Using (2.17) and Theorem 2.2, pseudospectra of (2.6) can be computed by evaluating

$$\left\| \left(\sum_{i=0}^{m} A_i p_i(\lambda) \right)^{-1} \right\|_{\alpha} \|w(\lambda)\|_{\beta}$$

for λ on a grid over a region of the complex plane. By using a contour plotter to view the results, the boundaries of ϵ-pseudospectra are then identified. Notice that for $\alpha = 2$, the left term can be computed as the inverse of the smallest singular value of $\sum_{i=0}^{m} A_i p_i(\lambda)$. Analogously, from Corollary 2.4 the complex stability radius can be computed using a grid, laid on the boundary of the stability region. Such an approach is taken for the numerical examples of Section 2.4.

It is important to mention that the computational complexity is similar to the complexity of computing pseudospectra of an n-by-n matrix (see formula (2.2)), since the characteristic matrix F has dimensions n by n. In particular, when the nonlinear eigenvalue problem stems from an infinite-dimensional linear system (such as the time-delay systems discussed in the next paragraph), a "lifting" to a first-order representation would have led to a formula of the form (2.2), with $\mathcal{A}$ an infinite-dimensional operator.

To compute pseudospectra of specific problems (for example, large matrices with a special structure) and, in particular, to solve optimization problems related to pseudospectra, efficiency can often be improved by exploiting properties of the problem under consideration. See, for instance, [87, Section 6] and the references therein for an efficient algorithm to compute stability radii of polynomial matrices and [35] for the efficient computation and optimization of so-called pseudospectral abscissa of matrices. However, this is beyond the scope of this chapter, where generality is the main concern.

To conclude this section, Table 2.1 lists publications where some of the results stated in Theorem 2.2 or Corollary 2.4 were obtained. Note that we have restricted ourselves to nonscalar problems (results for scalar polynomials are, for instance, described in [116, section 5.4.2] and the references therein).

Table 2.1. *Special cases of Theorem 2.2/ Corollary 2.4, treated in the literature.*

Reference	Problem	Perturbation measure	Weights
[306]	matrix pencil	(2.12)	/
[87]	polynomial matrices	(2.12), (2.14) with $p_2 = \infty$	/
[250]	polynomial matrices	(2.12), (2.14) with $p_2 = \infty$	yes
[297]	polynomial matrices	(2.12), (2.14) with $p_2 = \infty$	yes
[94]	time-delay systems	(2.14) with $p_1 = 2$ and $p_2 = \infty$	yes
[183]	general case		

2.2.4 Application to time-delay systems

We apply the results of Section 2.2 to linear DDEs of the form

$$\dot{x}(t) = A_0 x(t) + \sum_{i=1}^{m} A_i x(t - \tau_i), \tag{2.20}$$

where we assume that $0 < \tau_1 < \cdots < \tau_m$ and that the system matrices $A_i \in \mathbb{R}^{n\times n}$, $i = 0, \ldots, m$, are uncertain. In this particular case we have

$$F(\lambda) = \lambda I - A - \sum_{i=1}^{m} A_i e^{-\lambda \tau_i}. \tag{2.21}$$

Expressions

Pseudospectra and stability radii of (2.20), following the general definitions (2.8) and (2.15), can be computed as follows.

Theorem 2.6. *For perturbations $\delta A_i \in \mathbb{C}^{n\times n}$, $i = 0, \ldots, m$, measured by (2.12)–(2.14), the pseudospectrum Λ_ϵ of (2.20) satisfies*

$$\begin{aligned}\Lambda_\epsilon(F;\ \|\cdot\|_{\text{glob}}) = \Lambda(F) \cup \\ \left\{\lambda \in \mathbb{C} : \left\|\left(\lambda I - A_0 - \sum_{i=1}^{m} A_i e^{-\lambda\tau_i}\right)^{-1}\right\|_\alpha \cdot \|w(\lambda)\|_\beta > \epsilon^{-1}\right\}\end{aligned} \tag{2.22}$$

and, if the zero solution of (2.20) is asymptotically stable, then the associated stability radius satisfies

$$\begin{aligned} r_{\mathbb{C}}(F;\ \mathbb{C}_-, \|\cdot\|_{\text{glob}}) = \\ \left(\left(\sum_{i=0}^{m} w_i^{-\beta}\right)^{\frac{1}{\beta}} \sup_{\omega\geq 0} \left\|\left(j\omega I - A_0 - \sum_{i=1}^{m} A_i e^{-j\omega\tau_i}\right)^{-1}\right\|_\alpha\right)^{-1},\end{aligned} \tag{2.23}$$

where $w(\lambda) = \left[\frac{1}{w_0}\ \frac{e^{-\lambda\tau_1}}{w_1} \ldots \frac{e^{-\lambda\tau_m}}{w_m}\right]^T$ and α and β are defined as in Theorem 2.2.

Remark 2.7. *For the system*

$$\dot{x}(t) = (A + \delta A)x(t), \tag{2.24}$$

with $\|\Delta\|_{\text{glob}} = \|\delta A\|_2$, expression (2.22) simplifies to

$$\Lambda_\epsilon(F;\ \|\cdot\|_{\text{glob}}) = \Lambda(F) \cup \left\{\lambda \in \mathbb{C} : \|\mathcal{R}(\lambda, A)\|_2 > \epsilon^{-1}\right\}, \tag{2.25}$$

where $\mathcal{R}(\lambda, A) = (\lambda I - A)^{-1}$ is the resolvent *of A. As mentioned in the introduction, the right-hand side of (2.25) can also be considered as a definition for the ϵ-pseudospectrum of (2.24).*

In general, one can formulate (2.20) as an abstract evolution equation over the Hilbert space $X := \mathbb{C}^n \times \mathcal{L}_2([-\tau_m, 0], \mathbb{C}^n)$, equipped with the usual inner product

$$\langle (y_0, y_1), (z_0, z_1)\rangle_X = \langle y_0, z_0\rangle_{\mathbb{C}^n} + \langle y_1, z_1\rangle_{\mathcal{L}_2},$$

namely

$$\frac{d}{dt} z(t) = \mathcal{A}\, z(t), \tag{2.26}$$

where

$$\begin{aligned} D(\mathcal{A}) = \{z = (z_0, z_1) \in X : &\ z_1 \text{ is absolutely continuous on } [-\tau_m, 0], \\ &\tfrac{dz_1}{d\theta}(\cdot) \in \mathcal{L}_2([-\tau_m, 0], \mathbb{C}^n),\ z_0 = z_1(0)\big\}, \\ \mathcal{A}\, z = &\begin{pmatrix} A_0 z_0 + \sum_{i=1}^{m} A_i z_1(-\tau_i) \\ \frac{dz_1}{d\theta}(\cdot) \end{pmatrix}, \quad z \in D(\mathcal{A}). \end{aligned}$$

Furthermore, the solutions of (2.26) and (2.20) are connected by the relation $z_0(t) \equiv x(t)$, $z_1(t) \equiv x(t+\theta)$, $\theta \in [-\tau_m, 0]$. In this way, one can alternatively define the ϵ-pseudospectrum of (2.20) as the set

$$\Lambda(F) \cup \left\{\lambda \in \mathbb{C} :\ \|\mathcal{R}(\lambda, \mathcal{A})\| > \epsilon^{-1}\right\}. \tag{2.27}$$

See, for example, [57] for further discussions and remarks related to the abstract formulation of differential delay equations on a Hilbert space and [85] for pseudospectra of closed linear operators on complex Banach spaces.

Definition (2.27) is related to the effect of unstructured perturbations of the operator $\mathcal{A}$ *on stability. In this chapter we have chosen a more practical definition, by directly relating pseudospectra to concrete perturbations on the system matrices. Notice that such a practical definition is typically used also for polynomial equations [297, 250, 87].*

Remark 2.8. *When* $\alpha = 2$*, computing the complex stability radius (2.23) requires the computation of the* $\mathcal{H}_\infty$*-norm of the transfer function*

$$H(j\omega) := \left(j\omega I - A_0 - \sum_{k=1}^{m} A_k e^{-\lambda\tau_k} \right)^{-1}, \tag{2.28}$$

and for this a frequency grid can be used, as outlined in Section 2.2.3. The fast iterative methods to calculate $\mathcal{L}_\infty$*-norms of transfer functions or stability radii for finite-dimensional systems, as described in [38; 21; 22; 87; 86, Section 6], and the references therein, are not directly applicable to the delay case. The underlying principle, namely the fact that the frequencies* ω *where one of the singular values of* $H(j\omega)$ *equals a given constant value coincide with the imaginary eigenvalues of a certain Hamiltonian system, remains valid. However, for the delay case this Hamiltonian system is also infinite-dimensional and described by a FDE with both delayed and advanced terms. For a given dimension* n*, it can even have arbitrarily many eigenvalues on the imaginary axis, which prevents a fast direct calculation. This is illustrated with the scalar example*

$$h(j\omega) = \left(j\omega + 1 + \frac{1}{2} e^{-j\omega\tau} \right)^{-1}. \tag{2.29}$$

One easily shows that for $\tau \geq 10N\pi$*, there are more than* N *different values of* ω *in the interval* $\left[0, \frac{1}{10}\right]$*, where* $|h(j\omega)| = 1$*.*

However, despite the infinite-dimensional nature of equation (2.20), it describes an evolution in the finite-dimensional space $\mathbb{R}^n$*, and the dimension of (2.28) is* $n \times n$*. So if* n *is small, then using a simple grid to scan the imaginary axis is not a computational burden.*

Effect of weighting

Applying different weights to the system matrices A_i of 2.20, $i = 1, \ldots, m$, leads to changes in the pseudospectra. This can be understood by investigating the weighting function $w(\lambda) = w(\sigma + j\omega)$, where

$$\|w(\sigma + j\omega)\|_\beta = \left\| \left[\frac{1}{w_0}, \frac{e^{-\sigma\tau_1}}{w_1}, \ldots, \frac{e^{-\sigma\tau_m}}{w_m} \right]^T \right\|_\beta \quad \forall \sigma, \omega \in \mathbb{R}. \tag{2.30}$$

Note that $w(\lambda)$ only depends on the real part σ, that is, $w(\lambda) \equiv w(\sigma)$. From (2.30) the following conclusions can be drawn:

1. Characteristic roots in the right half-plane are more sensitive to perturbations of the nondelayed term A_0.

2. Characteristic roots in the left half-plane are more sensitive to perturbations of the delayed terms A_i, $i = 1, \ldots, m$.

3. Furthermore, the intersection of an ϵ-pseudospectrum contour with the imaginary axis is independent of the weights, provided that the β-norm of $w(\lambda) = w(0)$ is constant. As a consequence, under this condition, the stability radius is also independent of the weights.

Asymptotic properties

In order to characterize boundedness properties of pseudospectra, we investigate the behavior of

$$f(\lambda; \|\cdot\|_{\text{glob}}) := \begin{cases} \left\| \left(\lambda I - A_0 - \sum_{i=1}^{m} A_i e^{-\lambda\tau_i}\right)^{-1} \right\|_{\alpha} \|w(\lambda)\|_{\beta}, & \lambda \notin \Lambda(F), \\ +\infty, & \lambda \in \Lambda(F), \end{cases}$$

as $|\lambda| \to \infty$. This leads to the following two results from [183, Section 3.3].

Proposition 2.9. *For all* $\mu \in \mathbb{R}$,

$$\lim_{R\to\infty} \inf \left\{ f(\lambda, \|\cdot\|_{\text{glob}})^{-1} : \Re(\lambda) > \mu, \ |\lambda| > R \right\} = \infty. \tag{2.31}$$

As a consequence, the cross section between any pseudospectrum and any right half-plane is bounded.

Proposition 2.10. *Let*

$$\Psi_\gamma := \left\{ \lambda \in \mathbb{C} : \Re(\lambda) < -\gamma, \ |\lambda| < e^{-(\Re(\lambda)+\gamma)\tau_m} \right\}. \tag{2.32}$$

If A_m *is regular, then*

$$\begin{aligned} &\forall \epsilon \in \left(0, \tfrac{w_m}{\|A_m^{-1}\|_\alpha}\right), \ \exists \gamma > 0 \text{ such that } \Psi_\gamma \cap \Lambda_\epsilon(F; \|\cdot\|_{\text{glob}}) = \phi, \\ &\forall \epsilon > \tfrac{w_m}{\|A_m^{-1}\|_\alpha}, \ \exists \gamma > 0 \text{ such that } \Psi_\gamma \subset \Lambda_\epsilon(F; \|\cdot\|_{\text{glob}}). \end{aligned} \tag{2.33}$$

If A_m *is singular, then*

$$\forall \epsilon > 0, \ \exists \gamma > 0 \text{ such that } \Psi_\gamma \subset \Lambda_\epsilon(F; \|\cdot\|_{\text{glob}}). \tag{2.34}$$

In the case of a singular A_m, the pseudospectrum Λ_ϵ thus stretches out along the negative real axis, for *any* value of $\epsilon > 0$. Conversely, for the case of a regular A_m, this only happens for $\epsilon > w_m/\|A_m^{-1}\|_\alpha$. As a consequence, *infinitesimal* perturbations may result in the introduction of characteristic roots with *small* imaginary parts (but large negative real parts).

The two cases detailed above are connected as follows: when the matrix A_m is regular, we have

$$\inf_{\delta A_m \in \mathbb{C}^{n\times n}} \{\|\delta A_m\|_\alpha : \det(A_m + \delta A_m) = 0\} = 1/\|A_m^{-1}\|_\alpha;$$

that is, the smallest rank reducing perturbation has size $1/\|A_m^{-1}\|_\alpha$. Furthermore, the smallest perturbation $\Delta = (\delta A_0, \ldots, \delta A_m)$ on the delay equation (2.20), which introduces a characteristic root with a predetermined very large negative real part but small imaginary part, can be decomposed into a minimal size perturbation $\Delta_c = (0, \ldots, 0, \delta A_m)$ which makes A_m singular (due to the weights we have $\|\Delta_c\|_{\text{glob}} = w_m/\|A_m^{-1}\|_\alpha$), together with a very small perturbation to place the characteristic root, according to (2.34).

2.3 Structured pseudospectra for nonlinear eigenvalue problems

Structured pseudospectra of (2.6) are defined and motivated. Next, various computational expressions are presented and discussed. Finally, the results are applied to the time-delay system (2.20).

2.3.1 Exploiting the system's structure

A number of stability and robustness problems for linear systems lead to the study of the eigenvalues of a matrix A with a certain structure (for example, a block structure), which should be respected in the sensitivity analysis [303]. For this, perturbations of the form $A + DPE$ have been considered in [115], where the fixed matrices D and E describe the perturbation structure and P is a complex perturbation matrix. This approach has been further developed in [320] for perturbations of the form $A + \sum D_i P_i E_i$, which, in particular, allow one to deal with elementwise perturbations. On the other hand, specific classes of systems, like higher-order systems or systems with time-delays, lead to the study of the zeros of nonlinear eigenvalue problems of the form (2.5), as we have seen in the previous section. In this section we *combine* these two approaches for exploiting a system's structure. In light of this, we define pseudospectra for the nonlinear eigenvalue problem (2.5) and derive computable formulae, where, in addition to exploiting the form of the nonlinear eigenvalue problem, a particular structure can be imposed on the perturbations of the individual coefficient matrices A_i. This is necessary because in a lot of applications the different coefficient matrices have a certain structure that should naturally be respected in a sensitivity analysis, since unstructured perturbations may lead to irrelevant or nonphysical effects. One example is discussed in [94], where the emergence of unbounded pseudospectra of a delay system in certain directions is explained by nonphysical perturbations that destroy an intrinsic property, namely the singular nature, of one of the coefficient matrices. Another example from laser physics will be discussed in Section 2.4.2.

2.3.2 Definition and expressions

The definition (2.15) for the ϵ-pseudospectrum of the nonlinear eigenvalue problem (2.5), for complex perturbations measured with (2.14), where $p_1 = 2$ and $p_2 = \infty$, is equivalent to

$$\Lambda_\epsilon(F) := \Big\{\lambda \in \mathbb{C} : \det\Big(\sum_{i=0}^m (A_i + \delta A_i) p_i(\lambda)\Big) = 0 \text{ for some } \delta A_i \in \mathbb{C}^{n\times n} \text{ with } w_i\|\delta A_i\|_2 < \epsilon,\ 0 \le i \le m\Big\}. \tag{2.35}$$

Observe that the perturbations δA_i considered in (2.35) lead to an additive uncertainty on the characteristic matrix (2.6) given by

$$\delta F(\lambda) := \sum_{i=0}^{m} \delta A_i \, p_i(\lambda). \tag{2.36}$$

Although the structure of the expression (2.6) is explicitly taken into account in the definition (2.35), the perturbations δA_i applied to the different matrices A_i are unstructured. In other words, the elementwise structure of A_i is not taken into account when using the corresponding perturbation δA_i. We now present a framework for the definition and computation of pseudospectra, in which various types of structures on the perturbation matrices can be imposed, too. For this we assume a more general additive uncertainty on (2.6) than what (2.36) allows. This uncertainty takes the form

$$\delta F(\lambda) := \sum_{j=0}^{f} D_j(\lambda)\Delta_j E_j(\lambda) + \sum_{j=0}^{s} d_j G_j(\lambda) H_j(\lambda). \tag{2.37}$$

In this expression $\Delta_j \in \mathbb{C}^{k_j \times l_j}$ and $d_j \in \mathbb{C}$ denote the underlying unstructured perturbations, and $D_j \in \mathbb{C}^{n \times k_j}$, $E_j \in \mathbb{C}^{l_j \times n}$, $G_j \in \mathbb{C}^{n \times m_j}$, and $H_j \in \mathbb{C}^{m_j \times n}$ are appropriate shape matrices, whose elements are entire functions. We further assume that $m_j \geq 2$, G_j has full column rank and H_j has full row rank, for all $j = 0, \ldots, s$. We define the *uncertainty set* $\boldsymbol{\Delta}$ as

$$\begin{aligned} \boldsymbol{\Delta} \; := \; \Big\{ \mathrm{diag}(\Delta_0, \ldots, \Delta_f, d_0 I_{m_0}, \ldots, d_s I_{m_s}) : \; & \Delta_i \in \mathbb{C}^{k_i \times l_i}, \; d_j \in \mathbb{C}, \\ & 0 \leq i \leq f, \; 0 \leq j \leq s \Big\}, \end{aligned} \tag{2.38}$$

endowed with the following norm:

$$\begin{aligned} \|\Delta\|_{\mathrm{glob}} = \max & \left\{ \|\Delta_0\|_2, \ldots, \|\Delta_f\|, |d_0|, \ldots, |d_s| \right\}, \\ & \Delta = \mathrm{diag}(\Delta_0, \ldots, \Delta_f, d_0 I_{m_0}, \ldots, d_s I_{m_s}) \in \boldsymbol{\Delta}. \end{aligned}$$

The structured ϵ-pseudospectrum $\Lambda_\epsilon^{\mathrm{str}}(F)$ of F with respect to the uncertainty (2.37) can then be defined as follows.

Definition 2.11.

$$\begin{aligned} \Lambda_\epsilon^{\mathrm{str}}(F;\; \boldsymbol{\Delta}) := \{ & \lambda \in \mathbb{C} : \det(F(\lambda) + \delta F(\lambda)) = 0 \text{ for some } \delta F \text{ as in} \\ & (2.37) \text{ with } \|\Delta_j\|_2 < \epsilon, \; 0 \leq j \leq f, \text{ and } |d_j| < \epsilon, \; 0 \leq j \leq s \}. \end{aligned} \tag{2.39}$$

Let us make a comparison with Definition 2.1. In addition to the fact that the structure of (2.37) is exploited in (2.39), we notice that

- The underlying norm used in Definition 2.11 is of mixed type and similar to (2.14) with $p_1 = 2$ and $p_2 = \infty$. Although other types of norms can be used without any problem, we restrict ourselves to this type because it is most relevant from an application point of view and it gives rise to directly computable expressions in terms of structured singular values, as we shall see.

- Scalar weights of the perturbations are not used in Definition 2.11, as they can always be absorbed in the shape matrices in (2.37).

In a similar way the complex structured stability radius of (2.6) w.r.t. the perturbations (2.37)–(2.38) and w.r.t. the desired "stability" region $\mathbb{C}_d$ is given by the following definition.

Definition 2.12.

$$r_{\mathbb{C}}^{\mathrm{str}}(F;\ \mathbb{C}_d, \Delta) := \inf_{\lambda \in \mathbb{C}_u} \inf_{\epsilon > 0} \Big\{ \epsilon :\ \det\left(F(\lambda) + \delta F(\lambda)\right) = 0 \text{ for some } \delta F \text{ as in } (2.37) \text{ with } \|\Delta_j\|_2 < \epsilon,\ 0 \le j \le f, \text{ and } |d_j| < \epsilon,\ 0 \le j \le s \Big\}.$$

Several problems encountered in the literature fit into the general framework presented above. For instance,

- [250] deals with perturbations of matrix polynomials (2.6) of the form

$$\sum_{i=0}^{m} \left(A_i + \sum_{j=0}^{f} D_j \Delta_j E_{ij} \right) p_i(\lambda), \tag{2.40}$$

  in the context of stability radii for polynomial matrices. The shape matrices D_j and E_{ij} in (2.40) can be used to perturb only a submatrix of A_i, to assign weights to perturbations of rows, columns, or elements of each A_i, and to weight the perturbations applied to the matrices $A_0, \ldots, A_m$ with respect to each other. The uncertainty in (2.40) leads to the additive perturbation

$$\delta F(\lambda) = \sum_{j=0}^{f} D_j \Delta_j \left(\sum_{i=0}^{m} E_{ij} p_i(\lambda) \right).$$

- In many applications the characteristic matrix of an uncertain system is given by $\sum_{i=0}^{m} \tilde{A}_i p_i(\lambda)$, where the matrices $\tilde{A}_i$ depend linearly on a number of uncertain scalar parameters, say

$$\tilde{A}_i = A_i + \sum_j \theta_j P_{ij},$$

  with $\theta_j \in \mathbb{C}$ describing the uncertainties on these parameters. Assume that one wishes to investigate the possible positions of the characteristic roots when $|\theta_j| \le \epsilon$ for all j. It follows that we are in the framework of (2.6), (2.37), and (2.39), as we can express

$$\begin{aligned} \textstyle\sum_{i=0}^{m} \tilde{A}_i p_i(\lambda) &= F(\lambda) + \textstyle\sum_j \theta_j \left(\sum_{i=1, P_{ij} \neq 0}^{m} U_{ij} V_{ij}^* \, p_i(\lambda) \right) \\ &= F(\lambda) + \textstyle\sum_j \theta_j \left[\cdots U_{ij} \cdots \right] \left[\cdots V_{ij} \, \bar{p}_i(\lambda) \cdots \right]^*, \end{aligned} \tag{2.41}$$

  where each U_{ij} has full column rank and U_{ij} and V_{ij} can be computed, for instance, from a singular value decomposition of P_{ij}. Notice that (2.41) leads to $s > 0$ in the general expression (2.37) if and only if one of the matrices P_{ij} has rank larger than one, or if one of the parameters explicitly appears in different matrices $\tilde{A}_i$.

- Structured perturbations of the form (2.37) can sometimes be used for classes of systems with a nonlinear dependence on the parameters. As an illustration, the uncertain system

$$\dot{x}(t) = (A + \delta A)x(t) + (B + \delta B)(C + \delta C)x(t - \tau)$$

can be rewritten in a descriptor form as

$$\begin{aligned} \dot{x}(t) &= (A + \delta A)x(t) + (B + \delta B)y(t), \\ 0 &= (C + \delta C)x(t - \tau) - y(t); \end{aligned}$$

see also [83] and the references therein. This system has a nominal characteristic matrix

$$F(\lambda) = \begin{bmatrix} \lambda I - A & -B \\ Ce^{-\lambda\tau} & -I \end{bmatrix},$$

to which we may apply the structured perturbations

$$\delta F(\lambda) = -\begin{bmatrix} I \\ 0 \end{bmatrix} \delta A \,[I \;\; 0] - \begin{bmatrix} I \\ 0 \end{bmatrix} \delta B \,[0 \;\; I] - \begin{bmatrix} 0 \\ I \end{bmatrix} \delta C \left[e^{-\lambda\tau} I \;\; 0\right].$$

A reformulation of (2.39) in terms of *structured singular values* (ssvs) gives rise to a computable form for $\Lambda_\epsilon^{\text{str}}(F;\ \Delta)$ (the formula for $r_{\mathbb{C}}(F;\ \mathbb{C}_d,\ \Delta)$ follows straightforwardly). See Section A.2 of the appendix for a short introduction to the concept and computation of ssv.

Theorem 2.13. *Consider the characteristic matrix (2.6) with additive uncertainty (2.37)–(2.38). Defining*

$$\begin{aligned} M(\lambda) &:= \left[E_1(\lambda)^T \ldots E_f(\lambda)^T \; H_1(\lambda)^T \ldots H_s(\lambda)\right]^T, \\ N(\lambda) &:= [D_1(\lambda) \cdots D_f(\lambda) \; G_1(\lambda) \cdots G_s(\lambda)], \end{aligned} \tag{2.42}$$

and

$$T(\lambda) := M(\lambda)F(\lambda)^{-1}N(\lambda),$$

we have

$$\Lambda_\epsilon^{\text{str}}(F;\ \Delta) = \Lambda \cup \left\{\lambda \in \mathbb{C} :\ \mu_\Delta(T(\lambda)) > \epsilon^{-1}\right\}, \tag{2.43}$$

where $\mu_\Delta(\cdot)$ is the ssv with respect to the uncertainty set Δ.

The proof is based on an appropriate feedback interpretation of the perturbed system and can be found in [321].

2.3.3 Computational issues and special cases

From (2.43) the boundaries of structured ϵ-pseudospectra can be defined as level sets of the function

$$\mu_\Delta(T(\lambda)), \quad \lambda \in \mathbb{C}. \tag{2.44}$$

In general the computation of the ssv of a matrix with respect to the uncertainty set (2.38) is a difficult problem. Indeed, it is known to belong to the class of NP hard problems [299], which makes it inefficient for large problems. In many cases, however, the ssv can be computed efficiently by exploiting the structure of $T(\lambda)$. Some results in this sense will be presented below. Furthermore, for the examples presented in [321, 319], a good performance was found for numerical algorithms that approximate the ssv. There, lower and upper bounds on the ssv were computed by solving eigenvalue optimization problems, based on the ideas outlined in Section A.2. Such bounds are sharp in many cases. Note, in particular, that if the additional restriction $f + 2s \leq 3$ holds for the uncertainty set (2.38) and the full blocks are square, then an exact computation of $\mu_\Delta(\cdot)$ is always possible by solving a convex optimization problem. Hence, we propose the use of the ssv for computing structured pseudospectra. Note also that the computation of upper bounds for the ssv in (2.44) gives lower bounds for the ϵ-values, for which a point lies in the structured ϵ-pseudospectrum. This is in agreement with the common use of pseudospectra (or directly related values like stability radii) in a worst case analysis. In this context the above approach can be used to give rigorous bounds for the behavior of eigenvalues under perturbations.

In some cases the particular structure of $T(\lambda)$ can be exploited for an efficient computation of the ssv in (2.44). This is illustrated by the following result of [321], which slightly generalizes one of the assertions of Theorem 2.2 and is also related to Proposition 3.4 of [250].

Theorem 2.14. *We consider the characteristic matrix (2.6) with uncertainty (2.37). Furthermore, we assume that $s = 0$, and that there exist matrices D and E and functions $q_j : \mathbb{C} \to \mathbb{C}$ such that*

$$D_j(\lambda) = D(\lambda),$$
$$E_j(\lambda) = E(\lambda)\, q_j(\lambda), \quad 1 \leq j \leq f.$$

By defining $T(\lambda)$ and Δ as in Theorem 2.2, the following holds:

$$\mu_\Delta(T(\lambda)) = \left\| E(\lambda) F^{-1}(\lambda) D(\lambda) \right\|_2 \left(\sum_{j=0}^{f} |q_j(\lambda)| \right), \quad \lambda \notin \Lambda. \tag{2.45}$$

Note that, in addition to the availability of a directly computable formula, the dimensions of $E(\lambda)F^{-1}(\lambda)D(\lambda)$ are f *times smaller* than the dimensions of $T(\lambda)$.

To conclude this section we give an overview of special types of problems encountered in the pseudospectra literature for which the combination of Theorems 2.13 and 2.14 ensures an efficient computation of pseudospectra:

- Unstructured pseudospectra, in the sense of Definition 2.1, for the measure (2.14) with $p_1 = 2,\ p_2 = \infty$, and unit weights.

  With $F(\lambda) + \delta F(\lambda) = \sum_{i=0}^{m} (A_i + \delta A_i) p_i(\lambda)$ we have

$$\delta F(\lambda) = \sum_{i=0}^{m} \underbrace{I}_{D_i(\lambda)} \underbrace{\delta A_i}_{\Delta_i} \underbrace{p_i(\lambda) I}_{E_i(\lambda)}.$$

An application of Theorem 2.14 yields

$$\mu_\Delta(T(\lambda)) = \left\| F(\lambda)^{-1} \right\|_2 \sum_{i=0}^{m} |p_i(\lambda)|. \tag{2.46}$$

In combination with Theorem 2.13 one of the results of Theorem 2.2 is recovered.

- Pseudospectra in the sense of [297].

  With $F(\lambda) + \delta F(\lambda) = \sum_{i=0}^{m} (A_i + D\delta A_i E_i) p_i(\lambda)$ we get

$$\delta F(\lambda) = \underbrace{D}_{D(\lambda)} \underbrace{[\delta A_1 \dots \delta A_m]}_{\Delta} \underbrace{\left[E_1^T p_1(\lambda) \cdots E_m^T p_m(\lambda) \right]^T}_{E(\lambda)},$$

  and by Theorem 2.14

$$\mu_\Delta(T(\lambda)) = \left\| \begin{bmatrix} p_1(\lambda) E_1 \\ \vdots \\ p_m(\lambda) E_m \end{bmatrix} F(\lambda)^{-1} D \right\|_2 .$$

  For the special case $E_i = E,\ i = 1, \dots, m$, this expression can also be simplified to

$$\mu_\Delta(T(\lambda)) = \|EF^{-1}(\lambda) D\|_2 \sqrt{\sum_{i=0}^{m} |p_i(\lambda)|^2}. \tag{2.47}$$

  Furthermore, if we consider the dual problem, $F(\lambda) + \delta F(\lambda) = \sum_{i=0}^{m} (A_i + D_i \delta A_i E) p_i(\lambda)$, instead, we have

$$\delta F(\lambda) = \underbrace{[D_1 p_1(\lambda) \cdots D_m p_m(\lambda)]}_{D(\lambda)} \underbrace{[\delta A_1^T \cdots \delta A_m^T]^T}_{\Delta} \underbrace{E}_{E(\lambda)}$$

  and

$$\mu_\Delta(T(\lambda)) = \left\| EF(\lambda)^{-1} [p_1(\lambda) D_1 \cdots p_m(\lambda) D_m] \right\|_2 ,$$

  which can for the special case $D_i = D,\ i = 1, \dots, m$, be simplified to (2.47), as expected.

Notice that expression (2.47) does not reduce to expression (2.46) if $D = I$ and $E = I$, although in that case they concern the same additive perturbation on F. This is explained by the fact that both formulae are based on a different way of measuring the perturbations: $\max_{1 \le i \le m} \|\delta A_i\|_2$ for (2.46) and $\|[\delta A_1 \dots \delta A_m]\|_2$ for (2.47).

2.3.4 Application to time-delay systems

Rephrasing Theorems 2.13 and 2.14 for the time-delay system (2.20) yields the following theorem.

Theorem 2.15. *Consider the characteristic matrix (2.6) with additive uncertainty (2.37). Let F be defined as in (2.21) and let the uncertainty set Δ be as (2.38). Then*

$$\Lambda_\epsilon^{\mathrm{str}}(F;\ \Delta) = \Lambda(F) \cup \left\{\lambda \in \mathbb{C} : \ \mu_\Delta\left(M(\lambda)\left(\lambda I - A_0 - \textstyle\sum_{i=1}^m A_i e^{-\lambda\tau_i}\right)^{-1} N(\lambda)\right) > \epsilon^{-1}\right\},$$

where M and N are given by (2.42) and $\mu_\Delta(\cdot)$ is the ssv with respect to the uncertainty set Δ. If the zero solution of (2.20) is asymptotically stable, then the associated complex stability radius satisfies

$$r_{\mathbb{C}}^{\mathrm{str}}(F;\ \mathbb{C}_-, \Delta) = \left\{\sup_{\omega\geq 0} \mu_\Delta\left(M(j\omega)\left(j\omega I - A_0 - \sum_{i=1}^m A_i e^{-j\omega\tau_i}\right)^{-1} N(j\omega)\right)\right\}^{-1}.$$

If, furthermore, $s = 0$, and there exist matrices D and E and functions $q_j : \mathbb{C} \to \mathbb{C}$ such that

$$\begin{aligned} D_j(\lambda) &= D(\lambda), \\ E_j(\lambda) &= E(\lambda)\, q_j(\lambda), \quad 1 \leq j \leq f, \end{aligned}$$

then

$$\Lambda_\epsilon^{\mathrm{str}}(F;\ \Delta) = \Lambda(F) \cup \left\{\lambda \in \mathbb{C} : \ \left\|E(\lambda)\left(\lambda I - A - \textstyle\sum_{i=1}^m A_i e^{-\lambda\tau_i}\right)^{-1} D(\lambda)\right\|_2 \left(\textstyle\sum_{j=0}^f |q_j(\lambda)|\right) > \epsilon^{-1}\right\}$$

and

$$r_{\mathbb{C}}^{\mathrm{str}}(F;\ \mathbb{C}_-, \Delta) = \left\{\sup_{\omega\geq 0} \left\|E(j\omega)\left(j\omega I - A - \textstyle\sum_{i=1}^m A_i e^{-j\omega\tau_i}\right)^{-1} D(j\omega)\right\|_2 \left(\textstyle\sum_{j=0}^f |q_j(j\omega)|\right)\right\}^{-1}.$$

In Chapter 9 we will describe an algorithm for the related synthesis problem of optimizing complex stability radii as a function of system or controller parameters.

2.4 Illustrative examples

We illustrate the results obtained in this chapter by means of the time-delay system

$$\dot{x}(t) = A_0 x(t) + A_1 x(t-1), \tag{2.48}$$

for which we have

$$F(\lambda) = \lambda I - A_0 - A_1 e^{-\lambda}.$$

We take two different sets of parameters values. With the first set we illustrate the results of Section 2.2. With the second set, which comes from a laser physics application, we illustrate the results of Section 2.3.

2.4.1 Second-order system

Consider the system (2.48), where

$$A_0 = \begin{bmatrix} -5 & 1 \\ 2 & -6 \end{bmatrix} \quad \text{and} \quad A_1 = \begin{bmatrix} -2 & 1 \\ 4 & -1 \end{bmatrix}. \tag{2.49}$$

Figure 2.1(a) shows the rightmost characteristic roots of (2.48)–(2.49). The system is shown to be stable with all characteristic roots confined to the open left half-plane. To investigate how stability may change under perturbations of the matrices A_0 and A_1 we need to compute the corresponding pseudospectra. To this end, we consider unstructured perturbations of A_0 and A_1 using the global measure (2.14) with $p_1 = 2$ and $p_2 = \infty$. Pseudospectra can then be computed using Theorems 2.6 and 2.2 with $\alpha = 2$ and $\beta = 1$. Specifically (for $\lambda \notin \Lambda$),

$$f(\lambda;\ \|\cdot\|_{\text{glob}}) = \left\| \left(\lambda I - A_0 - A_1 e^{-\lambda}\right)^{-1} \right\|_2 \left(\frac{1}{w_0} + \frac{|e^{-\lambda}|}{w_1} \right). \tag{2.50}$$

By evaluating f on a grid over a region of the complex plane, and by using a contour plotter, we have identified the boundaries of ϵ-pseudospectra $\Lambda_\epsilon(F,\ \|\cdot\|_{\text{glob}})$.

Figures 2.1(b)–(d) show the ϵ-pseudospectra of (2.49) where different weights have been applied to A_0 and A_1. Specifically, $(w_0, w_1) = (\infty, 1)$ (b), $(w_0, w_1) = (2, 2)$ (c), and $(w_0, w_1) = (1, \infty)$ (d). In each panel, from outermost to innermost (or rightmost to leftmost if the curve is not closed), the curves correspond to boundaries of ϵ-pseudospectra with $\epsilon = 10^{1.25}$, $10^{1.0}$, $10^{0.75}$, $10^{0.5}$, $10^{0.25}$, 10^0, and $10^{-0.5}$. It can be seen that the conclusions drawn in Section 2.2.4 on the effect of weighting hold; that is, perturbations of A_0 stretch pseudospectra lying in the right half-plane (d), while perturbations applied to A_1 stretch the pseudospectra lying in the left half-plane (b). Furthermore, Figure 2.2 shows the intersection of ϵ-pseudospectrum curves with the imaginary axis. In each panel, the darkest curve corresponds to an ϵ-pseudospectrum curve of Figure 2.1(a), the next to a curve of Figure 2.1(b), and the lightest curve corresponds to an ϵ-pseudospectrum curve of Figure 2.1(c). Specifically, Figure 2.2(a) shows the intersection of the three curves for $\epsilon = 10^{1.25}$, Figure 2.2(b) for $\epsilon = 10^{1.0}$, and Figure 2.2(c) for $\epsilon = 10^{0.75}$. For a given ϵ, these curves are seen to intersect the imaginary axis at the same point, independent of the weighting applied to the system matrices, thus demonstrating the third conclusion.

Figure 2.3 shows which ϵ-pseudospectrum curve intersects the imaginary axis at $\lambda = j\omega$, that is $f^{-1}(j\omega,\ \|\cdot\|_{\text{glob}})$, for each $\omega \in [-50,\ 50]$. The minimum of this curve represents the complex stability radius of the system,

$$r_{\mathbb{C}}(F;\ \mathbb{C}_-,\ \|\cdot\|_{\text{glob}}) \approx 3.28011.$$

Since the minimum is reached for $\omega = 0$, the smallest destabilizing perturbations shift a characteristic root to the origin.

In Figure 2.4(a) we show ϵ-pseudospectra for the weights $(w_0, w_1) = (\infty, 1)$ and $\epsilon = 0.1,\ 0.2,\ 0.3,\ 0.4,\ 0.5$. In Figure 2.4(b) we take $(w_0, w_1) = (2, 2)$ and $\epsilon = 0.2,\ 0.4,\ 0.6,\ 0.8,\ 1$. In both cases only the ϵ-pseudospectrum for the largest value of ϵ stretches out infinitely far along the negative real axis. This is in accordance with Proposition 2.10 since $w_1/\|A_1^{-1}\|_2 \approx 0.4282\ w_1$.

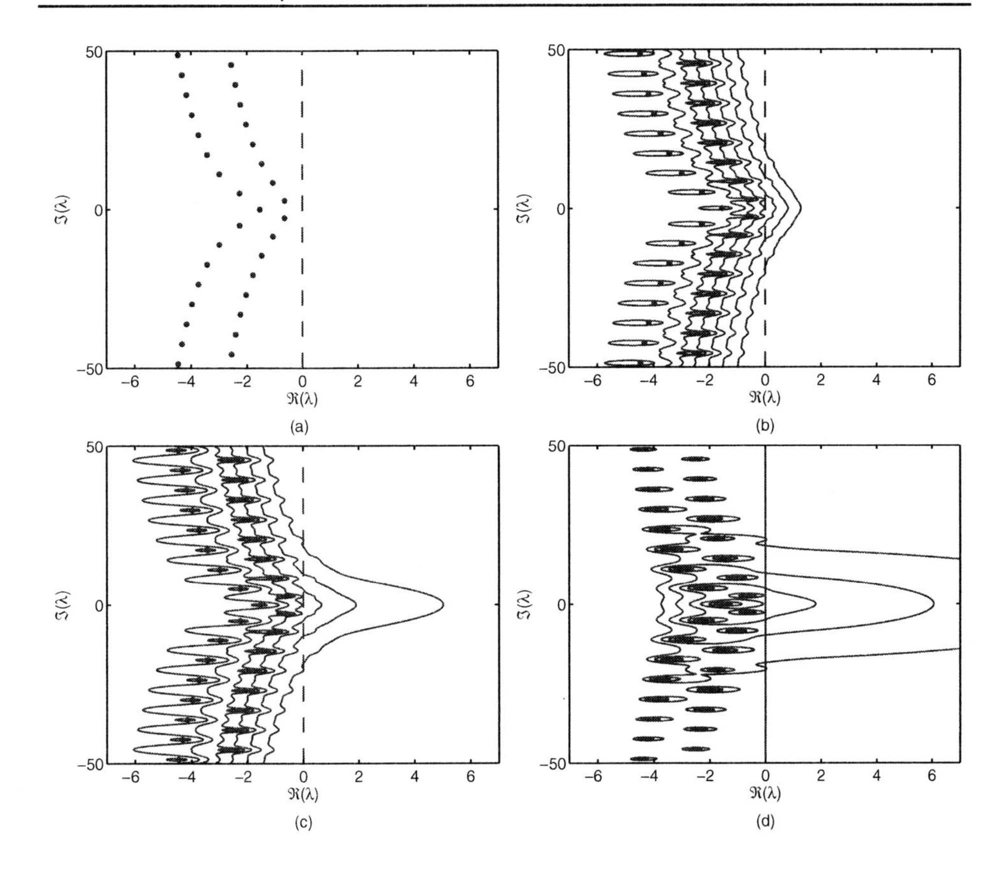

Figure 2.1. *Weighted pseudospectra of (2.48)–(2.49). Panel (a) shows the spectrum of the unperturbed problem. In all other panels, from rightmost to leftmost, the contours correspond to* $\epsilon = 10^{1.25}, 10^{1.0}, 10^{0.75}, 10^{0.5}, 10^{0.25}, 10^{0}$, *and* $10^{-0.5}$. *From (b) to (d), the weights* w_0 *and* w_1 *applied to the* A_0 *and* A_1 *matrices were* $(w_0, w_1) = (\infty, 1)$, $(w_0, w_1) = (2, 2)$, *and* $(w_0, w_1) = (1, \infty)$, *respectively.*

2.4.2 Feedback controlled semiconductor laser

In [94] pseudospectra were applied to the analysis of the robustness of stability of a model for a semiconductor laser subject to optical feedback. For certain fixed model parameters, the problem leads to the study of the DDE (2.48), where

$$A_0 = \begin{bmatrix} -0.8498 & 0.1479 & 44.37 \\ 0.003756 & -0.2805 & -229.2 \\ -0.1754 & 0.02296 & -0.3608 \end{bmatrix}, \quad A_1 = \begin{bmatrix} 0.28 & 0 & 0 \\ 0 & -0.28 & 0 \\ 0 & 0 & 0 \end{bmatrix}. \tag{2.51}$$

We investigate the effect of an uncertainty on specific elements of A_0 and A_1 on the characteristic roots by computing structured pseudospectra. From physical considerations an important requirement on the uncertainty is that in A_1 only the elements on positions (1,1)

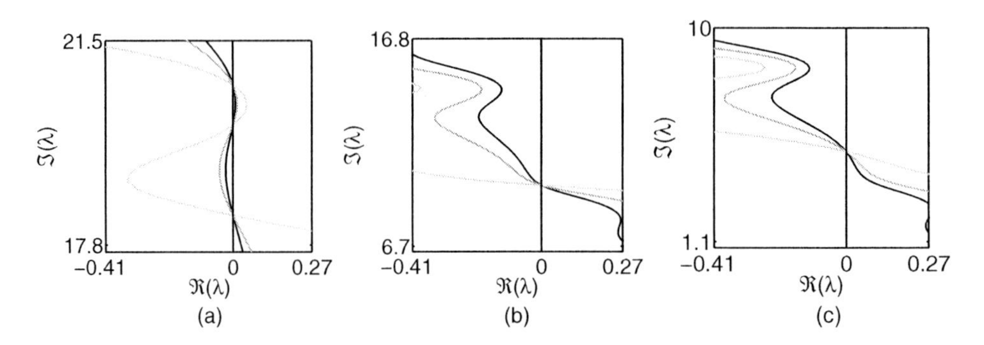

Figure 2.2. *Crossings of ϵ-pseudospectrum curves with the imaginary axis, for $\epsilon = 10^{1.25}$ (a), $\epsilon = 10$ (b), and $\epsilon = 10^{0.75}$ (c). In the three cases the darkest contour corresponds to the weights $(w_0, w_1) = (\infty, 1)$, the middle curve to $(2, 2)$, and the lightest curve to $(1, \infty)$.*

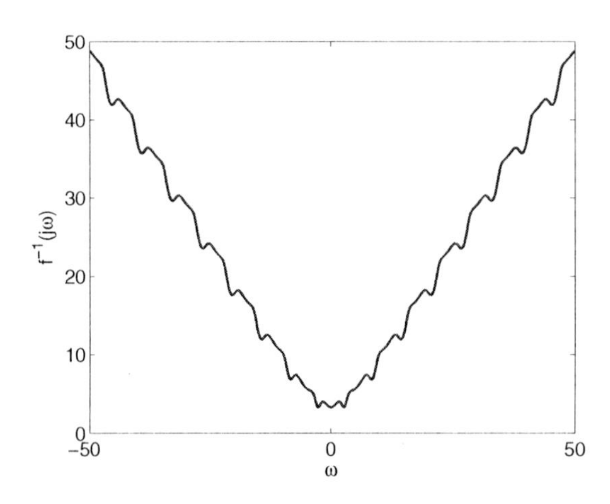

Figure 2.3. *The function $\omega \to f^{-1}(j\omega;\ \|\cdot\|_{\text{glob}})$. The minimum is the complex stability radius $r_{\mathbb{C}}(F;\ \mathbb{C}_-, \|\cdot\|_{\text{glob}})$.*

and (2,2) are nonzero and remain opposite to each other. Physically, these elements describe the feedback process of the laser; see [307] for the details. We can take this structure into account by considering perturbations on A_1 of the form $\text{diag}(\delta a, -\delta a, 0)$, with $\delta a \in \mathbb{C}$, in addition to unstructured perturbations on A_0. The resulting additive uncertainty on F has the general form (2.37), namely

$$\delta F(\lambda) = \underbrace{-I_3}_{D_1(\lambda)}\ \delta A_0\ \underbrace{I_3}_{E_1(\lambda)} + \delta a \underbrace{\begin{bmatrix} -1 & 0 \\ 0 & 1 \\ 0 & 0 \end{bmatrix}}_{G_1(\lambda)} \underbrace{\begin{bmatrix} 1 & 0 & 0 \\ 0 & 1 & 0 \end{bmatrix} e^{-\lambda}}_{H_1(\lambda)}, \tag{2.52}$$

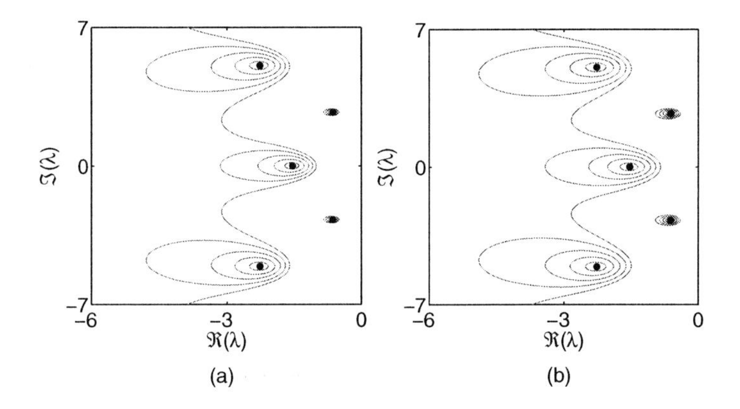

Figure 2.4. *(a) ϵ-pseudospectrum curves for $(w_0, w_1) = (\infty, 1)$ and $\epsilon = 0.1,\ 0.2,\ 0.3,\ 0.4,\ 0.5$. (b) ϵ-pseudospectrum curves for $(w_0, w_1) = (2, 2)$ and $\epsilon = 0.2,\ 0.4,\ 0.6,\ 0.8,\ 1$.*

and the uncertainty set Δ is the set of complex block-diagonal 5×5 matrices with one full 3×3 block and one repeated scalar 2×2 block. An application of Theorem 2.15 yields

$$\Lambda_\epsilon^{\text{str}}(F;\ \Delta) = \left\{ \lambda \in \mathbb{C} \ :\ \mu_\Delta \left(\left[\begin{array}{ccc} & I_3 & \\ \hline e^{-\lambda} & 0 & 0 \\ 0 & e^{-\lambda} & 0 \end{array} \right] F(\lambda)^{-1} \left[\begin{array}{c|cc} & -1 & 0 \\ -I_3 & 0 & 1 \\ & 0 & 0 \end{array} \right] \right) > \frac{1}{\epsilon} \right\}.$$

In this case ($f = s = 1$), the ssv can be computed exactly as the solution of a convex optimization problem; see Appendix A.2. We have combined the `mussv` routine of MATLAB with a contour plotter to visualize the structured pseudospectra and the results are shown in Figure 2.5 (left).

For comparison, unstructured pseudospectra of (2.48) and (2.51) are shown in Figure 2.5 (right), corresponding to the measure (2.14) with $p_1 = 2$, $p_2 = \infty$, and unity weights. An application of Theorem 2.6 then yields

$$\Lambda_\epsilon(F;\ \|\cdot\|_{\text{glob}}) = \left\{ \lambda \in \mathbb{C} :\ \|F(\lambda)^{-1}\|_2 \left(1 + \left|e^{-\lambda}\right|\right) > \frac{1}{\epsilon} \right\}.$$

As a significant qualitative difference, the (unstructured) ϵ-pseudospectra stretch out infinitely far along the negative real axis, even for arbitrarily small values of ϵ, following from the singularity of A_1 and Proposition 2.10. In Section 2.4, this phenomenon is related to the behavior of characteristic roots, which are introduced by perturbations that make the matrix A_1 nonsingular. Such perturbations are, however, nonphysical and, as we have demonstrated, can be excluded by applying the structured uncertainty (2.52).

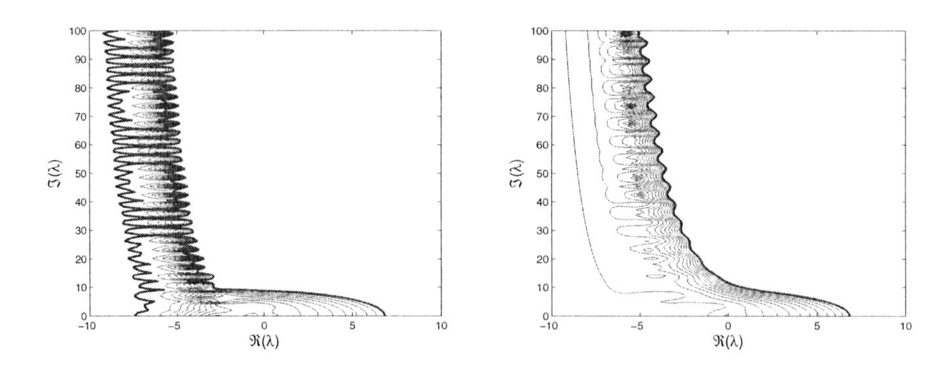

Figure 2.5. *Structured (left) and unstructured (right) pseudospectra for the system (2.48) and (2.51). The contours corresponding to* $\epsilon = 0.001$ *to* $\epsilon = 0.27$*, in intervals of 0.01, are depicted. The contours corresponding* $\epsilon = 0.27$ *are shown in bold.*

2.5 Notes and references

Inspired by the observation that the eigenvalue problems encountered in the study of linear systems described by higher-order differential equations, differential algebraic equations, and delay differential (algebraic) equations have a similar structure, we first presented a unifying treatment of pseudospectra and stability radii of nonlinear eigenvalue problems, where the structure of the problem is fully exploited. Various perturbation measures were considered and efficient formulae for the computation of both pseudospectra and stability radii were derived, whose computational complexity is similar to the complexity of computing the pseudospectra of an n-by-n matrix, with n-by-n being the dimension of the characteristic matrix. The results were applied to classes of time-delay systems, which are inherently infinite-dimensional. Qualitative properties of the pseudospectra of such systems were investigated, with the emphasis on boundedness properties and the effect of the weight factors in the definition.

Next, we presented an approach for computing structured pseudospectra and stability radii of nonlinear eigenvalue problems, and applied the results to time-delay systems. The proposed method allows one to direct perturbations to specific elements (or groups of elements) of the individual coefficient matrices of the eigenvalue problem under consideration, and thus to exploit the structure of these matrices, in addition to exploiting the structure of the eigenvalue problem itself. A general formula was presented which is based on the computation of appropriately defined ssvs. It was outlined how, for special cases, the computational efficiency can be improved by exploiting additional information on the system or the uncertainty set, based on the ideas presented in the first part. In fact, this reveals an inherent trade-off between the extent to which one wants to impose structure on the perturbations on the one hand, and the computational complexity of the resulting pseudospectra computations on the other hand.

Two illustrative examples were presented, including an example from laser physics, where structured pseudospectra were necessary to exclude physically nonrealistic perturbations, which would otherwise have a considerable impact on the pseudospectra.

In this chapter we have restricted ourselves to complex values perturbations. Some extensions to the case where only real valued perturbations are allowed can be found in [87, 192, 254, 282, 116]. Although the extension of Theorem 2.13 to real valued perturbations, based on a feedback interconnection point of view and appropriately defined ssvs, is trivial from a conceptional point of view, the complexity of computing or approximating the involved ssvs typically increases considerably. We recall that considering a larger class of perturbations leads to upper bounds on the sensitivity of the characteristic roots and lower bounds on the stability radii (which correspond to sufficient stability conditions). This is in agreement with the common use of pseudospectra and stability radii in a worst case analysis.

The presented results are based on [183, 321, 319] and the references therein.

Chapter 3

Computation of stability regions in parameter spaces

3.1 Introduction

As in the finite-dimensional case, the stability of a linear time-delay system is given by its spectrum location. The next important step is represented by the characterization of *spectrum behavior* as a function of the parameters' variation. To perform such an analysis, three ingredients are essential:

(a) the continuity property of the spectrum with respect to the parameters;

(b) the detection, and the explicit computation (if any!) of the characteristic roots located on the imaginary axis;

(c) the behavior of the characteristic roots located on the imaginary axis in a very small neighborhood if one or several parameters vary.

The first problem was largely discussed in the first chapter, and the main ideas are clear (continuity properties of the rightmost characteristic roots with respect to all the parameters of the system including also the delay parameters). The next two problems represent the main topics in what follows. More precisely, we start by presenting several approaches and methodologies to describe the characteristic roots behavior when they are located on the imaginary axis. Some basic ideas concerning the detection and the explicit computation of the characteristic roots located on the imaginary axis will be also presented and discussed in some particular cases. A deeper analysis will be proposed in the next chapters (stability in the delay-parameter space, and the characterization of delay-interference phenomena).

We start by analyzing the properties of some generic characteristic root located on the imaginary axis as a function of its parameters. Such a method is certainly not new, and it proved its interest in the case of finite-dimensional systems. Next, the crossing direction characterization w.r.t. the parameters is largely treated, and several intuitive approaches are presented (Jacobi's formulae for computing the derivative of a determinant, perturbation analysis). We focus only on the simple and semisimple characteristic root cases, but the general ideas for the analysis of multiple nonsemisimple characteristic roots are pointed out. Finally, explicit crossing direction computation in some particular cases that are useful in the next chapters are also proposed.

The chapter is organized as follows. Section 3.2 includes some basic notions and definitions. Next, Section 3.3 is devoted to various existing methods and approaches for deriving the stability crossing boundaries. The presentation is intuitive and starts with some historical perspective concerning the D-subdivision method with an exemplification in the scalar case. Next, we present the "dual" τ-decomposition method, and still use the scalar system as illustrative example. The difference between these approaches mainly lies in the way in which the delays are treated. Finally, we discuss numerical continuation methods. Roughly speaking, the idea behind the numerical continuation method in our case is to find a one-dimensional curve in the parameter space and to "follow" (or "continue") it in the parameter space starting from one or several "first-points" found on the curve. The next section is devoted to the crossing direction characterization and related remarks for the cases of simple and semisimple characteristic root crossings. The main ideas for treating the remaining cases complete the section. Finally, the computation of the crossing direction taking into account some particular delay interdependence is also proposed. The corresponding results will prove their utility in the forthcoming chapters.

Some notes and comments end the chapter, as well as a list of references related to the topics treated in the chapter.

3.2 Basic notions and definitions

Consider the following class of delay systems:

$$\dot{x}(t) = A_0(p)x(t) + \sum_{i=1}^{m} A_i(p)x(t - \tau_i(p)), \tag{3.1}$$

under appropriate initial conditions, with

$$p := (p_1, \ldots, p_{n_p}) \in \mathcal{D} \subset \mathbb{R}^{n_p}$$

being the parameters, under the constraint that

$$\tau_i(p) \geq 0, \qquad \forall p \in \mathcal{D},$$

and such that there exists at least one point $\tilde{p}$ in the parameter set $\mathcal{D}$, and at least one positive integer i_0, $1 \leq i_0 \leq n_p$, such that

$$\tau_{i_0}(\tilde{p}) > 0, \quad A_{i_0} \neq 0.$$

In other words, we assume that (3.1) is always a time-delay system in some neighborhood of $\tilde{p}$ in the corresponding parameter space. We also assume that the matrices A_i and delays τ_l smoothly depend on p.

Next, define the characteristic matrix associated to (3.1) as the mapping $M : \mathbb{C} \times \mathcal{D} \mapsto \mathbb{C}^{n \times n}$ given by

$$M(\lambda;\ p) := \lambda I - A_0(p) - \sum_{i=1}^{m} A_i(p)e^{-\lambda\tau_i(p)}, \tag{3.2}$$

and the characteristic function as the mapping $H : \mathbb{C} \times \mathcal{D} \mapsto \mathbb{C}$ given by

$$H(\lambda;\ p) := \det\left(\lambda I - A_0(p) - \sum_{i=1}^{m} A_i(p)e^{-\lambda\tau_i(p)}\right). \tag{3.3}$$

As defined in the previous chapters, the roots of the characteristic equation $H(\lambda;\ p) = 0$ will be called *characteristic roots* of the delay system under consideration.

We introduce the following notions for stability domains (or regions) and stability crossing boundaries.

Definition 3.1.

(1) *The set of values $p \in \mathcal{D}$ of the parameter space $\mathbb{R}^{n_p}$ such that the delay system (3.1) is exponentially stable is called the* stability domain *or* stability region.

(2) *The set of values $p \in \mathcal{D}$ of the parameter space $\mathbb{R}^{n_p}$ is called the* stability crossing boundary *of the delay system (3.1) if for any point p_0 of the boundary, the characteristic function $H(\lambda;\ p_0)$ has at least one root on the imaginary axis.*

The continuity properties discussed in the previous chapter allow us to divide the parameter space into domains characterized by the same number of characteristic roots in the open right half-plane for all the parameters points of the domain. The "separation" between domains with a distinct number of unstable characteristic roots is given by the stability crossing boundaries. This is mainly due to the fact that the characteristic roots are discrete, that is, there no accumulation points of characteristic roots in any finite domain. Such an idea is exploited by the D-subdivision and τ-decomposition methods that are discussed in the next sections.

3.3 From D-subdivision to numerical continuation

With the notations, definitions, and comments above, we first concentrate on the way in which stability crossing boundaries are defined in the parameter space, and on the corresponding crossing direction from one domain to another. We start with a brief discussion on the D-subdivision method. Next, we present the so-called τ-decomposition method, and we discuss the differences between these two methods. Finally, we focus on numerical continuation.

3.3.1 D-subdivision and stability crossing boundaries

To the best of the authors' knowledge, the first method for characterizing the stability regions of time-delay systems in a parameter space was proposed by Neimark [220] for quasipolynomials by the end of the 1940s. It consists of computing a particular decomposition of the parameter space in regions such that the number of unstable characteristic roots is invariant with respect to all the points of the parameter space inside the region, and such that for each point of the boundaries the corresponding characteristic equation has at least one root on the imaginary axis. This method is also known as the *D-subdivision* (or *D-decomposition*) *method*. In order to describe the algorithm, some supplementary assumptions are needed. More precisely, we assume that the delays $\tau_i, i = 1, \ldots, m$, do *not depend* on the parameters $p \in \mathbb{R}^{n_p}$, and furthermore they are considered as *fixed*. In other words, our system rewrites as

$$\dot{x}(t) = A_0(p)x(t) + \sum_{i=1}^{m} A_i(p)x(t - \tau_i),$$

under appropriate initial conditions. The characteristic function becomes

$$H(\lambda;\ p) = \det\left(\lambda I - A_0(p) - \sum_{i=1}^{m} A_i(p)e^{-\lambda\tau_i}\right).$$

With the notations and remarks above, the algorithm of the D-subdivision method can be summarized as follows:

(i) First, solve the equation

$$H(j\omega;\ p) = 0,$$

for p as a function of $j\omega$ (including the origin of the complex plane) in order to find (stability crossing) surfaces in the parameter space $\mathbb{R}^{n_p}$ such that for each p on such a surface, there exists at least one characteristic root on the imaginary axis.

(ii) Second, these surfaces divide the parameter space into several regions and sometimes it is possible to conclude, by using appropriate additional arguments, for which region the stability is guaranteed. As additional arguments, we can find, for example, a particular point (on some of the axis of the parameter space) for which the stability analysis becomes easier to perform (finite-dimensional systems, eventually). Each region derived in this way is characterized by the same number of strictly unstable characteristic roots for all the points of the corresponding domain. Such a number is also called the *instability degree* of the domain or region (see also [138]).

Such a method is quite computationally involved, and allows us to derive a complete description of the stability regions in the parameter space for relative simple systems. However, for high-order systems including a large number of parameters, the method loses its efficiency.

D-subdivision applied to a scalar system including one delay

Consider now the simplest example of a time-delay system, that is, the first-order system including a single delay,

$$\dot{x}(t) = -ax(t) - bx(t - \tau), \tag{3.4}$$

under appropriate initial conditions with $(a, b) \in \mathbb{R}^2$, and $\tau > 0$. In the D-subdivision method, the delay parameter needs to be *fixed*, and the analysis is done with respect to the other parameters of DDE. The characteristic equation of our scalar system simply writes as

$$H(\lambda;\ a, b) = \lambda + a + b^{-\lambda\tau} = 0. \tag{3.5}$$

In the parameter space Oab, the characteristic root 0 defines the line: $a + b = 0$. Such a line is the *only* crossing curve separating the stable domain from the unstable one of the system free of delay, that is, of the system $\dot{x}(t) = -(a + b)x(t)$.

Next, for $\omega \neq 0$, $H(j\omega, a, b) = 0$ if and only if the following conditions are satisfied simultaneously:

$$\begin{cases} a + b\cos(\omega\tau) = 0, \\ b\sin(\omega\tau) = \omega, \end{cases} \tag{3.6}$$

which simply leads to the following parameterization of the (a, b)-curves:

$$\begin{cases} a = -\dfrac{\omega\cos(\omega\tau)}{\sin(\omega\tau)}, \\ b = \dfrac{\omega}{\sin(\omega\tau)}. \end{cases} \tag{3.7}$$

Without any loss of generality, we can restrict our analysis only to $\omega \in \mathbb{R}_+$. Next, the parameters a and b as functions of ω, that is $a = a(\omega)$ and $b = b(\omega)$, are correctly defined if and only if $\omega \neq k\pi$, with $k \in \mathbb{N}$. Thus, we can define the set of crossing frequencies

$$\Omega = \bigcup_{k\in\mathbb{N}} \Omega_k = \bigcup_{k\in\mathbb{N}} (k\pi, (k + 1)\pi),$$

and the mapping $p : \Omega \mapsto \mathbb{R}^2$ given by the relations (3.7) defines all the (stability crossing) curves in the parameter space Oab. Denote with C_k the corresponding curve defined on the frequency interval Ω_k, where k is a positive integer.

Based on the construction above, we need to detect whether the curves C_k intersect the line $a+b = 0$. Since $b(\omega)$ can be prolonged by continuity at $\omega = 0$, and the corresponding limit value is $b(0) = 1/\tau$, it follows that the point $(-1/\tau, 1/\tau)$ corresponds to the intersection of the line $a + b = 0$ with C_0. For all $k \geq 1$, the curves C_k do not intersect this line. It is important to point out that the point $(-1/\tau, 1/\tau)$ corresponds to a double characteristic root at zero. Along the curve C_0 we have a pair of complex conjugate characteristic roots located on the imaginary axis, and along the line $a + b = 0$ we have a characteristic root at zero.

To summarize, the only curve intersecting $a + b = 0$ is C_0. Next, it is not difficult to see that the curves C_k never intersect Oa. However, the curves C_k always intersect the axis Ob, for all positive integer k. Next, based on the sinus sign on each interval, it is easy to see that the curves C_{2k} intersects Ob for positive b, and C_{2k+1} intersects Ob for negative b. Furthermore, two independent crossing curves C_{k_1} and C_{k_2} ($k_1, k_2 \in \mathbb{N}$, $k_1 \neq k_2$) defined on the disjoint frequency sets Ω_{k_1} and Ω_{k_2}, respectively, do *not intersect* each other. Finally, the distance between this intersection point ($C_{2k} \cap Ob$ and, respectively, $C_{2k+1} \cap Ob$) and the origin increases when k increases.

In this way, we have been able to define a *first domain* $\mathcal{D}_0$ which is bounded by the line $a+b = 0$, and by the curve C_0 given by (3.7) evaluated on Ω_0. The remaining domains are defined by the curves C_k corresponding to the frequency intervals Ω_k, $k \in \mathbb{N}$ ($k \neq 0$), and the line $a + b = 0$. Thus, we performed step (i) of the procedure above.

Consider for instance $\mathcal{D}_0$. It is easy to see that such a domain includes the (positive) axis, Oa. Now, since for $b = 0$ the system simply becomes a finite-dimensional system of the form $\dot{x}(t) = -ax(t)$, the situation where $a > 0$ corresponds to a stable system, it follows

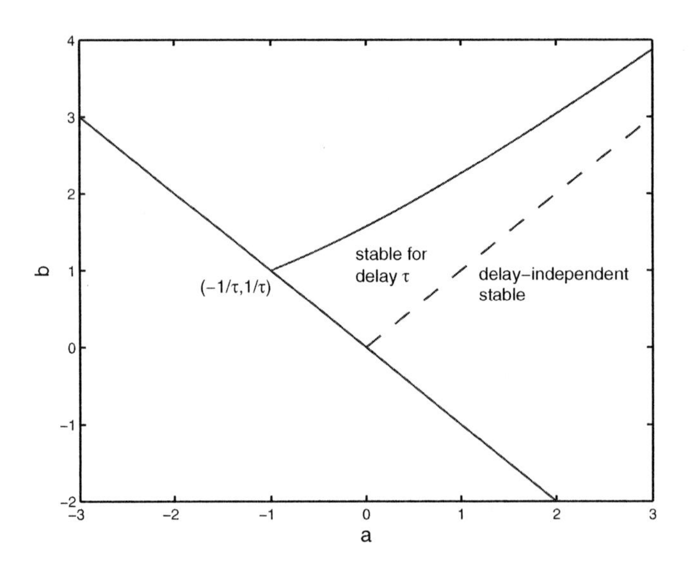

Figure 3.1. *The stability regions in the parameter space Oab of the scalar system* $\dot{x}(t) = -ax(t) - bx(t-\tau)$ *for the fixed delay value* $\tau = 1$.

that the domain $\mathcal{D}_0$ constructed above is a *stability domain*. Finally, it is important to point out that this region is the *only stability domain* for the corresponding delay system above. Starting from this point, one needs to analyze first the domains having a common boundary with $\mathcal{D}_0$. One of the approaches that can be used in such a case consists of analyzing the way in which the characteristic roots cross the imaginary axis if some small variations are applied to the parameters. All these aspects will be discussed in the next paragraphs.

Note also that we make a distinction between the cases with and without a characteristic root at zero. The explanation for such a distinction is easy to understand, since a characteristic root at zero is invariant under delay changes. Such an *invariance property* will play an important role in stability/instability characterization with respect to delay parameters, as we shall discuss in depth in the next chapter.

In Figure 3.1 we show the stability regions in the (a, b)-parameter space for the system (3.4) with $\tau = 1$.

Further examples on different applications of the method for low-order systems can be found in [286, 139, 138]. In particular, in Chapter 13 we will discuss the stability regions in some appropriate parameter space of a second-order system encountered in modeling congestion in high-performance networks by using fluid approximations.

Rightmost characteristic roots and stability/instability issues in the scalar case. Let us focus on a particular region of the parameter space defined by $a \geq |b|$, $a + b > 0$. As discussed above, by using the D-subdivision method this particular region corresponds to a stable system. Consider now the *rightmost characteristic root* of the equation $H(\lambda;\ a, b) = 0$, and denote it as λ_r. Such a root is characterized by the fact that its real part equals the spectral abscissa, that is, $\lambda_r = \alpha + j\omega_r$, with

$$\alpha = \max\{\Re(\lambda): \quad \lambda \in \mathbb{C}, \quad H(\lambda;\ a, b) = 0\}.$$

In what follows, we shall develop a simplified argument for proving by contradiction that $\lambda_r \in \mathbb{C}_-$ in the case under consideration. In this sense, assume $\lambda_r \in \mathbb{C}_+ \cup j\mathbb{R}$. Then $r \geq 0$. Since λ_r is a characteristic root, it should satisfy the following (real, imaginary part) conditions:

$$\begin{cases} r + a + be^{-r\tau}\cos(\omega_r\tau) = 0, \\ \omega - be^{-r\tau}\sin(\omega_r\tau) = 0. \end{cases} \tag{3.8}$$

Simple computations lead to the following relation:

$$(r+a)^2 + \omega_r^2 = b^2 e^{-2r\tau}. \tag{3.9}$$

Let us consider first the case $r > 0$. Then

$$a^2 < (r+a)^2 + \omega_r^2 = b^2 e^{-2r\tau} < b^2,$$

since a is positive by hypothesis, and $e^{-2r\tau} < 1$ due to the assumption that $r > 0$ ($\tau > 0$). In conclusion, we arrive to the condition $a^2 < b^2$, which cannot be satisfied for the region we considered. In conclusion, we need $r \leq 0$. Let us analyze now the case $r = 0$, that is, at least one root of the characteristic equation on the imaginary axis. Note that $\omega_r \neq 0$, since $\lambda = 0$ is not a characteristic root. Using the same argument as in the previous case, we arrive at

$$a^2 < a^2 + \omega_r^2 = b^2, \tag{3.10}$$

since $|\ \omega_r\ | > 0$, and $e^{-2r\tau} = 1$ ($r = 0$) in (3.9). Once again, we arrive at $a^2 < b^2$, which contradicts the initial hypothesis.

In conclusion, all the roots of the characteristic equation $H(\lambda;\ a, b) = 0$ should be located in $\mathbb{C}_-$ if (a, b) satisfies the condition $a \geq |\ b\ |, a+b > 0$. Furthermore, the condition does *not* include any information on the *delay* size, which is equivalent to saying that the stability is of *delay-independent* type.

Next, consider the case where $|\ a\ | < |\ b\ |$. Then the scalar system is *unstable* for sufficiently large delays. The corresponding bounds as a function of the parameters (a, b) will be detailed in the next section. Here we are interested in computing some (first) estimation of the domain of the complex plane in $\mathbb{C}_+$ where the rightmost root is located. Assume that for some $\tau = \tau_0$ the scalar system is unstable. Then, without any loss of generality, we can consider $r > 0$, and $\omega_r > 0$. Conditions (3.8) lead to the estimates

$$0 \leq \Re(\lambda_r) = r \leq -a + |\ b\ |,$$
$$0 \leq \Im(\lambda_r) = \omega_r \leq |\ b\ |,$$

by using the properties of sinus, cosinus, and exponential functions for real arguments. In conclusion, the rightmost root is located in some *appropriate rectangle* in $\mathbb{C}_+$, with the imaginary axis as a left boundary. As expected from the theory presented in Chapter 1, the upper limit for the real part of the rightmost root ($\Re(\lambda_r)$) is *finite*. In a more general setting (state-space representation), the construction above still works by using appropriate matrix measures (see, for instance, the last section of the chapter for further comments).

Finally, consider the particular situation where $a < 0$, with $a+b > 0$, and let us focus on the existence of some particular unstable roots: *positive* characteristic roots. In such a case, λ_r rewrites as $\lambda_r = r$, and the condition (3.8) reduces to

$$r + a + b^{-r\tau} = 0. \tag{3.11}$$

Since $a < 0$ and $a + b > 0$, it is easy to see that $f_\tau(0) < 0$ and $f_\tau(+\infty) = +\infty$, where the function f_τ is defined by $f_\tau : \mathbb{R}_+ \mapsto \mathbb{R}$, $f_\tau(x) = x + a + be^{-x\tau}$. Since f_τ is a continuous mapping, the conditions above simply say that its graphics should intersect the real axis for some *finite* r-value. In conclusion, the above choice of parameters always leads to some instability conditions, and this property holds *independently* of the *delay* value. In other words, we will have a particular delay-independent instability property. Such an idea holds for more general quasipolynomials, and will be discussed later in a different framework.

3.3.2 τ-decomposition and delay stability intervals

To the best of the authors' knowledge such a method has its origin in the works of Sokolov and Miasnikov in the 1950s and is discussed in Popov's book [252] (see also [140] for further discussions). The method is devoted to the stability analysis only with respect to the *delay parameter* by assuming that all the other parameters of the system are fixed.

Without any loss of generality, the method can be summarized as follows:

(i) First, "decompose" the delay axis $O\tau$ into (delay) intervals

$$\mathbb{R}_+^* = \bigcup_{k\in\mathbb{N}} (\tau_k, \tau_{k+1}),$$

with $\tau_0 = 0$, such that within each interval the number of unstable characteristic roots is invariant for all the points of the interval, that is, the same instability degree for all the delay values inside the corresponding interval.

(ii) Second, investigate the change of the number of roots at the end points of the delay intervals computed at the previous step.

This method can be seen as "dual" to the D-subdivision method discussed in the previous section. Indeed, here we concentrate our attention on only one parameter, and the domains (or regions) are reduced to intervals such that each end point of the interval corresponds to a characteristic root crossing with respect to the imaginary axis. The only point for which such a crossing does not necessarily exist is the origin of the delay axis $\tau_0 = 0$. Indeed, for $\tau_0 = 0$, two situations may occur, depending on the presence or absence of characteristic roots on the imaginary axis. The first situation fits well within the description above. In the second case, the first delay interval is of the form $[0, \tau_1)$, and its stability (instability) is given by the stability (instability) of the system free of delay.

τ-decomposition applied to the scalar case

In order to illustrate the method we reconsider the scalar system including a single (constant) delay,

$$\dot{x}(t) = -ax(t) - bx(t-\tau), \tag{3.12}$$

under appropriate initial conditions with $a, b \in \mathbb{R}$, and $\tau > 0$. As mentioned above, a and b should be fixed, and the only "free" parameter is the delay τ. The characteristic equation $\lambda + a + be^{-\lambda\tau} = 0$ admits a solution $\lambda = j\omega$ on the imaginary axis if and only if the following conditions are satisfied simultaneously:

$$\begin{cases} \omega^2 + a^2 = b^2, \\ a + b\cos(\omega\tau) = 0. \end{cases}$$

The first condition is nothing other than the modulus condition, and it has an interesting property in the sense that we can derive the crossing frequency $j\omega$ *independently* on the delay value. Such an idea can be applied to high-order systems also, and it can be generalized to delay systems in a state-space representation (using matrix pencil techniques).

Depending on the values of a and b, three situations may occur:

(a) $\mid a \mid > \mid b \mid$. In such a case, there does not exist any frequency ω satisfying the modulus condition mentioned above. In conclusion, there is no crossing with respect to the imaginary axis, that is, the scalar system is stable or unstable depending on the stability (or instability) of the corresponding system free of delay. In other words, the (stability, instability) property is of *delay-independent* type. If there exists at least one unstable characteristic root, that is, the instability degree is strictly positive, then the system is called *hyperbolic* (see, for instance, [103, 223, 222] for further discussions).

(b) $\mid a \mid = \mid b \mid \neq 0$, that is $a = \pm b$. If $a = -b$, it easy to see that $\lambda = 0$ is a solution of the characteristic equation for $\tau = 0$ and, furthermore, such a solution is *invariant* with respect to the delay τ (since $e^{0\cdot\tau} = 1$, for all $\tau \geq 0$). In conclusion, the system cannot be asymptotically stable due to the characteristic root at zero. Now, if $a = b$, then it follows that the pair (ω, τ) should satisfy simultaneously $\omega\tau = (2k+1)\pi$, for some positive integer k, and $\omega = 0$, conditions which are impossible for finite delays. Thus, the stability (instability) of the scalar system is still defined in this case by the stability (or instability) property of the system free of delay. In other words, the (stability, instability) property is still of *delay-independent* type. The main difference with respect to the previous situation is that the line $a = b$ defines the *boundary* of the *delay-independent stability (instability) domain* with respect to the dual one, the so-called *delay-dependent stability (instability) domain*. Indeed, it is easy to see that there exists some perturbation on the pair (a, a) such that the delay-independent stability (instability) is lost at $(a, a+\varepsilon)$ for ε sufficiently small, where $\operatorname{sign}(a) = \operatorname{sign}(\varepsilon)$.

(c) $\mid a \mid < \mid b \mid$. In this case, crossing frequency exists; that is, characteristic roots will cross the imaginary axis if the delay is increased starting from 0.

The situations described above have a very simple graphical representation by using the modulus condition above. Define the transfer

$$H_{yu}(\lambda) := -\frac{b}{\lambda + a},$$

which corresponds to a first-order system, and next represent the intersections between the graph of $H_{yu}(j\omega)$ with the unit circle $\mathcal{C}(0,1)$ in the complex plane for $\omega \geq 0$. Since $\lim_{\omega\to\infty} \mid H_{yu}(j\omega) \mid = 0$, and using the monotonicity of $\mid H_{yu}(j\omega) \mid$ as a function of ω for $\omega \in \mathbb{R}_+$, it follows that explicit crossing exists if and only if $\mid a \mid < \mid b \mid$. Case (b) above corresponds to the situation when the graph starts tangentially from the point $(-1, 0)$ of the unit circle, and excepting this (particular) point the whole graph is *inside* the unit circle.

Crossing direction characterization. Let us detail the τ-decomposition method in this last case since crossing with respect to the imaginary axis exists. More precisely, there exists only *one (crossing) frequency* that is given by

$$\omega_c = \sqrt{b^2 - a^2}.$$

This is also confirmed by the graphical representation in Figure 3.2. The corresponding delays for which characteristic roots cross the imaginary axis are given by the relation

$$\tau_c = \min_{\ell\in\mathbb{N}} \left\{ \frac{1}{\sqrt{b^2 - a^2}} \left[\cos^{-1}\left(-\frac{a}{b}\right) + 2\ell\pi \right] > 0 \right\},$$

$$\tau_{k+1} = \tau_c + \frac{2k\pi}{\sqrt{b^2 - a^2}}, \qquad k \in \mathbb{N}.$$

We need to analyze the behavior of the crossing root $j\omega_c$ when τ is increased from τ_c to $\tau_c + \varepsilon$, where $\varepsilon > 0$, but very small. Based on the developments in Chapter 1, it follows that the mapping $\lambda : \mathbb{R}_+ \mapsto \mathbb{C}$, defined by $\lambda = \lambda(\tau)$, is a differentiable function. Furthermore, since the root $\lambda = j\omega_c$ is *simple*, the derivative $d\lambda/d\tau$ evaluated at $\lambda = j\omega_c$, $\tau = \tau_1$ exists, and it is different from 0. Let us differentiate the corresponding characteristic function:

$$\begin{aligned} \frac{d}{d\tau} H(\lambda(\tau);\ \tau) &= \frac{d}{d\tau}\lambda + be^{-\lambda\tau}\left(-\tau\frac{d}{d\tau}\lambda - \lambda\right) \\ &= \left(1 - b\tau e^{-\lambda\tau}\right)\frac{d}{d\tau}\lambda - b\lambda e^{-\lambda\tau}. \end{aligned} \tag{3.13}$$

Since at $\tau = \tau_1$ the critical root $\lambda = j\omega_c$ is *simple*, it follows that evaluating (3.13) at $\tau = \tau_1$ will lead to

$$\left(1 - be^{-j\omega_c\tau_1}\right)\frac{d}{d\tau}\lambda(\tau_1) = j\omega_c be^{-j\omega_c\tau_1},$$

which simply rewrites as

$$\left[\frac{d}{d\tau}\lambda(\tau_1)\right]^{-1} = -\frac{\tau_1}{j\omega_c} - \frac{1}{j\omega_c be^{-j\omega_c\tau_1}},$$

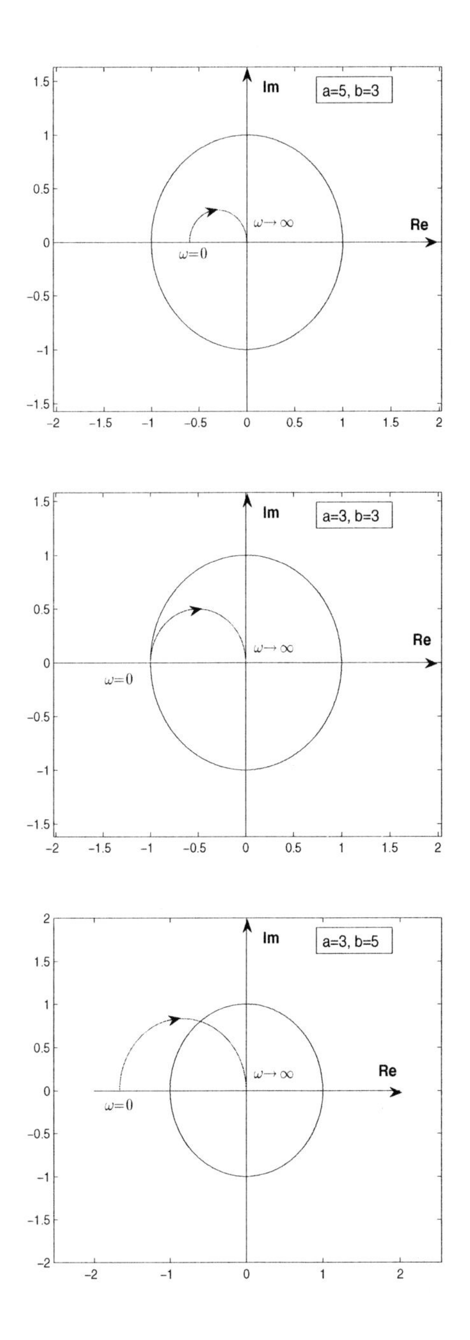

Figure 3.2. *The intersection in the complex plane of the transfer* $H_{yu}(\lambda) = -b/(\lambda + a)$ *evaluated on the imaginary axis* $j\mathbb{R}_+$ *with the unit circle* $\mathcal{C}(0, 1)$ *for different values of the parameters a and b covering all the possible situations.*

because $d\lambda/d\tau$ exists at $\tau = \tau_1$ and it is different from 0 since the corresponding critical root $j\omega_c$ on the imaginary axis is simple. It is important to note that $\omega_c \neq 0$ since $a+b \neq 0$. Next,

$$e^{-j\omega_c\tau_1} = -\frac{j\omega_c + a}{b},$$

since $j\omega_c$ is a characteristic root, and thus we get the following evaluation of the derivative at $\tau = \tau_1$:

$$\left[\frac{d}{d\tau}\lambda(\tau_1)\right]^{-1} = -\frac{\tau_1}{j\omega_c} + \frac{\omega_c + ja}{\omega_c(a^2 + \omega_c^2)}. \tag{3.14}$$

It is easy to see that the crossing direction given by an increment $\varepsilon > 0$ on $\tau = \tau_1$ is given by the sign of $\Re[\frac{d}{d\tau}\lambda(\tau)]_{\tau=\tau_1}$. In our case, one needs to evaluate the real part of the quantities in (3.14), that is,

$$\Re\left[\frac{d}{d\tau}\lambda(\tau_1)\right]^{-1} = \frac{1}{a^2 + \omega_c^2}, \tag{3.15}$$

a quantity which always *exists*, and it is always *positive*. Furthermore, this evaluation does *not depend* on the *critical delay* value τ_1, but only on the critical characteristic root value.

In other words, for all the critical delays τ_k, $k \geq 1$, the *crossing direction is always toward instability*. In conclusion, the only delay interval guaranteeing (asymptotic) stability is the *first delay interval* $[0, \tau_1)$ if the system free of delays is asymptotically stable. The bound τ_1 is also known as the *delay margin*. In such a situation, increasing the delay value does not improve the stability of the system, but it induces more roots crossing the imaginary axis toward instability. Such an *effect* is usually called *delay-induced instability* or *destabilizing effect* of the delay.

From Figure 3.2, it follows that the crossing directions at the end delay points of all intervals should be in the *same* direction, a fact that is confirmed by the relation above, which proves that the crossing direction is *invariant* with respect to the delay crossing value. Indeed, a change of crossing direction should be associated with the existence of at least *one more* critical characteristic root on the imaginary axis, a fact which is impossible in the scalar case (first-order polynomial in ω^2). However, the existence of at least two critical roots on the imaginary axis does not mean that the increase of the delay parameter will necessarily induce stability. These aspects will be discussed in the next paragraphs.

Remark 3.2. *Brauer [24] has also analyzed the stability of the linear scalar single delay case with respect to the delay term τ, by defining a "special" quantity called* characteristic return time τ_c *to the equilibrium $x \equiv 0$:*

$$\tau_c = \frac{-\tau}{\alpha},$$

where α is the spectral abscissa. Next, by analyzing the monotonicity properties of this "quantity" (characteristic return time) with respect to τ, we may decide whether there exist other stability regions by increasing the delay parameter τ for the case $b >| a |$. This leads to the following simple properties:

(i) If $b < 0$ and $a + b > 0$, then the characteristic return time is a monotone increasing function of τ.

(ii) If $b > 0$, then the characteristic return time is a decreasing function of τ, for all $\tau \in [0, \tau_c)$, where τ_c is defined by

$$b\tau^* e^{a\tau_c} = e^{-1}.$$

Note that it is an increasing function of τ for $\tau > \tau_c$ and it remains finite for all τ for which the zero solution is asymptotically stable.

Although the idea to analyze the behavior of such stability regions is quite attractive, it seems difficult to generalize it for more complicated (high-order) systems.

3.3.3 Numerical continuation

We present the main ideas behind the numerical continuation approach for the automatic computation of branches in a two-parameter space that correspond to characteristic roots on the imaginary axis.

Note that (3.1) has a characteristic root at $j\omega$, $\omega > 0$, if and only if

$$\begin{cases} \Re(M(j\omega;\ p)v) = 0, \\ \Im(M(j\omega;\ p)v) = 0, \\ \Re(n(v)) = 0, \\ \Im(n(v)) = 0, \end{cases} \tag{3.16}$$

where $v \in \mathbb{C}^{n\times 1}$ is the right eigenvector of M (more precisely, $ve^{\lambda\theta}$, $\theta \in [-\tau_m, 0]$ is the right eigenfunction of the infinitesimal generator of the solution operator corresponding to (3.1); see Chapter 1), and $n(v) = 0$ is a *normalizing* condition for the eigenvector. For instance, one can take $n(v) = a^T v = 0$ with $a \in \mathbb{C}^n$. Assume that there are two 2 parameters, that is, $p = (p_1, p_2)$. Then (3.16) consists of $2n + 2$ equations in the $2n + 3$ unknowns (p_1, p_2, ω, v) (note that both $\Re(v)$ and $\Im(v)$ have n components). So, if p_1 and p_2 are free, then (3.16) locally defines a branch under the assumptions of the implicit function theorem. For the problem under consideration, the purpose of numerical continuation is to compute a (discrete approximation of a) branch of solutions by starting from some point(s) on the branch and following the branch. The starting point(s) are typically obtained by freezing one parameter or adding an extra condition, such that the number of unknowns becomes equal to the number of parameters, and solving the resulting system using Newton's method.

The so-called *predictor-corrector methods* essentially boil down to the following main steps:

- the predictor step, consisting of estimating a new point on the branch;
- the correction step, consisting of computing a new point on the branch by correcting the predicted point.

We illustrate this with the commonly used *secant predictor*, combined with a correction based on an *arclength* or *pseudo-arclength* parameterization of the branch.

Assume that we have computed two nearby points on the branch, namely $P^{(j-1)}$ and $P^{(j)}$,

$$\begin{aligned} P^{(j-1)} &: (p_1^{(j-1)}, p_2^{(j-1)}, \omega^{(j-1)}, v^{(j-1)}), \\ P^{(j)} &: (p_1^{(j)}, p_2^{(j)}, \omega^{(j)}, v^{(j)}), \end{aligned}$$

where the superscript refers to point number. Then a prediction $\hat{P}^{(j+1)}$ of a new point $P^{(j+1)}$ can be obtained from the linear approximation of the branch through $P^{(j-1)}$ and $P^{(j)}$ as follows:

$$\hat{P}^{(j+1)} : (\hat{p}_1^{(j+1)}, \hat{p}_2^{(j+1)}, \hat{\omega}^{(j+1)}, \hat{v}^{(j+1)}),$$

where

$$\begin{aligned} \hat{p}_1^{(j+1)} &= p_1^{(j)} + \tfrac{\epsilon^{(j)}}{\|p^{(j)}-p^{(j-1)}\|_2} (p_1^{(j)} - p_1^{(j-1)}), \\ \hat{p}_2^{(j+1)} &= p_2^{(j)} + \tfrac{\epsilon^{(j)}}{\|p^{(j)}-p^{(j-1)}\|_2} (p_2^{(j)} - p_2^{(j-1)}), \\ \hat{\omega}^{(j+1)} &= \omega^{(j)} + \tfrac{\epsilon^{(j)}}{\|p^{(j)}-p^{(j-1)}\|_2} (\omega^{(j)} - \omega^{(j-1)}), \\ \hat{v}^{(j+1)} &= v^{(j)} + \tfrac{\epsilon^{(j)}}{\|p^{(j)}-p^{(j-1)}\|_2} (v^{(j)} - v^{(j-1)}), \end{aligned} \tag{3.17}$$

and $\epsilon^{(j)}$ is the steplength taken in the jth step. Note that, as it appears in (3.17), the latter corresponds to the distance between (the projections of) $\hat{P}^{(j+1)}$ and $P^{(j)}$ in the (p_1, p_2)-plane.

The next step consists of correcting $\hat{P}^{(j+1)}$ to a new point $P^{(j+1)}$ on the branch. This can be done by applying Newton's method to (3.16), with (3.17) as starting values. However, since p_1 and p_2 are freed, the solutions of (3.16) define a branch, as we have seen. Therefore, an extra equation has to be added first to (3.17), in order to uniquely specify the solution. One possibility consists of adding

$$\left(p_1 - p_1^{(j)}\right)^2 + \left(p_2 - p_2^{(j)}\right)^2 = \left(\varepsilon^{(j)}\right)^2. \tag{3.18}$$

This is called an arclength condition, as it expresses that the new point $P^{(j+1)}$ must lie at a distance $\varepsilon^{(j)}$ from $P^{(j)}$ in the (p_1, p_2)-plane and $\sum_j \varepsilon^{(j)}$ can be seen as an approximation of the parameter corresponding to an arclength parameterization of the branch in the (p_1, p_2)-plane. In combination with (3.17), it is natural to take

$$\varepsilon^{(j)} = \epsilon^{(j)}. \tag{3.19}$$

Because the condition (3.18) is nonlinear in the parameters and it contains no information about the direction in which the new point must be found along the branch, usually a linearization of (3.18) is employed, leading to a so-called pseudo-arclength condition. For instance, one can replace (3.18)–(3.19) with

$$\left\langle p^{(j)} - p^{(j-1)}, p - \hat{p}^{(j+1)} \right\rangle = 0,$$

which expresses that the new point lies on a line through the prediction $\hat{p}^{(j+1)}$ and perpendicular to the approximation of the tangent vector at $p^{(j)}$, given by $p^{(j)} - p^{(j-1)}$.

For more details on numerical computation, refer to [270, 145] and the references therein.

3.4 Computing the crossing direction of characteristic roots

Assume that $\lambda_0 = j\omega_0 \in \sigma(M)$ is a characteristic root on the imaginary axis with $\omega_0 \neq 0$ for some parameter vector $p = p^0 \in \mathbb{R}^{n_p}$, that is, $\lambda_0 = \lambda(p^0)$. Assume also that the functions $\tau_i : \mathbb{R}^{n_p} \mapsto \mathbb{R}_+$, $i = 1, \ldots, m$, and the matrix functions $A_i : \mathbb{R}^{n_p} \mapsto \mathbb{R}^{n\times n}$ are continuously differentiable functions of the parameters $p := (p_1\ p_2\ \ldots\ p_{n_p}) \in \mathbb{R}^{n_p}$.

3.4.1 Simple crossing characteristic roots

We start with the assumption that the characteristic roots on the imaginary axis are *simple*. Due to the differentiability property of $A_0(\cdot)$, $A_1(\cdot)$, ..., $A_m(\cdot)$, $\tau_1(\cdot)$, $\tau_2(\cdot)$, ..., $\tau_{n_p}(\cdot)$ it follows that $\lambda_0 = j\omega_0 = \lambda(p)$ is also a differentiable function, and its derivative exists in some neighborhood of p^0.

There are several ways to compute the derivative of the function, of $p \mapsto \lambda(p)$ evaluated at $p = p^0$ with respect to any of the directions p_i, $i = 1, \ldots, n_p$, in the parameter space. Formally, simple computations lead to

$$0 = \frac{\partial}{\partial p_i} H(\lambda(p);\ p) = \frac{\partial}{\partial \lambda} H(\lambda;\ p) \cdot \frac{\partial \lambda}{\partial p_i} + \frac{\partial}{\partial p_i} H(\lambda;\ p)$$

evaluated at (λ_0, p^0), and we get (using the implicit function theorem) that, *locally*, in some neighborhood of the point $p = p^0$ such that $\lambda_0 = \lambda(p^0)$, the condition

$$\frac{\partial \lambda}{\partial p_i} := -\frac{\frac{\partial}{\partial p_i} H(\lambda;\ p)}{\frac{\partial}{\partial \lambda} H(\lambda;\ p)} \tag{3.20}$$

holds, evaluated at the same point. It is important to note that $\frac{\partial}{\partial \lambda} H(\lambda;\ p) \mid_{\lambda=\lambda_0, p=p^0} \neq 0$ since the corresponding characteristic root is simple.

In conclusion, in order to evaluate explicitly this derivative, one needs to evaluate the derivative of the determinant $H(\lambda;\ \tau) = 0$ at $\lambda = \lambda_0$, and $\tau = \tau(p^0)$. In what follows, we shall propose two independent approaches for estimating such a crossing characteristic root derivative: one based on Jacobi's formula, and one based on the perturbation theory for matrices.

Jacobi's formula based approach

Consider also the corresponding right (left) eigenvectors u_0 (v_0^*) associated with the characteristic root $\lambda_0 = j\omega_0$. If $\lambda_0 = \lambda(p^0)$ is a *simple characteristic root*, then $\lambda = \lambda(p)$ is a differentiable function around $p = p^0$, and furthermore the eigenvectors are also differentiable. We can apply this last argument in order to derive the corresponding derivative. However, we shall present a different method based on the Jacobi's formula for the determinant's derivative.

We have the following result.

Proposition 3.3. *If the characteristic root $\lambda_0 = j\omega_0 = \lambda(p^0) \neq 0$ is simple, then for any positive integer i, $1 \leq i \leq m$, the following holds:*

$$\frac{\partial \lambda}{\partial p_i} = -\frac{v_0^* \dfrac{\partial M}{\partial p_i} u_0}{v_0^* \dfrac{\partial M}{\partial \lambda} u_0}, \tag{3.21}$$

where the partial derivatives of the characteristic matrix $M(\lambda;\ p)$ are evaluated at $\lambda = \lambda_0$, and $p = p^0$. The vectors v_0^* and u_0 are the left and right eigenvectors[8] *corresponding to characteristic root λ_0.*

Proof. From the definition of the characteristic function, it follows that $\lambda = \lambda_0$ is a simple root if and only if $\operatorname{rank}(M(\lambda_0, p^0)) = n - 1$. The use of Jacobi's formula

$$d \det(M) = Tr \left(\operatorname{Adj}(M) dM\right) \tag{3.22}$$

leads to the following relation:

$$\begin{aligned}
\frac{\partial}{\partial p_i} \det(M(\lambda;\ p)) &= Tr \left(\operatorname{Adj}(M(\lambda;\ p)) \frac{\partial}{\partial p_i} M(\lambda;\ p) \right) \\
&= \operatorname{Tr} \left[u_0 v_0^* \left(\frac{\partial}{\partial \lambda} M(\lambda;\ p) \frac{\partial \lambda}{\partial p_i} + \frac{\partial}{\partial p_i} M(\lambda;\ p) \right) \right] \\
&= v_0^* \left(\frac{\partial}{\partial \lambda} M(\lambda;\ p) \frac{\partial \lambda}{\partial p_i} + \frac{\partial}{\partial p_i} M(\lambda;\ p) \right) u_0,
\end{aligned} \tag{3.23}$$

where we used the standard property of the adjoint of a matrix $M \in \mathbb{C}^{n \times n}$ of rank $(n-1)$ to be a matrix of rank one [120], expressed as a function of its left and right null vectors as $u_0 v_0^*$.

Next, using the same argument as in deriving (3.20), the condition (3.21) evaluated at $\lambda = \lambda_0$ and $p = p^0$ follows straightforwardly from (3.23), and $v_0^* \frac{\partial M}{\partial \lambda} u_0 \neq 0$ at $\lambda = \lambda_0$ and $p = p^0$ since the characteristic root is simple. $\square$

An explicit computation of the derivative leads to the formula

$$\frac{\partial \lambda}{\partial p_i} = \frac{v_0^* \left(\dfrac{\partial A_0(p)}{\partial p_i} + \displaystyle\sum_{k=1}^{m} \frac{\partial A_k(p)}{\partial p_i} e^{-\lambda \tau_k(p)} - \lambda \sum_{k=1}^{m} A_k(p) \frac{\partial \tau_k(p)}{\partial p_i} e^{-\lambda \tau_k(p)} \right) u_0}{v_0^* \left(I_n + \displaystyle\sum_{k=1}^{m} A_k \tau_k(p) e^{-\lambda \tau_k(p)} \right) u_0}, \tag{3.24}$$

evaluated at $\lambda = \lambda_0$ and $p = p^0$.

---

[8] More precisely, the function segments $u_0 e^{\lambda_0 \theta}$, $\theta \in [-\tau, 0]$, and $v_0^* e^{\bar{\lambda}\theta}$, $\theta \in [0, \tau]$, are right and left eigenfunctions of the corresponding infinitesimal generator.

Perturbation theory based approach

As mentioned earlier, there exist several ways to compute the derivative of the function $\lambda(p)$ evaluated at $p = p^0$. Here, we focus on a different way to derive such a derivative by using the *perturbation theory* for (analytic) operators (see, for instance, [133]).

Assume now that the function $p : \mathbb{R} \mapsto \mathbb{R}^{n_p}$, $\epsilon \mapsto p(\varepsilon)$, is sufficiently smooth in some neighborhood of $p = p^0$, and assume that $p(0) = p^0$. Let $d \in \mathbb{R}^{n_p}$ be the *direction vector* in the parameter space defined as follows:

$$d := \frac{d}{d\varepsilon}p(0) = \left[\begin{array}{cccc} \frac{d}{d\varepsilon}p_1(0) & \frac{d}{d\varepsilon}p_2(0) & \dots & \frac{d}{d\varepsilon}p_{n_p}(0) \end{array} \right]. \tag{3.25}$$

Without any loss of generality, we can use the following normalization condition:

$$\|d\|_2 := \sqrt{\sum_{k=1}^{n_p} d_k^2} = 1.$$

Using the smoothness of $p(\varepsilon)$, it follows that the characteristic matrix $M(\lambda;\ p)$ rewrites as

$$M(\lambda;\ p(\varepsilon)) = M_{1,0}(\lambda) + \sum_{i=1}^{r} \varepsilon^i M_{i,1}(\lambda) + \dots, \tag{3.26}$$

for some positive integer $r > 1$. The construction of the matrices $M_{1,0}$ and $M_{i,1}$, $1 \leq i \leq r$, follows straightforwardly. For instance,

$$\begin{aligned} M_{1,0}(\lambda) &= M(\lambda;\ p^0), \\ M_{1,1}(\lambda) &= \sum_{k=1}^{n_p} \frac{\partial}{\partial p_k} M(\lambda;\ p^0) d_k \\ &= -\sum_{k=1}^{n_p} \frac{\partial}{\partial p_k} A_0(p^0) d_k - \sum_{l=1}^{m} \sum_{k=1}^{n_p} \frac{\partial}{\partial p_k} A_l(p^0) d_k e^{-\lambda \tau_l(p^0)} \\ &\quad + \sum_{l=1}^{m} \sum_{k=1}^{n_p} \lambda d_k \frac{\partial}{\partial p_k} \tau_l(p^0) A_l(p^0) e^{-\lambda \tau_l(p^0)}. \end{aligned}$$

In the case of simple characteristic roots, the *Puiseux formulae* (see, for instance, [133]) lead to the following expansion of the characteristic root λ, and of the corresponding right eigenvector u:

$$\lambda = \lambda_0 + \sum_{i=1}^{r} \varepsilon^i \lambda_i + \dots, \tag{3.27}$$

$$u = u_0 + \sum_{i=1}^{r} \varepsilon^i u_i + \dots. \tag{3.28}$$

Assume now that λ_0 is located on the imaginary axis ($\lambda_0 = j\omega_0 \neq 0$). The way in which the characteristic root is moving with respect to the imaginary axis (that is, toward stability/instability regions) is given by the complex λ_1, that needs an explicit evaluation.

More precisely,

(i) If $\Re(\lambda_1) > 0$, then the crossing direction at $\varepsilon > 0$ is *toward instability*.

(ii) If $\Re(\lambda_1) < 0$, then the crossing direction at $\varepsilon > 0$ is *toward stability*.

(iii) Finally, if $\Re(\lambda_1) = 0$, then we need a *second-order analysis* in order to see if the corresponding characteristic root will cross the imaginary axis or will stay in the same half-plane. In other words, we need the explicit computation of λ_2 in the corresponding λ-expansion.

In order to compute λ_1, we need a further expansion of (3.26) in some neighborhood of $\lambda_0 = j\omega_0 = \lambda(p^0) = \lambda(p(0))$, by using the mapping $\lambda \circ p : \mathbb{R} \mapsto \mathbb{C}$ and defined by $(\lambda \circ p)(\varepsilon) = \lambda(p(\varepsilon))$. For the sake of brevity, we shall "identify" $\lambda(p(\varepsilon))$ as $\lambda(\varepsilon)$. It is easy to see that (3.26) rewrites as

$$M(\lambda(\varepsilon);\ p(\varepsilon)) = M_0 + \sum_{i=1}^{r} \varepsilon^i M_i + \dots, \tag{3.29}$$

with

$$M_0 = M(\lambda_0, p^0).$$

Simple algebraic manipulations lead to

$$\begin{aligned} M_1 = {} & \lambda_1 \left(I_n + \sum_{l=1}^{m} \tau_l(p^0) A_l e^{-\lambda_0 \tau_l(p^0)} \right) \\ & - \sum_{k=1}^{n_p} \left(\frac{\partial}{\partial p_k} A_0(p^0) d_k + \sum_{l=1}^{m} \frac{\partial}{\partial p_k} A_l(p^0) d_k e^{-\lambda_0 \tau_l(p^0)} \right) \\ & + \sum_{k=1}^{n_p} \sum_{l=1}^{m} \lambda_0 d_k A_l(p^0) \frac{\partial}{\partial p_k} \tau_l(p^0) e^{-\lambda_0 \tau_l(p^0)}. \end{aligned} \tag{3.30}$$

Since the characteristic root $\lambda_0 = j\omega_0$ is *simple*, it follows that

$$M_0 u_0 = 0, \quad v_0^* M_0 = 0, \quad v_0^* u_0 \neq 0.$$

The last condition is a consequence of the fact that the eigenvalue 0 of the complex matrix M_0 is (algebraically) simple. Without any loss of generality, we can use the following normalization condition:

$$v_0^* u_0 = 1.$$

Next, by identifying the coefficients, we arrive at the following proposition.

Proposition 3.4. *For a given direction vector* $d \in \mathbb{R}^{n_p}$ *with* $\|d\|_2 = 1$*, the characteristic root* $\lambda = \lambda(p)$ *crosses the imaginary axis at* $p = p^0$ *toward instability (stability) with respect to the direction* d *if* $\Re(\lambda_1) > 0$ ($\Re(\lambda_1) < 0$)*, where*

$$\lambda_1 = \sum_{k=1}^{n_p} \frac{\partial}{\partial p_k} \lambda(p^0) d_k, \tag{3.31}$$

where $\partial\lambda/\partial p_i$*, for* $i = 1, \ldots n_p$*, are given by (3.24).*

There is still one case to be discussed, that is, the case where $\Re(\lambda_1) = 0$. Then we have two distinct situations depending on whether $\lambda_1 = 0$:

(i) The first case ($\lambda_1 = 0$) corresponds to the situation where the direction vector d is tangent to the "surface" in the parameter space that corresponds to the presence of characteristic roots on the imaginary axis.

(ii) The other case ($\lambda_1 \neq 0$) corresponds to the situation where the characteristic root arrives *tangentially* to the imaginary axis (one contact point in the sense mentioned by [29]) and, as mentioned above, a second-order analysis is needed, that is, an explicit computation of $\Re(\lambda_2)$. As expected, two situations are possible:

  (ii.1) the characteristic root will stay in the same domain, that is, the corresponding half-plane of $\mathbb{C}$ (the so-called mirror effect), or

  (ii.2) it will cross the imaginary axis depending on the sign of $\Re(\lambda_2)$.

Finally, if $\Re(\lambda_2) = 0$, the procedure should follow the same steps as above, and a new iteration will be needed.

For the brevity of the chapter, we do not discuss the second-order analysis in the general case. However, we mention the existence of such an analysis in the commensurate delay case as reported in [84].

3.4.2 Semisimple characteristic roots

Consider now the case where the characteristic root $\lambda_0 = j\omega_0 \neq 0$ is *not simple*, and assume for the sake of simplicity that its algebraic multiplicity μ_0 is equal to 2. In such a situation, the characteristic function rewrites as follows:

$$H(\lambda;\ p^0) = (\lambda - \lambda_0)^2 H_1(\lambda;\ p^0),$$

with $H_1(\lambda_0;\ p^0) \neq 0$.

Two situations may appear depending on the geometric multiplicity of the characteristic root as an eigenvector of the corresponding infinitesimal generator (number of linearly independent eigenfunctions):

(i) The characteristic root λ_0 is *double semisimple*; that is, there exist two linearly independent eigenvectors. In other words, the Riesz index $\gamma_0 = 1$ (see [123]);

(ii) The characteristic root λ_0 is *double, nonsemisimple*; that is, there exists only one linearly independent eigenvector.

Consider now the first case, that is the double semisimple characteristic root. Some ideas for treating the remaining case (double eigenvalue, nonsemisimple) will be considered in the next section.

If the root is double and semisimple, then the normalized left $(v_{0,1}^*, v_{0,2}^*)$ and right $(u_{0,1}, u_{0,2})$ eigenvectors should verify the following equations (see, for instance, [317] for general results and [216] for some discussions in the matrix case):

$$\begin{cases} M(\lambda_0;\ p^0)u_{0,l} = 0, \\ v_{0k}^* M(\lambda_0;\ p^0) = 0, \\ v_{0,k}^* u_{0,l} = \delta_{k,l}, \quad \forall k, l \in \{1, 2\}, \end{cases} \tag{3.32}$$

where δ_{kl} is the Kronecker symbol (1 if $k = l$, and 0 otherwise).

In this case, the Puiseux formulae rewrite as

$$\lambda = \lambda_0 + \sum_{i=1}^{r} \varepsilon^i \lambda_i + \dots, \tag{3.33}$$

$$u = u_0 + \sum_{i=1}^{r} \varepsilon^i u_i + \dots. \tag{3.34}$$

By substituting (3.33)–(3.34) and (3.32) in the equation

$$M(\lambda;\ p(\epsilon)),$$

some simple but tedious algebraic manipulations lead to the vector

$$u_0 = \gamma_1 u_{0,1} + \gamma_2 u_{0,2}, \tag{3.35}$$

where γ_1 and γ_2 should verify the following conditions:

$$\left(\lambda_1 \begin{bmatrix} v_{0,1}^* M_1^0 u_{0,1} & v_{0,1}^* M_1^0 u_{0,2} \\ v_{0,2}^* M_1^0 u_{0,1} & v_{0,2}^* M_1^0 u_{0,2} \end{bmatrix} + \begin{bmatrix} v_{0,1}^* M_1^1 u_{0,1} & v_{0,1}^* M_1^1 u_{0,2} \\ v_{0,2}^* M_1^1 u_{0,1} & v_{0,2}^* M_1^1 u_{0,2} \end{bmatrix}\right) \begin{bmatrix} \gamma_1 \\ \gamma_2 \end{bmatrix} = 0, \tag{3.36}$$

where M_1^i, $i = 0, 1$, are given by

$$M_1^0 = I_n + \sum_{l=1}^{m} \tau_l(p^0) A_l e^{-\lambda_0 \tau_l(p^0)},$$

$$M_1^1 = -\sum_{k=1}^{n_p} \left(\frac{\partial}{\partial p_k} A_0(p^0) d_k + \sum_{l=1}^{m} \frac{\partial}{\partial p_k} A_l(p^0) d_k e^{-\lambda_0 \tau_l(p^0)} \right)$$
$$+ \sum_{k=1}^{n_p} \sum_{l=1}^{m} \lambda_0 d_k A_l(p^0) \frac{\partial}{\partial p_k} \tau_l(p^0) e^{-\lambda_0 \tau_l(p^0)};$$

that is, the formula (3.30) of M_1 in the expansion (3.26) is given by

$$M_1 = \lambda_1 M_1^0 + M_1^1.$$

In other words, the *(double) semisimple* case is reduced to solving the *generalized eigenvalue problem* (3.36) where the eigenvector $[\gamma_1\ \gamma_2]^T$ will define the term u_0 in the expansion of the corresponding right eigenvector u. It is clear that the generalized eigenvalue problem above may have a finite number of solutions under the assumption of *regularity* (see, e.g., [86] for a definition) of the matrix pencil: $\lambda\mathcal{M} + \mathcal{N}$, where $\mathcal{M}$ and $\mathcal{N}$ are identified by the corresponding 2×2 matrices in formula (3.36):

$$\mathcal{M} = \begin{bmatrix} v_{0,1}^* M_1^0 u_{0,1} & v_{0,1}^* M_1^0 u_{0,2} \\ v_{0,2}^* M_1^0 u_{0,1} & v_{0,2}^* M_1^0 u_{0,2} \end{bmatrix},$$

$$\mathcal{N} = \begin{bmatrix} v_{0,1}^* M_1^1 u_{0,1} & v_{0,1}^* M_1^1 u_{0,2} \\ v_{0,2}^* M_1^1 u_{0,1} & v_{0,2}^* M_1^1 u_{0,2} \end{bmatrix}.$$

More precisely, the cardinality of the spectrum of the matrix pencil above is given by the rank of the matrices $\mathcal{N}$ and $\mathcal{M}$. We can have a finite number (0, 1, or 2) or an infinite number of generalized eigenvalues. Since the rank of the matrices $\mathcal{M}$ and $\mathcal{N}$ is strongly dependent on the direction vector d, some "small" perturbations on d will lead to $\mathcal{M}$ and $\mathcal{N}$ of full rank, that leads to *two generalized eigenvalues* of the matrix pencil under consideration.

Based on the remarks above, consider now the case of $\det(\mathcal{N}) \neq 0$. In such a situation, some simple computations lead to the fact that the corresponding generalized eigenvalue problem reduces to solving the second-order equation:

$$\frac{\det(\mathcal{M})}{\det(\mathcal{N})}\lambda^2 + \mathrm{Tr}(\mathcal{M}\mathcal{N}^{-1})\lambda + 1 = 0.$$

Now, depending on the value of $\det(\mathcal{M})$, we have three possibilities:

- $\det(\mathcal{M}) = 0$, and $Tr(\mathcal{M}\mathcal{N}^{-1}) = 0$. In such a case, the eigenvalue problem has no solution.
- $\det(\mathcal{M}) = 0$, and $Tr(\mathcal{M}\mathcal{N}^{-1}) \neq 0$. It is easy to see that the eigenvalue problem has only one finite root, given by $\lambda = -1/Tr(\mathcal{M}\mathcal{N}^{-1})$.
- $\det(\mathcal{M}) \neq 0$. In such a case, we will have *two (finite) roots*, not necessarily distinct.

In conclusion, consider now that the generic case of interest, the matrix pencil $\lambda\mathcal{M} + \mathcal{N}$, has *two generalized eigenvalues*, denoted $\lambda_1^{(1)}$ and $\lambda_1^{(2)}$. The corresponding *regularity condition* is

$$\det(\mathcal{M}) \cdot \det(\mathcal{N}) \neq 0. \tag{3.37}$$

With the remarks, comments, and notations above, we have the following result.

Proposition 3.5. *Consider a direction vector $d \in \mathbb{R}^{n_p}$ with $\|d\|_2 = 1$, and assume that the characteristic root $\lambda = \lambda(p)$ on the imaginary axis is (double) semisimple at $p = p^0$. Assume furthermore that the regularity condition (3.37) is satisfied.*

Then the double semisimple characteristic root $\lambda = \lambda(p)$ at $p = p^0$ splits up into two simple characteristic roots with respect to the direction d if and only if

$$\mathrm{Tr}(\mathcal{M}\mathcal{N}^{-1}) \neq \pm 2\sqrt{\det(\mathcal{M})\det(\mathcal{N})}. \tag{3.38}$$

Furthermore, the corresponding crossing directions toward instability (stability) are given by $\Re(\lambda_1^{(1)}) > 0(< 0)$ and $\Re(\lambda_1^{(2)}) > 0(< 0)$, where $\lambda_1^{(1)}$ and $\lambda_1^{(2)}$ are given by

$$\lambda_1^{(1)} = -\frac{\det(\mathcal{M})}{2\det(\mathcal{N})} + \frac{\sqrt{\mathrm{Tr}(\mathcal{M}\mathcal{N}^{-1})^2 - 4\det(\mathcal{M})\det(\mathcal{N})}}{2\det(\mathcal{N})}, \tag{3.39}$$

$$\lambda_1^{(2)} = -\frac{\det(\mathcal{M})}{2\det(\mathcal{N})} - \frac{\sqrt{\mathrm{Tr}(\mathcal{M}\mathcal{N}^{-1})^2 - 4\det(\mathcal{M})\det(\mathcal{N})}}{2\det(\mathcal{N})}. \tag{3.40}$$

Proof. The condition (3.38) simply says that, under the regularity assumption (3.37), the matrix pencil $\det(\lambda\mathcal{M} + \mathcal{N})$ has two *distinct* roots. Next, the corresponding $\lambda_1^{(1)}$ and $\lambda_1^{(2)}$ are obtained by directly computing the roots of the corresponding second-order equation. □

Remark 3.6 (Weak interactions). *The proposition above simply describes the way in which a (double) semisimple characteristic root on the imaginary axis will split by a variation of the parameters defined by the direction d. The behavior above is nothing other than the behavior of two ("almost-independent") simple crossing roots, and it is also known as the* weak interaction *of the characteristic roots; see, for instance, the work of Seyranian and Mailybaev [271] for finite-dimensional systems.*

3.4.3 Further analysis: basic ideas

Let us consider now the case where the characteristic root λ_0 is *double, nonsemisimple.* In other words, there exists only one linearly independent eigenvector. Denote by u_0 (and v_0) the corresponding right (left) eigenvector.

The fact that the eigenvalue λ_0 is double (nonsemisimple) implies that there exist vectors u_0, v_0, u_1, and v_1 such that the following conditions hold simultaneously:

$$\begin{cases} \left(\lambda_0 I_n - A_0(p^0) - \sum_{i=1}^{m} A_i(p)e^{-\lambda_0\tau_i(p^0)}\right)u_0 = 0, \\ \left(\lambda_0 I_n - A_0(p^0) - \sum_{i=1}^{m} A_i(p)e^{-\lambda_0\tau_i(p^0)}\right)u_1 + u_0 = 0, \end{cases} \tag{3.41}$$

and

$$\begin{cases} v_0^*\left(\lambda_0 I_n - A_0(p^0) - \sum_{i=1}^{m} A_i(p)e^{-\lambda_0\tau_i(p^0)}\right) = 0, \\ v_1^*\left(\lambda_0 I_n - A_0(p^0) - \sum_{i=1}^{m} A_i(p)e^{-\lambda_0\tau_i(p^0)}\right) + v_0^* = 0, \end{cases} \tag{3.42}$$

respectively. Furthermore, u_0, v_0, u_1, and v_1 should satisfy the conditions

$$v_0^* u_0 = 0, \qquad v_1^* u_0 = v_0^* u_1 \neq 0.$$

As in the double semisimple characteristic root case, and without any loss of generality, we can impose the following regularity conditions:

$$v_1^* u_1 = 0, \qquad v_0^* u_1 = 1. \tag{3.43}$$

In this case of double nonsemisimple characteristic roots, the Puiseux formulae (see, for instance, [133]) lead to the following expansion of λ, and of the corresponding right eigenvector u:

$$\lambda = \lambda_0 + \sum_{i=1}^{r} \varepsilon^{\frac{i}{2}} \lambda_i + \dots, \tag{3.44}$$

$$u = u_0 + \sum_{i=1}^{r} \varepsilon^{\frac{i}{2}} w_i + \dots, \tag{3.45}$$

where λ_i and w_i, for all $i = 1, 2, \dots$, need be computed. Since the characteristic root λ and the right eigenvector v satisfy

$$M(\lambda(\varepsilon);\ p(\varepsilon))u(\varepsilon) = 0,$$

some simple but tedious computations lead us to the following first condition concerning λ_1, λ_2, w_1, and w_2:

$$M_{1/2} u_0 + M_0 w_1 = 0, \tag{3.46}$$

$$M_1 u_0 + M_{1/2} w_1 + M_0 w_2 = 0. \tag{3.47}$$

Thus, the first two terms of the expansion of M (that is, M_0 and $M_{1/2}$) are given by

$$M_0 := M(\lambda_0, p^0) = \lambda_0 I - A_0(p^0) - \sum_{i=1}^{m} A_i(p^0) e^{-\lambda_0 \tau_i(p^0)}, \tag{3.48}$$

$$M_{1/2} := \lambda_1 \left(I_n + \sum_{l=1}^{m} \tau_l(p^0) A_i(p^0) e^{-\lambda_0 \tau_l(p^0)} \right). \tag{3.49}$$

The conditions (3.46)–(3.47), combined with the conditions (3.48)–(3.49), (3.41)–(3.42), and using the regularity conditions (3.43), will help to derive the formulae of λ_1 and of w_1 in the corresponding expansions.

Notice that we preferred to present only some extremely simple and *intuitive* ideas of the perturbation based theory approach for deriving such a crossing characterization in the double nonsemisimple characteristic root. A different angle for handling the problem in the case of commensurate delay was proposed and developed in [45]. Further comments on the corresponding results are included in the next chapter.

3.4.4 Delay interdependence and crossing direction evaluation

Consider now the delay dependence on the parameters. Recall that the mapping $\tau : \mathbb{R}^{n_p} \mapsto \mathbb{R}^m$ was assumed continuous and differentiable, and subjected to the constraint $\tau_i(p) \geq 0$, for all $1 \leq i \leq m$ and all $p \in \mathcal{D} \subset \mathbb{R}^{n_p}$. As discussed at the beginning of this chapter, there exist some simple cases of *delay interdependence* for which the conditions above are automatically satisfied. Since the interdependence of the delays plays an important role in computing stability crossing boundaries, and related stability regions, we need to have the explicit expressions of the corresponding crossing direction.

Corollary 3.7 (Delays interdependence: crossing direction evaluation). *Under the assumptions of Proposition 3.3 the following expressions hold with respect to the delays:*

(i) [Independent delays: $n_p = m$, $\tau_i = p_i$, $i = 1, \ldots, m$] The crossing direction with respect to the delay τ_i is given by

$$\frac{\partial \lambda}{\partial \tau_i} = \frac{v_0^* \left(\dfrac{\partial A_0(\tau)}{\partial \tau_i} + \displaystyle\sum_{k=1}^{m} \dfrac{\partial A_k(\tau)}{\partial \tau_i} e^{-\lambda \tau_k} - \lambda A_i(\tau) e^{-\lambda \tau_i} \right) u_0}{v_0^* \left(I_n + \displaystyle\sum_{k=1}^{m} A_k \tau_k e^{-\lambda \tau_k} \right) u_0}, \tag{3.50}$$

evaluated at $\lambda = \lambda_0$, and $\tau_i = \tau_i^0$, for all $i = 1, \ldots, m$.

(ii) [Dependent delay: $n_p < m$, $\tau_i = \sum_{k=1}^{n_p} \gamma_{ik} p_k$, $i = 1, \ldots, m$] The crossing direction with respect to the delay p_i is given by

$$\frac{\partial \lambda}{\partial p_i} = \frac{v_0^* \left(\dfrac{\partial A_0(p)}{\partial p_i} + \displaystyle\sum_{k=1}^{m} \dfrac{\partial A_k(p)}{\partial p_i} e^{-\lambda \tau_k(p)} - \lambda \sum_{k=1}^{m} \gamma_{ki} A_k(p) e^{-\lambda \tau_k(p)} \right) u_0}{v_0^* \left(I_n + \displaystyle\sum_{k=1}^{m} A_k \tau_k(p) e^{-\lambda \tau_k(p)} \right) u_0}, \tag{3.51}$$

evaluated at $\lambda = \lambda_0$ and $p = p^0$.

(iii) [Commensurate delays: $\tau_k = k\tau$, $k = 1, \ldots, m$] The crossing direction with respect to the delay τ is given by

$$\frac{d\lambda}{d\tau} = \frac{v_0^* \left(\dfrac{dA_0(\tau)}{d\tau} + \displaystyle\sum_{k=1}^{m} \dfrac{dA_k(\tau)}{d\tau} e^{-k\lambda\tau} - \lambda \sum_{k=1}^{m} k A_k(\tau) e^{-k\lambda\tau} \right) u_0}{v_0^* \left(I_n + \tau \displaystyle\sum_{k=1}^{m} k A_k e^{-k\lambda\tau} \right) u_0}, \tag{3.52}$$

evaluated at $\lambda = \lambda_0$ and $\tau = \tau^0$.

The main difference between the formulae above and the general (3.24) lies in the different forms of the corresponding numerators. As we shall see in the next chapter, in the cases when the matrices A_0 and A_k, $k = 1, \ldots, m$, do not explicitly depend on the parameters defined by the vector p, the crossing direction evaluations will have some interesting *invariance properties* (with respect to the delays) that will be exploited in defining the stability regions and the corresponding stability crossing boundaries.

3.5 Notes and references

The characterization of stability domains by using the D-subdivision method has been largely presented in the literature starting with the 1950s. The presentation here follows the lines considered by [223, 67]. For the brevity of this part, we decided to present only the essential facts, results, and ideas. An interesting discussion can be found in the almost forgotten monograph of Pinney [251]. In the case of scalar systems including one delay, the complete characterization (for fixed delay) of all crossing curves in the parameter space can be found, for instance, in [67].

Next, the estimation of some domain in $\mathbb{C}_+$ where the rightmost characteristic root of the scalar system should be located, or the property of existence of positive (real) characteristic roots are valid for more general systems than the scalar case under consideration.

The ideas for developing simple estimates for the stability domain of delay systems in state-space representation can be found in [215] (and the references therein). Their idea simply makes use of the maximum principle of a harmonic (or subharmonic) complex function. Without entering into details, their stability results can be summarized as follows.

Consider the system

$$\dot{x}(t) = A_0 x(t) + A_1 x(t - \tau)$$

under appropriate initial conditions, and let $\mu(A_0)$ be the corresponding matrix measure with respect to some p-(matrix) norm ($p = 1, 2, \ldots, \infty$). If $\mu(A_0) + \|A_1\| < 0$, the system above is stable independently of the delay size. Assume now that $\mu(A_0) + \|A_1\| > 0$. Then, as suggested by [215], the system is still asymptotically stable if there does not exist any root located in the Γ rectangle bounded by the imaginary axis, by the line $\Re(\lambda) = \mu(A_0) + \|A_1\|$, and the lines $\Im(\lambda) = \pm(\mu(-jA_0) + |A_1\|)$. However, the construction of Mori and Kokame is more general in the sense that if the system is unstable, the rightmost root together with all other unstable characteristic roots should be located in the *domain* Γ defined above.

Next, the property concerning the existence of strictly unstable positive roots for scalar systems for some particular choice of systems parameters also holds for more general quasipolynomials. To the best of the authors' knowledge, such a property was first mentioned in the literature in the 1960s and can be found in [92, 286] (and the references therein). Some remarks in this sense can be found also in the Applications part, in particular in the chapter devoted to the delayed output feedback control problem for single input single output (SISO) linear systems. Without any loss of generality, let us consider the following analytic function ($\tau > 0$),

$$p(\lambda;\ \tau) := Q(\lambda) + P(\lambda)e^{-\lambda\tau},$$

with P and Q analytic, and assume that

$$\operatorname{sign}\left[(P(0)+Q(0))Q(+\infty)\right] = -1,$$

that is, $P(0)+Q(0)$ and $Q(+\infty)$ have opposite signs. Then, using the same argument as in the scalar case, it simply follows that the characteristic equation $p(\lambda;\ \tau) = 0$ always has a characteristic root located on the *positive real axis*, and this property holds independently of the delay size; that is, we have a *delay-independent unstable* system. In the case where P and Q are polynomials with $\deg(P) < \deg(Q)$, the condition above simply says that the coefficient of the dominant term of the polynomial Q should have an opposite sign to $P(0)+Q(0)$, as discussed in [286].

The "dual" method of D-subdivision, the τ-decomposition method, is largely discussed by Lee and Hsu [147] in the context of quasipolynomials including one delay parameter. As in the presentation of the previous method, we focused more on ideas and essential facts by adding, in particular, some simple geometric interpretations that are useful for the other developments of the chapter. The characterization of the only (stability) delay interval follows the lines in [223], where the reader can find further discussions on the boundary of the delay-independent stability domain together with the relations existing with strong/weak delay-independent stability.

The τ-decomposition in the most general case for analytical functions with respect to only one delay parameter can be found in the paper of Cooke and van den Driessche [54], which is at the origin of various research results in control literature published in the last years. Cooke and van den Driessche [54] considered the following analytic function ($\tau > 0$),

$$p(\lambda;\ \tau) := Q(\lambda) + P(\lambda)e^{-\lambda\tau},$$

with P and Q analytic, and focused on the first-order analysis of the crossing roots with respect to the delay parameter, that is, the definition of all delay intervals for which stability/instability is guaranteed. Independently, a similar analysis in the case of polynomials P and Q was proposed in the control literature by Walton and Marshall [322] one year later. Corrections to [54] for the case of neutral systems (see Chapter 1 for definitions and related stability results) have been reported by Boese [18] (and mentioned first by Kuang [144] in his monograph published a couple of years before).

Although it seems that almost everything was said for the scalar system including one constant delay, we would like to mention two particular approaches that are quite simple to apply in the stability analysis: the *pseudodelay technique* and the *Lambert functions* based approach.

Consider now the following quasipolynomial:

$$p(\lambda;\ \tau) := Q(\lambda) + P(\lambda)e^{-\lambda\tau}.$$

The idea of the first approach (pseudodelay technique) consists of *finding* two parameter-dependent polynomials $A_0(\lambda T)$ and $A_1(\lambda T)$ such that the intersection in the complex plane of the ratio curve $-Q(j\omega)/P(j\omega)$ with the unit circle (described by $e^{-j\omega\tau}$) is *reduced* to the roots location of the parameter-dependent polynomial

$$p(\lambda;\ T) := Q(\lambda)A_1(\lambda T) + P(\lambda)A_0(\lambda T),$$

for $\lambda = j\omega$, when T increases from 0 to ∞. The last polynomial was derived by "replacing" the delay element $e^{-\lambda\tau}$ by $A_0(\lambda T)/A_1(\lambda T)$ in the quasipolynomial $p(\lambda;\ \tau)$. The "quantity" or the "parameter" T is called *pseudodelay*. In conclusion, the analysis of the roots distribution with respect to the imaginary axis of the corresponding characteristic equation is reduced to the analysis of the roots of some polynomials of higher order.

Such a transformation allows us to reformulate the *crossing roots* characterization of quasipolynomials as a *crossing roots* analysis for some parameter-dependent polynomials. Hence, the stability analysis of a delay systems is reduced to the analysis of the roots distribution of some appropriate parameter-dependent polynomials. To the best of the authors' knowledge, such an idea was first used in control by Rekasius in 1980 [257] (see also the comments in [294]). The result of Rekasius [257] makes use of the transformation $(1 - j\omega T)/(1 + j\omega T)$, which maps $[0, \infty)$ into a semicircle. Various related results (as well as some appropriate corrections) can be found in [114, 165, 322, 295], etc. Note that similar (bilinear) transformations are used in signal processing (see, for instance, [242] and the references therein). Further comments on the method can be found in MacDonald's monograph devoted to the stability analysis of some biological models [168]. Finally, a deeper analysis of this approach and various extensions are proposed by Sipahi [276] (see also [241]).

The second approach mentioned above can be found in [55, 127] (see also [48, 169] for some applications), where the authors used an *analytical* method based on Lambert functions for solving the corresponding transcendental characteristic equation. Recall that a Lambert function is a complex function $L(\lambda)$ satisfying

$$L(\lambda)e^{L(\lambda)} = \lambda.$$

The corresponding characteristic equation in the single delay case can be rewritten in this form. The next step is to compute the (infinite) branches of the complex function $L(\lambda)$. Note that the (first) "critical" roots will correspond to the principal branch (see, e.g., [169] and the references therein).

The continuation method presented in this chapter is inspired by [270, 74, 69]. We refer to [270, 145] for the general theory of numerical continuation and to the work of Engelborghs and co-workers for an application to time-delay systems. See also [74] for a software package for continuation and bifurcation analysis of time-delay systems (a brief description can be found in Section A.5 of the appendix).

Concerning the computation of the crossing direction of the characteristic roots, the proof idea in Proposition 3.3 based on the use of Jacobi's formula for the derivative of a determinant was first used in the context of commensurate delays by [229]. Next, the perturbation theory approach follows the classical lines of the approach (see, for instance, [133], but also [275] for a comprehensive introduction). The case of simple characteristic roots was considered in the literature by various authors [241, 229], mainly in the commensurate delays case or in the delay-parameter space [276]. The approach presented here generalizes the ideas in [229]. The characterization of the semisimple characteristic root case in such a general framework is completely new, and it is inspired by the recent developments in [45]. However, the arguments are independent of the methods and ideas in [45]. The double (nonsemisimple) eigenvalue case was not explicitly treated, but the main ideas were presented.

We note that for the case of proportional-integral-derivative (PID) controlled plants, specific methods for computing stability regions in the controller parameter space, based on Pontryagin-type criteria [139], can be found in [274]. The approach proposed here can be seen as an alternative to these methods.

Finally, various other results needed in the next chapters, such as the computation of the crossing direction taking into account delay interdependence information, completed this chapter.

Chapter 4

Stability regions in delay-parameter spaces

4.1 Introduction

In the previous chapters, we focused on the behavior of the system

$$\dot{x}(t) = A_0(p)x(t) + \sum_{i=1}^{m} A_i(p)x(t - \tau_i(p)), \tag{4.1}$$

in some neighborhood of some (generic) stability crossing boundary (that is, the set of parameters such that the corresponding characteristic equation has at least one root on the imaginary axis) in the parameter space $\mathcal{D} \subset \mathbb{R}^{n_p}$. Some properties of the crossing direction of the characteristic roots on the imaginary axis were presented and discussed.

In this chapter, we concentrate on the particular case where the parameter space is defined *only* by the delays. We start by presenting some important *invariance* properties concerning the characteristic roots on the imaginary axis and/or concerning the crossing direction. Next, we discuss in detail the stability in the delay-parameter space for a particular ray defined by *commensurate* delays. We shall see that the particular structure of the delay system allows us to use some elimination principle that leads to an explicit computation of the delay intervals guaranteeing stability. Emphasis will be put on the use of matrix pencil techniques. The presentation will be completed with the analysis of some particular problems such as hyperbolicity (no characteristic roots on the imaginary axis for all positive delays), and the stability characterization of a particular class of quasipolynomials.

In the second part of the chapter a full characterization of the stability crossing curves will be made for a class of quasipolynomials including two distinct delays. Such an approach will be adopted in later chapters to the stability analysis of immune dynamics models in leukemia and to the delay sensitivity analysis of Smith predictors.

The analysis is based on the duality between geometric and algebraic approaches. This choice is explained by some facility to understand the methodologies and techniques that we are describing. It is important to point out that the delay ratio sensitivity analysis, that is, the analysis of perturbed rays in the delay-parameter space, is the main subject of the next chapter.

The chapter is organized as follows: some important invariance properties of characteristic roots are presented and discussed in Section 4.2. The stability analysis in the commensurate delay case is presented in Section 4.3. Section 4.4 is devoted to the geometric approach applied to quasipolynomials with two distinct delays. Some notes and comments end the chapter, including a long list of references related to the topics treated in the chapter.

4.2 Invariance properties

We focus on the interactions between characteristic roots on the imaginary axis and the delays. More precisely, we discuss two problems: the invariance of characteristic roots on the imaginary axis w.r.t. particular delay shifts, and the crossing direction invariance in the commensurate case or in the case where we have independent delays with one varying and the other ones fixed.

Consider the system (4.1) and assume that the delay vector $\vec{\tau}$ defines the parameter space $\mathcal{D} \subset \mathbb{R}^{n_p}$, that is, $n_p = m$ and $\mathcal{D} = \mathbb{R}_+^m$. Then the system rewrites as follows:

$$\dot{x}(t) = A_0 x(t) + \sum_{i=1}^{m} A_i x(t - \tau_i). \tag{4.2}$$

Define the following *shift function*: $f_s : \mathbb{R}^m \times \mathbb{Z}^m \times \mathbb{R} \to \mathbb{R}^m$,

$$(\vec{\alpha}, \vec{\ell}, r) \mapsto f_s(\vec{\alpha}, \vec{\ell}, r) := (\alpha_1 + \ell_1 r, \alpha_2 + \ell_2 r, \dots, \alpha_m + \ell_m r). \tag{4.3}$$

Such a shift function allows us to associate a given vector $\vec{\alpha} \in \mathbb{R}^m$ with another vector in $\mathbb{R}^m$ situated on the ray defined by the direction $\vec{\ell} \in \mathbb{Z}^m$ and the point $\vec{\alpha}$.

4.2.1 Delay shifts and characteristic roots

Let $\lambda = \lambda_0 := j\omega_0 \neq 0$ be a zero of the characteristic function

$$p(\lambda;\ \vec{\tau}) := \det\left(\lambda I_n - A_0 - \sum_{i=1}^{m} A_i e^{-\lambda \tau_i}\right),$$

for some delays $\vec{\tau} = \vec{\tau}_0$, and let u_0 (v_0) be the corresponding right (left) eigenvector. Then we have the following result.

Proposition 4.1. *For any integer vector $\vec{\ell} \in \mathbb{Z}^m$, a zero $\lambda = j\omega_0 \neq 0$ of the characteristic function $p(\lambda;\ \vec{\tau})$ corresponding to some delays $\vec{\tau}_0 \in \mathbb{R}_+^m$ is invariant under the delay shift:*

$$\vec{\tau} := f_s\left(\vec{\tau}_0, \vec{\ell}, \frac{2\pi}{\omega_0}\right). \tag{4.4}$$

Proof. The result straightforwardly follows from the invariance of the exponential function

$$\alpha \in \mathbb{R} \mapsto e^{-j\alpha}$$

under the shift $\alpha \leftarrow \alpha + 2\pi \ell_k$, with $\ell_k \in \mathbb{Z}$. □

4.2.2 Crossing direction invariance

Using the developments of the previous chapter, it follows that $\partial\lambda/\partial\tau_k$, evaluated at a simple root $\lambda = \lambda_0$ and $\vec{\tau} = \vec{\tau}_0$, satisfies

$$\frac{\partial\lambda}{\partial\tau_k} = -\lambda_0 \frac{v_0^* A_k u_0 e^{-\lambda_0\tau_{0,k}}}{v_0^*\left(I_n + \sum_{i=1}^m A_i \tau_{0,i} e^{-\lambda_0\tau_{0,i}}\right) u_0}, \tag{4.5}$$

where u_0 (v_0) is the right (left) eigenvector corresponding to the characteristic root λ_0. It is easy to see that, under the assumption of a root on the imaginary axis, $\lambda_0 = j\omega_0$, the condition (4.5) leads to

$$\left[\frac{\partial\lambda}{\partial\tau_k}\right]^{-1} = -\frac{v_0^* u_0}{j\omega_0 v_0^* A_k u_0 e^{-j\omega_0\tau_{0,k}}} - \sum_{i=1,i\neq k}^{m} \frac{\tau_{0,i}}{j\omega_0}\cdot\frac{v_0^* A_i u_0}{v_0^* A_k u_0} e^{-j\omega_0(\tau_{0,i}-\tau_{0,k})}$$
$$-\frac{\tau_{0,k}}{j\omega_0}. \tag{4.6}$$

Since the crossing direction is given by the sign of $\Re(\partial\lambda/\partial\tau_k)$, evaluated at $\lambda = j\omega_0$ and $\vec{\tau} = \vec{\tau}_0$, we arrive at the following.

Proposition 4.2. *Under the assumption of a simple crossing, the crossing direction at $\lambda = \lambda_0 := j\omega_0 \neq 0$ and $\vec{\tau} = \vec{\tau}_0$ w.r.t. the delay parameter τ_k, $k \in \{1, 2, \ldots, m\}$, is invariant under the delay shift*

$$\vec{\tau} := f_s\left(\vec{\tau}_0, he_k, \frac{2\pi}{\omega_0}\right), \tag{4.7}$$

where e_k is the kth row of I_n and $h \in \mathbb{Z}$.

Proof. The result follows from the observation that the delay shift f_s described by (4.7) does not affect the real part of the quantity (4.6). Indeed, the quantities involving $\tau_{0,k}$ are on the imaginary axis (for instance, $-\tau_{0,k}/(j\omega_0)$) or appear in some exponential terms, $e^{-j\omega_0(\tau_{0,i}-\tau_{0,k})}$, for which the considered delay shift does not change the value. $\square$

Consider next a particular case that will be largely treated in what follows: *commensurate* delays. If the delays satisfy $\tau_i := i\tau$, $i = 1, \ldots, m$, with $\tau \geq 0$, then the crossing direction $\partial\lambda/\partial\tau$, evaluated at a simple characteristic root $\lambda = \lambda_0$ and $\tau = \tau_0$, becomes

$$\frac{\partial\lambda}{\partial\tau} := -\lambda_0 \frac{\sum_{i=1}^m i v_0^* A_i u_0 e^{-i\lambda_0\tau_0}}{v_0^*\left(I_n + \sum_{i=1}^m A_i i\tau_0 e^{-i\lambda_0\tau_0}\right) u_0}, \tag{4.8}$$

where u_0 (v_0) is the right (left) eigenvector corresponding to the characteristic root λ_0. If $\lambda_0 = j\omega_0$, $\omega_0 > 0$, then the condition (4.8) leads to

$$\left[\frac{\partial\lambda}{\partial\tau}\right]^{-1} = -\frac{v_0^* u_0}{j\omega_0 v_0^* \sum_{i=1}^m i A_i u_0 e^{-j\omega_0 i\tau_{0,k}}} - \frac{\tau_0}{j\omega_0}. \tag{4.9}$$

In conclusion, we have the following result.

Corollary 4.3. *Assume that the delays satisfy* $\tau_i = i\tau$, $i = 1, 2, \ldots, m$. *Assume that* $\lambda = \lambda_0 := j\omega_0 \neq 0$ *is a simple characteristic root for* $\tau = \tau_0$. *The crossing direction w.r.t. the delay parameter* τ *is invariant under the delay shift*

$$\tau := \tau_0 + h\frac{2\pi}{\omega_0}, \tag{4.10}$$

where $h \in \mathbb{Z}$.

4.3 Algebraic methods

Throughout this section we assume that the delays τ_i, $i = 1, \ldots, m$, are commensurate. Without losing generality we may assume that

$$\tau_i = i\tau, \ i = 1, \ldots, m,$$

with $\tau \geq 0$. Then the system (4.1) rewrites as

$$\dot{x}(t) = A_0 x(t) + \sum_{i=1}^{m} A_i x(t - i\tau), \tag{4.11}$$

and the corresponding characteristic function becomes

$$p(\lambda;\ \tau) := \det\left(\lambda I_n - A_0 - \sum_{i=1}^{m} A_i e^{-\lambda i \tau}\right). \tag{4.12}$$

4.3.1 Elimination principle: basic ideas

We briefly discuss some methodological ideas for the computation and/or detection of characteristic roots on the imaginary axis. More precisely, we consider approaches based on *two variables* and approaches based on *bilinear transformations* (including two-dimensional (2-D) polynomials and the *pseudodelay* technique).

Two variable-based approaches

We start by rewriting the characteristic function (4.12) as

$$p_1(\lambda;\ z) := \det\left(\lambda I_n - A_0 - \sum_{i=1}^{m} A_i z^i\right), \tag{4.13}$$

where we have formally replaced $e^{-\lambda i \tau}$ by z^i, since λ and $e^{-\lambda\tau}$ can be seen as algebraically independent variables if $\lambda \neq 0$. The function p_1 is a *bivariate* polynomial. The relation between the "original" characteristic function p and the "associated" function p_1 becomes clear when we consider the case of characteristic roots on the imaginary axis. Indeed, let $j\omega_s \in j\mathbb{R}^*$ be a zero of p_1 for some $z_s \in \mathcal{C}(0, 1)$; that is, $p_1(j\omega_s;\ z_s) = 0$. Then it is easy to see that $j\omega_s$ is a zero of the characteristic function p of the original delay system (4.11) if the delay τ belongs to the set $\mathcal{T}_{\omega_s}$ given by

$$\mathcal{T}_{\omega_s} := \left\{ \frac{1}{\omega_s} \left[\angle(\bar{z}) + 2\ell\pi \right] > 0, \quad \ell \in \mathbb{Z} \right\}.$$

By reciprocity, a zero $j\omega_s \in j\mathbb{R}^*$ of the characteristic function p is a zero of the characteristic function p_1 for $z_s = e^{-j\omega_s}$.

The particular relationship between the zeros of p and p_1 suggests the idea of computing the characteristic roots on the imaginary axis by *exploiting* the particular *form, structure,* and *dependence* of the characteristic function p_1 with respect to the two *variables*: $j\omega$ on the imaginary axis, and z on the unit circle. More precisely, the idea is to *eliminate* one of the variables, leading to two types of solutions: μ-analysis, and matrix pencil based solutions. Notes and comments on the first solution can be found in [223] (see also [96] for an analysis for both delay-independent and delay-dependent stability cases). The matrix pencil approach will be presented to some extent since it leads to explicit algorithms for the computation of stability regions in the delay-parameter space.

Bilinear transformations and related methods

The explicit computation of the characteristic function leads to an expression of the form

$$p(\lambda;\ \tau) = \sum_{i=0}^{n_d} p_i(\lambda) e^{-i\lambda\tau}, \tag{4.14}$$

where p_i are polynomials of degree at most n, and $n_d \geq m$ represents the number of commensurate delays in the quasipolynomial representation. As in the previous paragraph, we can interpret (4.14) as a bivariate polynomial p_1:

$$p_1(\lambda;\ z) := \sum_{i=0}^{n_d} p_i(\lambda) z^i, \tag{4.15}$$

where z formally replaces $e^{-\lambda\tau}$.

2-D polynomials. The representation of the characteristic function p as a bivariate polynomial leads to a relatively simple idea to analyze the stability of p in terms of a 2-D polynomial. We consider the bilinear transformation

$$\lambda := \frac{1+w}{1-w},$$

which maps the open unit disk onto the open left half-plane. Next, we construct the 2-D polynomial:

$$p_2(w;\ z) := (1-w)^n p_1\left(\frac{1+w}{1-w};\ z\right).$$

It is evident that p_1 has zeros on $j\mathbb{R} \times \mathcal{C}(0, 1)$ if and only if p_2 has zeros on $\mathcal{C}(0, 1) \times \mathcal{C}(0, 1)$. Further comments and discussions can be found in [96].

Pseudodelay technique. This approach was already commented on in the notes and references in the previous chapter. Although it is not considered in detail in this monograph, the main lines are presented since there exist natural connections with the generalized eigenvalue distribution of some appropriate matrix pencils, as we shall see later. Based on the bilinear transformation

$$z := \frac{1 - \lambda T}{1 + \lambda T}, \qquad T \geq 0,$$

we construct the parameter-dependent polynomial

$$p_3(\lambda;\ T) := (1 + \lambda T)^{n_d} p_1\left(\lambda;\ \frac{1 - \lambda T}{1 + \lambda T}\right)$$
$$= \sum_{i=0}^{n_d} p_i(\lambda)(1 - T\lambda)^i (1 + T\lambda)^{n_d - i}. \tag{4.16}$$

We have the following result (see, e.g., [294]).

Proposition 4.4. *The quasipolynomial $p(\lambda;\ \tau)$ has zero $\lambda = j\omega$, $\omega > 0$, for some delay value $\tau \geq 0$, if and only if the parameter-dependent polynomial $p_3(\lambda;\ T)$ has zero $j\omega$ for some $T \geq 0$.*

Further comments and discussions can be found in [276] (see also Section 4.5).

4.3.2 Matrix pencil approach and crossing characterization

Recall the system (4.11) and its characteristic function (4.12). Introduce the following matrix pencil [46, 222]:

$$\Lambda(z) := zM + N, \tag{4.17}$$

where $M,\ N \in \mathbb{R}^{(2mn^2)\times(2mn^2)}$ are given by

$$M := \begin{bmatrix} I_{n^2} & 0 & \dots & 0 & 0 \\ 0 & I_{n^2} & \dots & 0 & 0 \\ & & \ddots & & \\ 0 & 0 & \dots & I_{n^2} & 0 \\ 0 & 0 & \dots & 0 & B_m \end{bmatrix}, \tag{4.18}$$

$$N := \begin{bmatrix} 0 & -I_{n^2} & 0 & \dots & 0 \\ 0 & 0 & -I_{n^2} & \dots & 0 \\ & & & \ddots & \\ 0 & 0 & 0 & \dots & -I_{n^2} \\ B_{-m} & B_{-m+1} & B_{-m+2} & \dots & B_{m-1} \end{bmatrix}, \tag{4.19}$$

and B_{-k} $(k = 1, \dots, m)$, B_i $(i = 1, \dots, m)$ are defined as

$$B_{-k} = I_n \otimes A_k^T, \qquad B_i = A_i \otimes I_n, \qquad B_0 = A \oplus A^T.$$

The operators $\otimes$ and $\oplus$ denote the Kronecker product and sum (see, e.g., [93]).

Frequency crossing set characterization

The following proposition gives a complete description of characteristic roots on the imaginary axis in terms of the generalized eigenvalue distribution of the matrix pencil Λ. The proof closely follows the ideas mentioned in [46, 222] and was presented in [229].

Proposition 4.5. *Assume that the matrix pencil Λ is regular. The characteristic equation $p(\lambda;\ \tau) = 0$ has a root $j\omega_0$, $\omega_0 > 0$, for some positive delay value τ if and only if there exists a complex number*

$$z_0 \in \sigma(\Lambda) \cap \mathcal{C}(0, 1) \tag{4.20}$$

such that

$$j\omega_0 \in \sigma\left(A + \sum_{i=1}^{m} A_i z_0^i\right). \tag{4.21}$$

Furthermore, the corresponding delay values are given by

$$\begin{aligned}\mathcal{T}_{\omega_0} = \Big\{ \tfrac{\angle \overline{z}_0}{\omega_0} + \tfrac{2\pi\ell}{\omega_0} > 0 :\ & z_0 \in \sigma(\Lambda) \cap \mathcal{C}(0, 1), \\ & j\omega_0 \in \sigma\left(A_0 + \textstyle\sum_{i=1}^{m} A_i z_0^i\right),\ \ell \in \mathbb{Z}\Big\}.\end{aligned} \tag{4.22}$$

Proof. We start with the first assertion.

$\Leftarrow$ Let (ω_0, z_0) satisfy (4.21). It remains to prove that there exists at least one positive delay value τ_0 such that

$$z_0 = e^{-j\omega_0\tau_0}.$$

This is true since the general solution of this equation is given by

$$\tau = \frac{\mathrm{Log}(\overline{z_0})}{j\omega_0} + \frac{2\pi\ell}{\omega_0},$$

with $\ell \in \mathbb{Z}$, and we can always take ℓ such that τ is positive.

$\Rightarrow$ Assume that the characteristic function $p(\lambda;\ \tau)$ has at least one (nonzero) root on the imaginary axis, $j\omega_0$, for some delay $\tau_0 > 0$; that is, $p(j\omega_0;\ \tau_0) = 0$. Let z_0 be defined as $z_0 = e^{-j\omega_0\tau_0}$. It is clear that

$$\det\left(j\omega_0 I_n - A - \sum_{i=1}^{m} A_i z_0^i\right) = 0 \tag{4.23}$$

and

$$\det\left(-j\omega_0 I_n - A^T - \sum_{i=1}^{m} A_i^T \overline{z}_0^i\right) = 0. \tag{4.24}$$

Expressions (4.23)–(4.24) imply that

$$D(z_0) := \det\left[\left(A + \sum_{i=1}^{m} A_i z_0^i\right) \oplus \left(A + \sum_{i=1}^{m} A_i z_0^i\right)^*\right] = 0. \tag{4.25}$$

Some simple computations lead us to

$$D(z_0) = z_0^m \det(z_0 M + N),$$

with M, N given by (4.18)–(4.19). Since $z_0 \neq 0$, Λ has at least one generalized eigenvalue on the unit circle z_0, for which (4.21) is satisfied.

To prove the second assertion we have to find all values of τ such that $p(j\omega_0;\ \tau) = 0$. From the arguments spelled out in the proof of the first assumption (more precisely, the implication $\Rightarrow$) it follows that $e^{-j\omega_0\tau} \in \sigma(\Lambda) \cap \mathcal{C}(0, 1)$. Expression (4.22) then follows straightforwardly. $\square$

The following definitions are related with the elimination technique used in the proof of Proposition 4.5.

Definition 4.6. *A complex number z_0 satisfying (4.20) and for which (4.21) is satisfied for some $\omega_0 > 0$ is called a crossing generator. The set of all crossing generators is denoted with σ_g. The frequency crossing set Ω is defined as*

$$\Omega := \left\{ \omega \in \mathbb{R}_+^* : j\omega \in \sigma\left(A_0 + \sum_{i=1}^{m} A_i z_0^i\right), \quad z_0 \in \sigma_g \right\} \tag{4.26}$$

and the set

$$\mathcal{T} := \bigcup_{\omega \in \Omega} \mathcal{T}_\omega \tag{4.27}$$

is called the delay crossing set.

In terms of Definition 4.6, Proposition 4.5 says that the existence of characteristic roots on the imaginary axis for some delay values is equivalent to the property that the frequency crossing set is not empty. By construction, Ω is not empty if and only if the set σ_g of all crossing generators is not empty.

Remark 4.7. Not all *the generalized eigenvalues of the pencil Λ on the unit circle correspond to* crossing generators. *Indeed, by the properties of the Kronecker sum, the solutions z on the unit circle of the equation*

$$\det\left[\left(A_0 + \sum_{i=1}^{m} A_i z^i\right) \oplus \left(A_0 + \sum_{i=1}^{m} A_i z^i\right)^*\right] = 0$$

include not only the crossing generators, but all the values of z on the unit circle for which the matrix

$$A_0 + \sum_{i=1}^{m} A_i z^i$$

has some eigenvalues symmetric w.r.t. the imaginary axis (which is more general than having eigenvalues on *the positive imaginary axis) [46, 222].*

Proposition 4.5 directly leads to the following algorithm for the computation of the delay crossing set.

ALGORITHM 4.1.

Computation of delay crossing set

A. Compute the generalized eigenvalues of the matrix pencil Λ on the unit circle of the complex plane (under the assumption of regularity).

B. For each generalized eigenvalue $z \in \mathcal{C}(0,1) \cap \sigma(\Lambda)$, compute the eigenvalues on the imaginary axis of the complex matrix $A_0 + \sum_{i=1}^{m} A_i z^i$.

C. Compute the frequency crossing set Ω from (4.26) and the delay crossing set $\mathcal{T}$ from (4.27) and (4.22).

Crossing direction characterization: simple characteristic roots

We first consider the characterization of a *simple* root crossing. The multiple semisimple crossing case will be considered separately. We have the following result.

Proposition 4.8. *Let $j\omega_0$, $\omega_0 > 0$, be a characteristic root of the system (4.11) for some delay value τ_0. Let $z_0 \in \sigma_g$ be the corresponding crossing generator. If the delay is increased, then the characteristic root crosses the imaginary axis toward instability (stability) if and only if*

$$\Re\left\{\frac{\omega_0}{jv_0^* u_0}\sum_{i=1}^{m} i z_0^i v_0^* A_i u_0\right\} > 0(< 0), \tag{4.28}$$

where v_0 and u_0 are the left and right eigenvectors of the complex matrix

$$A_0 + \sum_{i=1}^{m} A_i z_0^i,$$

corresponding to the eigenvalue $j\omega_0$.

The proof is omitted since it follows straightforwardly from Corollary 3.7.

Crossing direction characterization: semisimple characteristic roots

The result below is taken from [45], where the authors handled the problem by using a slightly different argument than the one proposed in the previous chapter.

Proposition 4.9. *Let $j\omega_0$, $\omega_0 > 0$, be a semisimple characteristic root of the system (4.11) for some delay value τ_0, with multiplicity equal to q. Let $z_0 \in \sigma_g$ be a corresponding* *crossing generator. For τ sufficiently close to τ_0, the characteristic roots corresponding to $j\omega_0$ can be expanded as*

$$j\omega_0 - \lambda_k \left(R^* \left(j\omega_0 \sum_{i=1}^{m} iA_i z_0^i\right) Q\right)(\tau - \tau_0) + O\left((\tau - \tau_0)^2\right), \quad k = 1, 2, \ldots, q, \tag{4.29}$$

where $Q = [u_1\ u_2\ \cdots\ u_q]$, $R = [v_1\ v_2\ \cdots\ v_q]$, and the notation $\lambda_k(\cdot)$ stands for the kth eigenvalue. The vectors u_i and v_i, $i = 1, 2, \ldots, q$, constitute a normalized set of right and left eigenvectors of the matrix $A_0 + \sum_{i=1}^{m} A_i z_0^i$, corresponding to the eigenvalue $j\omega_0$.

Thus, for τ sufficiently close to τ_0 but $\tau > \tau_0$, there are at least ℓ ($\ell \leq q$) characteristic roots in the open right half-plane if one of the following inequalities is satisfied:

$$\Re\left\{\lambda_k \left[R^* \left(j\omega_0 \sum_{i=1}^{m} iA_i z_0^i\right) Q\right]\right\} < 0, \quad k = 1, \ldots, q. \tag{4.30}$$

4.3.3 Particular cases and other elimination techniques

In what follows, we consider some special cases for which the analysis presented in the previous section simplifies significantly.

Hyperbolicity and delay-independent stability

If there are no characteristic roots on the imaginary axis, for any delay value, then the system is called *hyperbolic*, in the sense mentioned by Hale, Infante, and Tsen in [103]. In other words, the number of strictly unstable characteristic roots is *constant* for all delay values, including also the case free of delay. We have the following result [222].

Proposition 4.10. *Assume that the system (4.11) free of delays has no characteristic roots on the imaginary axis. Then it is hyperbolic if and only if the frequency crossing set Ω is empty.*

The characterization of *delay-independent* asymptotic *stability* is similar. The only difference is that asymptotic stability of the delay-free system needs to be imposed instead of having no characteristic roots on the imaginary axis.

Rank-one matrices and frequency-sweeping tests

In the final chapter on applications in biosciences, we will encounter models of the form (4.11), where the matrices $A_1, \ldots, A_m$ have rank one. In what follows, we present the main ideas for analyzing such problems by means of the single delay case. Therefore, we study the system

$$\dot{x}(t) = A_0 x(t) + BC^T x(t - \tau), \tag{4.31}$$

where B and C are column matrices. For the sake of conciseness, we assume that the system free of delay is asymptotically stable and that the matrix A_0 has no eigenvalues on the imaginary axis.

Define the analytic function

$$p_a(\lambda;\ \tau) := 1 - a(\lambda)e^{-\lambda\tau}, \tag{4.32}$$

where a is given by

$$a(\lambda) := C^T(\lambda I - A_0)^{-1}B.$$

The particular structure of the rank-one matrix BC^T allows us to derive the following result.

Proposition 4.11. *The characteristic function of (4.31) and the function $p_a(\lambda;\ \tau)$ have the same zeros in a neighborhood $\mathcal{V}_\delta$ of the imaginary axis, where*

$$\mathcal{V}_\delta := \{\lambda \in \mathbb{C} : \quad \delta \geq \Re(\lambda) > -\delta\},$$

for some $\delta > 0$.

Proof. Since A_0 has no eigenvalues on the imaginary axis, the continuity property of the roots of the corresponding characteristic equation leads to the existence of some $\delta > 0$ such that $\lambda I - A_0$ is invertible in $\mathcal{V}_\delta$. Next, for all $\lambda \in \mathcal{V}_\delta$, we have

$$\begin{aligned}\det(\lambda I_n - A_0 - BC^T e^{-\lambda\tau}) &= \det(\lambda I_n - A_0)\det(I_n - (\lambda I_n - A_0)^{-1}BC^T e^{-\lambda\tau}) \\ &= p_a(\lambda;\ \tau)\det(\lambda I_n - A_0),\end{aligned}$$

where we used the properties of the Schur complement. $\square$

If p_a has a zero $j\omega_0$, $\omega_0 > 0$, for some delay value τ_0, then

$$a(j\omega_0) = e^{j\omega_0\tau}, \tag{4.33}$$

which implies that the frequency crossing set Ω consists of the strictly positive zeros of the function

$$f : \mathbb{R}_+ \to \mathbb{R},\ \omega \mapsto f(\omega) = 1 - \mid a(j\omega) \mid^2, \tag{4.34}$$

while the delay crossing set can be computed from the phase information of (4.33). Furthermore, if we differentiate

$$p_a(\lambda(\tau);\ \tau) = 0$$

w.r.t. τ (under the assumption of a simple root) we get

$$\left[\frac{d\lambda}{d\tau}\right]^{-1} = -\frac{\tau}{\lambda} + \frac{a'(\lambda)}{\lambda a(\lambda)},$$

which in the case $\lambda = j\omega_0$, $\omega_0 \geq 0$, yields

$$\Re\left[\frac{d\lambda}{d\tau}\right]^{-1} = \Re\frac{a'(j\omega_0)}{j\omega_0 a(j\omega_0)} = \frac{1}{2\omega_0|a(j\omega_0)|^2}f'(\omega_0).$$

In this way, we arrive at the following result.

Proposition 4.12. *Consider the system (4.31). Assume that A_0 has no eigenvalues on the imaginary axis and that $1 - a(0) \neq 0$. Assume further that the zeros of f are simple. The system (4.31) has a characteristic root $j\omega_0$, $\omega_0 > 0$, for some delay value if and only if $f(\omega_0) = 0$. Furthermore, the set of corresponding delay values is given by*

$$\mathcal{T}_{\omega_0} = \left\{ \frac{1}{\omega_0} [-j\mathrm{Log}(a(j\omega_0)) + 2\pi l] \geq 0, \ \ l \in \mathbb{Z} \right\}. \tag{4.35}$$

When increasing the delay, the corresponding crossing direction of the characteristic root is toward instability (stability) if $f'(\omega) > 0$ (< 0).

Quasipolynomials with one delay. Consider the following class of quasipolynomials including one delay:

$$p(\lambda;\ \tau) := Q(\lambda) + P(\lambda)e^{-\lambda\tau}, \tag{4.36}$$

where P and Q are coprime polynomials with real coefficients such that $\deg(Q) > \deg(P)$. Such a form is inspired by some stabilization problems for linear systems using delayed output feedback, as we shall see in Chapter 11. It is easy to see that (4.36) can be rewritten as

$$p(\lambda;\ \tau) = Q(\lambda)\left(1 + \frac{P(\lambda)}{Q(\lambda)}e^{-\lambda\tau}\right).$$

Hence, under the assumption that Q has no zeros on the imaginary axis, the stability can be analyzed as outlined previously. In what follows we present an alternative method that does not make such an assumption.

If p has a zero $j\omega_0$, $\omega_0 > 0$, for some delay value τ_0, then

$$Q(j\omega_0) = -P(j\omega_0)e^{-j\omega_0\tau_0}, \tag{4.37}$$

which implies that the frequency crossing set Ω consists of the strictly positive zeros of the function

$$F : \mathbb{R}_+ \to \mathbb{R}, \ \ \omega \mapsto F(\omega) := \mid Q(j\omega) \mid^2 - \mid P(j\omega) \mid^2, \tag{4.38}$$

while the delay crossing set can be calculated from the phase information of (4.37). Furthermore, in [233] it is shown that under the assumption that the roots of F are simple, the crossing direction information can be obtained from the derivative of F, which leads us to the following result.

Proposition 4.13. *Assume that the zeros of F are simple. The characteristic function p has a zero $j\omega_0$, $\omega_0 > 0$, for some delay value τ_0 if and only if*

$$F(\omega_0) = 0. \tag{4.39}$$

Furthermore, for any ω_0 satisfying (4.39) the set of corresponding delay values is given by

$$\mathcal{T}_{\omega_0} = \left\{ \frac{1}{\omega_0}\left[-j\mathrm{Log}\left(-\frac{P(j\omega_0)}{Q(j\omega_0)} \right) + 2\pi l \right] \geq 0, \ \ l \in \mathbb{Z} \right\}. \tag{4.40}$$

When increasing the delay the corresponding crossing direction of the zero is toward instability (stability) if $F'(\omega_0) > 0$ (< 0).

Remark 4.14. *The methodology considered here for the characterization of the delay crossing set is sometimes called the* direct approach *(see, e.g., [96]). To the best of our knowledge, the crossing characterization in terms of the function F was proposed by Cooke and van den Driessche [54], generalizing the second-order characterization proposed in [53].*

Pseudodelay and matrix pencils

Recall that the pseudodelay technique allows us to detect characteristic roots on the imaginary axis using the bilinear transformation

$$z = \frac{1 - \lambda T}{1 + \lambda T}, \qquad T \geq 0,$$

which puts the quasipolynomial $p(\lambda;\ \tau)$, defined in (4.14), into the form

$$p_3(\lambda;\ T) = \sum_{i=0}^{n_d} p_i(\lambda)(1 - T\lambda)^i(1 + T\lambda)^{n_d - i} := \sum_{i=0}^{n_d} q_i(\lambda)T^i,$$

with $q_0, \ldots, q_{n_d}$ polynomials. More precisely, Proposition 4.4 establishes a connection between the characteristic function $p(\lambda;\ \tau)$ and the parameter-dependent polynomial $p_3(\lambda;\ T)$ in terms of zeros on the imaginary axis. Thus, the detection of such zeros for particular delay values is reduced to finding parameters T such that p_3 has zeros on the imaginary axis. We outline how this problem can be solved by computing the generalized eigenvalues of an appropriately defined matrix pencil.

With an arbitrary polynomial

$$a(\lambda) := \sum_{i=0}^{n_a} a_i \lambda^{n_a - i}$$

we can associate a matrix $H(a)$, defined as

$$H(a) := \begin{bmatrix} a_1 & a_3 & a_5 & \ldots & a_{2n_a-1} \\ a_0 & a_2 & a_4 & \ldots & a_{2n_a-2} \\ 0 & a_1 & a_3 & \ldots & a_{2n_a-3} \\ 0 & a_0 & a_2 & \ldots & a_{2n_a-4} \\ \vdots & & & \ddots & \vdots \\ 0 & 0 & 0 & \ldots & a_{n_a} \end{bmatrix} \in \mathbb{R}^{n_a \times n_a}, \tag{4.41}$$

where the coefficients $a_l = 0$ are assumed zero for $l > n_a$. Next, we introduce the matrix pencil,

$$\Gamma(z) := zU + V,$$

with U, V given by

$$U = \begin{bmatrix} I & & & \\ & \ddots & & \\ & & I & \\ & & & H(q_{n_d}) \end{bmatrix}, V = \begin{bmatrix} 0 & -I & \cdots & 0 \\ \vdots & \vdots & \ddots & \vdots \\ 0 & 0 & \cdots & -I \\ H(q_0) & H(q_1) & \cdots & H(q_{n_d-1}) \end{bmatrix}, \quad (4.42)$$

where the identity and the zero-block matrices have appropriate dimensions.

The following result gives a characterization of the zeros of p_3 on the imaginary axis as a function of T, and generalizes the matrix pencil solution proposed in [44] in the context of static output feedback.

Proposition 4.15. *Let* $0 < \lambda_1 < \lambda_2 < \cdots < \lambda_h$, *with* $h \leq nn_d$, *be the real eigenvalues of the matrix pencil* Γ. *The parameter-dependent polynomial* p_3 *has a zero on the imaginary axis if and only if* $T \in \{\lambda_1, \ldots, \lambda_h\}$. *Furthermore, if there are r unstable zeros for* $T = T^*$, $T^* \in (\lambda_i, \lambda_{i+1})$, *then there are r unstable roots for all* $T \in (\lambda_i, \lambda_{i+1})$. *In other words, the instability degree of* p_3 *remains constant as T varies within each interval* $(\lambda_i, \lambda_{i+1})$. *The same property holds for the intervals* $(0, \lambda_1)$ *and* (λ_h, ∞).

In conclusion, the generalized eigenvalues of Γ on the real axis define the values of the parameter T for which the polynomial p_3 has zeros on the imaginary axis. For all $T \in \{\lambda_1, \ldots, \lambda_h\}$ one can compute these zeros on the imaginary axis, whose imaginary parts define the *frequency crossing set* Ω. Next, for every frequency $\omega \in \Omega$ and corresponding $T \in \{\lambda_1, \ldots, \lambda_n\}$, the set $\mathcal{T}_\omega$ can be derived by solving the equation

$$e^{-j\omega\tau} = \frac{1 - j\omega T}{1 + j\omega T}.$$

It is important to point out that this procedure for detection the crossing frequency set Ω is quite distinct to the procedures proposed in [241, 276].

4.4 Geometric methods

In this section, we focus on a particular problem—the characterization of the stability regions and of their boundaries in the delay-parameter space—by considering a simplified characteristic function including two independent delays and of the form

$$p(\lambda;\ \tau_1, \tau_2) := p_0(\lambda) + p_1(\lambda)e^{-\lambda\tau_1} + p_2(\lambda)e^{-\lambda\tau_2}, \quad (4.43)$$

where

$$p_l(\lambda) := \sum_{i=1}^{n_l} p_{li}\lambda^i, \qquad i = 0, 1, 2,$$

with $n_0 > n_1, n_2$. This condition implies that the corresponding system is of *retarded* type. The analysis of the neutral case is omitted, but it can be treated in a similar way under some appropriate assumptions on the difference operator (see, for instance, [98]).

Our first objective is to identify the regions of (τ_1, τ_2) in $\mathbb{R}_+^2$ such that $p(\lambda;\ \tau_1, \tau_2)$ has zeros on the imaginary axis. We first exclude some simple trivial cases and restrict the analysis to the cases when $p(\lambda;\ \tau_1, \tau_2)$ satisfies the following conditions:

I. Zero frequency,
$$p_0(0) + p_1(0) + p_2(0) \neq 0; \tag{4.44}$$

II. The polynomials $p_0(\lambda)$, $p_1(\lambda)$, and $p_2(\lambda)$ do not have any common zeros.

Let us comment on these conditions. If condition I is not satisfied, then 0 is a zero of $p(\lambda;\ \tau_1, \tau_2)$ for any $(\tau_1, \tau_2) \in \mathbb{R}_+^2$, and therefore it can never be stable. Next, condition II is natural. If it is not satisfied, there exists a common factor $c(\lambda) \neq$ constant such that $p_l(\lambda) = c(\lambda)q_l(\lambda), l = 0, 1, 2$. Let $c(\lambda)$ be the highest possible order; then $q_l(\lambda), l = 0, 1, 2$, do not have any common zeros, and the underlying DDE can be decomposed to an ODE with characteristic polynomial $c(\lambda)$ and a DDE with characteristic quasipolynomial
$$q_0(\lambda) + q_1(\lambda)e^{-\lambda\tau_1} + q_2(\lambda)e^{-\lambda\tau_2}, \tag{4.45}$$
which satisfies condition II.

Due to the continuity, given $\tau_1 = \tau_1^0$ and $\tau_2 = \tau_2^0$, in principle we may find the number of zeros of $p(\lambda;\ \tau_1, \tau_2)$ on $\mathbb{C}_+$ using the following simplified *algorithm*: first, find the number of unstable roots in the case free of delays, that is the polynomial $p(\lambda : 0, 0)$; second, form a curve in the τ_1-τ_2-plane within $\mathbb{R}_+^2$ initiating from the origin and ending at the desired point (τ_1^0, τ_2^0); finally, find all the points of (τ_1, τ_2) in the curve such that there are zeros of $p(\lambda; \tau_1, \tau_2)$ crossing the imaginary axis, and find the directions of crossing (from left to right, or the other way) as one moves along the curve. In the remainder of the section we will identify these crossing points and the corresponding stability crossing curves. We will need the following definition [98].

Definition 4.16. *Let $\mathcal{C}_k : [a, b] \to \mathbb{R}^2$, $k = 1, 2, \ldots,$ be a series of curves satisfying*
$$\mathcal{C}_k(b) - \mathcal{C}_k(a) = A,\ k = 1, 2, \ldots, \tag{4.46}$$
where $A \in \mathbb{R}^2$ is a constant two-dimensional vector independent of k, and
$$\mathcal{C}_{k+1}(a) = \mathcal{C}_k(b). \tag{4.47}$$
Then, the curve $\mathcal{C}$ formed by connecting all the curves $\mathcal{C}_k$, $k = 1, 2, \ldots,$
$$\mathcal{C} = \bigcup_{k=1}^{\infty} \mathcal{C}_k, \tag{4.48}$$
is known as a spiral-like curve, *and A is known as its* axis. *If in addition*
$$\mathcal{C}_{k+1}(\xi) = \mathcal{C}_k(\xi) + A \forall \xi \in [a, b], \tag{4.49}$$
then $\mathcal{C}$ is known as a spiral.

The definition above simply says that a spiral is formed by connecting identical curves head to tail. It is important to point out that the composite curves in a spiral-like curve do not have to be identical. In what follows, in the spiral-like curves case, $\mathcal{C}_{k+1}$ can often be viewed as formed from $\mathcal{C}_k$ with a small deformation, which justifies the term "spiral-like curve."

4.4.1 Identification of crossing points

Let us adapt the notations from the commensurate and single delay cases to the delay-parameter space defined by τ_1 and τ_2. Thus, $\mathcal{T}$ denotes the set of all the points of (τ_1, τ_2) in $\mathbb{R}_+^2$ such that $p(\lambda;\ \tau_1, \tau_2)$ has at least one zero on the imaginary axis. Any $(\tau_1, \tau_2) \in \mathcal{T}$ is known as a *crossing point*. The set $\mathcal{T}$, which is the collection of all the crossing points, is called the *stability crossing curves*.

Define now $a_l(\lambda) = p_l(\lambda)/p_0(\lambda)$, $l = 1, 2$, and

$$a(\lambda;\ \tau_1, \tau_2) := 1 + a_1(\lambda)e^{-\lambda\tau_1} + a_2(\lambda)e^{-\lambda\tau_2}. \tag{4.50}$$

Using a similar argument to the one proposed in the rank-one delayed matrix case, for any given τ_1 and τ_2, as long as $p_0(\lambda)$ does not have characteristic roots on the imaginary axis, $p(\lambda;\ \tau_1, \tau_2)$ and $a(\lambda;\ \tau_1, \tau_2)$ *share* all the zeros in some neighborhood of the imaginary axis. Therefore, in general, we may obtain all the crossing points and directions of crossing from

$$a(\lambda;\ \tau_1, \tau_2) = 0 \tag{4.51}$$

instead of $p(\lambda;\ \tau_1, \tau_2) = 0$.

For each given $\lambda = j\omega$, $\omega > 0$, we may consider the three terms in $a(j\omega;\ \tau_1, \tau_2)$ as three vectors in the complex plane, with the magnitudes 1, $|a_1(j\omega)|$, and $|a_2(j\omega)|$, respectively. Furthermore, if we adjust the values of τ_1 and τ_2, we may arbitrarily adjust the directions of the vectors represented by the second and third terms. Equation (4.51) means that if we put these vectors head to tail, they form a triangle, as illustrated in Figure 4.1. This allows us to conclude the following proposition (see [98] for a complete proof).

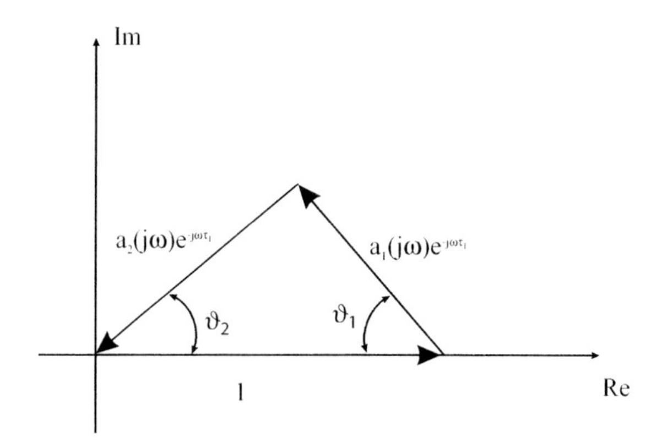

Figure 4.1. *Triangle formed by* 1, $|a_1(j\omega)|$, *and* $|a_2(j\omega)|$.

Proposition 4.17. *For each* ω, $\omega \neq 0$, $p_0(j\omega) \neq 0$, $\lambda = j\omega$ *can be a solution of* $p(\lambda;\ \tau_1, \tau_2) = 0$ *for some* $(\tau_1, \tau_2) \in \mathbb{R}_+^2$ *if and only if*

$$|a_1(j\omega)| + |a_2(j\omega)| \geq 1, \tag{4.52}$$

$$-1 \leq |a_1(j\omega)| - |a_2(j\omega)| \leq 1. \tag{4.53}$$

For $\omega \neq 0$ satisfying $p_0(j\omega) = 0$, $\lambda = j\omega$ can be a zero of $p(\lambda;\ \tau_1, \tau_2)$ for some $(\tau_1, \tau_2) \in \mathbb{R}_+^2$ if and only if

$$|p_1(j\omega)| = |p_2(j\omega)|. \tag{4.54}$$

Let Ω be the set of all $\omega > 0$ which satisfy (4.52) and (4.53) if $p_0(j\omega) \neq 0$ and (4.54) if $p_0(j\omega) = 0$. Similar to the commensurate and single delay cases, we will refer to Ω as the *frequency crossing set*. It contains all the ω such that some zero of $p(\lambda;\ \tau_1, \tau_2)$ may cross the imaginary axis at $j\omega$. Then, for any given $\omega \in \Omega$, $p_l(j\omega) \neq 0$, $l = 0, 1, 2$, one may easily find all the pairs of (τ_1, τ_2) satisfying (4.51) as follows.

$$\tau_1 = \tau_1^{u\pm}(\omega) = \frac{\angle a_1(j\omega) + (2u-1)\pi \pm \theta_1}{\omega} \geq 0,\ u = u_0^{\pm}, u_0^{\pm}+1, u_0^{\pm}+2, \ldots, \tag{4.55}$$

$$\tau_2 = \tau_2^{v\pm}(\omega) = \frac{\angle a_2(j\omega) + (2v-1)\pi \mp \theta_2}{\omega} \geq 0,\ v = v_0^{\pm}, v_0^{\pm}+1, v_0^{\pm}+2, \ldots, \tag{4.56}$$

where $\theta_1, \theta_2 \in [0, \pi]$ are the internal angles of the triangle in Figure 4.1, and can be calculated by the cosine law as

$$\theta_1 = \cos^{-1}\left(\frac{1 + |a_1(j\omega)|^2 - |a_2(j\omega)|^2}{2|a_1(j\omega)|}\right), \tag{4.57}$$

$$\theta_2 = \cos^{-1}\left(\frac{1 + |a_2(j\omega)|^2 - |a_1(j\omega)|^2}{2|a_2(j\omega)|}\right), \tag{4.58}$$

and $u_0^+, u_0^-, v_0^+, v_0^-$ are the smallest possible integers (may be negative and may depend on ω) such that the corresponding $\tau_1^{u_0^+ +}, \tau_1^{u_0^- -}, \tau_2^{v_0^+ +}, \tau_2^{v_0^- -}$ calculated are nonnegative. Notice that $u_0^+ \leq u_0^-$, $v_0^+ \geq v_0^-$. The position in Figure 4.1 corresponds to $(\tau_1^{u+}, \tau_2^{v+})$. The position corresponding to $(\tau_1^{u-}, \tau_2^{v-})$ is its mirror image about the real axis. Next, let $\mathcal{T}_{\omega,u,v}^+$ and $\mathcal{T}_{\omega,u,v}^-$ be the singletons defined by

$$\mathcal{T}_{\omega,u,v}^{\pm} = \{(\tau_1^{u\pm}(\omega), \tau_2^{v\pm}(\omega))\},$$

and define

$$\mathcal{T}_\omega = \left(\bigcup_{u \geq u_0^+, v \geq v_0^+} \mathcal{T}_{\omega,u,v}^+\right) \bigcup \left(\bigcup_{u \geq u_0^-, v \geq v_0^-} \mathcal{T}_{\omega,u,v}^-\right),$$

which generalizes the delay crossing set notion encountered in the commensurate or single delay cases. Here, $\mathcal{T}_\omega$ represents the set of all (τ_1, τ_2) such that $p(\lambda;\ \tau_1, \tau_2)$ has a zero at $\lambda = j\omega$. In the following remark, we will discuss the degenerate cases of $p_i(j\omega) = 0$ for at least one i $(= 1, 2)$.

Remark 4.18. *If $p_0(j\omega) = 0$, $\omega \in \Omega$. Then $p(j\omega) = 0$ and assumption II implies $|p_1(j\omega)| = |p_2(j\omega)| \neq 0$. In this case, $\mathcal{T}_\omega$ consists of the solutions of*

$$\angle p_1(j\omega) - \omega\tau_1 + 2\pi u = \angle p_2(j\omega) - \omega\tau_2 + 2\pi v + \pi$$

in $\mathbb{R}_+^2$ for integers u, v. Instead of isolated points, $\mathcal{T}_\omega$ now consists of an infinite number of straight lines of slope 1 of equal distance.

On the other hand, if $p_0(j\omega) \neq 0$, $\omega \in \Omega$, *and* $p_1(j\omega) = 0$, *then* $a_1(j\omega) = 0$ *and* $|a_2(j\omega)| = 1$*; we have* $\theta_2 = 0$, *and* θ_1 *can assume all the values in* $[0, \pi]$*; and* $\mathcal{T}^{\pm}_{\omega,u,v}$ *contains all the points calculated by (4.55) and (4.56) with* $\theta_1 \in [0, \pi]$, $\theta_2 = 0$. *The corresponding* $\mathcal{T}_\omega$ *is a series of horizontal lines. Similarly, for* $\omega \in \Omega$ *satisfying* $p_0(j\omega) \neq 0$, $p_2(j\omega) = 0$, *the corresponding* $\mathcal{T}^{\pm}_{\omega,u,v}$ *contains all the points calculated by (4.55) and (4.56) with* $\theta_1 = 0$, $\theta_2 \in [0, \pi]$, *and* $\mathcal{T}_\omega$ *is a series of vertical lines.*

Obviously,

$$\mathcal{T} = \{\mathcal{T}_\omega \mid \omega \in \Omega\}. \tag{4.59}$$

Since the behavior of the degenerate cases discussed in the above remark is easily understood, for brevity we will exclude these degenerate situations in what follows by imposing the following last assumption:

III. Nondegeneracy,

$$p_l(j\omega) \neq 0 \quad \forall \omega \in \Omega \text{ and } l = 0, 1, 2. \tag{4.60}$$

4.4.2 Stability crossing curves

In this section, we will give the complete characterization of the crossing set Ω and of the stability crossing curves $\mathcal{T}$. The presentation follows [98] closely. We have the following result.

Proposition 4.19. *The crossing set* Ω *consists of a finite number of intervals of finite length, including the cases which may violate (4.60).*

From Propositions 4.17 and 4.19, it follows that the end intervals frequency points ω of the crossing set Ω should satisfy one of the following conditions:

$$|a_1(j\omega)| + |a_2(j\omega)| = 1, \tag{4.61}$$

or

$$|a_1(j\omega)| - |a_2(j\omega)| = 1, \tag{4.62}$$

or

$$|a_2(j\omega)| - |a_1(j\omega)| = 1. \tag{4.63}$$

Let these intervals be Ω_k, $k = 1, 2, ..., N$, arranged in such an order that the left end point of Ω_k increases with increasing k. Then

$$\Omega := \bigcup_{k=1}^{N} \Omega_k. \tag{4.64}$$

It is worth clarifying that $0 \notin \Omega$ by definition even if $\omega = 0$ satisfies (4.52) and (4.53). Indeed, if (4.52) and (4.53) are satisfied for $\omega = 0$ and sufficiently small positive values of ω, then $\Omega_1 = (0, \omega_1^r]$, and we will let $\omega_1^l = 0$ in this case. Otherwise, $\Omega_1 = [\omega_1^l, \omega_1^r]$, $\omega_1^l \neq 0$. For $k \geq 2$, $\Omega_k = [\omega_k^l, \omega_k^r]$. We will subdivide the intervals if necessary so that for any $\omega \in (\omega_k^l, \omega_k^r)$, none of the three equations (4.61), (4.62), and (4.63) is satisfied.

Let

$$\mathcal{T}_{u,v}^{\pm k} = \bigcup_{\omega \in \Omega_k} \mathcal{T}_{\omega,u,v}^{\pm} = \{(\tau_1^{u\pm}(\omega), \tau_2^{v\pm}(\omega)) \mid \omega \in \Omega_k\},$$

and

$$\begin{aligned} \mathcal{T}^k &= \bigcup_{u=-\infty}^{\infty} \bigcup_{v=-\infty}^{\infty} (\mathcal{T}_{u,v}^{+k} \cup \mathcal{T}_{u,v}^{-k}) \cap \mathbb{R}_+^2 \\ &= \bigcup_{\omega \in \Omega_k} \mathcal{T}_\omega. \end{aligned} \tag{4.65}$$

Then,

$$\mathcal{T} = \bigcup_{k=1}^{N} \mathcal{T}^k.$$

Note that we allow part of $\mathcal{T}_{u,v}^{+k}$ or $\mathcal{T}_{u,v}^{-k}$ to be outside of $\mathbb{R}_+^2$ in some cases for the convenience of discussions. We should, however, keep in mind that the part of $\mathcal{T}_{u,v}^{+k}$ or $\mathcal{T}_{u,v}^{-k}$ outside of $\mathbb{R}_+^2$ no longer represents the boundary of a meaningful change of the number of right half-plane (RHP) zeros of $p(\lambda;\ \tau_1, \tau_2)$. As is well known, $p(\lambda;\ \tau_1, \tau_2)$ has an infinite number of RHP zeros if τ_1 or τ_2 assumes a negative value [13].

We will not restrict $\angle a_l(j\omega)$ to be within a range of 2π but make it a continuous function of ω within each Ω_k. This is always possible due to the way Ω_k is defined. As a result, for a fixed pair of integers (u, v), each $\mathcal{T}_{u,v}^{+k}$ or $\mathcal{T}_{u,v}^{-k}$ is a continuous curve.

To study how each $\mathcal{T}_{u,v}^{+k}$ or $\mathcal{T}_{u,v}^{-k}$ is connected in $\mathcal{T}^k$ at the ends of Ω_k, we make the following observation—under our standing nondegenerate assumption (4.60), the end points of the intervals, ω_k^l, $k = 2, 3, \ldots$, and ω_k^r, $k = 1, 2, \ldots$, must satisfy one and only one of the three equations (4.61), (4.62), and (4.63). Accordingly, we can classify these end points into three types according to which equation $\omega = \omega_k^l$ or $\omega = \omega_k^r$ satisfies. The left end of Ω_1 may have an additional type if $\omega_1^l = 0$. A careful examination of the equations (4.55) and (4.56) allows us to arrive at the following list [98]:

Type 1. (4.62) is satisfied. In this case, $\theta_1 = 0$, $\theta_2 = \pi$, and $\mathcal{T}_{u,v}^{+k}$ is connected with $\mathcal{T}_{u,v-1}^{-k}$ at this end.

Type 2. (4.63) is satisfied. In this case, $\theta_1 = \pi$, $\theta_2 = 0$, and $\mathcal{T}_{u,v}^{+k}$ is connected with $\mathcal{T}_{u+1,v}^{-k}$ at this end.

Type 3. (4.61) is satisfied. In this case, $\theta_1 = \theta_2 = 0$, and $\mathcal{T}_{u,v}^{+k}$ is connected with $\mathcal{T}_{u,v}^{-k}$ at this end.

Type 0. $\omega_k^l = 0$. This requires that $\omega = 0$ satisfy (4.52) and (4.53). In this case, as $\omega \to 0$, $\mathcal{T}_{u,v}^{+k}$ and $\mathcal{T}_{u,v}^{-k}$ approach ∞ with asymptotes passing through the points $(\hat{a}_1 \pm \hat{\theta}_1, \hat{a}_2 \mp \hat{\theta}_2)$ with slopes of

$$\frac{\tau_2^{v\pm}}{\tau_1^{u\pm}} \to \kappa_{u,v}^{\pm} = \frac{\angle a_2(0) + (2v-1)\pi \mp \theta_2(0)}{\angle a_1(0) + (2u-1)\pi \pm \theta_1(0)}, \tag{4.66}$$

where $\theta_1(0)$ and $\theta_2(0)$ are evaluated by (4.57) and (4.58) using $a_1(0)$ and $a_2(0)$, respectively, and

$$\hat{a}_l = \frac{d}{d\omega}[\angle a_l(j\omega)]_{\omega=0}, \tag{4.67}$$

$$\hat{\theta}_l = \frac{d}{d\omega}\theta_l(j\omega)|_{\omega=0}. \tag{4.68}$$

Correspondingly, we say an interval Ω_k is of type lr if the left end of Ω_k is of type l and its right end is of type r. There are a total of $4 \times 3 = 12$ possible types of such intervals. Let us consider one of the cases.

Example 4.20. *Consider a second-order system including two delays such that*

$$a_1(\lambda) = \frac{3}{\lambda^2 + 2\lambda + 1}, \tag{4.69}$$

$$a_2(\lambda) = \frac{9\lambda + 1}{\lambda^2 + 2\lambda + 1}. \tag{4.70}$$

In this case, Ω *contains two intervals*

$$\Omega_1 = [0.188, 0.453], \text{ of type 12,}$$
$$\Omega_2 = [8.532, 9.217], \text{ of type 23.}$$

Figure 4.2 plots $|a_1(j\omega)| + |a_2(j\omega)|$ *and* $|a_1(j\omega)| - |a_2(j\omega)|$ *against* ω.

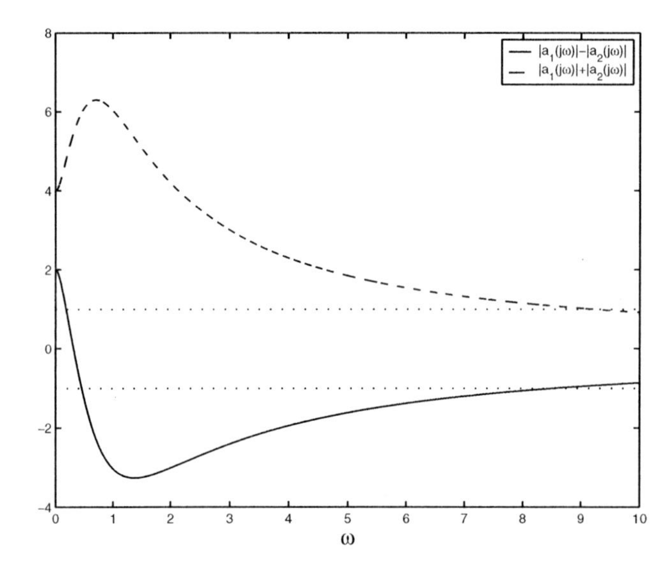

Figure 4.2. $|a_1(j\omega)| \pm |a_2(j\omega)|$ *versus* ω *for system represented by (4.69) and (4.70).*

Further examples covering most of the situations can be found in [98] (see also Section 4.5).

According to the types of Ω_k, $\mathcal{T}^k$ may have different shapes, as specified in the following proposition.

Proposition 4.21. *Under the standing assumption (4.60), the stability crossing curves $\mathcal{T}^k$ corresponding to Ω_k must be an intersection of $\mathbb{R}_+^2$ with a series of curves belonging to one of the following categories:*

A. *A series of closed curves;*

B. *A series of spiral-like curves with axes oriented either horizontally, vertically, or diagonally;*

C. *A series of open-ended curves with both ends approaching ∞.*

The rest of this section is devoted to showing the validity of the above proposition via a detailed list of scenarios (see, for instance, [98] for more details and illustrative examples corresponding to each situation).

Closed curves

As an illustration of this situation, examine first $\mathcal{T}^k$ corresponding to Ω_k of type 11. In this case, for given u and v, $\mathcal{T}_{u,v}^{+k}$ and $\mathcal{T}_{u,v-1}^{-k}$ are connected on both ends to form a closed curve. As u and v vary, a series of deformed versions of such closed curves are generated along the horizontal and vertical directions. $\mathcal{T}^k$ is the intersection of $\mathbb{R}_+^2$ with this series of closed curves. Similarly, it is easily shown that a $\mathcal{T}^k$ corresponding to Ω_k of type 22 or type 33 also forms a similar series of closed curves. In the case of type 22, a closed curve is formed by connecting both ends of $\mathcal{T}_{u,v}^{+k}$ and $\mathcal{T}_{u+1,v}^{-k}$. For type 33, a closed curve is formed by connecting both ends of $\mathcal{T}_{u,v}^{+k}$ and $\mathcal{T}_{u,v}^{-k}$.

Spiral-like curves

Several situations occur, as follows.

Spiral-like curves with axes oriented diagonally. To illustrate such a case, consider $\mathcal{T}^k$ corresponding to Ω_k of type 12. In this case, $\mathcal{T}_{u,v}^{+k}$ is connected to $\mathcal{T}_{u+1,v}^{-k}$ at ω_k^r, and the other end of $\mathcal{T}_{u+1,v}^{-k}$ is connected to $\mathcal{T}_{u+1,v+1}^{+k}$ at ω_k^l, which is again connected to $\mathcal{T}_{u+2,v+2}^{-k}$ at ω_k^r, and so on. According to Definition 4.16, with $\mathcal{C}_k = \mathcal{T}_{u,v}^{+k} \cup \mathcal{T}_{u+1,v}^{-k}$, it can be easily verified that this forms a spiral-like curve with the axis

$$\begin{aligned} A &= \left(\tau_1^{u+1-}(\omega_k^l) - \tau_1^{u+}(\omega_k^l), \tau_2^{v-}(\omega_k^l) - \tau_2^{v+}(\omega_k^l)\right) \\ &= \left(\frac{2p}{\omega_k^l}, \frac{2p}{\omega_k^l}\right), \text{ independent of } u \text{ (or } v), \end{aligned}$$

forming a 45° angle from the horizontal. This spiral is repeated an infinite number of times in a deformed form as the difference between u and v changes. Shown in Figure 4.3 is $\mathcal{T}^1$ for the system in Example 4.20.

We can observe that a $\mathcal{T}^k$ corresponding to Ω_k of type 21 also forms such a series of spiral-like curves with axes oriented diagonally. In this case $\mathcal{T}_{u,v}^{+k}$ is connected to $\mathcal{T}_{u+1,v}^{-k}$ at ω_k^l instead, $\mathcal{T}_{u+1,v}^{-k}$ is connected to $\mathcal{T}_{u+1,v+1}^{+k}$ at ω_k^r, and so on.

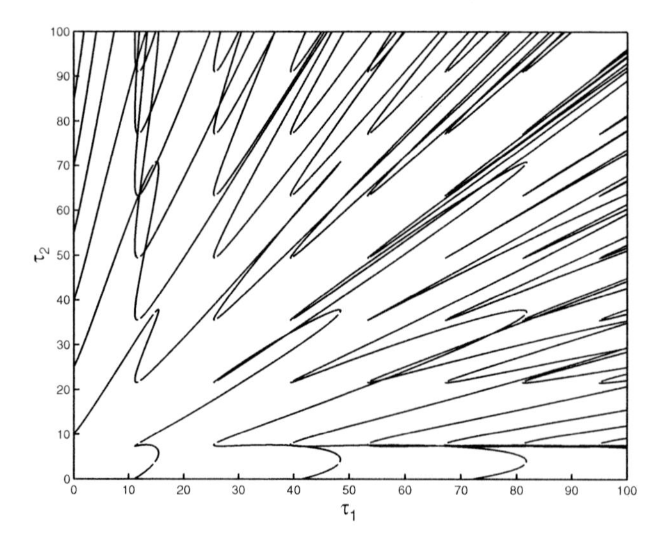

Figure 4.3. $\mathcal{T}^1$ *of the system in Example 4.20.*

Spiral-like curves with vertical axes. To illustrate such a case, consider $\mathcal{T}^k$ corresponding to Ω_k of type 13. In this case, $\mathcal{T}_{u,v}^{+k}$ is connected to $\mathcal{T}_{u,v}^{-k}$ at ω_k^r, the other end of $\mathcal{T}_{u,v}^{-k}$ is connected to $\mathcal{T}_{u,v+1}^{+k}$ at ω_k^l, and so on. This forms a spiral-like curve with a vertical axis. This spiral-like curve is repeated in deformed form along the horizontal direction as u changes.

It is easily shown that $\mathcal{T}^k$ corresponding to Ω_k of type 31 is also in the form of a series of vertically oriented spiral-like curves, with $\mathcal{T}_{u,v}^{+k}$ and $\mathcal{T}_{u,v}^{-k}$ connected at ω_k^l, $\mathcal{T}_{u,v}^{-k}$ and $\mathcal{T}_{u,v+1}^{+k}$ connected at ω_k^r, and so on.

Spiral-like curves with horizontal axes. Finally, the curves of $\mathcal{T}^k$ corresponding to Ω_k of types 23 and 32 are in the form of a series of *spiral-like curves with horizontal axes*. For type 23, $\mathcal{T}_{u,v}^{+k}$ is connected to $\mathcal{T}_{u+1,v}^{-k}$ at ω_k^l, the other end of $\mathcal{T}_{u+1,v}^{-k}$ is connected to $\mathcal{T}_{u+1,v}^{+k}$ at ω_k^r, and so on. For type 32, $\mathcal{T}_{u,v}^{+k}$ and $\mathcal{T}_{u+1,v}^{-k}$ are connected at ω_k^r, $\mathcal{T}_{u+1,v}^{-k}$ and $\mathcal{T}_{u+1,v}^{+k}$ are connected at ω_k^l, and so on.

Open-ended curves

Corresponding to $\Omega_1 = (0, \omega_1^r]$, $\mathcal{T}^1$ is a series of *open-ended curves*. For the type 01, $\mathcal{T}_{u,v}^{-1}$ and $\mathcal{T}_{u,v+1}^{+1}$ are connected at ω_1^r. The other end of $\mathcal{T}_{u,v}^{-1}$ extends to infinity with asymptote of a slope $\kappa_{u,v}^-$ passing through the point $(\hat{a}_1 - \hat{\theta}_1, \hat{a}_2 + \hat{\theta}_2)$. The other end of $\mathcal{T}_{u,v+1}^{+1}$ extends to infinity with asymptote of a slope $\kappa_{u,v+1}^+$ passing through the point $(\hat{a}_1 + \hat{\theta}_1, \hat{a}_2 - \hat{\theta}_2)$. This pattern is repeated in a deformed form in both horizontal and vertical directions. Note also that the slopes also change for different u and v.

It is easy to show that $\mathcal{T}^1$ corresponding to Ω_1 of type 02 and type 03 also forms open-ended curves. For type 02, $\mathcal{T}_{u,v}^{+1}$ and $\mathcal{T}_{u+1,v}^{-1}$ are connected at ω_1^r. The other ends of

$\mathcal{T}_{u,v}^{+1}$ and $\mathcal{T}_{u+1,v}^{-1}$ extend to infinity with slopes $\kappa_{u,v}^{+}$ and $\kappa_{u+1,v}^{-}$, respectively. For type 03, $\mathcal{T}_{u,v}^{+1}$ and $\mathcal{T}_{u,v}^{-1}$ are connected at ω_1^r. The other ends of $\mathcal{T}_{u,v}^{+1}$ and $\mathcal{T}_{u,v}^{-1}$ extend to infinity with slopes $\kappa_{u,v}^{+}$ and $\kappa_{u,v}^{-}$, respectively.

Thus far, we have exhausted all 12 types of Ω_k.

4.4.3 Tangents, smoothness, and crossing direction

We discuss in detail the smoothness of the stability crossing curves, characterize their tangents, and derive expressions for the corresponding direction of crossing.

Tangents and smoothness

For a given k, we will discuss the smoothness of the curves in $\mathcal{T}^k$ and thus of $\mathcal{T}$. For this purpose, we consider τ_1 and τ_2 as implicit functions of $\lambda = j\omega$ defined by (4.51). As λ moves along $j\mathbb{R}$, $(\tau_1,\tau_2) = (\tau_1^{u\pm}(\omega), \tau_2^{v\pm}(\omega))$ moves along $\mathcal{T}^k$. For a given $\omega \in \Omega_k$, let

$$
\begin{aligned}
R_0 &= \Re\left(\frac{j}{\lambda}\frac{\partial a(\lambda;\ \tau_1,\tau_2)}{\partial\lambda}\right)_{\lambda=j\omega} \\
&= \frac{1}{\omega}\Re\left(\left[a_1'(j\omega) - \tau_1 a_1(j\omega)\right]e^{-j\tau_1\omega} + [a_2'(j\omega) - \tau_2 a_2(j\omega)]e^{-j\tau_2\omega}\right), \qquad (4.71)\\
I_0 &= \Im\left(\frac{j}{\lambda}\frac{\partial a(\lambda;\ \tau_1,\tau_2)}{\partial\lambda}\right)_{\lambda=j\omega} \\
&= \frac{1}{\omega}\Im\left(\left[a_1'(j\omega) - \tau_1 a_1(j\omega)\right]e^{-j\tau_1\omega} + [a_2'(j\omega) - \tau_2 a_2(j\omega)]e^{-j\tau_2\omega}\right), \qquad (4.72)
\end{aligned}
$$

and

$$
R_l = -\Re\left(\frac{1}{\lambda}\frac{\partial a(\lambda;\ \tau_1,\tau_2)}{\partial\tau_k}\right)_{\lambda=j\omega} = \Re\left(a_k(j\omega)e^{-j\tau_k\omega}\right), \qquad (4.73)
$$

$$
I_l = -\Im\left(\frac{1}{\lambda}\frac{\partial a(\lambda;\ \tau_1,\tau_2)}{\partial\tau_k}\right)_{\lambda=j\omega} = \Im\left(a_k(j\omega)e^{-j\tau_k\omega}\right), \qquad (4.74)
$$

for $l = 1, 2$. Then, since $a(\lambda;\ \tau_1, \tau_2)$ is an analytic function of λ, τ_1, and τ_2, the implicit function theorem indicates that the tangent of $\mathcal{T}^k$ can be expressed as

$$
\begin{aligned}
\begin{pmatrix} \frac{d\tau_1}{d\omega} \\ \frac{d\tau_2}{d\omega} \end{pmatrix} &= \begin{pmatrix} R_1 & R_2 \\ I_1 & I_2 \end{pmatrix}^{-1} \begin{pmatrix} R_0 \\ I_0 \end{pmatrix} \\
&= \frac{1}{R_1 I_2 - R_2 I_1}\begin{pmatrix} R_0 I_2 - I_0 R_2 \\ I_0 R_1 - R_0 I_1 \end{pmatrix}, \qquad (4.75)
\end{aligned}
$$

provided that

$$
R_1 I_2 - R_2 I_1 \neq 0. \qquad (4.76)
$$

It follows from a well-known result [29] that $\mathcal{T}^k$ is smooth everywhere except possibly at the points where either (4.76) is not satisfied or

$$
\frac{d\tau_1}{d\omega} = \frac{d\tau_2}{d\omega} = 0. \qquad (4.77)
$$

A careful examination of these cases allows us to conclude with the following (see also [98]).

Proposition 4.22. *Under the standing assumptions including (4.60), the curves in $\mathcal{T}^k$ are smooth everywhere except possibly at the degenerate points corresponding to ω in any one of the following three cases:*

(1) $\lambda = j\omega$ is a multiple solution of $a(\lambda;\ \tau_1, \tau_2) = 0$.

(2) ω is a type 3 end point of Ω_k, and

$$\frac{d}{d\omega}\left(|a_1(j\omega)| + |a_2(j\omega)|\right) = 0.$$

(3) ω is a type 1 or type 2 end point of Ω_k, and

$$\frac{d}{d\omega}\left(|a_1(j\omega)| - |a_2(j\omega)|\right) = 0.$$

Furthermore, if the point is not among the three cases, then the tangents of the curves in $\mathcal{T}^k$ can be expressed as

$$\frac{d\tau_2}{d\tau_1} = \begin{cases} \dfrac{1/\tan\varphi_0 - 1/\tan\varphi_1}{1/\tan\varphi_0 - 1/\tan\varphi_2}, & \omega \in (\omega_k^l, \omega_k^r), \\[2ex] -\dfrac{|a_1(j\omega)|}{|a_2(j\omega)|}, & \omega \text{ is a type 3 end point of } \Omega_k, \\[2ex] \dfrac{|a_1(j\omega)|}{|a_2(j\omega)|}, & \omega \text{ is a type 1 or 2 end point of } \Omega_k, \end{cases} \tag{4.78}$$

where

$$\varphi_0 = \angle\left(\left[a_1'(j\omega) - \tau_1 a_1(j\omega)\right] e^{-j\tau_1\omega} + [a_2'(j\omega) - \tau_2 a_2(j\omega)] e^{-j\tau_2\omega}\right),$$
$$\varphi_k = \angle\left(a_k(j\omega) e^{-j\tau_k\omega}\right),\ k = 1, 2.$$

Direction of crossing

Next, we will discuss the direction in which the solutions of (4.51) cross the imaginary axis as (τ_1, τ_2) deviates from a curve in $\mathcal{T}^k$. We will call the direction of the curve that corresponds to increasing ω the *positive direction.* Notice that as the curve passes through the points corresponding to the end points of Ω_k, the positive direction is reversed. We will also call the region on the left-hand side as we head in the positive direction of the curve *the region on the left.* Again, due to the possible reversion of parameterization, the same region may be considered on the left with respect to one point of the curve, and be considered as on the right on another point of the curve.

For the purpose of discussing the direction of crossing, we need to consider τ_1 and τ_2 as functions of

$$\lambda = \sigma + j\omega,$$

i.e., functions of two real variables σ and ω, and partial derivative notation needs to be adopted instead. Since the tangent of $\mathcal{T}^k$ along the positive direction is $(\partial\tau_1/\partial\omega, \partial\tau_2/\partial\omega)$, the normal to $\mathcal{T}^k$ pointing to the left-hand side of the positive direction is $(-\partial\tau_2/\partial\omega, \partial\tau_1/\partial\omega)$.

Also, as a pair of complex conjugate solutions of (4.51) cross the imaginary axis to the RHP, (τ_1, τ_2) moves along the direction $(\partial\tau_1/\partial\sigma, \partial\tau_2/\partial\sigma)$. We can therefore conclude that if the inner product of these two vectors is positive, i.e.,

$$\left[\frac{\partial\tau_1}{\partial\omega}\frac{\partial\tau_2}{\partial\sigma} - \frac{\partial\tau_2}{\partial\omega}\frac{\partial\tau_1}{\partial\sigma}\right]_{s=j\omega} > 0, \tag{4.79}$$

the region on the left of $\mathcal{T}^k$ at ω has two more solutions on the RHP. On the other hand, if the inequality in (4.79) is reversed, then the region on the left of $\mathcal{T}^k$ has two fewer solutions on the right-hand side of the complex plane. We can very easily express, parallel to (4.75), that

$$\begin{aligned}\begin{pmatrix}\frac{\partial\tau_1}{\partial\sigma}\\ \frac{\partial\tau_2}{\partial\sigma}\end{pmatrix}_{s=j\omega} &= \begin{pmatrix}R_1 & R_2\\ I_1 & I_2\end{pmatrix}^{-1}\begin{pmatrix}I_0\\ -R_0\end{pmatrix}\\ &= \frac{1}{R_1I_2 - R_2I_1}\begin{pmatrix}R_0R_2 + I_0I_2\\ -R_0R_1 - I_0I_1\end{pmatrix},\end{aligned} \tag{4.80}$$

where R_l and I_l, $l = 0, 1, 2$, are defined in (4.71) to (4.74). This allows us to conclude with the following.

Proposition 4.23. *Let $\omega \in (\omega_k^l, \omega_k^r)$ and $(\tau_1, \tau_2) \in \mathcal{T}^k$ such that $j\omega$ is a simple solution of $a(j\omega;\ \tau_1, \tau_2) = 0$, and*

$$a(j\omega';\ \tau_1, \tau_2) \neq 0 \text{ for any } \omega' > 0,\ \omega' \neq \omega. \tag{4.81}$$

Then as (τ_1, τ_2) moves from the region on the right to the region on the left of the corresponding curve in $\mathcal{T}^k$, a pair of solutions of (4.51) cross the imaginary axis to the right if

$$\Im(a_1(j\omega)a_2(-j\omega)e^{j\omega(\tau_2-\tau_1)}) = R_2I_1 - R_1I_2 > 0. \tag{4.82}$$

The crossing is in the opposite direction if the inequality is reversed.

The condition (4.81) simply means that (τ_1, τ_2) is not an intersection point of two curves or different sections of a single curve in $\mathcal{T}$.

Finally, any given direction, $d = (d_1, d_2)$, with $\|d\|_2 = 1$, is to the left-hand side of the curve if its inner product with the left-hand side normal $(-\partial\tau_2/\partial\omega, \partial\tau_1/\partial\omega)$ is positive, i.e.,

$$-d_1\partial\tau_2/\partial\omega + d_2\partial\tau_1/\partial\omega > 0, \tag{4.83}$$

from which we have the following corollary.

Corollary 4.24. *Let ω, τ_1, and τ_2 satisfy the same condition as Proposition 4.23. Then as (τ_1, τ_2) crosses the curve along the direction (d_1, d_2), a pair of solutions of (4.51) cross the imaginary axis to the right if*

$$d_1(R_0I_1 - I_0R_1) + d_2(R_0I_2 - I_0R_2) > 0. \tag{4.84}$$

The crossing is in the opposite direction if the inequality is reversed.

4.5 Notes and references

This chapter addressed the stability problem in the delay-parameter space. More precisely, we discussed and presented various algebraic tests for characterizing the crossing existence together with corresponding delay intervals in the case of linear systems in state-space representation and including commensurate delays. We also presented the geometry of the stability crossing curves for some quasipolynomials including two independent delays. Some connection between these independent approaches will be considered in the next chapter, where we will focus on the delay ratio sensitivity and delay-interference phenomena.

The analysis of the crossing roots and the corresponding stability characterization of linear systems with commensurate delays presented here follows the lines of [229] with some complements taken from [223, 96]. It is important to point out that, to the best of our knowledge, the interpretation (in the control area) of the characteristic equation of a linear system with commensurate delays as an equation including two variables—one on the imaginary axis and one on the unit circle—goes back to the work of Kamen [131, 132] in the 1980s in the context of *delay-independent* (asymptotic) *stability*. Such an idea was largely exploited in the literature in the 1980s (quasipolynomials) and in the 1990s (state-space representation). A more extensive overview of such results can be found in [223, 96]. The bilinear transformation based approaches can be found in [223, 96]. For the 2-D polynomial approach, we mention the test proposed by [28, 50]. The pseudodelay technique goes back to the work of Rekasius [257] and Thowsen [294, 295] (delay-independent analysis). The complete characterization of the commensurate delays case can be found in [241], and several extensions of the method for dealing with neutral and multiple delays can be found in [276].

Proposition 4.8 is described in [229] and represents a natural extension of the root crossing characterization proposed by [54, 53] to the state-space representation of the delay system. The characterization of the crossing direction in the case of semisimple characteristic roots is taken from [84]. The particular cases covering the characterizations of hyperbolicity and of delay-independent stability can be found in [223] and have been revisited in [229]. The rank-one delayed matrix presentation is new, and will be useful in the stability analysis of some models from the biosciences in the last chapter. A thorough discussion on frequency-sweeping tests can be found in [47] (see also [96]). The stability crossing characterization of quasipolynomials including one delay was considered first in [54] and [322].

The connection between pseudodelays and the computation of generalized eigenvalues of some appropriate matrix pencils is new, and is inspired by some characterization proposed in the context of distributed delays in [209]. Such an idea will be adopted later to the case of delayed output feedback stabilization of SISO systems.

Next, a geometric approach was used for the characterization of the stability crossing curves in the delay-parameter space of quasipolynomials with two delays. The presentation closely followed [98]. More precisely, we proved that the frequency crossing set can be expressed by three constraints and consists of a finite number of intervals of finite length. Except in a few degenerate cases, the stability crossing curves are smooth. These curves may be closed, open-ended, or spiral-like with axis in the horizontal, vertical, or diagonal directions. The classification of the curves is determined by the constraints which are violated at the end points of the frequency crossing set. Finally, smoothness properties of the stability crossing curves were investigated and a characterization of the crossing direction was made. We used an approach similar to the one described in Chapter 11 of [67], based on the implicit function theorem.

This geometric approach will be particularly exploited in the next chapters for giving some insights in the case of a Smith predictor subjected to delay uncertainty, the analysis of some immune dynamics models in chronic leukemia, and the delayed output feedback stabilization problem of SISO systems.

Chapter 5

Stability of delay rays and delay-interference

5.1 Introduction

The stability analysis in the delay-parameter space was the main objective of the previous chapter, and several algebraic and geometric based methods and algorithms have been proposed for performing such an analysis. The first part of the chapter was mainly devoted to presenting efficient, computationally tractable methods for characterizing stability/instability regions and evaluating the crossing direction of characteristic roots for linear systems including commensurate delays. More precisely, the use of some appropriate elimination techniques helped us handle the asymptotic stability in the delay-parameter space of the linear system

$$\dot{x}(t) = A_0 x(t) + \sum_{i=1}^{m} A_i x(t - \tau_i), \tag{5.1}$$

where $A_i \in \mathbb{R}^{n \times n}$, $i = 0, \ldots, m$, are given system matrices and $\tau_i \geq 0$, $i = 1, \ldots, m$, are delay constants, but under the constraint that the delays are restricted to be multiples of the same number τ_0. The latter implies a relation of the form

$$(\tau_1, \ldots, \tau_m) = \tau_0\,(n_1, \ldots, n_m), \quad n_i \in \mathbb{N},\ i = 1, \ldots, m, \tag{5.2}$$

where $n_1, \ldots, n_m$ are fixed positive integers, and $\tau_0 \in \mathbb{R}_+$ is the free parameter. Thus, the stability was reduced to a one-parameter analysis problem, in which the commensurate character of the delays is essential. Geometrically, the parametrization (5.2) above corresponds to a particular *ray* or direction in the complete delay-parameter space.

Next, in the second part of the chapter, the delay ray constraint (5.2) above was relaxed in the quasipolynomial case including two distinct and independent delays, and we discussed in detail some geometric based methods for characterizing the stability crossing curves together with some appropriate algorithms for the computation of the corresponding stability regions in the delay-parameter space. Some extensions to multiple delays cases were also outlined.

The main objective of this chapter is to investigate the way in which the stability properties are affected by the ratio between the delays. More precisely, we will consider a particular stability property, the *delay-independent* stability along some ray in the delay-parameter space. As a byproduct of this analysis, we will also give a characterization of the so-called delay-interference phenomena.

Inspired by these algebraic and geometric approaches briefly mentioned above, we start the chapter by studying the eigenvalue distribution of some complex matrix-valued functions, derived from the delay system (5.1). As a consequence of this study, new necessary and sufficient conditions for *delay-independent stability* are obtained. The derived criteria cover both the case where all delays vary independently of each other, and where they are restricted to *an arbitrary* ray in the delay-parameter space, more exactly,

$$(\tau_1, \ldots, \tau_m) = \tau_0 \,(r_1, \ldots, r_m), \quad r_i \in \mathbb{R}_+, \; i = 1, \ldots, m, \tag{5.3}$$

with $\tau_0 \in \mathbb{R}_+$ being once again the free parameter.

The relaxation from (5.2) to (5.3) has the following important consequences:

1. It allows us to make assertions about *delay-independent stability* when the ray under consideration consists of delay values with *any type of interdependence* (commensurate, rationally (in)dependent delays). As mentioned above, this interdependency turns out to be important.

2. A study of the *sensitivity* of delay-independent stability *along a ray* w.r.t. changes of the direction (determined by the r_i in (5.3)) becomes possible. Among others, it leads us to a complete characterization of the so-called *delay-interference phenomenon*, that is, the presence of delay-independent stability along a particular ray, which is not robust against arbitrarily small perturbations of the direction of the ray.

Due to the fragility of delay-independent stability properties (note that they involve a *noncompact* set of delay values), these issues are nontrivial and deserve special attention, as we shall see in what follows. They have, however, barely been treated in the literature, excepting some contributions on the interference phenomenon and related topics [166, 167, 60, 159, 225].

To the best of the authors' knowledge, the notion of delay-interference[9] was first mentioned by MacDonald [167] (see also [166]), where a second-order system including two delays is shown to be subjected to delay-interference if it has the following property: delay-independent stability if the delays are equal, and delay-dependent stability with respect to each delay if the other is equal to zero. A further example (scalar system including two delays) of delay-interference can be found in [60]. This example will be briefly discussed in the next section, and reconsidered as a particular illustrative example later.

The chapter is organized as follows: definitions, assumptions, and an introductory example are presented in the next section. Section 5.3 contains preliminary results concerning the properties of the eigenvalues of some associated matrix-valued functions, in support of Section 5.4, which is devoted to the main results (delay-independent stability and interference characterizations). In Section 5.5 some illustrative examples are presented and discussed. Some notes and references complete the chapter.

[9] In Physics, interference represents the combination of two or more wave motions to form a resultant wave in which the displacement is reinforced or canceled.

5.2 Preliminary results

In what follows, we present the definition of a ray together with appropriate delay-independent stability and delay-interference notions. Finally, a technical assumption needed to avoid the existence of an invariant root at the origin is considered. An introductory illustrative example of a scalar DDE including two delays ends the section.

5.2.1 Definitions and assumptions

Given a direction $(r_1, \ldots, r_m) \in \mathcal{B}^m_+$ in the delay-parameter space, we define the associated *ray*, $\mathcal{T}(\vec{r})$, as follows.

Definition 5.1. *For* $\vec{r} \in \mathcal{B}^m_+$, *let* $\mathcal{T}(\vec{r}) := \{\tau_0 \, \vec{r} : \; \tau_0 \in \mathbb{R}_+\}$.

As we address delay-independent stability properties for both cases where all delays vary independently of each other, and where they are restricted to a particular ray, we will from now on use the following terminology, in order to avoid confusion.

Definition 5.2. *The system (5.1) is delay-independent stable if and only if its zero solution is asymptotically stable for all* $\vec{\tau} \in \mathbb{R}^m_+$.

Definition 5.3. *The ray* $\mathcal{T}(\vec{r})$ *is stable if and only if the zero solution of (5.1) is asymptotically stable for all* $\vec{\tau} \in \mathcal{T}(\vec{r})$.

Regarding delay-interference we have the following definition.

Definition 5.4. *A stable ray* $\mathcal{T}(\vec{r})$ *is subjected to the delay-interference phenomenon if and only if for all* $\epsilon > 0$ *there exists a* $\vec{s} \in \mathcal{B}^m_+$ *with* $\|\vec{r} - \vec{s}\| < \epsilon$ *such that the ray* $\mathcal{T}(\vec{s})$ *is not stable.*

Recall that the zero solution of (5.1) is asymptotically stable if and only if all its *characteristic roots*, the zeros of

$$H(\lambda; \; \vec{\tau}) = \det\left(\lambda I - A_0 - \sum_{i=1}^{m} A_i e^{-\lambda \tau_i}\right),$$

are in the open LHP (left half-plane). See Chapter 1 for stability definitions and their relation with the location of the characteristic roots.

The following technical assumption will be made throughout this chapter.

Assumption 5.5.

$$\det\left(A_0 + \sum_{i=1}^{m} A_i\right) \neq 0.$$

It excludes the presence of a characteristic root of (5.1) at zero for *all* delays values, implying that the asymptotic stability is not possible. Note that if this assumption is not satisfied, we have

$$\forall \vec{\tau} \in \mathbb{R}_+^m \;\; H(0;\; \vec{\tau}) = 0,$$

which implies that the zero solution is not asymptotically stable, whatever the values of the delays. Since this chapter is devoted to the characterization of delay-independent stability, the assumption above can be made without losing generality.

5.2.2 Introductory example

Consider the following scalar system including two distinct delays:

$$\dot{x}(t) = -x(t) - x(t-\tau_1) - \frac{1}{2}x(t-\tau_2), \tag{5.4}$$

where $(\tau_1, \tau_2) \in \mathbb{R}_+ \times \mathbb{R}_+$. This example is taken from [60]. The corresponding characteristic function is

$$p(\lambda;\; \tau_1, \tau_2) := \lambda + 1 + e^{-\lambda\tau_1} + \frac{1}{2}e^{-\lambda\tau_2}.$$

The geometric approach proposed in Chapter 4 for the computation of the stability crossing curves applies to the system (5.4), and we can derive without any difficulty all the stability regions in the delay-parameter set. It is easy to see that the quasipolynomial p and the analytic function q given by

$$q(\lambda;\; \tau_1, \tau_2) := 1 + a(\lambda)e^{-\lambda\tau_1} + \frac{1}{2}a(\lambda)e^{-\lambda\tau_2},$$

with $a(\lambda) = 1/(\lambda+1)$, share the same characteristic roots on the imaginary axis. More precisely, the crossing set Ω is given by all the frequencies $\omega \geq 0$ such that

$$\frac{2}{3} \leq \mid a(j\omega) \mid \leq 2.$$

Simple computations prove that Ω has the form $\Omega = [0, \sqrt{5}/2]$, where $\tilde{\omega} = \sqrt{5}/2$ is the unique positive solution of $\mid a(j\tilde{\omega}) \mid = 2/3$, and the stability crossing curves characterization follows.

However, we will concentrate on the analysis of the asymptotic stability on some particular ray in the delay-parameter space, and on some of its perturbations. More precisely, we consider the following ray:

$$(\tau_1, \tau_2) = \tau(1, 2), \;\; \tau \geq 0.$$

Thus, (5.4) rewrites as follows:

$$\dot{x}(t) = -x(t) - x(t-\tau) - \frac{1}{2}x(t-2\tau), \tag{5.5}$$

where τ is constant. Using the methods proposed in Chapter 4, it is easy to see that the system (5.5) is *delay-independent* asymptotically stable.

Indeed, the system free of delay is asymptotically stable, and the corresponding characteristic function

$$p(\lambda;\ \tau) := \lambda + 1 + e^{-\lambda\tau} + \frac{1}{2}e^{-2\lambda\tau}$$

has no roots on the imaginary axis for all delay $\tau \geq 0$. The last point can be checked easily by using the elimination technique discussed in Chapter 4. As suggested there, we may concentrate on analyzing the existence of solutions $(\omega, z) \in \mathbb{R}_+^* \times \mathcal{C}(0, 1)$ of the 2-D equation:

$$j\omega + 1 + z + \frac{1}{2}z^2 = 0. \tag{5.6}$$

Some algebraic manipulations (by eliminating the variable on the imaginary axis) lead to the following conclusion: the characteristic function $p(\lambda;\ \tau)$ has roots on the imaginary axis for some positive τ if and only if the equation

$$2 + z + \frac{1}{2}z^2 + \bar{z} + \frac{1}{2}\bar{z}^2 = 0 \tag{5.7}$$

has solutions on the unit circle, and these solutions generate crossing frequencies. There are two different ways to see if (5.7) has roots on the unit circle—by direct calculations, or by assuming the existence of at least one root, and next arriving at some contradiction.

Let us use the second argument. If there exists any root z_0 on the unit circle, this root should be of the form $z_0 = e^{-j\theta_0} = \cos(\theta_0) - j\sin(\theta_0)$, for some $\theta_0 \in (0, 2\pi)$, since $z_0 = 1$ is no solution of the equation above. Thus, (5.7) leads to the following second-order equation in $\cos(\theta_0)$:

$$2\cos(\theta_0)^2 + 2\cos(\theta_0) + 1 = 0,$$

which does not have any real solution in the interval $[-1, 1]$. So, we arrive at a contradiction and, thus, there does not exist any crossing frequency for all positive delays $\tau \geq 0$.

In conclusion, the scalar system including two distinct delays (5.4) is asymptotically stable for all delays (τ_1, τ_2) on the ray

$$(\tau_1, \tau_2) = \tau(1, 2).$$

Furthermore, it is easy to see that (5.4) is also *delay-independent* asymptotically stable if $\tau_1 = 0$ and $\tau_2 \neq 0$, or if $\tau_1 \neq 0$ and $\tau_2 = 0$, that is, both delay axes are also *stable rays*.

Consider now some perturbation of the ray of the form

$$(\tau_1, \tau_2) = \tau(1, 2 + \varepsilon) \tag{5.8}$$

for some $\varepsilon > 0$, but sufficiently small. Since the system (5.4) is *not* asymptotically stable for *all positive delays* τ_1 and τ_2, the natural question is to know *if such a ray will intersect some stability crossing curve in the domain*. This is indeed the case, because Datko [60] has shown that there exists a sequence $\{\varepsilon_n\}_{n\geq 1}$, where

$$\varepsilon_n = \frac{1}{2(2n + 1)}$$

such that the ray (5.8) with $\varepsilon = \varepsilon_n$ is not stable. More precisely, for some delay values $\tau > 2(2n + 1)\pi$, the system becomes unstable on the ray corresponding $\epsilon = \varepsilon_n$. Hence, the ray $(1, 2)$ is subjected to the so-called *delay-interference* as specified by Definition 5.4.

More discussions on this example can be found in Section 5.5.

5.3 Properties of some associated matrix-valued functions

Consider now the following matrix-valued functions:

- $L_1 : [0,\ 2\pi]^m \mapsto \mathbb{C}^{n\times n}$, given by

$$L_1(\vec{\theta}) := L_1(\theta_1, \ldots \theta_m) = A_0 + \sum_{i=1}^{m} A_i e^{-j\theta_i}, \tag{5.9}$$

- $L_2 : \mathbb{R}_+ \times \mathbb{R}_+^m \mapsto \mathbb{C}^{n\times n}$, given by

$$L_2(\theta, \vec{r}) := L_2(\theta, r_1, \ldots, r_m) = A_0 + \sum_{i=1}^{m} A_i e^{-j\theta r_i}. \tag{5.10}$$

The following sets and quantities will play a major role in the characterization of delay-independent stability and the interference phenomenon.

Definition 5.6. *Let*

$$\mathbb{W} = \bigcup_{\vec{\theta}\in[0,\ 2\pi]^m} \sigma\left(L_1(\vec{\theta})\right),$$

$$\alpha_0 = \sup\left\{\Re(\lambda) : \lambda \in \mathbb{W}\right\},$$

and for $\vec{r} \in \mathcal{B}_+^m$ let

$$\mathbb{V}(\vec{r}) = \bigcup_{\theta\geq 0} \sigma\left(L_2(\theta, \vec{r})\right),$$

$$\alpha(\vec{r}) = \sup\left\{\Re(\lambda) : \lambda \in \mathbb{V}(\vec{r})\right\}.$$

It directly follows that

$$\forall \vec{r} \in \mathbb{B}_+^m \quad \begin{array}{l} \overline{\mathbb{V}(\vec{r})} \subseteq \mathbb{W}, \\ \alpha(\vec{r}) \leq \alpha_0. \end{array} \tag{5.11}$$

As we shall see the relations (5.11) can be strict ($\subset$, resp., $<$). However, if the components of $\vec{r}$ are rationally independent,[10] then the next proposition applies.

Proposition 5.7. *If the components of $\vec{r}$ are rationally independent, then*

$$\overline{\mathbb{V}(\vec{r})} = \mathbb{W}, \quad \alpha(\vec{r}) = \alpha_0.$$

Sketch of the proof. The proof idea can be summarized as follows: since $\mathbb{V}(\vec{r}) \subseteq \mathbb{W}$ and $\mathbb{W}$ is closed, it remains to prove for the first statement that every element of W can be approximated arbitrarily well with elements from $\mathbb{V}(\vec{r})$. Such a conclusion follows by applying Kronecker's theorem [110, Theorem 444], combined with a Rouché-type result (see Section A.1 of the appendix). A complete proof can be found in [190].

[10] See Section A.4 in the appendix for definitions regarding the interdependency of numbers.

Next, with $\mathcal{B}_+^m$ equipped with the Euclidean norm and $\mathcal{P}(\mathbb{C})$ with the Hausdorff metric, we address the continuity of the functions $\mathbb{V}(\cdot)$ and $\alpha(\cdot)$; see Section A.3 for precise definitions. As we illustrate at the end of the section, these functions are in general *not* continuous at each point. However, the following weaker property holds (see [190] for the proof, and see Section A.3 for the semicontinuity definition).

Proposition 5.8. *The function* $\mathbb{V}: \mathcal{B}_+^m \to \mathcal{P}(\mathbb{C}),\ \vec{r} \mapsto \mathbb{V}(\vec{r})$ *is lower semicontinuous at each* $\vec{r} \in \mathcal{B}_+^m$.

Corollary 5.9. *The function* $\alpha: \mathcal{B}_+^m \to \mathbb{R},\ \vec{r} \mapsto \alpha(\vec{r})$ *is lower semicontinuous at each* $\vec{r} \in \mathcal{B}_+^m$.

By combining Propositions 5.7 and 5.8 with relations (5.11) we arrive at a stronger result at rationally independent $\vec{r}$.

Proposition 5.10. *The function* $\mathbb{V}: \mathcal{B}_+^m \to \mathcal{P}(\mathbb{C}),\ \vec{r} \mapsto \mathbb{V}(\vec{r})$ *is continuous at rationally independent* $\vec{r}$. *The function* $\alpha: \mathcal{B}_+^m \to \mathbb{R},\ \vec{r} \mapsto \alpha(\vec{r})$ *is continuous at rationally independent* $\vec{r}$.

We illustrate the above concepts and properties with the equation

$$\dot{x}(t) = -1.3x(t) - x(t-\tau_1) - \frac{1}{2}x(t-\tau_2), \tag{5.12}$$

an example that slightly modifies the introductory example briefly discussed in the previous section. We have

$$\begin{aligned} \mathbb{W} &= \textstyle\bigcup_{(\theta_1,\theta_2)\in[0,\ 2\pi]^2} \left(-1.3 - e^{-j\theta_1} - \frac{1}{2}e^{-j\theta_2}\right) = \left\{\lambda \in \mathbb{C}:\ \frac{1}{2} \le |\lambda + 1.3| \le \frac{3}{2}\right\}, \\ \mathbb{V}(\vec{r}) &= \textstyle\bigcup_{\theta\ge 0} \left(-1.3 - e^{-jr_1\theta} - \frac{1}{2}e^{-jr_2\theta}\right). \end{aligned}$$

In Figure 5.1 we have shown the set $\mathbb{V}(\vec{r})$ for some values of $\vec{r}$. We have also indicated the boundaries of $\mathbb{W}$. When the components of $\vec{r}$ are commensurate, $\mathbb{V}(\vec{r})$ forms a closed curve. In all the cases displayed we have $\alpha(\vec{r}) < \alpha_0$ and $\overline{\mathbb{V}(\vec{r})} \subset \mathbb{W}$ (strict relations). *Discontinuities* of the function $\vec{r} \mapsto \mathbb{V}(\vec{r})$ occur at all rationally dependent $\vec{r}$. To illustrate the mechanism, let us first compare cases (a) and (f). Ratio $r_2/r_1 = 1.9$ can be seen as a perturbation of $r_2/r_1 = 2$ and, although the components of $\vec{r}$ are still commensurate, they are "more independent," since the coprime numbers 19 and 10 (having ratio 1.9) are larger than 2 and 1 (having ratio 2). As a consequence, the curve

$$\theta \ge 0 \mapsto -1.3 - e^{-j\theta} - \frac{1}{2}e^{-j\frac{r_2}{r_1}\theta} \tag{5.13}$$

with $r_2/r_1 = 1.9$ only closes at $\theta = 20\pi$ when θ is increased from zero (instead of $\theta = 2\pi$ for $r_2/r_1 = 2$), and a "larger portion" of $\mathbb{W}$ is covered by $\mathbb{V}(\vec{r})$. This is clearly visible in subplot (f), where we have also depicted the values of (5.13) for $\theta \in [0,\ 2\pi]$ (bold part of curve). Next, if $r_2/r_1 = 2$ would be perturbed instead to an irrational value, such as $2 - \pi/n,\ n \ge 2$, then the corresponding curve $\mathbb{V}(\vec{r})$ would never close and its points would densely fill $\mathbb{W}$, as follows from Proposition 5.7. Since n can be taken arbitrarily large, the perturbation can be taken arbitrarily small.

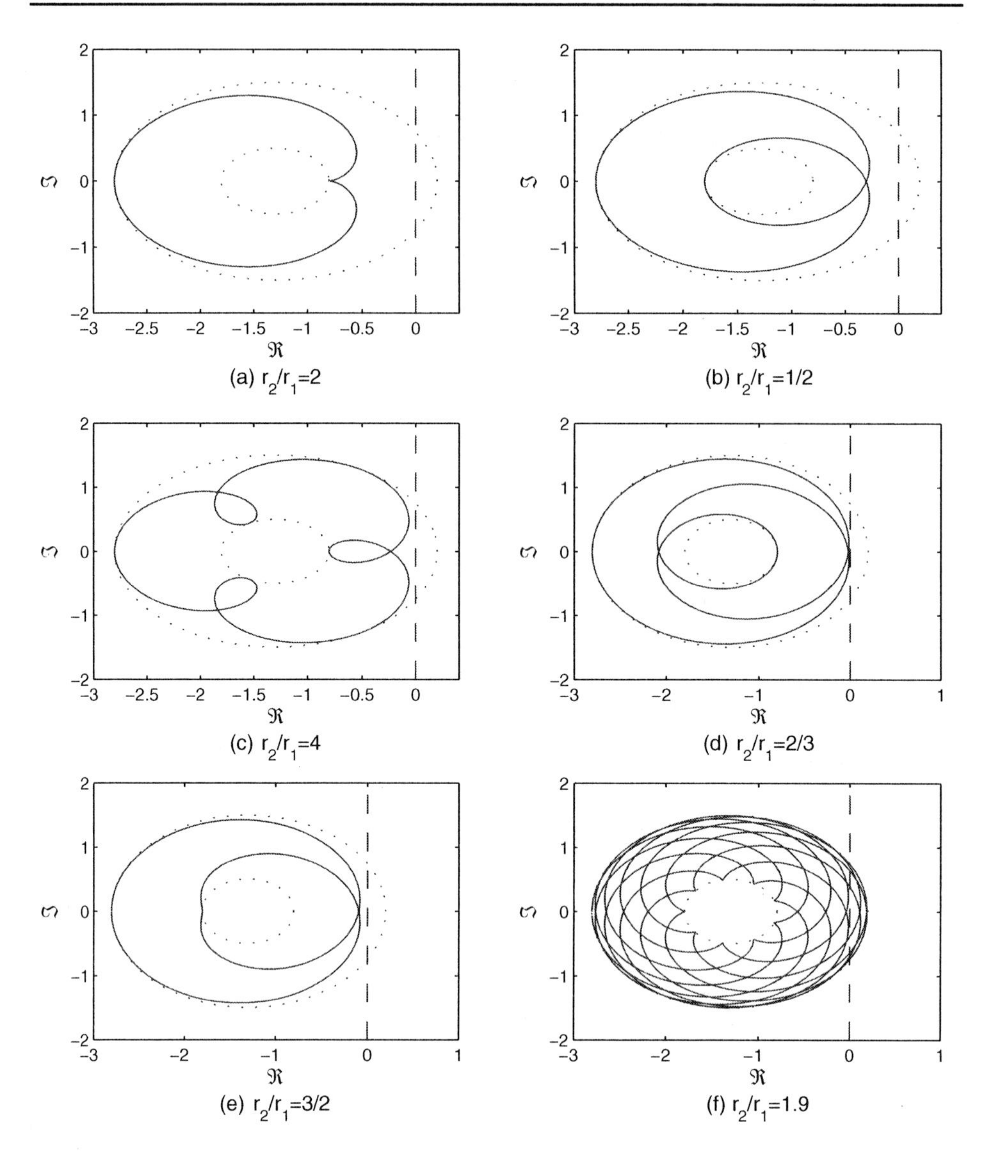

Figure 5.1. *The set $\mathbb{V}(\vec{r})$ for the system (5.12), for different values of $\vec{r}$ (solid curves). The dotted curves are the boundaries of $\mathbb{W}$.*

5.4 Delay-independent stability and delay-interference phenomena

Step by step we characterize the stability of the zero solution of (5.1) in the delay parameters, in terms of the sets $\mathbb{V}(\cdot)$ and $\mathbb{W}$ defined in the previous section. The main results will be stated in Theorems 5.11, 5.15, and 5.17, and they will be summarized in Table 5.2.

5.4.1 Delay-independent stability characterization

We first recall some properties of the characteristic roots of (5.1), which are needed in the subsequent analysis (see also Chapter 1). The number of characteristic roots in any RHP is finite. Furthermore, the function

$$g : \mathbb{R}_+^m \to \mathbb{R}, \ \vec{\tau} \mapsto g(\vec{r}) = \max_{\lambda \in \mathbb{C}} \{\Re(\lambda) : \ H(\lambda; \ \vec{\tau}) = 0\} \tag{5.14}$$

is continuous [223]. As a consequence, if the zero solution of (5.1) is asymptotically stable for some delay values but the system is not delay-independent stable, then there exist delay values for which the rightmost characteristic roots are *on the imaginary axis*.

Necessary and sufficient conditions for delay-independent stability are expressed by the following theorem (see [190] for a complete proof, where both implications are proved by contradiction).

Theorem 5.11. *The system (5.1) is delay-independent stable if and only if*

$$\mathbb{W} \subset (\mathbb{C}_- \cup \{0\}) \,. \tag{5.15}$$

Remark 5.12. *For a delay-independent stable system, the condition* $0 \notin \mathbb{W}$*, respectively* $0 \in \mathbb{W}$*, is referred to in the literature as* strong, *respectively* weak, *delay-independent stability (see [17, 223] and the references therein), notions which are related to stability properties of some associated 2-D system. Further comments on these notions can be found in Chapter 4.*

5.4.2 Delay-interference characterization

The stability of a ray $\mathcal{T}(\vec{r})$ can be characterized in an analogous way to Theorem 5.11 and it is omitted.

Proposition 5.13. *The ray* $\mathcal{T}(\vec{r})$ *is stable if and only if*

$$\mathbb{V}(\vec{r}) \subset (\mathbb{C}_- \cup \{0\}) \,. \tag{5.16}$$

The next proposition states that if the system (5.1) is not delay-independent stable, then it is always prone to the delay-interference phenomenon.

Proposition 5.14. *Assume that* $\mathbb{W} \not\subset (\mathbb{C}_- \cup \{0\})$*. If a ray* $\mathcal{T}(\vec{r})$ *is stable, then it is subjected to the delay-interference phenomenon.*

Proof. From the assumption $\mathbb{W} \not\subset (\mathbb{C}_- \cup \{0\})$ it follows that (5.1) is not delay-independent stable. So there exist delay values $\vec{\tau} \in \mathbb{R}_+^m$ and a frequency $\omega > 0$ such that $H(j\omega; \vec{\tau}) = 0$. For any given $\vec{k} \in \mathbb{N}^m$, $\vec{k} \neq 0$, this implies that $H(j\omega; \vec{\tau}(n)) = 0$ for all $n \in \mathbb{N}$, where

$$\vec{\tau}(n) = \left(\tau_1 + k_1 n \, \frac{2\pi}{\omega}, \ldots, \tau_m + k_m n \, \frac{2\pi}{\omega} \right). \tag{5.17}$$

Therefore, the ray

$$\mathcal{T}\left(\frac{\vec{\tau}(n)}{\|\vec{\tau}(n)\|}\right)$$

is unstable for all $n \in \mathbb{N}$. From

$$\lim_{n\to\infty} \frac{\vec{\tau}(n)}{\|\vec{\tau}(n)\|} = \frac{\vec{k}}{\|\vec{k}\|}$$

and the fact that this analysis can be repeated for every nonzero $\vec{k} \in \mathbb{N}^m$, it follows that every ray $\mathcal{T}(\vec{r})$, where $\vec{r}$ has *commensurate* components, gets unstable when applying certain infinitesimal perturbations to $\vec{r}$.

If the components of $\vec{r}$ are noncommensurate, then there exist infinitesimal $\delta\vec{r}$ such that $\vec{r} + \delta\vec{r}$ has commensurate components ($\mathbb{Q}^m$ is dense in $\mathbb{R}^m$), and the above arguments can be repeated. This completes the proof. □

Combining the previous results for the case where $\mathbb{W}$ has a nonempty intersection with the open RHP yields the following.

Theorem 5.15. *Assume that* $\mathbb{W} \cap \mathbb{C}_+ \neq \phi$. *Then the following holds:*

1. *If the components of* $\vec{r}$ *are rationally independent, then the ray* $\mathcal{T}(\vec{r})$ *is unstable.*
2. *If the ray* $\mathcal{T}(\vec{r})$ *is stable, then it is subjected to the delay-interference phenomenon.*
3. *The set* $\left\{\vec{r} \in \mathcal{B}_+^m : \mathcal{T}(\vec{r}) \text{ stable}\right\}$ *is nowhere dense in* $\mathcal{B}_+^m$.

Proof. If the components of $\vec{r}$ are rationally independent, then $\overline{\mathbb{V}(\vec{r})} = \mathbb{W}$ by Proposition 5.7. A combination with the assumption of the theorem leads to $\mathbb{V}(\vec{r}) \cap \mathbb{C}_+ \neq \phi$. Applying Proposition 5.13 then yields the first assertion.

The second assertion follows from Proposition 5.14, whose proof mainly relies on the invariance property (5.17).

Finally, the third assertion follows from Propositions 5.7–5.8 and 5.13. □

Under the assumption of the previous theorem, we comment on the detection of *stable* rays $\mathcal{T}(\vec{r})$ (if any), *without* the explicit computation of stability/instability regions of (5.1) in the delay-parameter space. From Theorem 5.15, the components of $\vec{r}$ cannot be rationally independent. By Proposition 5.13, we have to search for values for which $\mathbb{V}(\vec{r}) \subset (\mathbb{C}_- \cup \{0\})$, whereas we have $\mathbb{W} \cap \mathbb{C}_+ \neq \phi$ and $\mathbb{V}(\vec{r}) \subseteq \mathbb{W}$. Thus, values of $\vec{r}$, for which $\mathbb{V}(\vec{r})$ "doesn't cover $\mathbb{W}$ very well," are good candidates. Following from Propositions 5.7 and 5.8, such values must be characterized by a "large dependence," for instance, $\vec{r} = \vec{n}/\|\vec{n}\|$, with $\vec{n} \in \mathbb{N}^m$ and $\|\vec{n}\|$ small. To fix the ideas, we pick up example (5.12), which satisfies $\mathbb{W} \cap \mathbb{C}_+ \neq \phi$ since $\alpha_0 = 0.2$. Table 5.1 displays the corresponding values of $\alpha(\vec{r})$, for rationally dependent $\vec{r}$. It becomes apparent that stable rays correspond to

$$(n_1, n_2) \in \{(1,0),\ (0,1),\ (1,2),\ (1,4),\ (2,1),\ (2,3),\ (3,2)\}. \tag{5.18}$$

Table 5.1. *Values of* $\alpha(\vec{r})$, *where* $\vec{r} = (n_1,\ n_2)/\sqrt{n_1^2 + n_2^2}$, *computed for the system (5.12).*

$\alpha(\vec{r})$	0	1	2	3	4	5	6	7	8	$\cdots\ n_1$
0		−0.8								
1	−1.8	0.20	−0.27	0.20	0.06	0.20	0.13	0.20	0.16	
2		−0.55		−0.01		0.11		0.15		
3		0.20	−0.08		0.08	0.20		0.20	0.16	
4		−0.06		0.06		0.13		0.16		
5		0.20	0.05	0.20	0.11		0.15	0.20	0.17	
6		0.07				0.14		0.16		
7		0.20	0.11	0.20	0.14	0.20	0.16		0.17	
8		0.13		0.14		0.16		0.17		
$\vdots$										
n_2										

Notice that for five of these cases the set $\mathbb{V}(\vec{r})$ is depicted in Figure 5.1, subplots (a) to (e).

Finally, we look at the special case, where $\mathbb{W} \not\subset (\mathbb{C}_- \cup \{0\})$, but still lies in the *closed* LHP. Inspired by the proof of Lemma 2.2 in [106], and using an approximation and a continuation argument, we have the following result (see, for instance, [190] for a complete proof).

Lemma 5.16. *Assume that (5.1) has a characteristic root in* $\mathbb{C}_+$ *for some delay values. Then* $\mathbb{W} \cap \mathbb{C}_+ \neq \phi$.

Finally, the combination of Theorem 5.11, Proposition 5.14, and Lemma 5.16 results in the following theorem.

Theorem 5.17. *Assume that* $\mathbb{W} \cap \mathbb{C}_+ = \phi$ *and* $\mathbb{W} \not\subset (\mathbb{C}_- \cup \{0\})$. *Then there exist delay values for which (5.1) has characteristic roots on the imaginary axis, while there are no delay values for which (5.1) has characteristic roots in* $\mathbb{C}_+$. *Every stable ray is subjected to the delay-interference phenomenon.*

The main results of this section are summarized in Table 5.2. The cases $\mathbb{W} \cap \mathbb{C}_+ \neq \phi$ and $\mathbb{W} \subset (\mathbb{C}_- \cup \{0\})$ with $0 \notin \mathbb{W}$ are generic. The other cases, where the rightmost elements of $\mathbb{W}$ are on the imaginary axis, characterize situations where a system is on the edge of losing or acquiring delay-independent stability. Recall that this analysis has been performed under Assumption 5.5, which can be checked a priori (if the assumption is not satisfied, then there exist no values of $\vec{\tau}$ for which the zero solution of (5.1) is asymptotically stable).

Further remarks on delay-interference

To conclude, we briefly discuss some existing results related to the interference phenomenon. In [159] it is shown that delay-independent stability is equivalent to asymptotic stability for

Table 5.2. *Characterization of stability of the steady state solution of (5.1) along rays in the delay-parameter space, as a function of* $\mathbb{W}$ *and* $\mathbb{V}(\cdot)$.

$\mathbb{W} \subset (\mathbb{C}_- \cup \{0\})$	$\mathbb{W} \not\subset (\mathbb{C}_- \cup \{0\})$ and $\mathbb{W} \cap \mathbb{C}_+ = \phi$	$\mathbb{W} \cap \mathbb{C}_+ \neq \phi$
delay-independent stable:	not delay-independent stable ray $\mathcal{T}(\vec{r})$ stable if and only if $\mathbb{V}(\vec{r}) \subset (\mathbb{C}_- \cup \{0\})$	
$0 \notin \mathbb{W}$: strong	stable ray (if any) subjected to the delay-interference phenomenon	
$0 \in \mathbb{W}$: weak	characteristic roots in $\mathbb{C}_+$ not possible	ray $\mathcal{T}(\vec{r})$ unstable for $\vec{r}$ rat. independent

all delay values lying in a nontrivial sector in the delay-parameter space, and to the robustness of stability of a ray, consisting of commensurate delay values, w.r.t. small perturbations of the direction (see [159] for precise formulations). Note that the latter two statements imply the existence of a stable ray, which is not subjected to the interference phenomenon. Given a stable ray, the additional condition to have interference in [167] (delay-dependent stability when one of the (two) delays is set to zero) and in [225] (frequency-sweeping test) are in fact conditions for the presence of characteristic roots on the imaginary axis for some delay values, and thus conditions for *not* having delay-independent stability. In this way, these cited results are a direct corollary of the fact that a system is *either* delay-independent stable *or* every stable ray is subjected to the delay-interference phenomenon; see Table 5.2.

5.5 Illustrative examples

We present two examples, which together illustrate all results and phenomena described in the previous sections. Besides the delays, the examples will exhibit another parameter. This allows us to illustrate, in addition, two scenarios for the transition between delay-independent stability and delay-dependent stability: one via weak delay-independent stability, the other via the case $\mathbb{W} \not\subset (\mathbb{C}_- \cup \{0\})$, $\mathbb{W} \cap \mathbb{C}_+ = \phi$.

5.5.1 Interference in parameterized scalar delay systems

As a first example we consider

$$\dot{x}(t) = -ax(t) - x(t-\tau_1) - \frac{1}{2}x(t-\tau_2), \tag{5.19}$$

where $a \in \mathbb{R}$ is a parameter. This example represents a natural generalization of the illustrative examples presented in the previous sections of the chapters (that correspond to the cases $a = 1$ and $a = 1.3$, respectively).

We have

$$\mathbb{W} = \left\{ \lambda \in \mathbb{C} : \ \frac{1}{2} \leq |\lambda + a| \leq \frac{3}{2} \right\}.$$

According to Table 5.2 the zero solution of (5.19) is delay-independent stable if and only if $a \geq \frac{3}{2}$. For $a > \frac{3}{2}$ we have strong delay-independent stability, and for $a = \frac{3}{2}$ we have weak delay-independent stability. For $a < \frac{3}{2}$ we are directly in the case $\mathbb{W} \cap \mathbb{C}_+ \neq \phi$.

For $a = 1.3$ the system reduces to (5.12), for which the analysis in the previous section and Table 5.1 show that there are seven stable rays, characterized by $\vec{r} = \vec{n}/\|\vec{n}\|$, with $\vec{n}$ given by (5.18). Such an analysis can easily be repeated for other values of a, since a change of this parameter, say δa, only affects the sets $\mathbb{W}$ and $\mathbb{V}(\vec{r})$ by a shift along the real axis of $-\delta a$, while the numerical values of Table 5.1 change accordingly with $-\delta a$. For $a = 1$, there are only three stable rays left, characterized by

$$(n_1, n_2) \in \{(1,0),\ (0,1),\ (1,2)\}, \tag{5.20}$$

and for $a < -0.5$ all rays are unstable.

In Figure 5.2 we show the complete stability/instability regions of (5.19) in the delay-parameter space,[11] for $a = 1$ and $a = 1.3$. The solid lines correspond to delay values for which there are characteristic roots on the imaginary axis. The dashed lines indicate the stable rays. Notice that small perturbations of their slope lead to intersections with the solid curves, a consequence of the delay-interference phenomenon. As $a \to 1.5$, we have in fact a scenario toward delay-independent stability, characterized by an increase in the number of stable rays. That number becomes arbitrarily large when getting arbitrarily closed to the "bifurcation value" $a = 1.5$, for which we have weak delay-independent stability.

5.5.2 Delay rays and second-order delay systems

Next, we analyze the system

$$\dot{x}(t) = A_0 x(t) + A_1 x(t - \tau_1) + A_2 x(t - \tau_2), \tag{5.21}$$

where

$$A_0 = \begin{bmatrix} 0 & 1 \\ a - a^2 - \frac{5}{4} & 2a - 1 \end{bmatrix}, \quad A_1 = \begin{bmatrix} 0 & 0 \\ \frac{1}{5} & 0 \end{bmatrix}, \quad A_2 = \begin{bmatrix} 0 & 0 \\ -\frac{4}{5} & 0 \end{bmatrix}, \tag{5.22}$$

and $a \in \mathbb{R}$ is a parameter. The corresponding set $\mathbb{W}$ is depicted in Figure 5.3. A change of parameter a results once again in a shift of this set along the real axis.

For $a < 0$ the system is strongly delay-independent stable, and for $a > 0$; we have $\mathbb{W} \cap \mathbb{C}_+ \neq \phi$; see Table 5.2. For the intermediate value, $a = 0$, we are in the special case where the assumptions of Theorem 5.17 are satisfied.

In Figure 5.4, we plot the stability region in the (τ_1, τ_2)-space for $a = 1/16$ and $a = 0$. In the first case there are three stable rays, determined by $\vec{r} = \vec{n}/\|\vec{n}\|$, with

$$\vec{n} \in \{(1,0),\ (1,1),\ (4,1)\}.$$

These directions can also be obtained directly by constructing a table similar to Table 5.1. As $a \to 0+$, all closed curves in the (τ_1, τ_2)-space corresponding to characteristic roots

[11] The stability crossing curves were computed as Hopf bifurcation curves with the package DDE-BIFTOOL [74], thereby exploiting (frequency-dependent) invariance properties w.r.t. delay shifts. Other approaches are described in [98] and [279].

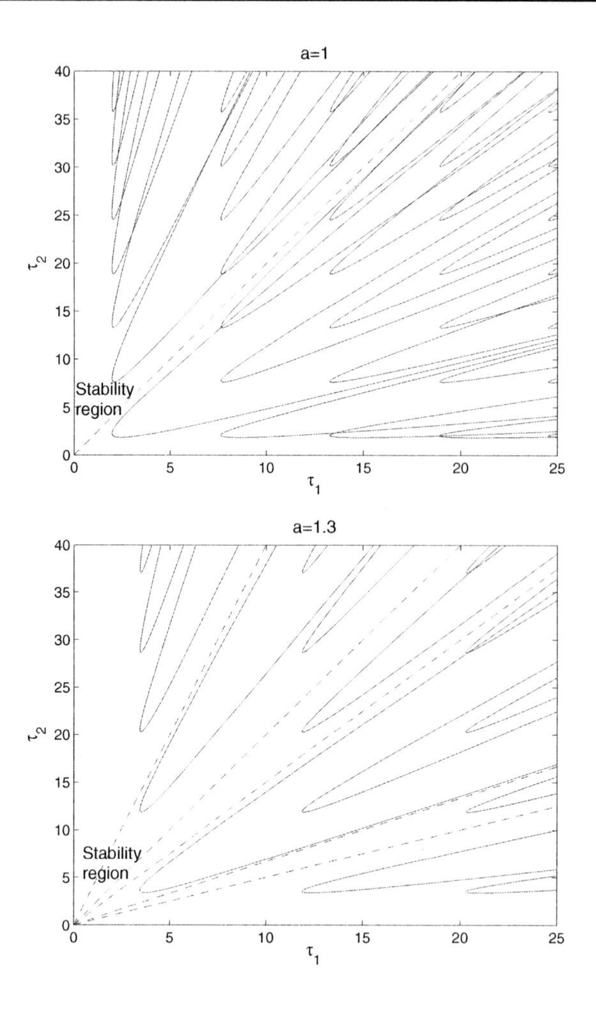

Figure 5.2. *Stability/instability regions of the zero solution of (5.19) in the (τ_1, τ_2)-space, for $a = 1$ (top) and $a = 1.3$ (bottom). For $a = 1$, the directions of the stable rays (dashed lines) are given by (5.20); for $a = 1.3$ by (5.18). For $a \geq 1.5$ the system is delay-independent stable.*

on the imaginary axis shrink, and at the limit $a = 0$, where Theorem 5.17 applies, they have collapsed to equally spaced points. For such delay values, which can be computed analytically as

$$(\tau_1, \tau_2) = \left(\frac{10\pi}{3\sqrt{3}} + k\frac{4\pi}{\sqrt{3}}, \frac{4\pi}{3\sqrt{3}} + l\frac{4\pi}{\sqrt{3}}\right), \quad k, l \in \mathbb{N}, \tag{5.23}$$

the system has characteristic roots on the imaginary axis (more precisely $\pm j\sqrt{3}/2$), but for all other delay values its zero solution is asymptotically stable. This is illustrated in

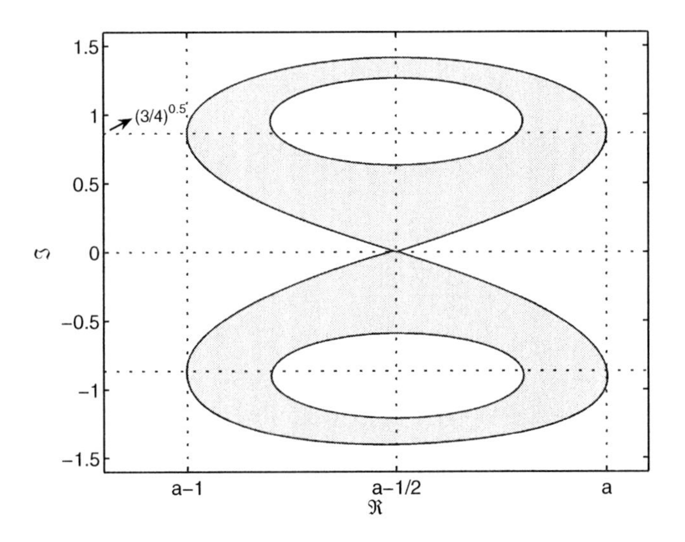

Figure 5.3. *The set $\mathbb{W}$, corresponding to the system (5.21)–(5.22), is shaded.*

Figure 5.5, where the real parts of the rightmost characteristic roots are displayed along a particular ray, containing some of the points (5.23).

Notice that for the limit case, $a = 0$, the number of unstable rays is infinite but *countable*, since the unstable rays have to contain at least one of the points (5.23). Hence, they are characterized by a slope,

$$\frac{r_2}{r_1} = \frac{2+6l}{5+6k}, \quad k, l \in \mathbb{N},$$

being rational numbers. This leads to a paradox: for $a = 0$ all rays $\mathcal{T}(\vec{r})$, with r_2/r_1 irrational, are stable, whereas for any $a > 0$, they are unstable (since $\mathbb{W} \cap \mathbb{C}_+ \neq \phi$ and Theorem 5.15 applies). The explanation is as follows: for all rationally independent $\vec{r}$ and every value of a, we have for the example

$$\alpha(\vec{r}) = \alpha_0,$$
$$\Re(\lambda) < \alpha_0, \quad \forall \lambda \in \mathbb{V}(\vec{r}).$$

Consequently, $\mathbb{V}(\vec{r})$ intersects the closed RHP if $\alpha_0 > 0$ but not if $\alpha_0 = 0$.

5.6 Notes and references

In this chapter, we proposed a complete characterization of delay-independent stability, stability of rays in the delay-parameter space, and the delay-interference phenomenon. The results were illustrated with numerical examples. In addition, scenarios for the transition between delay-dependent stability and delay-independent stability were shown. This chapter is based on the results proposed by the authors in [190].

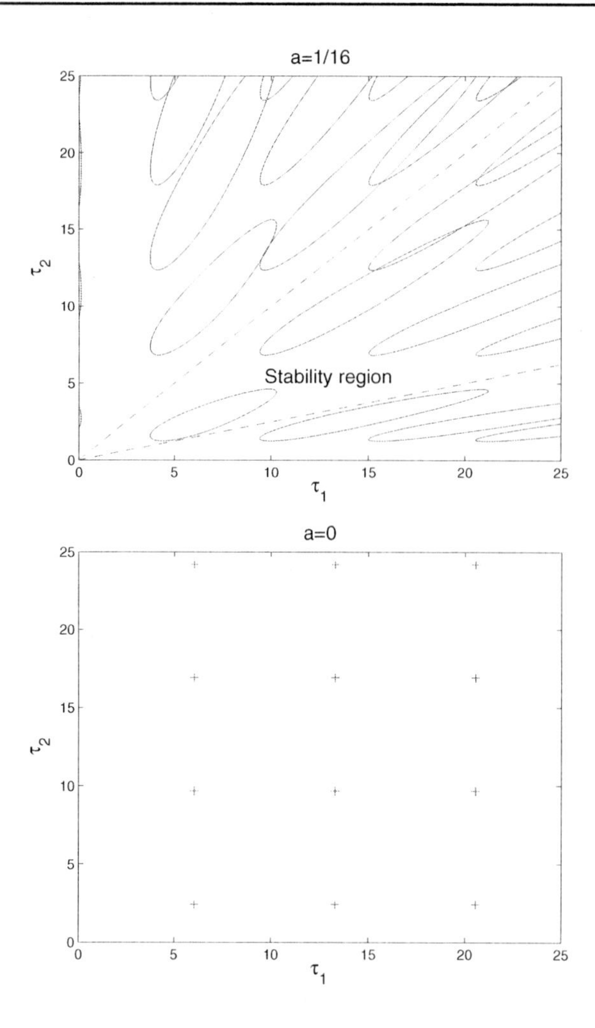

Figure 5.4. *Stability/instability regions of the zero solution of (5.21)–(5.22) in the* (τ_1, τ_2)*-space, for* $a = 1/16$ *(top) and* $a = 0$ *(bottom).*

The subject of delay-independent stability of the zero solution of (5.1), that is, asymptotic stability guaranteed for all delay values, received a lot of attention in the literature starting with the 1980s. Without being exhaustive, we cite some of the approaches proposed to handle the problem: two-variable criteria [131, 132], matrix pencil techniques [46, 222], pseudodelay techniques [241], frequency-sweeping tests [60, 47], and finite-dimensional LMI conditions derived by using some appropriate quadratic Lyapunov–Krasovskii functionals [17]. For further discussions and references, see [300, 96, 223, 70]. Essentially, in most of the existing literature for delay-independent stability and, more generally, in work on characterizing and computing stability regions in the delay-parameter space, the delays are *either* allowed to vary completely independently of each other, *or* they are restricted to

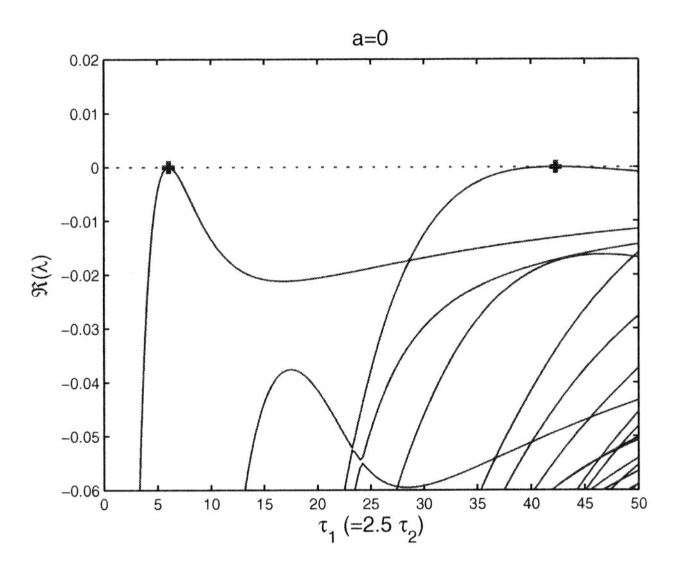

Figure 5.5. *Real parts of the rightmost characteristic roots of the system (5.21)–(5.22) with $a = 0$, for delay values along the ray $\mathcal{T}(5/\sqrt{29}, 2/\sqrt{29})$.*

be commensurate (multiples of the same number). Most of the works mentioned above did not pay a lot of attention to the delay ratio sensitivity and related problems.

As mentioned in the introduction of the chapter, to the best of our knowledge the notion of delay-interference was first mentioned by MacDonald [167] (see also [166]), where a particular second-order system including two delays encountered in biosciences is shown to be subjected to delay-interference if it has the following property: delay-independent stability if the delays are equal, and delay-dependent stability with respect to each delay if the other is equal to zero. A similar property holds for the scalar system including two delays considered as the introductory example in the chapter, and taken from [60]. This example was also discussed in [159] and [225], where some characterizations regarding interference are given (a frequency-sweeping test combined with a matrix pencil condition in [225], a sector characterization in [159]). As discussed at the end of Section 5.5, these results appear as corollaries of the general theory developed throughout the chapter.

It is worthwhile to mention that the stability results for rays in the delay-parameter space can easily be extended toward higher-dimensional subspaces. For instance, asymptotic stability for all delay values lying in a subspace spanned by $\vec{r}^{(1)} \in \mathcal{B}_+^m$ and $\vec{r}^{(2)} \in \mathcal{B}_+^m$ will be determined by the position of the set

$$\bigcup_{\theta_1, \theta_2 \geq 0} \sigma\left(A_0 + \sum_{i=1}^m A_i e^{-j\left(\theta_1 r_i^{(1)} + \theta_2 r_i^{(2)}\right)}\right) \subseteq \mathbb{W}.$$

Furthermore, the study of the sets $\mathbb{W}$ and $\mathbb{V}(\cdot)$ does not only lead to a powerful tool for characterizing delay-independent stability properties. Note for instance that the intersection of $\mathbb{W}$ with the imaginary axis contains all possible values where characteristic roots can cross the imaginary axis when the delays are varied. Actually, from these values and/or the

Table 5.3. *Conditions for the stability of the zero solution of (5.24). The set $\mathbb{W}$ is defined as in Definition 5.6, with A_0 set to zero.*

$\mathbb{W} \subset \{\lambda \in \mathbb{C} : \ \lvert\lambda\rvert < 1\}$	$\sup \{\lvert\lambda\rvert : \ \lambda \in \mathbb{W}\} > 1$
asympt. stable for all values of $\vec{r}$	not asympt. stable for all values of $\vec{r}$
	if $\vec{r}$ is stabilizing, then there exist infinitesimal perturbations on $\vec{r}$ which destroy stability $\equiv$ not strongly stable
	unstable for rat. independent $\vec{r}$

Table 5.4. *The similarity between the stability of the zero solution of (5.1) along rays in the delay space, and the stability of the zero solution of (5.24).*

Equation (5.1)	$\rightsquigarrow$	Equation (5.24)
relation of $\mathbb{W}$ w.r.t. imaginary axis	$\rightsquigarrow$	relation of $\mathbb{W}$ w.r.t. unit circle
direction $\vec{r}$ in delay space	$\rightsquigarrow$	nominal delay value $\vec{r}$
(un)stable ray $\mathcal{T}(\vec{r})$	$\rightsquigarrow$	(de)stabilizing delay value $\vec{r}$
stable ray subjected to the delay interference phenomenon	$\rightsquigarrow$	stability for nominal $\vec{r}$ destroyed by infinitesimal perturbations on $\vec{r}$

corresponding parameters, that is, $\vec{\theta}$ in Definition 5.6, detailed information can be retrieved about the geometry of the stability crossing curves/(hyper)surfaces in the delay-parameter space, thereby generalizing the results of [98].

Finally, we would like to point out the relation with the stability theory of continuous-time delay-difference equations. In this sense, we relate the results of the previous sections to stability conditions for continuous-time delay-difference equations of the form

$$x(t) - \sum_{i=1}^{m} A_i x(t - r_i) = 0, \tag{5.24}$$

which are discussed in, e.g., [106, 199, 181] and the references therein.

We assume that

$$\det\left(I - \sum_{i=1}^{m} A_i\right) \neq 0,$$

an assumption which prevents a characteristic root at zero for all $\vec{r} \in \mathbb{R}_{+}^{m}$. Then the stability conditions for the zero solution of (5.24) from [106, Theorem 2.2 and Corollary 2.2] can be rephrased as displayed in Table 5.3.

A comparison between Tables 5.2 and 5.3 reveals a strong correspondence between the stability of the zero solution of (5.1) along rays in the delay space, and the stability of the zero solution of (5.24). This correspondence is described in Table 5.4.

Chapter 6

Stability of linear periodic systems with delays

6.1 Introduction

We discuss mathematical and computational tools for the stability analysis of linear time-varying delay systems of the form

$$\dot{x}(t) = A(\omega t)\, x(t) + B(\omega t)\, x(t - \tau(t)), \tag{6.1}$$

$$\tau(t) = \tau_0 + \delta f(\Omega t), \tag{6.2}$$

under appropriate initial conditions. We assume that $x \in \mathbb{R}^n$, $A : \mathbb{R} \to \mathbb{R}^{n\times n}$, and $B : \mathbb{R} \to \mathbb{R}^{n\times n}$ are bounded periodic functions with period 2π, and that $f : \mathbb{R} \to [-1, 1]$ is a periodic function with zero mean and period 2π, $\max f = 1$ and $\min f = -1$. Further, $\delta, \tau_0, \omega, \Omega$ are strictly positive numbers satisfying $\delta \leq \tau_0$. In this way, the parameters δ and Ω determine the amplitude and frequency of the delay variation, while ω determines the frequency of the variation of A and B. We do not a priori require that ω and Ω are correlated, although such a correlation exists in many applications.

The problem under consideration is inspired by applications in mechanical engineering. More precisely, in the manufacturing literature one encounters models of the form (6.1) for the dynamics of rotating cutting and milling machines; see, e.g., [128, 283] and the references therein. In cutting machines, a workpiece rotates and the cutting inserts have a fixed position, while in milling machines the workpiece is fixed and the cutting inserts are mounted on a rotating axis. In both cases the time-delay represents the time taken for one revolution of the workpiece or cutting inserts. Therefore, it is proportional to the inverse of the rotational speed of the machine. In models for cutting machines the system matrices are typically constant, whereas they are periodically varying in models for milling machines, due to the varying angle between the cutting inserts and the workpiece. The nominal behavior of the machines corresponds to an asymptotically stable steady state solution. A loss of stability is undesired as it leads to chatter, that is, unwanted oscillations which cause irregularities in the surface of the workpiece to be processed. A typical approach used to enlarge the stability region of the steady state solution of such rotating machines in a relevant parameter space consists of *fast modulating* the speed around the nominal value [128, 269]. Such a modulation of the speed precisely corresponds to a modulation of the time-delay in the model (6.1) of the form (6.2). Note that an analysis of this stabilization approach calls for mathematical tools which are also able to characterize situations where a variation of a delay has a stabilizing effect.

If the time-variation of a delay is fast compared to the system's dynamics, then its distribution rather than its precise dependence upon time determines the stability properties of the system, as we shall see in Section 6.2. This makes some of the described results directly applicable to the emerging field of network controlled systems, since the varying delays in communication networks are typically of a stochastic nature, yet knowledge is available about their distribution; see, for instance, [260, 292] and the references therein.

This chapter is devoted to a presentation of *eigenvalue based techniques* for the stability analysis of the system (6.1)–(6.2). As an advantage w.r.t. most time-domain approaches, they lead to nonconservative results, in the sense that exact stability information is available (in terms of eigenvalues of appropriate operators), instead of sufficient stability conditions. Furthermore, the combination of detection of critical eigenvalues and a continuation facility allows us to compute the boundaries of stability regions in parameter spaces in a (numerically tractable) efficient and semiautomatic way. For systems with time varying delay an additional advantage lies in the fact that stabilizing effects of a delay variation can be investigated, which is not possible with approaches where a variation of the delay around a nominal value is explicitly or implicitly treated as uncertainty. Notice that time integration (simulation), see [12] for an overview of methods, can in principle also be used to determine stability of (6.1)–(6.2) and stability regions in parameter spaces, yet it is time consuming and boundaries of stability regions are hard to determine accurately.

The structure of the chapter is as follows: section 6.2 is devoted to the case where the time-variation of the periodic terms is fast compared to the system's dynamics, while Section 6.3 deals with the general case. Two practical examples are presented in section 6.4. Some notes and references complete the chapter.

6.2 Systems with fast varying coefficients

Using averaging techniques, first we show how the stability analysis problem of the system (6.1)–(6.2) for large values of ω and Ω can be reduced to the stability analysis of a *time-invariant* system exhibiting distributed delays. Next, computational and analytical tools for the averaged system are briefly discussed.

6.2.1 Averaging periodic systems

As an indication that the system (6.1)–(6.2) with parameters ω and Ω is suitable for averaging, observe that the upper bound in the estimate

$$\|\dot{x}(t)\| \leq \left(\max_{s} \|A(s)\| + \|B(s)\|\right) \max_{s\in[t-\tau_0-\delta,\ t]} \|x(s)\|$$

does not depend on the values of ω and Ω. This suggests that, on compact time-intervals, the trajectories of the system have a limit as ω, $\Omega \to \infty$. In fact, due to the filtering property of the integration process, the parameters ω and Ω regulate the separation between the time-scale associated with the periodically varying coefficients and the time-scale associated with the long-term behavior of the solutions, which forms the backbone of the averaging approach.

Averaging methods for periodic systems described by ODEs with a fast varying right-hand side are discussed in [265] and extended to DDEs with constant delays in [148, 104, 105]. Averaging methods for time-varying delays are treated in [197]. Their combination leads to the following result, which slightly generalizes [197, Theorem 1].

Theorem 6.1. *Consider the system (6.1) and (6.2). Let the integrable function*[12] w : $[-1,\ 1] \to \mathbb{R}_+$ *be defined by the relation*

$$\int_{-1}^{1} \alpha(t)w(t)dt = \frac{1}{2\pi}\int_{0}^{2\pi} \alpha(f(t))dt, \quad \forall \alpha \in \mathcal{C}([-1,\ 1],\ \mathbb{R}), \tag{6.3}$$

and let

$$\bar{A} = \frac{1}{2\pi}\int_{t}^{t+2\pi} A(s)\, ds, \ \bar{B} = \frac{1}{2\pi}\int_{t}^{t+2\pi} B(s)\, ds. \tag{6.4}$$

If the averaged system

$$\dot{x}(t) = \bar{A}x(t) + \bar{B}\int_{t-\tau_0-\delta}^{t-\tau_0+\delta} \frac{w((t-\tau_0-\theta)/\delta)}{\delta} x(\theta)d\theta \tag{6.5}$$

is asymptotically stable, then there exists a threshold ω_c such that the system (6.1) and (6.2) is globally uniformly asymptotically stable for all $\Omega > \omega_c$ and $\omega > \omega_c$.

Sketch of the proof. The existence of an integrable function w satisfying (6.3) follows from a change of measure and the Radon–Nikodym theorem (see [263]).

The proof of the stability assertion is based on an application of the *trajectory based proof technique*, developed in [211] for the stability analysis of ODEs and extended and applied to classes of DDEs in [212]. It relates closeness results for trajectories (in the sense of uniform convergence of trajectories on compact time-intervals) with stability results (which involve the behavior of trajectories on infinite time-intervals). The main steps are as follows:

1. One proves that trajectories of (6.1)–(6.2) and (6.5) with matching initial conditions uniformly converge on each other at *compact time-intervals* as the parameters ω and Ω tend to infinity. This is done by estimating the deviation between the solutions of (6.1)–(6.2) and (6.5) at time-instants later than the initial time, and involves the application of a generalization of the celebrated Gronwall Lemma.

2. This closeness result of trajectories of (6.5) and (6.1)–(6.2) is linked with stability assertions. By a slight generalization of [211, Theorem 1], one can conclude from the exponential stability of (6.5) and the closeness result that the null solution of system (6.1)–(6.2) is practically uniformly asymptotically stable (see [197] for a precise definition).

3. Practical uniform asymptotic stability of the null solution of (6.1)–(6.2) implies global uniform asymptotic stability by a scaling property of its solutions.

[12] More precisely, w represents a positive density measure.

Remark 6.2. *Under the conditions of the theorem the asymptotic stability of (6.1)–(6.2) is only guaranteed if Ω and ω are sufficiently large. An explicit bound ω_c may be obtained from theoretical considerations, see the discussion in [210], but such bounds are typically conservative. Therefore, it is advisable to determine a threshold based on numerical simulation (if desired), and to switch to the methods described in the next section if there are indications that the separation of time-scales may not be sufficient.*

Remark 6.3. *From the definition (6.3) the weight function w can be interpreted as the probability distribution of $f(\zeta)$, where ζ is uniformly distributed over the interval $[0,\ 2\pi]$. This interpretation is often very useful for computing the function w out of f and offers an alternative for using the definition (6.3) directly. It also lays the basis for an extension of the theorem to classes of systems where the time-delay is a random variable with a known probability density function.*

The importance of a reduction to a time-invariant system lies in the fact that frequency-domain techniques become applicable. For instance, the stability of the averaged system (6.5) is determined by the rightmost roots of the characteristic equation, which can be written in the form

$$\det\left(\lambda I - \bar{A} - \bar{B}e^{-\lambda\tau_0}g(\lambda\delta)\right) = 0, \tag{6.6}$$

where

$$g(\lambda) = \int_{-1}^{1} e^{-\lambda t}w(t)dt. \tag{6.7}$$

Notice that the $g(\lambda)$ can be interpreted as a correction of $e^{-\lambda\tau_0}$, corresponding to the mean delay value τ_0, and takes all the effects of the delay *variation* into account. As an illustration, several tuples (f, w, g), characterizing the varying part of the delay, are displayed in Table 6.1.

6.2.2 Computational tools

We give an overview of tools to compute the rightmost characteristic roots of integro-differential equations of retarded type, that is, DDEs of retarded type that contain terms of the form

$$\int_{\tau_1}^{\tau_2} K(\theta)x(t-\theta)\,\mathrm{d}\theta, \tag{6.8}$$

with measurable *kernel function* $K(\cdot) \in \mathbb{R}^{n\times n}$ and $\tau_1 \leq \tau_2$.

First, we consider the trivial case where the kernel is a constant matrix times the *Dirac impulse*, i.e., $K(\theta) \equiv Kh(\theta - \tau_0)$ with $\tau_1 \leq \tau_0 \leq \tau_2$. Then, the integral (6.8) equals $K\,x(t-\tau_0)$ and the integro-differential equation reduces to an equation with a pointwise delay. Various methods for determining the rightmost characteristic roots of systems with pointwise delays have been proposed in the literature. These include, but are not limited to, methods based on discretizing the solution operator associated with the equation (see, for instance, [77]) and methods based on discretizing its infinitesimal generator [25]. The

Table 6.1. *For three examples of f in (6.2), the corresponding weight function w of the distributed delay comparison system (6.5), as well as the correction term $g(s)$ in the characteristic equation (6.6) are shown. $J_0(.)$ denotes the Bessel function of the first kind of order zero, $h(.)$ is the Dirac impulse function. Notice that w can be seen as the probability density function of the image of f.*

$f(t)$	$w(t)$	$g(\lambda)$
$\begin{cases} \frac{2}{\pi}(t-\frac{\pi}{2}), & t \in [0,\ \pi), \\ \frac{2}{\pi}(\frac{3\pi}{2}-t), & t \in [\pi,\ 2\pi) \end{cases}$ (sawtooth)	$\frac{1}{2}$	$\begin{cases} \frac{\sinh\lambda}{\lambda}, & \lambda \neq 0 \\ 1, & \lambda = 0 \end{cases}$
$\sin(t)$	$\frac{1}{\pi\sqrt{1-t^2}}$	$J_0(j\lambda)$
$\begin{cases} 1, & t \in [0,\ \pi), \\ -1, & t \in [\pi,\ 2\pi) \end{cases}$ (square wave)	$\frac{h(t-1)+h(t+1)}{2}$	$\cosh\lambda$

stability routine for equilibria, contained in the software package DDE-BIFTOOL [78, 74], is based on the former approach.

Integro-differential equations with a constant kernel or a *gamma distribution* kernel, that is,

$$K(\theta) = K\theta^j e^{-\alpha\theta}, \tag{6.9}$$

where j is a positive integer, are treated in [162], which shows how a bifurcation analysis of this type of equation can be done using computational tools for pointwise delay equations. In [161, 163] the numerical stability analysis of scalar equations with more general bounded kernels is discussed. The equations are discretized using a linear multistep method and a quadrature method, based on Lagrange interpolation and a Gauss–Legendre quadrature rule.

Finally, we consider an integro-differential equation where the term (6.8) takes the form

$$K \int_{\tau_1}^{\tau_2} \frac{1}{\sqrt{1-(\theta-\tau_m)^2/\delta^2}} x(t-\theta)\, d\theta, \tag{6.10}$$

with $\tau_m := (\tau_1 + \tau_2)/2$ and $\delta := (\tau_2 - \tau_1)/2$. Note that the kernel function, which appears in the second row of Table 6.1, goes to infinity at the boundaries of the integration interval. A natural way to handle this problem is to expand the (perturbed) trajectories in terms of the polynomials that are orthogonal w.r.t. the weighted $\mathcal{L}_2$ inner product $\langle p_1, p_2\rangle := \int w_2 p_1 p_2$. In this way one arrives at the well-known *Chebyshev polynomials*, used frequently in spectral collocation methods [302].

6.2.3 Analytical tools

Determining analytically the sensitivity of characteristic roots with respect to parameters is a useful tool for characterizing stabilizability properties. Following this approach, the effects of constant delays on stability have been thoroughly analyzed in the literature; see, e.g., [53, 230, 223, 191]. Here, we examine the effects of a fast *variation* of parameters around a nominal value.

As follows from (6.4) and an application of the trajectory based proof technique, an exponentially unstable system of the form (6.1)–(6.2), with constant A, B, and τ, cannot be stabilized by introducing a fast modulation of the matrices A and B, as long as their mean value remains the same. Therefore, we restrict ourselves to the stabilizability by means of a fast delay variation. By virtue of Theorem 6.1, the latter is implied by the stabilizability by means of a delay "distribution." In this way, we arrive at investigating how the characteristic roots of (6.5) or, equivalently, the zeros of

$$H(\lambda, g(\lambda\delta)) := \det\left(\lambda I - \bar{A} - \bar{B}e^{-\lambda\tau_0} g(\lambda\delta)\right), \tag{6.11}$$

behave for small values of $\delta \geq 0$. For $\delta = 0$, this expression simplifies to

$$\det\left(\lambda I - \bar{A} - \bar{B}e^{-\lambda\tau_0}\right), \tag{6.12}$$

the characteristic quasipolynomial of the constant delay system.

Because g is a smooth function, the zeros of (6.11) are continuous at each value of parameter $\delta \geq 0$. For a given zero λ_0 of (6.12) with multiplicity one, this implies the existence of a root function $r(\delta)$ of (6.11), satisfying $r(0) = \lambda_0$ and

$$H(r(\delta), g(r(\delta)\delta)) = 0. \tag{6.13}$$

To compute the sensitivity of the zero λ_0 w.r.t. δ, we differentiate (6.13), leading to

$$r'(0) = 0, \tag{6.14}$$

$$r''(0) = -\frac{\frac{\partial H}{\partial g}(\lambda_0, 1)}{\frac{\partial H}{\partial \lambda}(\lambda_0, 1)}\, g''(0)\lambda_0^2, \tag{6.15}$$

where $g''(0) = \int_{-1}^{1} t^2 w(t)dt > 0$. A stability related corollary is the following.

Proposition 6.4. *Assume that the rightmost characteristic roots of (6.12) are simple and on the imaginary axis. Denote them by $j\omega_i$, $i = 1, m$. If*

$$\Re\left(\frac{\frac{\partial H}{\partial g}(j\omega_i, 1)}{\frac{\partial H}{\partial s}(j\omega_i, 1)}\right) < 0, \quad i = 1, \dots, m, \tag{6.16}$$

then the system (6.5) is asymptotically stable for small values of δ.

In Section 6.4 we will discuss a *parameterized* system, for which condition (6.16) is always satisfied[13] in cases of characteristic roots on the imaginary axis. This means that stability regions in the parameter space become *larger* when increasing δ from zero. Indeed, internal points of a stability region in the parameter space correspond to *asymptotic* stability of the system, which is preserved by increasing δ from zero (continuity argument), whereas points on the boundaries of a stability region, if any, correspond to a system having its

[13] Such a situation is likely to occur, as explained in [197].

rightmost characteristic roots on the imaginary axis, which becomes asymptotically stable under the conditions of Proposition 6.4. Notice that parameter values corresponding to a *zero* characteristic root are invariant w.r.t. δ, following from $g(0) = 1$.

Compared to the numerical tools described in the Section 6.2.2, which allow us to compute stability information of (6.5) for all values of the parameters, analytical tools such as the ones described above are less powerful in the sense that they are suitable for asymptotic results only (e.g., stability assertions for small δ in Proposition 6.4). However, they are more powerful in the sense that they allow us to make assertions about stabilizability by means of a delay variation, *independently* of the precise way in which the delay is varied (notice that the condition (6.16) is independent of the choice of g and, thus, of f). This illustrates the *complementarity* of analytical and numerical methods.

6.3 General case

In this section, no assumptions are made on the size of the frequencies ω and Ω in system (6.1)–(6.2). First, the *collocation* scheme for computing solutions of (6.1)–(6.2) is outlined. Next, the computation of stability determining characteristic roots corresponding to the zero solution is described, as well as the computation of the boundaries of stability regions. Then a special case is commented on, where a reduction to a time-invariant system is still possible. Finally, a comparison with the averaging based approach of the previous section is made.

For technical reasons it is assumed throughout the section that ω and Ω are rationally dependent, and that the functions f, A, and B in (6.1)–(6.2) are smooth.

6.3.1 Collocation scheme

By means of the system (6.1)–(6.2) we explain the collocation scheme to compute a trajectory of a periodic system with delays for a given initial condition. This collocation variant is based on [312, 311] (see also [73]).

Denote by T the least common multiple of $2\pi/\omega$ and $2\pi/\Omega$. Hence there exist integers k and K so that $T\omega = 2\pi k$ and $T\Omega = 2\pi K$. For simplicity, we call T the *period* of the system (6.1)–(6.2). Instead of solving the latter system, we rescale the time variable by $1/T$ such that the period is one in the transformed time. The transformed system is

$$\dot{x}(t) = TA(2\pi kt)x(t) + TB(2\pi kt)x(t - \sigma(t)), \tag{6.17}$$

$$\sigma(t) = \frac{\tau_0}{T} + \frac{\delta}{T} f(2\pi Kt). \tag{6.18}$$

We represent a trajectory $x(t)$ of (6.17)–(6.18) by a continuous piecewise polynomial, or *spline*. The mesh used for this *discrete* approximation is constructed as follows. First, let $\{0 = t_0 < t_1 < \cdots < t_m = 1\}$ be a mesh on $[0, 1]$ with m mesh intervals. Next, this mesh is periodically extended to the left to obtain a mesh on $[t_{-\ell}, 1]$ with $\ell + m$ intervals. For notational convenience, we use the variable $x(t)$ also for the piecewise polynomial approximation on this extended mesh. Since it is desirable that the trajectory $x(t)$ for $t > 0$ can be obtained by time-stepping if $x(t)$ on the initial interval $[t_{-\ell}, 0]$ is given, we choose ℓ such that $t_{-\ell} \leq \sigma_{\min} < t_{-\ell+1}$, where $\sigma_{\min}$ is the minimal value of $t - \sigma(t)$ for $t \in [0, 1]$.

Note that the discrete approximation has $n((\ell + m)d + 1)$ degrees of freedom, where d is the dimension of the spline.

For ease of implementation, the piecewise polynomial is represented as a linear combination of basis functions $\phi_{i+\frac{j}{d}}(t)$ that are only "locally" nonzero. More specifically, the restriction of the unknown trajectory to the interval $[t_i, t_{i+1}]$ can be written as

$$\sum_{j=0}^{d} c_{i+\frac{j}{d}} \phi_{i+\frac{j}{d}}(t), \tag{6.19}$$

with unknown coefficients $c_{i+\frac{j}{d}} \in \mathbb{R}^{n\times 1}$.

Next, we write down the conditions that determine the discrete approximation. First, *collocation requirements* are imposed, i.e., (6.17)–(6.18) have to be satisfied in a number of points. That is,

$$\dot{x}(c_{i,\nu}) = TA(2\pi k c_{i,\nu})x(t) + TB(2\pi k c_{i,\nu})x(c_{i,\nu} - \sigma(c_{i,\nu})), \tag{6.20}$$

$$\sigma(c_{i,\nu}) = \frac{\tau_0}{T} + \frac{\delta}{T} f(2\pi K c_{i,\nu}). \tag{6.21}$$

The so-called *collocation points*,

$$c_{i,\nu} := t_i + \Delta t_i z_\nu, \quad \text{for } i = 0, \ldots, m-1 \text{ and } \nu = 1, \ldots d, \tag{6.22}$$

where $\Delta t_i := t_{i+1} - t_i$, are obtained by scaling and shifting the set of *collocation parameters* $z_\nu \in [0, 1]$. In the case of *Gauss–Legendre collocation*, one chooses z_ν as the zeros of the *Legendre polynomial* of degree d on $[0, 1]$; see, e.g., the bifurcation packages AUTO [69] (for ODEs) and DDE-BIFTOOL [74, 78] (for DDEs). In our implementation, on the contrary, we choose the collocation parameters to be the zeros of the *Radau polynomial* of degree d, which is defined as the difference between the Legendre polynomials of degree d and degree $d - 1$. By construction, the Radau polynomial vanishes at the right end point, so that $z_d = 1$ is a collocation parameter. Consequently, (6.17)–(6.18) are satisfied at the mesh points t_i. Secondly, besides the md collocation requirements (6.20)–(6.21), $n(\ell d + 1)$ additional requirements are imposed. These can be used to specify the initial condition; see Sections 6.3.2 and 6.3.3.

The resulting nonlinear system of equations can be solved by using Newton iterations. In each iteration, a linearized system is solved. A typical structure for this matrix is shown in [312, Figure 1]. In general, the Newton iterations converge if the initial guess is "close enough" to the exact solution.

6.3.2 Computation of stability determining eigenvalues

A complex number λ is called an eigenvalue, or *(characteristic) root*, of (6.17)–(6.18) if and only if there exists a "perturbation" of the zero solution of the form

$$x(t) = e^{\lambda t} p(t), \tag{6.23}$$

where $p(t)$ is 1-periodic. This particular solution satisfies $x(t + 1) = e^{\lambda} x(t)$ for all t, where e^{λ} is called a *(characteristic) multiplier*. It follows that the multipliers are the eigenvalues of the linear map between the restriction of a trajectory $x(t)$ on the time intervals $[t_{-\ell}, 0]$ and $[t_{-\ell} + 1, 1]$. This map is called the *monodromy operator* if the system (6.17)–(6.18) is considered *before* discretization. *After* discretization, the map becomes finite-dimensional and is called the *monodromy matrix*.

The monodromy matrix of (6.17)–(6.18) is readily obtained from the collocation requirements (6.20)–(6.21). Indeed, the $\ell d + 1$ remaining degrees of freedom can be used to specify the solution on $[t_{-\ell}, 0]$. By solving the resulting system of equations, the corresponding solution on $[t_{-\ell} + 1, 1]$ is recovered.

The eigenvalues of the monodromy matrix can be computed by the QR algorithm, a reliable procedure that returns all eigenvalues of a matrix. Mostly, only the rightmost ones are good approximations for the exact eigenvalues of the system (but precisely these eigenvalues determine the stability of the system (6.1)–(6.2) before discretization). This is explained in [314, 312] and the references therein. The main idea is as follows. If the discretization is refined, then the approximate eigenvalues with *smallest modulus* first converge to the corresponding eigenvalues of the system, which is intuitively expected from the fact that the corresponding solutions of the form (6.23), with $|\lambda|$ small, have a nonsteep profile (for comparison, note also that in the frequency domain the error due to some type of discretization, for instance $|\lambda - (1 - e^{-\lambda h})/h|$ for a finite difference approximation of a derivative, is typically an unbounded function of $|\lambda|$, even if the discretization stepsize h is arbitrarily small). Next, because the set of eigenvalues of (6.1)–(6.2) has the same qualitative properties as the set of characteristic roots of delay equations of retarded type with constant delays, described in Chapter 1, the eigenvalues with the smallest modulus typically *correspond* to the *rightmost* eigenvalues.

If the monodromy matrix is "large," then it is much more efficient to use large-scale eigenvalue solvers, such as the *Arnoldi method* or the *Jacobi–Davidson method*, instead of the QR algorithm; see [6]. These methods rely on matrix-vector products only, which can be efficiently computed, and return only a small number of stability-determining eigenvalues. The monodromy matrix for the examples in Section 6.4 is of moderate size, so that the QR method could be used. The choice of the large-scale methods to be considered depends on different factors (e.g., the structure and spectrum of the matrix). Finally, note that other numerical methods to compute the eigenvalues are described in, e.g., [121, 37].

6.3.3 Computation of stability regions

When parameters are changes, a loss of stability of the zero solution of (6.17)–(6.18) is associated with an eigenvalue λ that crosses the imaginary axis. In that case, $|e^{\lambda}| = 1$, i.e., $\lambda = i\vartheta$ with $\vartheta \in \mathbb{R}$. Two cases can be distinguished:

- $|e^{i\vartheta}| = 1$ and ϑ/π is integer,
- $|e^{i\vartheta}| = 1$ but ϑ/π is *not* integer.

The latter case, which generically occurs, corresponds to a *torus bifurcation point*. This section discusses the computation of curves of torus bifurcation points in a parameter space. By using this procedure, stability regions can be traced efficiently and in a semiautomatic way.

First, we treat the computation of one torus bifurcation point, which requires that one model parameter, say η, is freed. The unknowns are the bifurcation value of η and the trajectory $x(t) := p(t)\exp(i\vartheta t)$ corresponding to the multiplier on the unit circle; see (6.23). Note that $p(t)$ is complex-valued, since the eigenvalue λ is complex. Hence $p(t)$ has $2n((\ell+m)d+1)$ real degrees of freedom and the total number of degrees of freedom is $2n((\ell+m)d+1)+2$. There are nmd complex (or $2nmd$ real) collocation requirements as per (6.20)–(6.21). Additionally, we impose $n(\ell d+1)$ complex (or $2n(\ell d+1)$ real) requirements that specify the boundary value problem, namely requirements that express the periodicity of $p(t)$. Finally, the *normalization* of the unknown $x(t)$ gives one complex requirement (or two real ones). Typically, one requires that the (complex) inner product of $x(t)$ and a given $c(t) \approx x(t)$ equals one. In summary, the total number of (real) requirements and unknowns are both $2n((\ell+m)d+1)+2$ and Newton's method can be used to solve this system of equations.

Next, we make free a second parameter, say ζ, and outline the continuation of a curve of torus bifurcations in the two-parameter space (η, ζ). To compute a new point of the curve, an extra condition has to be added to the defining system of equations in order to ensure uniqueness, and a starting value for the Newton iterations needs to be generated. A good starting value $(\hat{x}(t), \hat{\vartheta}, \hat{\eta}, \hat{\zeta})$ can be computed using the previous points on the curve. The easiest procedure to do so, the *secant predictor*, constructs an initial guess along the direction defined by the two previous points on the curve. Since it is desirable that the continuation procedure does not fail if the curve is not single-valued w.r.t. parameter η or ζ, e.g., a closed curve, we add a *pseudoarclength* steplength condition [270] to the equations to specify completely the new point search for

$$w_\vartheta(\vartheta-\hat{\vartheta})\dot{\vartheta} + w_p\begin{bmatrix} \eta-\hat{\eta} & \zeta-\hat{\zeta} \end{bmatrix}\begin{bmatrix} \dot{\eta} \\ \dot{\zeta} \end{bmatrix} = 0, \tag{6.24}$$

where $(\dot{\vartheta}, \dot{\eta}, \dot{\zeta})$ is the direction of the secant predictor. Here the *weights* w_ϑ and w_p are usually chosen to be one.

6.3.4 Special cases

If the time-dependence of $A(\cdot)$, $B(\cdot)$, and $\tau(\cdot)$ can be replaced by a dependence on a *finite* number of harmonics,

$$c_j := \cos\left(\frac{2\pi j}{T}t\right) \quad \text{and} \quad s_j := \sin\left(\frac{2\pi j}{T}t\right), \quad j = 1, \dots l, \tag{6.25}$$

then the system (6.1)–(6.2) can be transformed into a *time-invariant* system with state-dependent delay [305]. The latter takes the form (with a slight abuse of notation)

$$\begin{aligned} \dot{x}(t) &= A(c_1, s_1)\,x(t) + B(c_1, s_1)\,x(t-\tau(c_1, s_1)), \\ \dot{c}_1(t) &= -\tfrac{2\pi j}{T}s_1(t) + \gamma c_1(t)\left(1 - c_1(t)^2 - s_1(t)^2\right), \\ \dot{s}_1(t) &= \tfrac{2\pi j}{T}jc_1(t) + \gamma s_1(t)\left(1 - c_1(t)^2 - s_1(t)^2\right), \end{aligned} \tag{6.26}$$

where $\gamma > 0$ and (c_j, s_j), $j > 1$, are expressed as a function of c_1 and c_2 by means of trigonometric formulae.[14] In this way the stability analysis of the zero solution of (6.1)–(6.2) is reduced to the local stability analysis of the periodic solution

$$(x(t), c_1(t), s_1(t)) = \left(0, \cos\left(\frac{2\pi}{T}t\right), \sin\left(\frac{2\pi}{T}t\right)\right)$$

of (6.26), which can be performed directly with the package DDE-BIFTOOL [74]. Notice that stability is not affected by the transformation as the (decoupled) second and third equation of (6.26) generate a *stable* periodic solution, which corresponds to (6.25).

6.3.5 Comparison with the averaging based approach

The main advantages of the approach outlined in this section are its general applicability and the computation of exact stability information. No underlying assumptions on parameters need be taken, in contrast to the averaging based results of Section 6.2, which are only valid for a sufficiently high frequency of variation of the periodic coefficients. Notice that in general a threshold on this frequency depends on other systems parameters. As a consequence, the validity of the averaged model should always be questioned if parameters are changed (e.g., when computing stability regions in parameter spaces out of the averaged model).

On the other hand, since the averaging approach reduces the stability analysis problem to that of a steady state solution of a time-invariant system, the computational cost of the determination of stability information and the detection and continuation of bifurcations is reduced a lot, especially if a discretization of distributed delays in stability computations can be avoided (see Section 6.2.2). Furthermore, the classical frequency-domain controller design methods and the analytical approaches based on the sensitivity of the characteristic roots can easily be adapted to systems with a fast varying time-delay, because all the information about the varying part of a delay can be concentrated in a correction of the exponential term corresponding the mean delay value (see expression (6.6)).

6.4 Illustrative examples

The results presented in this chapter are applied to two examples from mechanical engineering. An example with a periodic delay function is discussed in Section 6.4.1, and an example with periodic system coefficients is in Section 6.4.2.

6.4.1 Variable spindle speed cutting machine

The equation

$$\ddot{x}(t) + 2\xi\omega_n\dot{x}(t) + \omega_n^2 x(t) = \frac{k}{m}\left(x(t-\tau(t)) - x(t)\right), \quad x \in \mathbb{R}, \tag{6.27}$$

[14]It is necessary to include these explicit links between all harmonics, since otherwise an arbitrary phase difference would occur.

taken from [128], models one mode of a mechanical rotational cutting process, where x represents the deflection of the machine tool and/or workpiece, ω_n the natural frequency, ξ the damping ratio, and m the modal mass. The term $k(x(t-\tau(t))-x(t))$, with $k>0$ the cutting force coefficient, models the cutting force, which depends on the time τ, taken by the cutting insert for one revolution of the workpiece. Clearly, the time-delay is inversely proportional to the rotational speed of the machine. In [128] one assumes that this speed is varied around a nominal value in a periodic way, which corresponds to a modulation of the time-delay in (6.27) of the form

$$\tau(t)=\tau_0+\delta f(\Omega t). \tag{6.28}$$

Based on the theory developed in the chapter and the model (6.27) we now give an explanation for the beneficial effect of a high-frequency modulating of the machine speed on increasing stability regions.

The characteristic equation of the averaged system is given by

$$H(\lambda,g(\lambda)):=\lambda^2+2\xi\omega_n\lambda+\omega_n^2+\frac{k}{m}(1-e^{-\lambda\tau}g(\lambda\delta))=0.$$

Applying the sensitivity formula (6.15) to an imaginary root $j\omega$ yields

$$\Re\left(r''(0)^{-1}\right)=-\frac{1}{g''(0)\,\omega^2}\left(\tau+2\xi\omega_n^2\frac{\omega^2+(\omega_n-k/m)^2}{(\omega^2-(\omega_n^2+k/m))^2+4\xi^2\omega_n^2\omega^2}\right), \tag{6.29}$$

which is *strictly negative* for *any* value of the system parameters and *any* delay forcing function f. Therefore, the stability region of the steady state solution can always be enlarged by "distributing" the pointwise delay over an interval, or, by virtue of Theorem 6.1, by modulating the pointwise delay (speed).

To illustrate this, we consider a delay modulation of the form

$$\tau(t)=\tau_0+\delta f_1(\Omega t), \tag{6.30}$$

where $f_1(t)$ is the sawtooth function described in Table 6.1, and use the tools of section 6.3 to compute the exact stability limits of the system (6.27)–(6.30) in the (τ_0,k)-plane, for $m=100$, $\omega_n=632.45$, $\xi=0.039585$, $\delta=0.05\tau_0$, and $\Omega=2\pi/\tau_0$. We use $m=12$ mesh intervals and a piecewise polynomial of degree $d=3$. The results are shown in Figure 6.1 (solid line). For comparison, this figure also shows the stability limits when the delay is fixed at its mean value, i.e., when $\delta=0$ in (6.30). In the latter case, the computations can be done using the DDE-BIFTOOL package [74]. Note that the parameter k depends on the width of the cutting tool and on the nominal depth of cut. Therefore, the variation of the rotating speed allows us to use a larger tool and/or to remove more material at once, especially at lower speed (i.e., larger delay).

The stability limits for the averaged system corresponding to (6.27)–(6.30) are shown in [197, Fig. 4]. With the delay function (6.30), the kernel function of the distributed delay of the averaged system is a constant. Consequently, as explained in [162], we can also use the DDE-BIFTOOL package for the computations. The resulting stability limits are very close to those of the original system (6.27) with (6.30).

Note that in a preliminary step before the actual computations, we rescaled system (6.27) in order to avoid numerical problems. In particular, the time variable was multiplied

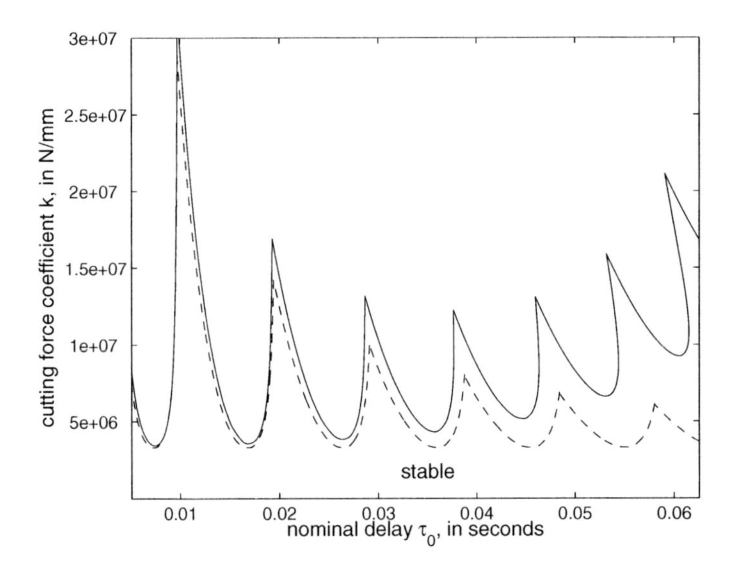

Figure 6.1. *The solid curve separates the stable and unstable regions in the* (τ_0, k)*-space for the system (6.27) and (6.30). The dashed curve separates the stable and unstable regions for the system with* $\delta = 0$.

by 10^2, so that the system could be transformed to a form with—among others—a "new" variable k that is 10^2 times smaller. Additionally, k and m were both divided by 10^3. In the postprocessing step after the computations, the numerical values were retransformed to restore the original meaning of the variables.

6.4.2 Forced elastic column

We consider a variation of the model of an elastic column studied in [312]. A column of height H is subjected to a time periodic force at the top and clamped at the bottom. Shear and inertia effects are neglected and the damping is assumed to be linear. We are interested in the local stability about the zero steady state and consider the following linear model for the displacement y at height h:

$$\alpha y_{hhhh}(h,t) + (\phi_1 + \phi_2 \cos(2\pi t)) y_{hh}(h,t) + y_{tt}(h,t) + \kappa y_t(h,t) = 0, \tag{6.31}$$

with boundary conditions

$$y(0,t) = 0, \quad y_h(0,t) = 0, \quad y_{hh}(H,t) = 0 \tag{6.32}$$

and

$$y_{hhh}(H,t) + (\phi_1 + \phi_2 \cos(2\pi t))(y_h(H,t) - \beta y_h(H, t-\tau)) = 0. \tag{6.33}$$

Note that here the delay only appears in the boundary condition (6.33).

In a preliminary step, this PDE is discretized in space to obtain a large system of DDEs. First, the column is divided into k intervals of height $\Delta h := H/k$, and $x(t) \in \mathbb{R}^{2k\times 1}$ is defined by

$$\begin{aligned} x_i(t) &:= y_t(i\Delta h, t) \quad \text{for } i = 1, \dots, k, \\ x_{k+i}(t) &:= y(i\Delta h, t) \quad \text{for } i = 1, \dots, k. \end{aligned} \tag{6.34}$$

Next, central finite difference formulae are used to approximate the spatial derivatives. The resulting system has the form

$$\dot{x}(t) = A(t)x(t) + B(t)x(t-\tau), \tag{6.35}$$

where

$$A(t) := \begin{bmatrix} -\kappa I_k & C(t) \\ I_k & 0 \end{bmatrix}, \quad B(t) := \begin{bmatrix} 0 & D(t) \\ 0 & 0 \end{bmatrix}. \tag{6.36}$$

The k-by-k matrices $C(t)$ and $D(t)$ take the form

$$C(t) = \begin{bmatrix} \bar{c}(t) & b(t) & a(t) & & & & \\ b(t) & c(t) & b(t) & a(t) & & & \\ a(t) & b(t) & c(t) & b(t) & a(t) & & \\ & \ddots & & \ddots & & \ddots & \\ & & a(t) & b(t) & c(t) & b(t) & a(t) \\ & & & a(t) & b(t) & \tilde{c}(t) & \tilde{b}(t) \\ & & & & \hat{a}(t) & \hat{b}(t) & \hat{c}(t) \end{bmatrix}, \tag{6.37}$$

$$D(t) = \begin{bmatrix} 0 & \cdots & 0 & 0 & 0 & 0 \\ \vdots & & & & & \vdots \\ 0 & \cdots & 0 & 0 & 0 & 0 \\ 0 & \cdots & 0 & d_1(t) & d_2(t) & d_3(t) \end{bmatrix}, \tag{6.38}$$

with

$$\begin{aligned} a(t) &= -\frac{\alpha}{\Delta h^4}, \\ b(t) &= \frac{4\alpha}{\Delta h^4} - \frac{\phi_1+\phi_2\cos(2\pi t)}{\Delta h^2}, \\ c(t) &= -\frac{6\alpha}{\Delta h^4} + 2\frac{\phi_1+\phi_2\cos(2\pi t)}{\Delta h^2}, \\ \bar{c}(t) &= -\frac{7\alpha}{\Delta h^4} + 2\frac{\phi_1+\phi_2\cos(2\pi t)}{\Delta h^2}, \\ \tilde{c}(t) &= -\frac{5\alpha}{\Delta h^4} + 2\frac{\phi_1+\phi_2\cos(2\pi t)}{\Delta h^2}, \\ \tilde{b}(t) &= \frac{2\alpha}{\Delta h^4} - \frac{\phi_1+\phi_2\cos(2\pi t)}{\Delta h^2}, \\ \hat{a}(t) &= \alpha\left(-\frac{2}{\Delta h^4} + \frac{\phi_1+\phi_2\cos(2\pi t)}{\Delta h^2}\right), \\ \hat{b}(t) &= 4\alpha\left(\frac{1}{\Delta h^4} - \frac{\phi_1+\phi_2\cos(2\pi t)}{\Delta h^2}\right), \\ \hat{c}(t) &= \alpha\left(-\frac{2}{\Delta h^4} + 3\frac{\phi_1+\phi_2\cos(2\pi t)}{\Delta h^2}\right), \\ d_1(t) &= -\frac{\alpha\beta(\phi_1+\phi_2\cos(2\pi t))}{\Delta h^2}, \\ d_2(t) &= \frac{4\alpha\beta(\phi_1+\phi_2\cos(2\pi t))}{\Delta h^2}, \\ d_3(t) &= -\frac{3\alpha\beta(\phi_1+\phi_2\cos(2\pi t))}{\Delta h^2}. \end{aligned} \tag{6.39}$$

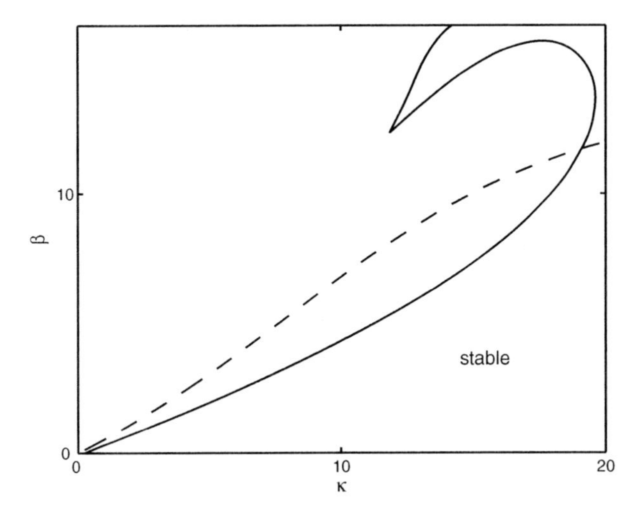

Figure 6.2. *The solid curve separates the stable and unstable regions in the* (κ, β)*-space for the forced system (6.31)–(6.33). The dashed curve separates the stable and unstable regions for the corresponding averaged system.*

Note that the ith equation of (6.35), with $1 \leq i \leq k$, corresponds to

$$y_{tt}(i\Delta h, t) = -\kappa y_t(i\Delta h, t) - \alpha y_{hhhh}(i\Delta h, t) - (\phi_1 + \phi_2 \cos(2\pi t))y_{hh}(i\Delta h, t),$$

where the spatial derivatives y_{hh} and y_{hhhh} are replaced by an approximation with central finite differences. For $i \in \{1, 2, k-1, k\}$, this requires function values at points which do not belong to the discretization points, namely $h = 0,\ h = -\Delta h,\ h = H + \Delta h,\ h = H + 2\Delta h$. The latter can, however, be eliminated using the three boundary conditions.

For the numerical example we choose $k = 32$; hence the size of the resulting system of DDEs is $n = 2k = 64$. Furthermore, we use $m = 12$ mesh intervals and a piecewise polynomial of degree $d = 3$, as for the previous example. Figure 6.2 shows the stability region of the zero solution in the (κ, β)-space for $\alpha = 1, \phi_1 = \phi_2 = 1, \tau = 0.4$, and $H = 1$, as well as the stability region of the corresponding averaged system (obtained by setting $\phi_2 = 0$).

6.5 Notes and references

Both analytical methods and computational tools for the stability analysis of periodic systems with time-varying delays were described and compared. Their applicability was illustrated with examples from mechanical engineering.

With respect to Sections 6.3.1 and 6.3.2, we notice that there are several other approaches for computing stability determining eigenvalues of linear periodic systems with delays, which we have not discussed. For instance, in [121, 122] the monodromy operator is discretized using the so-called semidescretization method, and in [37, 164] using a spectral method (see [302]) based on an expansion of the solutions in Chebychev polynomials.

The chapter is based on [197, 198, 305, 314, 312] and the references therein.

Part II

Stabilization and robust stabilization

Chapter 7
The continuous pole placement method

7.1 Introduction

The stabilization of linear time-delay systems has been studied extensively in the literature. Existing stabilization methods include those based on finite spectrum assignment [27, 173, 239, 323], time-domain approaches (which are usually Lyapunov-based and lead to stability conditions expressed by the solvability of algebraic Riccati equations (AREs) or the feasibility of linear matrix inequalities (LMIs); see [96] and the references therein), and variants of the Smith predictor (see [248] for an overview).

In this chapter we outline a stabilization procedure which is related to the classical pole placement method for ODEs [2]. The approach starts from the state-space description of the linear time-delay system. Because it is based on the continuous dependence of the characteristic roots on the controller parameters and the stabilization strategy consists of shifting the unstable characteristic roots to the LHP in a quasi-continuous way, it is called continuous pole placement.

As an illustration of the approach, we study the stabilization of the system

$$\dot{x}(t) = Ax(t) + Bu(t-\tau), \quad A \in \mathbb{R}^{n\times n}, \ B \in \mathbb{R}^{n\times 1}, \tag{7.1}$$

where $x(t) \in \mathbb{R}^n$ is the state at time t, $u \in \mathbb{R}$ is the input, and $\tau \geq 0$ represents an input delay. We assume that the pair (A, B) is controllable and use a linear static state feedback controller,

$$u(t) = K^T x(t), \quad K \in \mathbb{R}^{n\times 1}. \tag{7.2}$$

The form of the control law (7.2) is explained by the fact that it reveals the link between the proposed stabilization procedure and the classical pole placement method, and allows us to obtain some nice theoretical results. Furthermore, it will be shown that for the system (7.1), a static state feedback controller leads to stabilizability properties which are comparable to those of methods based on finite spectrum assignment, when small perturbations are taken into account.

The structure of the chapter is as follows. We first study some theoretical stability properties of the system (7.1)–(7.2) and motivate the use of the control law (7.2), thereby explaining the ideas behind the continuous pole placement method. Then we study the stabilization method in detail and apply it to a (numerical) example. We end with a discussion of dynamic feedback controllers and generalizations to multiple input, multiple output (MIMO) systems, as well as systems of neutral type.

Throughout the chapter we will use both the closed-loop description (7.1)–(7.2) and the description in the control canonical form,

$$\dot{z}(t) = A_c z(t) + B_c K_c^T z(t-\tau), \tag{7.3}$$

where $K_c = [k_1\, k_2 \cdots k_n]^T$, $B_c = [0 \cdots 0\ \ -1]^T$, and

$$A_c = \begin{bmatrix} 0 & 1 & & & \\ & 0 & 1 & & \\ & & \ddots & \ddots & \\ & & & 0 & 1 \\ -a_1 & -a_2 & \dots & -a_{n-1} & -a_n \end{bmatrix}. \tag{7.4}$$

Based on this transformation the characteristic equation of the closed-loop system can be written in the form

$$\begin{aligned} H(\lambda) := \lambda^n + (a_n + k_n e^{-\lambda\tau})\lambda^{n-1} + \cdots \\ +(a_2 + k_2 e^{-\lambda\tau})\lambda + (a_1 + k_1 e^{-\lambda\tau}) = 0. \end{aligned} \tag{7.5}$$

7.2 Motivation

The stabilization of system (7.1) with a feedback controller of the form (7.2) is a hard problem because its design involves the determination of only n parameters, while the closed-loop system has an infinite number of characteristic roots. On the other hand, the proposed controller structure is very simple and easy to implement, and stability of the closed-loop system is robust w.r.t. small perturbations on the parameters. Furthermore, it is shown in [72, 305, 203, 186] that more complex controller laws, which are based on prediction and only yield a finite number of closed-loop characteristic roots, may be sensitive to arbitrarily small implementation errors, and may therefore suffer from similar limitations as control law (7.2) from a practical point of view. This will be illustrated with a one-dimensional example in Section 7.2.3.

7.2.1 A finite-dimensional controller for an infinite-dimensional problem

The classical pole placement algorithm [2] cannot be applied directly to the time-delay case because the characteristic equation (7.5) has infinitely many solutions while the number of degrees of freedom in the controller is equal to the dimension n. A direct placement of n characteristic roots is always possible. This follows from the linearity of the characteristic equation (7.5) w.r.t. the components of K_c. Indeed, (7.5) can be rewritten as

$$\left[1\ \lambda\ \lambda^2\ \dots\ \lambda^{n-1}\right][k_1\ k_2\ \dots\ k_n]^T = -\bar{p}(\lambda)e^{\lambda\tau}, \tag{7.6}$$

where $\bar{p}(\lambda) = \sum_{i=1}^{n} a_i \lambda^{i-1} + \lambda^n$. When forcing n different numbers $\lambda_1, \ldots, \lambda_n$ to satisfy this equation, a linear system of equations with the components of K_c as unknowns and a regular Jacobian matrix (a Vandermonde matrix) is obtained. However, by placing n characteristic roots, which determines the complete spectral picture, control is lost over the position of the other ones and, as these may cause instability, an a posteriori stability check is necessary. See [331], where such an approach is worked out.

The continuous pole placement method is based on the fact that the number of unstable characteristic roots is finite [106, Lemma I.4.1] and on the availability of a software tool to calculate the rightmost characteristic roots; see Chapter 1. Once unstable characteristic roots are detected, the strategy consists of moving them to the LHP by applying small changes to the controller gain K, and meanwhile monitoring the other characteristic roots with a large real part. Because the characteristic roots move continuously w.r.t. changes in the feedback gain, the method is referred to as the *continuous* pole placement method. Before describing it in more detail in the next section, we introduce some theoretical results concerning the limitations and difficulties of linear state feedback control in the presence of input delays.

When the uncontrolled system (7.1) has characteristic roots in the closed RHP, the destabilizing effect of the time-delay is illustrated with the fact that no fixed feedback gain K is able to achieve stabilization for all values of the time-delay τ, a consequence of the following well-known result (see for instance [73, 223, 70]). By $\sigma(\cdot)$ we denote the spectrum and by $r_\sigma(\cdot)$ the spectral radius.

Theorem 7.1. *If the zero solution of $\dot{x}(t) = Ax(t) + A_d x(t-\tau)$ is asymptotically stable for all $\tau \geq 0$, then $\Re(\sigma(A)) < 0$ and $\sup_{\omega \in \mathbb{R}} r_\sigma((j\omega I - A)^{-1} A_d) \leq 1$.*

If A has eigenvalues in the open RHP, then this result can be strengthened.

Theorem 7.2. *If A has eigenvalues in the open RHP, then the system $\dot{x} = Ax + Bu(t-\tau)$ cannot be asymptotically stabilized semiglobally in the delay using state feedback $u(t) = K^T x(t)$; that is, there exists a $\bar{\tau} \geq 0$ such that*

$$\forall K \in \mathbb{R}^{n \times 1}\ \exists \tau \leq \bar{\tau} : \ \dot{x}(t) = Ax(t) + BK^T x(t-\tau) \text{ is not asymptotically stable.}$$

Proof. The proof is by contradiction. Suppose that the theorem does not hold, i.e., there exist sequences $\{K_n\}_{n \geq 1}$ and $\{\tau_n\}_{n \geq 1}$ with $\lim_{n \to \infty} \tau_n = \infty$ such that the feedback $u(t) = K_n^T x(t)$ asymptotically stabilizes the system for $0 \leq \tau \leq \tau_n$. We show that this always leads to a contradiction. We make distinction between two cases.

Case 1: The sequence $\{\|K_n\|\}_{n \geq 1}$ is bounded.

Because all elements of $\{K_n\}_{n \geq 1}$ belong to a compact region in $\mathbb{R}^n$, there exists a converging subsequence with limit K. Denote by λ_0 an eigenvalue of A in the open RHP and define the disk $D = \{\lambda : |\lambda - \lambda_0| \leq \frac{\Re(\lambda_0)}{2}\}$. The sequence of analytic functions

$$\{f(\lambda, \tau_n)\}_{n \geq 1} = \left\{\det(\lambda I - A - BK^T e^{-\lambda \tau_n})\right\}_{n \geq 1}$$

converges uniformly on the disk D to the function $f(\lambda) = \det(\lambda I - A)$. We can apply Corollary A.1, which states that for large n, the number of zeros of $f(\lambda, \tau_n)$ and $f(\lambda)$ in D are equal. As a consequence, the system $\dot{x}(t) = Ax(t) + BK^T x(t - \tau_n)$ has a characteristic

root in the open RHP for large values of n. Hence, there exists an integer $\hat{n}$ such that $\dot{x}(t) = Ax(t) + BK^T x(t - \tau_{\hat{n}})$ has a characteristic root in the open RHP, whereas $\dot{x}(t) = Ax(t) + BK_n^T x(t - \tau_{\hat{n}})$ is asymptotically stable for large n, since $\{K_n\}_{n\geq 1}$ asymptotically stabilizes the system for $\tau \in [0,\ \tau_n]$. Because $\{K_n\}_{n\geq 1}$ has a subsequence converging to K, this implies that an arbitrarily small change of the feedback gain K stabilizes the (unstable) system when $\tau = \tau_{\hat{n}}$, and we have a contradiction because the characteristic roots move continuously w.r.t. parameter changes.
Case 2: The sequence $\{\|K_n\|\}_{n\geq 1}$ is unbounded.
First note that with an arbitrary feedback gain K and for any $\omega > r_\sigma(A)$, the system $\dot{x}(t) = Ax(t) + BK^T x(t - \tau)$ has characteristic roots at $\pm j\omega$ if and only if

$$\det(I - (j\omega I - A)^{-1} BK^T e^{-j\omega\tau}) = 0,$$

meaning that the matrix $(j\omega I - A)^{-1} BK^T e^{-j\omega\tau}$ has a characteristic root 1. Since this is a rank-1 matrix, with its only possible nonzero characteristic root equal to $\lambda_{nz} = K^T(j\omega I - A)^{-1} Be^{-j\omega\tau}$, this is equivalent to $|\lambda_{nz}| = 1$ and $\Im(\lambda_{nz}) = 0$. Consequently, when $|K^T(j\omega I - A)^{-1}B| = 1$, we have characteristic roots at $\pm j\omega$ for a value of the delay τ satisfying $\tau \leq \frac{2\pi}{\omega}$. We will also use the fact that $\lim_{\omega\to\infty} K^T(j\omega I - A)^{-1}B = 0$.

There exists a converging subsequence $\{F_n\}_{n\geq 1}$ of $\{\frac{K_n}{\|K_n\|}\}_{n\geq 1}$ and let its limit be F. Consider the parameterized curve in the complex plane, $\omega \in (r_\sigma(A),\ \infty) \to F^T(j\omega I - A)^{-1}B$. It is impossible that this curve is identically zero for all $\omega > r_\sigma(A)$. Indeed, when $\omega > r_\sigma(A)$, we can expand $(j\omega I - A)^{-1} = \frac{1}{j\omega}\sum_{k=0}^{\infty}(\frac{A}{j\omega})^k$. Hence

$$\begin{aligned} 0 &= F^T(j\omega I - A)^{-1}B \\ &= \tfrac{1}{j\omega}\left[F^T B + \tfrac{1}{j\omega}F^T AB + \tfrac{1}{(j\omega)^2}F^T A^2 B \ldots\right] \quad \forall\omega > r_\sigma(A) \end{aligned}$$

would imply that $F^T A^k B = 0,\ k \in \mathbb{N}$, and thus $F^T[B\ AB \ldots\ A^{n-1}B] = 0$. Because the pair (A, B) is controllable, it follows that $F = 0$ and we have a contradiction since $\|F\| = 1$.

As a consequence, $\exists\alpha > 0$ and $\exists\bar{\omega} > r_\sigma(A)$ such that $|F^T(j\bar{\omega}I - A)^{-1}B| = \alpha$. Because $F_n \to F$ there exists an integer N such that for $n \geq N$, $\frac{3}{2}\alpha \geq |F_n^T(j\bar{\omega}I - A)^{-1}B| \geq \frac{\alpha}{2}$. With a feedback gain $kF_n,\ k > 0,\ n \geq N$, there are characteristic roots on the imaginary axis when $k \geq \frac{2}{\alpha}$ for a delay $\tau \leq \frac{2\pi}{\bar{\omega}}$, because then the curve $\omega \in (r_\sigma(A),\ \infty) \to kF_n^T(j\omega I - A)^{-1}B$ intersects the unit circle, and hence $|kF_n^T(j\omega I - A)^{-1}B| = 1$, for some $\omega \geq \bar{\omega}$. As a result, there is a subsequence of $\{K_n\}_{n\geq 0}$ which cannot asymptotically stabilize the system in delay intervals larger than $[0,\ \frac{2\pi}{\bar{\omega}}]$, and we have a contradiction. □

As illustrated with a feedback controlled multiple integrator [178, 193], semiglobal stabilization in the delay does not require that A is Hurwitz.

The next theorem illustrates an inherent difference between the ODE and the DDE case. When $\tau \neq 0$, the set of all stabilizing feedback gains is bounded. Therefore, it is impossible to apply a high-gain approach and move the characteristic roots as far away as desired into the LHP by increasing the controller gain.

Theorem 7.3. *Assume that the control law $u(t) = K^T x(t)$ asymptotically stabilizes the system $\dot{x}(t) = Ax(t) + Bu(t - \tau)$ for $\tau \in [0,\ r]$ with $r > 0$. Then there exist constants $\gamma, \delta > 0$ independent of K, such that $\|K\| \leq \gamma$ and, for each $\tau \in [0,\ r]$, $\inf_K \sup \Re(\lambda) \geq -\delta$.*

Proof. Denote by $\mathcal{B}$ the unit ball in $\mathbb{R}^n$ and define $\bar{\omega} = \max(\frac{2\pi}{r}, 2r_\sigma(A))$. For $K \in \mathcal{B}$, we have $\sup_{\omega \geq \bar{\omega}} |K^T(j\omega I - A)^{-1}B| > 0$, because of the controllability of (A, B); see the proof of Theorem 7.2. From the compactness of $\mathcal{B}$ it follows that

$$\xi := \inf_{K \in \mathcal{B}} \sup_{\omega \geq \bar{\omega}} |K^T(j\omega I - A)^{-1}B| > 0.$$

Therefore, when the feedback gain K satisfies $\|K\| \geq \frac{1}{\xi}$, there will be characteristic roots on the imaginary axis for values of the delay smaller than $\frac{2\pi}{\bar{\omega}} \leq r$, as follows from the arguments spelled out in the proof of Theorem 7.2. Hence the family of all asymptotically stabilizing controls for the delay interval $[0,\ r]$ can be embedded in the set $\{K : \|K\| \leq \frac{1}{\xi}\}$. Letting $\gamma = 1/\xi$ proves the first part of the theorem. The second statement follows from the fact that the spectral abscissa, $\sup \Re(\lambda)$, is continuous w.r.t. the components of K, and the fact that all stabilizing feedback gains belong to a compact set. $\square$

7.2.2 Methods based on prediction

Existing methods based on finite spectrum assignment (FSA) [173, 239, 281, 323] avoid at first sight the conflict between the finite-dimensional controller parameter space and the infinite-dimensional closed loop system. They use an infinite-dimensional controller, which makes the closed loop system finite-dimensional. This can be done by counteracting the effect of the delay with a prediction of the state variable over one delay interval. For the system (7.1), let $x_p(t_1, t_2)$ be the prediction of x at time t_2 based on its values for $t \leq t_1$. With the control law

$$\begin{aligned} u(t) &= K^T x_p(t, t+\tau) \\ &= K^T e^{A\tau} x(t) + K^T \int_0^\tau e^{A(\tau-\theta)} B u(t+\theta-\tau)\, d\theta, \end{aligned} \tag{7.7}$$

the characteristic equation of the closed-loop system is given by

$$\det(\lambda I - A - BK^T) = 0, \tag{7.8}$$

and under the controllability assumption of the pair (A, B), the n closed-loop characteristic roots can be assigned arbitrarily. However, in order to apply the control law (7.7), the integral term needs to be calculated online, and in [305, 72] it is shown that the stability of the closed-loop system (7.1)–(7.7) might not be robust against *arbitrarily* small implementation errors of the integral term, caused by an approximation with a finite sum of pointwise delays. The underlying reason is that the approximation error of the integral in (7.7) forms a noncompact perturbation of the solution semigroup associated with the closed-loop system. As shown in [72, 186], this perturbation introduces additional characteristic roots, whose real parts converge (as the approximation error tends to zero) to the real parts of the characteristic roots of

$$\dot{x}(t) = (A + BK^T)x(t) - BK^T e^{A\tau} x(t-\tau), \tag{7.9}$$

where the eigenvalues of A are removed. As a consequence, the location of the characteristic roots of (7.9) determines the robustness of the control law (7.7) w.r.t. small implementation errors of the distributed delays. In the next paragraph, we show that for a one-dimensional example, the *practical* stabilizability properties of (7.7) are qualitatively the same as those of the simple state feedback controller, $u(t) = K^T x(t)$. We refer to Chapter 15 for a more elaborate analysis of the effect on stability of the implementation of controllers based on FSA.

7.2.3 Scalar example

We investigate the stabilization of the system

$$\dot{x}(t) = ax(t) + u(t-\tau), \quad x \in \mathbb{R}. \tag{7.10}$$

Using the control law $u(t) = kx(t)$ we obtain the closed-loop system

$$\dot{x}(t) = ax(t) + kx(t-\tau), \tag{7.11}$$

which is also analyzed in [53, 70, 138] and corresponds to problem (3.4) of Section 3.2. Figure 7.1 shows the stability regions in the parameter space (a, k). The system is asymptotically stable for all values of the delay when $a < 0$ and $-|a| \le k < |a|$. Note that when the uncontrolled system is unstable ($a > 0$), delay-independent stability is not possible (Theorem 7.1). When only stability for all delays $\tau \le r$ is required, the stability region extends toward the curve through the point $P = (\frac{1}{r}, -\frac{1}{r})$, analytically given by

$$a = \frac{\omega}{\tan(\omega r)}, \quad k = -\frac{\omega}{\sin(\omega r)}, \quad \omega r \in (0, \pi).$$

Hence, it is impossible to stabilize the system for all $\tau \in [0, r]$ when $r > \frac{1}{a}$, whatever the value of the feedback gain k (Theorem 7.2). Note that the stabilizability condition is given by $ar < 1$, which can be interpreted as a trade-off between the instability of the uncontrolled system (a) and the destabilizing effect of the delay (r). By computing the sensitivity of characteristic roots w.r.t. the delay, one proves that for a particular value of the delay τ, a necessary and sufficient condition for asymptotic stabilizability of (7.10) is also given by

$$a\tau < 1. \tag{7.12}$$

For $a > 0$ and $\tau < \frac{1}{a}$, the stabilizing values of k lie in a compact interval (Theorem 7.3) and a right upper bound on the spectrum can never reach $-\infty$. A simple calculation shows that the value of k which minimizes the spectral abscissa, $\alpha = \sup \Re(\lambda)$, is given by $k = -\frac{1}{\tau} e^{a\tau - 1}$ and results in $c_1 = \inf_k \sup(\Re(\lambda)) = a - \frac{1}{\tau}$.

In [72] one considers the stabilization of the system (7.10) with $a = 1$ using the feedback $u(t) = -(1 - \lambda_d)(e^{\tau} x(t) + \int_0^{\tau} e^{\tau - \theta} u(t + \theta - \tau)\, d\theta)$, yielding one closed-loop characteristic root at λ_d. The robustness of the controller is investigated when the integral term is implemented with a numerical quadrature rule, by analyzing the spectrum of the corresponding equation (7.9). Since placing λ_d far away into the LHP increases the destabilizing influence of the implementation error, there also exists an optimal position of λ_d. In Figure 7.2, $c_2 = \inf_{\lambda_d} \sup \Re(\lambda)$ is depicted as a function of τ (solid line), taking into account the influence of arbitrarily small implementation errors. The dashed line shows the maximal exponential decay rate in the case of pure state feedback, $c_1 = \inf_k \sup \Re(\lambda)$. This figure reveals that, from a practical point of view, both approaches are comparable.

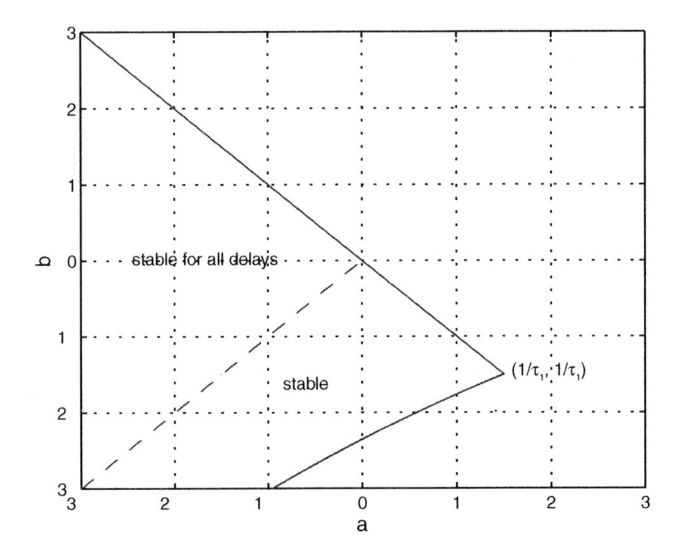

Figure 7.1. *Stability regions of* $\dot{x}(t) = ax(t) + kx(t - \tau)$.

For system (7.10) the (practical) stabilizability condition in case of FSA is given by

$$a\tau < 1.293, \tag{7.13}$$

which is only slightly better than (7.12). In Section 7.5.3, we will show that with a modification of the continuous pole placement method, namely an adaptation to the dynamic state feedback case, we are able to improve this stabilizability condition.

7.3 Continuous pole placement algorithm

7.3.1 Description of the algorithm

The idea behind the stabilization method is to shift the unstable characteristic roots to the LHP in a quasi-continuous way, by applying small changes to the feedback gain and monitoring the other characteristic roots with a large real part. It is based on the continuity of the characteristic roots w.r.t. the components of the feedback gain. The basic algorithm is as follows.

ALGORITHM 7.1.

The continuous pole placement method

A. Initialize $m = 1$.

B. Compute the rightmost characteristic roots for the nominal delay τ.

C. Compute the sensitivity of the m rightmost characteristic roots w.r.t. changes in the feedback gain K.

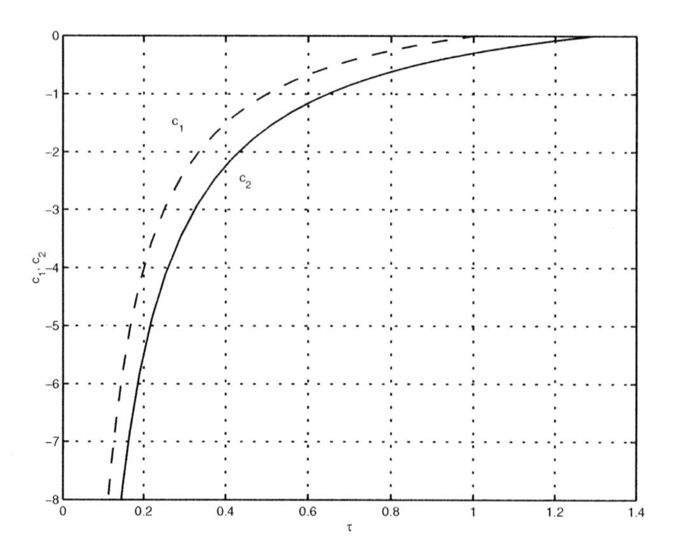

Figure 7.2. *Stabilizability limits of the system* $\dot{x}(t) = x(t) + u(t - \tau)$.

D. Move the m rightmost characteristic roots in the direction of the LHP by applying a small change to the feedback gain K, using the computed sensitivities.

E. Monitor the rightmost uncontrolled characteristic roots. If necessary, increase the number of controlled characteristic roots, m. Stop when stability is reached or when the available degrees of freedom in the controller do not allow us to further reduce $\sup \Re(\lambda)$. In the other case, go to step B.

We now describe the different steps of the algorithm in more detail.

Computation of the rightmost characteristic roots

We refer to Section 1.2.6.

Sensitivity of characteristic roots with respect to the feedback gain

When λ_i is a solution of the characteristic equation and $v_i e^{\lambda\theta}$, $\theta \in [-\tau, 0]$, its corresponding eigenfunction, we have

$$\begin{cases} \left(\lambda_i I - A - BK^T e^{-\lambda_i \tau}\right) v_i = 0, \\ n(v_i) = 0, \end{cases} \tag{7.14}$$

where $n(v_i)$ is a normalizing condition. Differentiating this equation w.r.t. a component k_j of K, we obtain a linear system of equations in the unknowns $\frac{\partial \lambda_i}{\partial k_j}$ and $\frac{\partial v_i}{\partial k_j}$,

$$\begin{bmatrix} \lambda_i I - A - BK^T e^{-\lambda_i \tau} & \left(I + BK^T \tau e^{-\lambda_i \tau}\right) v_i \\ \frac{dn}{dv_i}^T & 0 \end{bmatrix} \begin{bmatrix} \frac{\partial v_i}{\partial k_j} \\ \frac{\partial \lambda_i}{\partial k_j} \end{bmatrix} = \begin{bmatrix} B v_i^T e_j e^{-\lambda_i \tau} \\ 0 \end{bmatrix}, \tag{7.15}$$

with $e_j \in \mathbb{R}^{n \times 1}$ the jth unity vector. When the system is in the control canonical form (7.3), one can compute directly from (7.5)

$$\frac{\partial \lambda_i}{\partial k_j} = -\frac{e^{-\lambda_i \tau} \lambda_i^{j-1}}{\frac{dH}{d\lambda_i}}. \tag{7.16}$$

This result is useful for the derivation of theoretical properties of the method, but our implementation is based on (7.14), since (7.5) is not practical from a numerical point of view.

Continuation of characteristic roots as a function of the feedback gain K

Assume that $m \leq n$ characteristic roots $\lambda_1, \ldots, \lambda_m$ are controlled. When the "sensitivity" matrix S_m, defined by

$$S_m = [s_{i,j}] \in \mathbb{R}^{m \times n}, \quad \text{where } s_{i,j} = \frac{\partial \lambda_i}{\partial k_j}, \tag{7.17}$$

is of rank m and the desired (small) displacement of the controlled characteristic roots is given by $\Delta \Lambda_m^d = [\Delta \lambda_1^d \ldots \Delta \lambda_m^d]^T$, one can compute a change ΔK for K such that

$$S_m \Delta K = \Delta \Lambda_m^d. \tag{7.18}$$

When $m < n$ this equation has infinitely many solutions. One possibility to determine a unique solution consists of controlling the m characteristic roots using only m selected components of K and hence taking $n - m$ components of ΔK equal to zero. Another possibility, which we explain later on, consists of taking the solution with $\|\Delta K\|$ minimal. This solution is given by

$$\Delta K = S_m^{\dagger} \Delta \Lambda_m^d, \tag{7.19}$$

where $S_m^{\dagger}$ is the Moore–Penrose inverse of S_m; see [14, Chapter 3].

A physical constraint on the feedback gain is imposed by the fact that its components must be real, but this is assured by taking the components of $\Delta \Lambda_m^d$ in complex conjugate pairs.

With the new feedback gain $K + \Delta K$, the displacement of the controlled characteristic roots will generically not be equal to $\Delta \Lambda_m^d$, since (7.18) is based on linearization, and some correction is needed. However, both characteristic roots and eigenfunctions are continuous w.r.t. parameter changes and with, for instance, the predictor

$$\lambda_i^{(p)} = \lambda_i + \Delta\lambda_i^d, \quad v_i^{(p)} = \sum_{j=1}^{n} \frac{\partial v_i}{\partial k_j}\Delta k_j, \quad i = 1, m,$$

only a few Newton iterations on (7.14) are needed when ΔK is sufficiently small. Since it is also desirable to have the characteristic roots for the new feedback gain close to their predictions, (7.19) is generally a good choice for the solution of (7.18).

The n components of the controller gain allow us to control both real and imaginary parts of at most n characteristic roots, as illustrated by (7.18) or, when stabilization is of primary concern, only the real parts of possibly more than n characteristic roots. For instance, with $n = 2$ degrees of freedom in the controller, one can control either two real characteristic roots, or one real characteristic root and the real part of a complex pair, or the real parts of two complex pairs. For the latter case, notice that

$$\frac{\partial \Re(\lambda_i)}{\partial k_j} = \Re\left(\frac{\partial \lambda_i}{\partial k_j}\right).$$

This approach is applied to all examples in this chapter, combined with an adaptation of *all* controller parameters in an iteration step, as in (7.19). This leads to the adaptation formula

$$\Delta K = (\Re(S_m))^{\dagger} \Delta\Re(\Lambda_m^d), \tag{7.20}$$

where $\Delta\Re(\Lambda_m^d)$ is the desired displacement of the real parts of the controlled characteristic roots.

Increasing the number of controlled characteristic roots

In the examples of this chapter, where the focus lies on stabilization, we only control the real parts of the characteristic roots. We start by reducing the real part of the rightmost characteristic root. In this way control is lost over the other characteristic roots and when an interaction occurs, the number of controlled characteristic roots, m, generally needs to be increased to further reduce the spectral abscissa $\sup \Re(\lambda)$. This is now illustrated.

In Figure 7.3 we display two typical situations, where the real part of only the rightmost characteristic root is controlled and the number of controlled characteristic roots must be increased when it interacts with other characteristic roots. In the upper part the situation is shown where two real characteristic roots interact. When only the dominant characteristic root is controlled, no further reduction of $\sup \Re(\lambda)$ is possible after the interaction (left): when the two characteristic roots coincide, they first transform to a complex conjugate pair which does not move to the left in the complex plane with the next parameter change but immediately splits up again into two real characteristic roots (we control only the real part of the pair, and by the parameter change the imaginary part again becomes zero), and then the whole process repeats itself. By controlling the two real characteristic roots from iteration number 150 on, a further reduction of $\sup \Re(\lambda)$ is possible (right). In the lower part of Figure 7.3, a real characteristic root interacts with a complex pair of characteristic roots. A decrease of the real characteristic root leads to an increase of the real part of the complex pair and vice versa (left). This problem can be avoided by controlling both the real characteristic root and the real part of the complex pair (right).

A special situation also occurs when a complex pair, whose real part is controlled, splits up into two real characteristic roots. This is possible because its imaginary part is not controlled and may tend to zero during the stabilization procedure. After the splitting *both* real characteristic roots need to be controlled.

In the continuous pole placement algorithm, the number of controlled characteristic roots is manually increased, whenever stagnation of $\sup \Re(\lambda)$ or a splitting of a complex pair occurs. The method breaks down when no further reduction of $\sup \Re(\lambda)$ is possible, i.e., when the available degrees of freedom in the controller (n) are not sufficient to control the real parts of all dominant characteristic roots. This can happen when $n+1$ real parts need to be controlled to reduce $\sup \Re(\lambda)$ or when the matrix $\Re(S_m)$ in (7.20) has a rank strictly smaller than m.

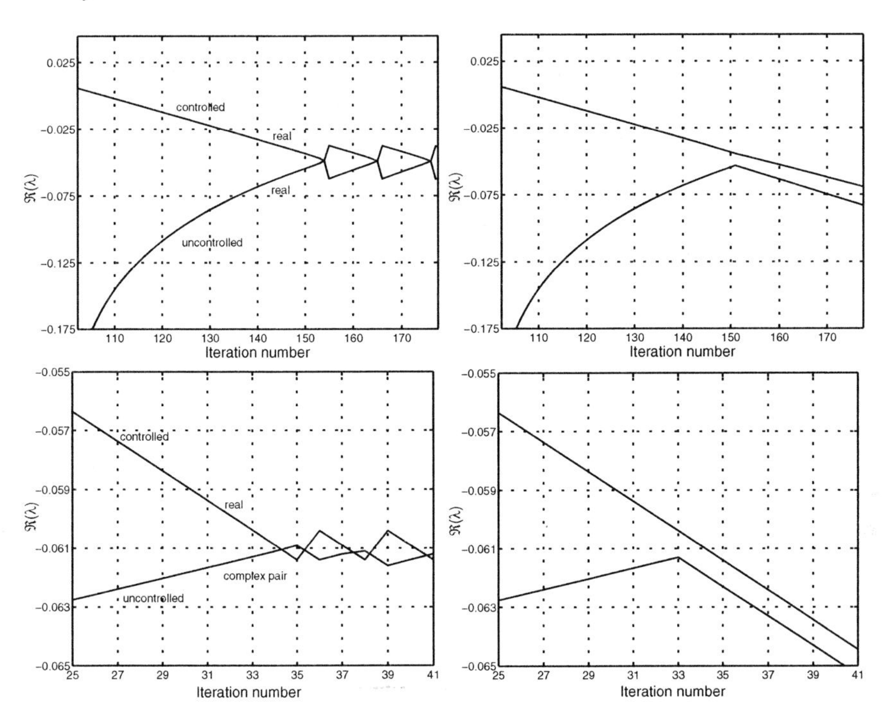

Figure 7.3. *Two typical situations where it is necessary to extend the number of controlled characteristic roots.*

7.3.2 Theoretical properties

In each step of the continuous pole placement algorithm, the sensitivity matrix S_m must be of rank m, if a full control over the changes of the m characteristic roots is desired. The next theorem states that when both real and imaginary parts of n isolated characteristic roots are controlled, the sensitivity matrix S_n is always regular.

Theorem 7.4. *If $\lambda_1, \ldots, \lambda_n$ are characteristic roots of multiplicity one, then the corresponding sensitivity matrix S_n is regular.*

Proof. Since the pair (A, B) is controllable, the system can be transformed into the control canonical form (7.3)–(7.5). Using (7.16), we obtain the following sensitivity matrix:

$$S_n = \begin{bmatrix} \frac{e^{-\lambda_1 \tau}}{\frac{dH}{d\lambda_1}} & & \\ & \ddots & \\ & & \frac{e^{-\lambda_n \tau}}{\frac{dH}{d\lambda_n}} \end{bmatrix} \begin{bmatrix} 1 & \lambda_1 & \ldots & \lambda_1^{n-1} \\ 1 & \lambda_2 & \ldots & \lambda_2^{n-1} \\ \vdots & \vdots & \ldots & \vdots \\ 1 & \lambda_n & \ldots & \lambda_n^{n-1} \end{bmatrix}. \tag{7.21}$$

This matrix is regular if no characteristic roots coincide, since $\frac{dH}{d\lambda_i} \neq 0$ and the Vandermonde matrix is regular. □

When two characteristic roots are close to each other, the sensitivity w.r.t. changes in the feedback gain is large, as shown in the next theorem. This result is related to the fact that large qualitative changes in the spectral picture may occur when characteristic roots coincide.

Theorem 7.5. *When two characteristic roots are brought together, the norm of the sensitivity matrix S_n becomes arbitrarily large.*

Proof. Start from (7.21) and consider the asymptotic case where $\lambda_j \to \lambda_i$ while the other characteristic roots are kept fixed. With the explicit expression for the determinant of a Vandermonde matrix, we have

$$\begin{aligned} \lim_{\lambda_j \to \lambda_i} \det(S_n) &= e^{-2\lambda_i \tau} \prod_{k=1,\, k \neq i,\, k \neq j}^{n} \left(\frac{e^{-\lambda_k \tau}}{\frac{dH}{d\lambda_k}} \right) \\ &\quad \lim_{\lambda_j \to \lambda_i} \left(\frac{1}{\frac{dH}{d\lambda_i} \frac{dH}{d\lambda_j}} \prod_{l>k=1}^{n} (\lambda_l - \lambda_k) \right) \\ &= \infty. \end{aligned}$$

The last step follows from the fact that $\frac{dH}{d\lambda_i} = O(\lambda_i - \lambda_j)$ and $\frac{dH}{d\lambda_j} = O(\lambda_i - \lambda_j)$ as $\lambda_j \to \lambda_i$. Finally, $\det(S_n) \to \infty$ implies $\|S_n\| \to \infty$ for any induced norm $\|.\|$. □

When only the real parts of characteristic roots are controlled, it is therefore recommended to keep the real parts separated in order to avoid highly sensitive sensitivity matrices leading to numerical problems. This is illustrated with the example in the next section.

7.3.3 Optimization point of view

The continuous pole placement method, as described in Algorithm 7.1, can be interpreted as a local strategy to solve or find a feasible (stabilizing) solution of the optimization problem

$$\min_K \alpha(K), \tag{7.22}$$

where the objective function is given by

$$\alpha(K) := \sup\left\{\Re(\lambda) : \det(\lambda I - A - BK^T e^{-\lambda\tau}) = 0\right\}. \tag{7.23}$$

We outline some properties of the optimization problem (7.22). The objective function (7.23) is not differentiable. Discontinuities in its derivatives may not only occur when the rightmost characteristic roots change (due to the supremum function), but also when some rightmost characteristic roots have a multiplicity larger than one (which may result in an infinitely high sensitivity w.r.t. parameter changes). Furthermore, the objective function is in general neither convex nor quasi-convex [23], the latter meaning that the sublevel sets

$$S_\gamma = \left\{K \in \mathbb{R}^{n\times 1} : \alpha(K) \leq \gamma\right\} \tag{7.24}$$

are also not convex. For instance, a feedback controlled triple integrator with input delay τ has a characteristic equation of the form

$$\lambda^3 + k_1 e^{-\lambda\tau}\lambda^2 + k_2 e^{-\lambda\tau}\lambda + k_3 e^{-\lambda\tau} = 0.$$

Substituting $\lambda = j\omega$ yields a set of stabilizing feedback gains given by $0 \leq k_3 \leq k_1 k_2$ for $\tau = 0$. For small values of the time-delay, the nonconvexity of this set is preserved.

Despite the complexity of the optimization problem, numerous experiments indicate that each sublevel set (7.24) is connected and, as a consequence, the optimization problem has only one minimum, yet a theoretical proof is missing.

The following theorem provides us with information on the configuration of the rightmost characteristic roots in a (the) minimum of (7.23).

Theorem 7.6. *Let $\bar{K}$ be a local minimum of the function $\alpha(K)$, defined in (7.23). Then for $K = \bar{K}$, there are at least $n+1$ characteristic roots with real part equal to $\alpha(\bar{K})$.*

Proof. Take any value of the feedback gain K and assume that there are at most n characteristic roots (multiplicity taken into account) with real part equal to $\sup \Re(\lambda)$. We prove that this situation does not correspond to a minimum.

Denote the n rightmost characteristic roots by $\lambda_i,\ i = 1, k$, each with a multiplicity m_i. As follows from (7.6), the feedback gain $K = [k_1 \ldots k_n]^T$ satisfies

$$\begin{bmatrix} 1 & \lambda_i & \lambda_i^2 & \cdots & \lambda_i^n \\ & 1 & 2\lambda_i & \cdots & n\lambda_i^{n-1} \\ & & \ddots & & \vdots \\ & & & & \frac{n!}{(n-m_i)!}\lambda_i^{n-m_i+1} \end{bmatrix} K = \begin{bmatrix} q(\lambda_i) \\ q'(\lambda_i) \\ \vdots \\ q^{m_i-1}(\lambda_i) \end{bmatrix}, \tag{7.25}$$

where $q(\lambda) = \bar{p}(\lambda)e^{\lambda\tau}$. When putting these equations together for each $i = 1, \ldots, k$, we obtain a set of n equations of the form

$$M(\lambda_1, \ldots, \lambda_k)K^c = Q(\lambda_i, \ldots, \lambda_k).$$

From the structure of (7.25) it follows that $M(\lambda_i, \ldots, \lambda_k)$ is regular when its arguments are different: note that the components of K can be considered as the coefficients of the Hermite interpolating polynomial of $q(\lambda)$ in $\lambda_i,\ i = 1, k$, which is unique.

Consider next the parameterized feedback gain K_ϵ which places n characteristic roots at the positions $\lambda_i - \epsilon$, i.e., it satisfies

$$M(\lambda_1 - \epsilon, \ldots, \lambda_k - \epsilon)K_\epsilon = Q(\lambda_1 - \epsilon, \ldots, \lambda_k - \epsilon). \tag{7.26}$$

Because $M(\lambda_1 - \epsilon, \ldots, \lambda_k - \epsilon)$ is continuous and invertible for each value of ϵ and the right-hand side of (7.26) is continuous in ϵ, we have that K_ϵ is continuous in ϵ, hence $K_\epsilon \to K$ as $\epsilon \to 0+$. Consequently, with an infinitesimal change of K, the characteristic roots with real part equal to $\sup \Re(\lambda)$ are shifted to the left while no other characteristic roots can become dominant; see Chapter 1. Hence, we are not in a minimum. $\square$

For second-order systems, a complete characterization of the configuration of the rightmost characteristic roots in the optimum will be provided in Chapter 8, which will illustrate that different configurations are possible, depending on the system matrices and the delay.

Finally, we note that in Chapter 10 another eigenvalue based stabilization procedure will be discussed, which is directly based on the above outlined optimization point of view toward stabilization.

7.4 Illustrative examples

7.4.1 Model problem: stabilizing a third-order system

We illustrate the continuous pole placement method and its theoretical properties by means of the system

$$\dot{x}(t) = Ax(t) + Bu(t - \tau), \quad u(t) = K^T x(t), \tag{7.27}$$

where

$$A = \begin{bmatrix} -0.08 & -0.03 & 0.2 \\ 0.2 & -0.04 & -0.005 \\ -0.06 & 0.2 & -0.07 \end{bmatrix}, \quad B = \begin{bmatrix} -0.1 \\ -0.2 \\ 0.1 \end{bmatrix}, \quad \tau = 5. \tag{7.28}$$

The uncontrolled system is unstable ($\sup \Re(\lambda) = 0.108$) and with the control law $u(t) = K^T x(t)$, $K = [0.719\ 1.04\ 1.29]^T$, the rightmost characteristic roots are shown in Figure 7.4. Calculations were made using the MATLAB package DDE-BIFTOOL [74]. Note that, as $\tau \to 0+$, the three dominant characteristic roots converge to the ODE characteristic roots while the others move off to $-\infty$. Although the particular control law achieves stability for $\tau = 0$, the system is unstable for the nominal delay $\tau = 5$. We now describe its stabilization with the continuous pole placement method.

In Figure 7.5, the rightmost characteristic roots are shown as a function of the number of iterations taken in Algorithm 7.1. The corresponding values of the feedback gain $K = [k_1\, k_2\, k_3]^T$ are displayed in Figure 7.6. First we only reduce the real part of the dominant pair of complex conjugate characteristic roots, which is already sufficient to achieve stabilization. Meanwhile, the real part of an uncontrolled pair grows and after 50 iterations we reduce the real part of the two complex pairs of characteristic roots. In this way we lose control over the imaginary parts and because we want to avoid coinciding characteristic roots, which

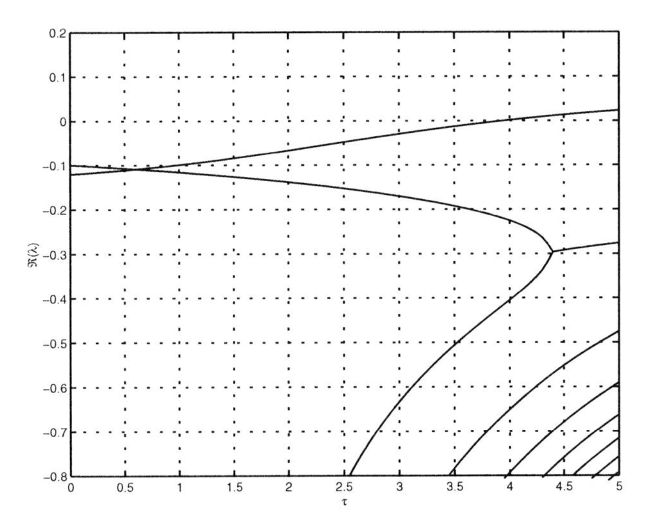

Figure 7.4. *Rightmost characteristic roots of the system (7.27)–(7.28) as a function of the delay, $K = [0.719\ 1.04\ 1.29]^T$.*

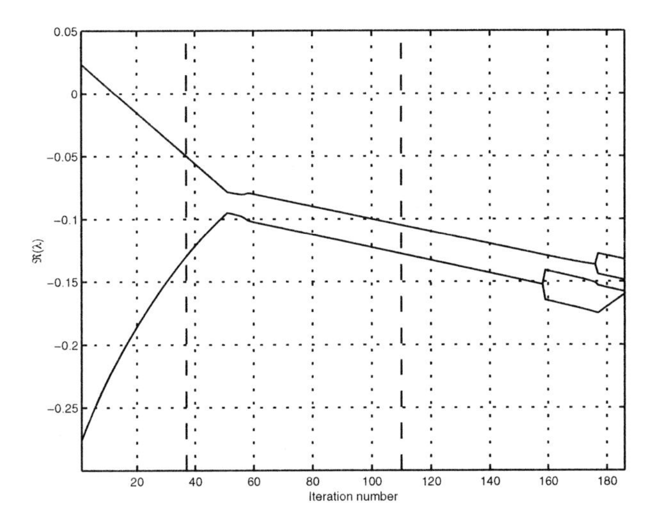

Figure 7.5. *Real parts of the rightmost characteristic roots of (7.27)–(7.28) with $\tau = 5$ as a function of the iterations of the continuous pole placement algorithm. For values of the feedback gain at iterations 37 and 110 (indicated by the dashed lines), and its value in the optimum, the rightmost characteristic roots are continued as a function of the delay in Figure 7.7.*

cause numerical problems (Theorem 7.5), we keep the real parts separated. However, around iteration 58, the characteristic roots are close to each other and the effect of a high sensitivity w.r.t. changes in the feedback gain is visible. At iteration 158, a complex pair splits into two real characteristic roots and all components of the feedback gain K are used to reduce

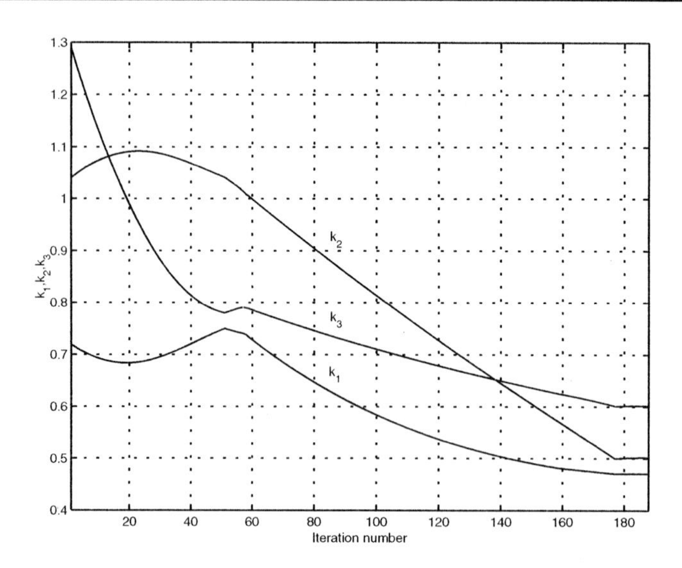

Figure 7.6. *Feedback gain $K = [k_1\ k_2\ k_3]^T$ as a function of the iterations of the continuous pole placement algorithm applied to (7.27)–(7.28).*

these real characteristic roots and the real part of the dominant pair. From iteration 176 on, where this pair also splits, we only control the three dominant real characteristic roots, and around iteration 190 no further reduction of $\sup \Re(\lambda)$ is possible and the method breaks down. At this point we are close to the minimum of $\sup \Re(\lambda)$ characterized by a rightmost characteristic root with multiplicity four at $\lambda = -0.150$. The corresponding feedback gain is given by $K = [0.471\ 0.504\ 0.602]^T$.

In Figure 7.7 the rightmost characteristic roots are displayed as a function of the delay at iterations 37 and 110, and in the optimum. This figure illustrates how the real parts of the dominant characteristic roots evolve toward the minimum for the nominal delay, characterized by characteristic roots with multiplicity larger than one and the resulting high sensitivity w.r.t. parameter changes (including the delay). This high sensitivity around the optimum is a property of the optimization problem under consideration, not of the algorithm. Our numerical experience indicates that for this type of problem, characteristic roots with a multiplicity larger than one are a general characteristic of the global minimum. Another illustration can be found in the next chapter, where all possible configurations of the rightmost characteristic roots in the global minimum described are given for second-order systems.

In our example, we faced the high sensitivity w.r.t. delay changes in the final iteration steps, because we executed Algorithm 7.1 until we reached the minimum. However, from a practical point of view, where also the robustness aspect is important, this is not appropriate. For instance, the feedback gain obtained at iteration number 37, see Figure 7.7, already achieves stability. The exponential decay rate of the closed loop solutions is smaller than for the optimum, yet less sensitive to delay changes.

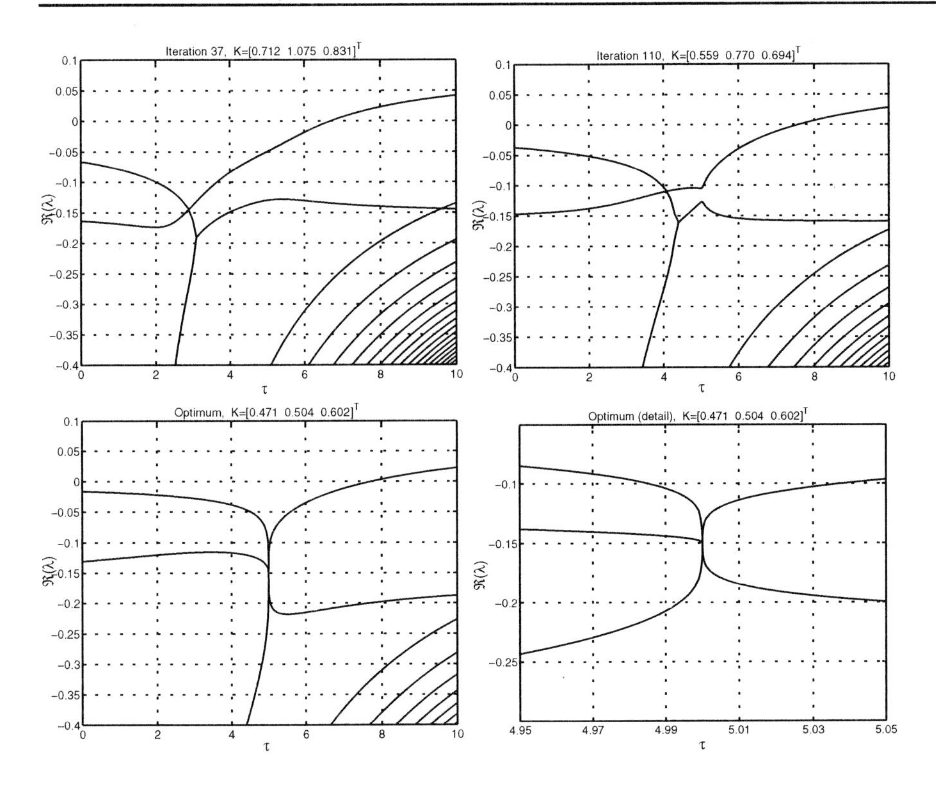

Figure 7.7. *Rightmost characteristic roots of (7.27)–(7.28) as a function of the delay, for values of K at iteration 37 (upper left), iteration 110 (upper right), and in the optimum (below). The corresponding values of the feedback gain are* $K = [0.712\ 1.075\ 0.831]^T$, $K = [0.559\ 0.770\ 0.694]^T$, *and* $K = [0.471\ 0.504\ 0.602]^T$.

7.4.2 General stabilization problems

The continuous pole placement method can be considered as a natural generalization of the classical pole placement algorithm to the input delay case. Since it is based on the continuous dependence of the rightmost characteristic roots on the controller parameters and because the algorithms for the computation of characteristic roots, described in Chapter 1, can also deal with equations with several discrete delays, it can easily be extended to more general types of linear delay equations involving delays in both state and control variables, and multiple feedback paths. Although theoretical stabilizability properties and convergence results, such as Theorem 7.4, are not generally valid and depend on the structure of the system under consideration, we will briefly illustrate with some examples the effectiveness of the method in solving more complicated stabilization problems.

As a first example consider the system

$$\begin{cases} \dot{x}(t) = A_1 x(t) + A_2 x(t-\tau_1) + B_1 u(t-\tau_2) + B_2 u(t-\tau_3), \\ u(t) = K^T x(t), \end{cases} \tag{7.29}$$

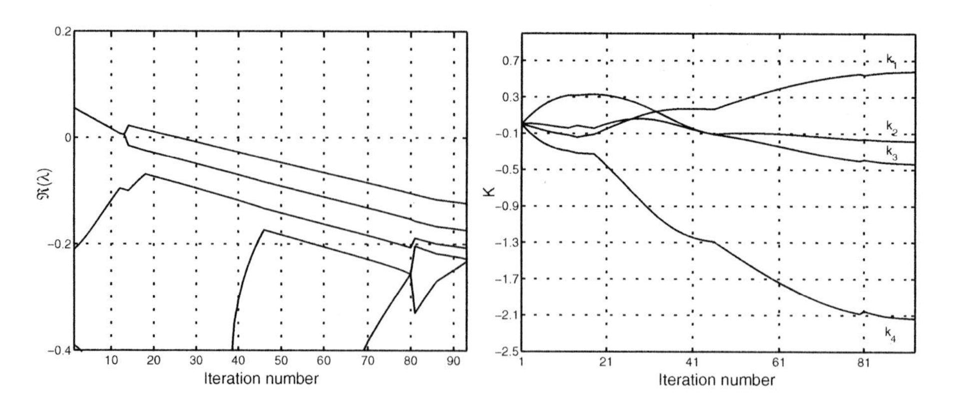

Figure 7.8. *Rightmost characteristic roots of (7.29) (left) and components of the feedback gain $K = [k_1\ k_2\ k_3\ k_4]^T$ (right) as a function of the iterations of the continuous pole placement algorithm.*

where

$$A_1 = \begin{bmatrix} 0.1 & -0.5 & -0.3 & 0.3 \\ 0 & 0 & 0.2 & -0.2 \\ 0 & 0 & -0.1 & 0 \\ 0 & 0.4 & 0.3 & -0.4 \end{bmatrix}, \quad A_2 = \begin{bmatrix} -0.2 & 0.3 & 0.2 & -0.5 \\ 0 & -0.2 & -0.2 & 0.2 \\ -0.1 & 0.1 & 0.1 & 0.1 \\ 0.1 & -0.6 & -0.6 & 0 \end{bmatrix},$$

$$B_1 = \begin{bmatrix} 0 \\ -0.1 \\ 0 \\ 0.1 \end{bmatrix}, \quad B_2 = \begin{bmatrix} -0.3 \\ 0 \\ 0 \\ 0 \end{bmatrix}, \quad \tau_1 = 2,\ \tau_2 = 1,\ \tau_3 = 3, \tag{7.30}$$

and a stabilizing feedback gain $K = [k_1\ k_2\ k_3\ k_4]^T$ needs to be determined. In Figure 7.8 the rightmost characteristic roots of (7.29) are shown as a function of the iterations of the continuous pole placement algorithm, as well as the components of the feedback gain. The initial value of the feedback gain is $[0\ 0\ 0\ 0]^T$. We have four control parameters and the method converges toward an optimum characterized by five coinciding real characteristic roots (in the last iteration step we already have a good approximation when taking into account the high sensitivity around the optimum).

Next we consider the case where the two inputs in (7.29) are independent,

$$\begin{cases} \dot{x}(t) = A_1 x(t) + A_2 x(t-\tau_1) + B_1 u_1(t-\tau_2) + B_2 u_2(u-\tau_3), \\ u_1(t) = K_1^T x(t), \quad u_2(t) = K_2^T x(t), \end{cases} \tag{7.31}$$

with the system parameters given by (7.30). In Figure 7.9, iterations of the continuous pole placement algorithm are shown. Compared to the system (7.29) a further reduction of the real parts of the dominant characteristic roots is possible, which is expected from the independent choice of both feedback gains. The algorithm also converges toward an optimum (w.r.t. all controller parameters), characterized by five coinciding real characteristic roots and hence, although there are eight controller parameters, it is not possible to further reduce

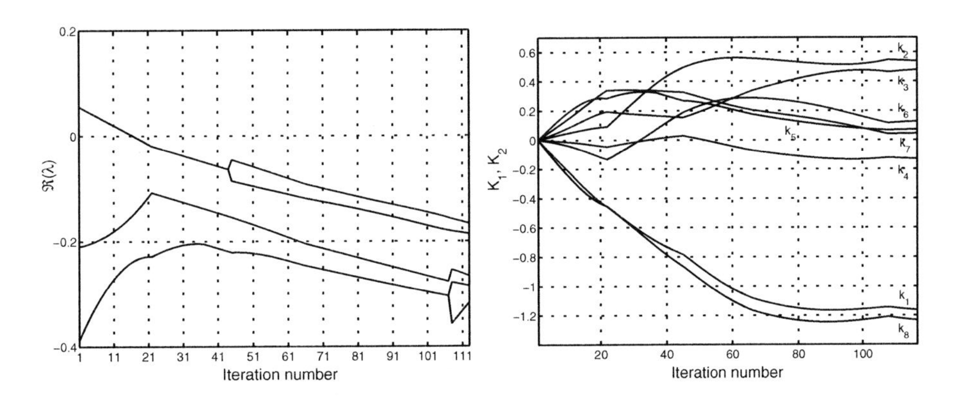

Figure 7.9. *Rightmost characteristic roots of (7.31) (left) and components of the feedback gains $K_1 = [k_1\ k_2\ k_3\ k_4]^T$ and $K_2 = [k_5\ k_6\ k_7\ k_8]^T$ (right) as a function of the iterations of the continuous pole placement algorithm.*

$\sup \Re(\lambda)$ by trying to control the real parts of more than four characteristic roots. This limitation is imposed by the structure of the controlled system, not by our algorithm. In this context we would like to emphasize first the danger of an "overparameterization" of the controller: having more degrees of freedom in the controller doesn't generally imply that more characteristic roots can be controlled or better stability results can be obtained. This is clearly illustrated in [195], where a stabilization problem with four controller parameters was studied, which could analytically be reduced to three effective controller parameters. Second, although the continuous pole placement algorithm uses a local strategy in each iteration step and in the beginning only a few characteristic roots are controlled, generically *all* controller parameters are adapted in an iteration step when using formula (7.20); see, e.g., Figure 7.9. Therefore the search space of the underlying optimization procedure (that is, finding a minimum of $\sup \Re(\lambda)$) is generally not limited to a subspace of the parameter space where, for instance, some controller parameters are not used and remain zero.

As a second example we consider the system

$$\begin{cases} \dot{x}(t) = A_1 x(t) + A_2 x(t-\tau_1) + \int_{t-\tau_1}^{t} A_3 x(s)ds + Bu(t-\tau_2), \\ u(t) = K^T x(t), \end{cases} \tag{7.32}$$

where

$$A_1 = \begin{bmatrix} 0.1 & 0 & 0 \\ 0.2 & 0 & -0.2 \\ 0.3 & 0.1 & -0.2 \end{bmatrix}, \quad A_2 = \begin{bmatrix} -0.2 & 0 & 0 \\ -0.4 & -0.2 & 0.4 \\ -0.4 & -0.1 & 0.2 \end{bmatrix},$$
$$A_3 = \begin{bmatrix} 0.1 & -0.2 & 0 \\ 0 & 0.1 & 0.1 \\ -0.1 & 0 & 0.1 \end{bmatrix}, \quad B = \begin{bmatrix} 0.1 \\ 0 \\ 0 \end{bmatrix}, \quad \tau_1 = 6, \quad \tau_2 = 1. \tag{7.33}$$

We can deal with the distributed delay term because differentiation of (7.32) leads to the

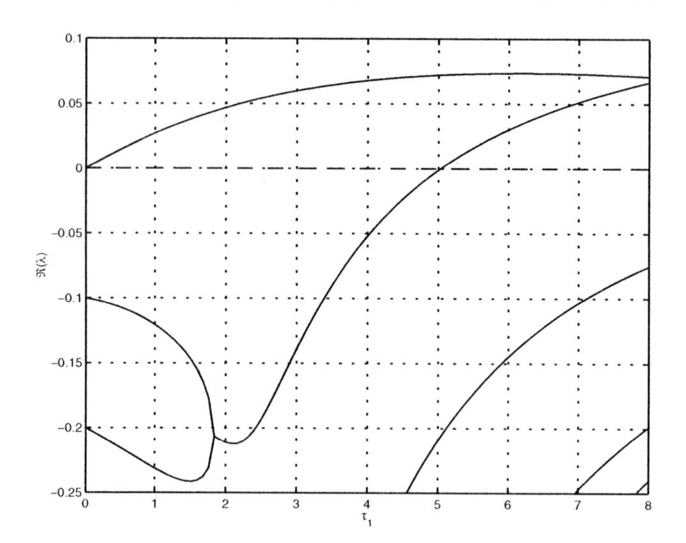

Figure 7.10. *Rightmost characteristic roots of the system (7.32)–(7.34) for $u = 0$ as a function of the delay τ_1. The solid lines correspond to the characteristic roots of (7.32). (7.34) has in addition a triple characteristic root at zero (dashed line).*

following equation with only discrete delays,

$$\dot{z} = \begin{bmatrix} 0 & I \\ A_3 & A_1 \end{bmatrix} z + \begin{bmatrix} 0 & 0 \\ -A_3 & A_2 \end{bmatrix} z(t-\tau_1) + \begin{bmatrix} 0 \\ BK^T \end{bmatrix} z(t-\tau_2), \tag{7.34}$$

where $z = [\ x\ \dot{x}\ ]^T$. It is easy to show that the transformation from (7.32) to (7.34) introduces n *additional* zero characteristic roots in the spectrum, with n the dimension of the system. However, the continuous pole placement method can cope with this problem because the zero characteristic roots can simply be removed after applying step B of Algorithm 7.1 to (7.34). In Figure 7.10 the characteristic roots of the uncontrolled system (7.32)–(7.34) are shown as a function of the delay τ_1. For the nominal delay $\tau_1 = 6$, there are three characteristic roots in the open RHP. Iterations of the continuous pole placement algorithm are shown in Figure 7.11. In Figure 7.12 we depict the characteristic roots of (7.32) for $K = 0$ and for the final value of the feedback gain. The continuous pole placement algorithm converges to an optimum characterized by five rightmost characteristic roots, a complex pair with multiplicity two and a real characteristic root. Such a situation also occurred in the three-parameter problem discussed in [195].

Remark 7.7. *The technique of transforming a system with distributed delays to a system with discrete delays at the price of the introduction of additional a priori known characteristic roots can be generalized to the case where the kernels of the distributed delays are gamma-distributions; see [162] and the references therein.*

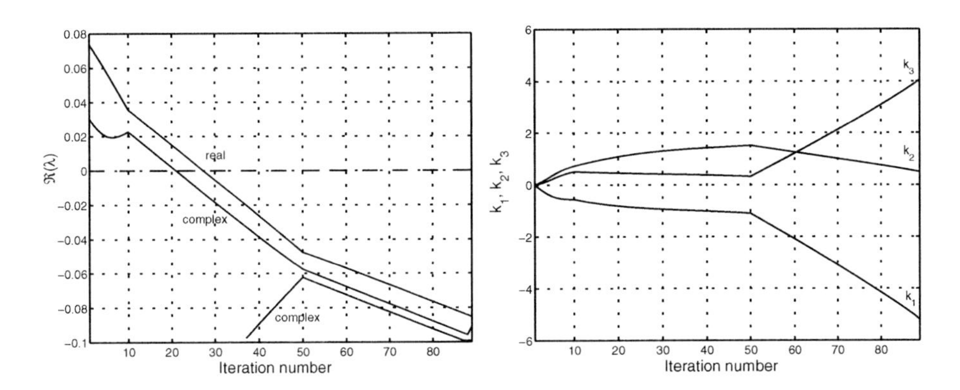

Figure 7.11. *Rightmost characteristic roots of (7.32)–(7.34) and components of the feedback gain $K = [k_1\ k_2\ k_3]^T$ (right) as a function of the iterations of the continuous pole placement algorithm.*

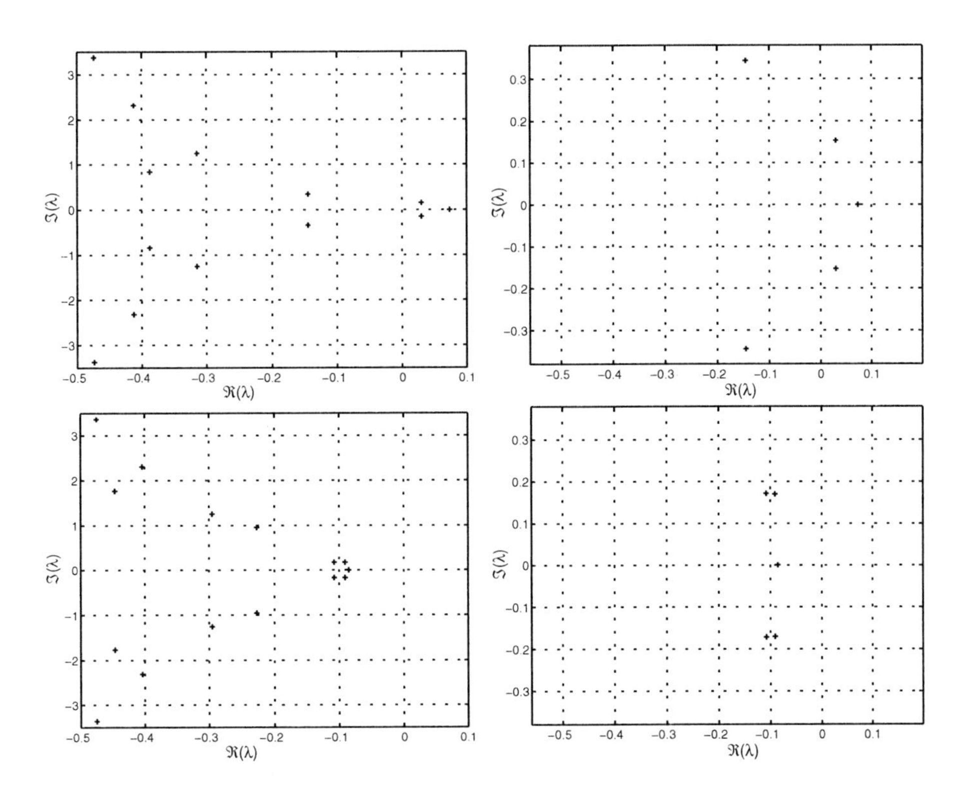

Figure 7.12. *Rightmost characteristic roots of (7.32) for $K = 0$ (above) and for the final value of the feedback gain (Iteration 89, $K = [-5.19\ 0.491\ 4.06]^T$) (below) on two different scales.*

7.5 Extensions of state feedback

7.5.1 Multiple input, multiple output systems

In this section we comment on the MIMO case. Because of the duality between controller and observer design, as spelled out in the previous section, we restrict ourselves to the multiple input case.

For the single input system (7.1), controllability of the pair (A, B) implies that A is cyclic, i.e., that the geometric multiplicity of each characteristic root is one; see [326, Chapter 1]. When the cyclic index of A, the maximal geometric multiplicity of its characteristic roots, is equal to k, at least k inputs are needed for controllability. In the ODE case, the full pole assignment procedure can then be split up into k single input problems. However, we illustrate that a complete decoupling induces additional restrictions on the characteristic roots in input delay cases, due to the infinite-dimensional nature of DDEs. Therefore, consider the multiple input system

$$\dot{x}(t) = Ax(t) + B\mathbf{u}(t-\tau), \quad A \in \mathbb{R}^{n\times n}, \quad B \in \mathbb{R}^{n\times m}, \tag{7.35}$$

where $\mathbf{u} = [u_1 \dots u_m]^T$ and (A, B) controllable. When the cyclic index of A is k, there always exist nonsingular transformations of the state, $z = T_x x$, and the input, $\mathbf{w} = T_u \mathbf{u}$, which transform (7.35) into

$$\dot{z}(t) = A_c z(t) + B_c \mathbf{w}(t-\tau), \tag{7.36}$$

where

$$A_c = \begin{bmatrix} A_{11} & & 0 \\ & \ddots & \\ 0 & & A_{kk} \end{bmatrix}, \quad B_c = \left[\begin{array}{ccc|ccc} B_{11} & \cdots & B_{1k} & \dots & & B_{1m} \\ & \ddots & \vdots & & & \vdots \\ 0 & & B_{kk} & \dots & & B_{km} \end{array}\right],$$

with A_{ii} cyclic and the pair (A_{ii}, B_{ii}) controllable for each $i = 1, k$ (see [326, p. 44]). Hence, a state feedback of the form $\mathbf{w} = K_c^T z$, where

$$K_c = \left[\begin{array}{ccc|c} K_{11} & & 0 & \\ & \ddots & & 0 \\ 0 & & K_{kk} & \end{array}\right], \tag{7.37}$$

yields a triangular closed-loop system. The characteristic roots are determined by the k (single input) systems

$$\dot{z}_i(t) = A_{ii} z_i(t) + B_{ii} K_{ii}^T z_i(t-\tau), \quad i = 1, \dots, k, \tag{7.38}$$

and the continuous pole placement can be applied to each subsystem to calculate the controller parameters.

Although for $\tau = 0$ the pole assignment problem is completely solved with the feedback (7.37), we now show that in the DDE case better stability results can be obtained when using all the components of the feedback gain in the continuous pole placement procedure and, hence, a design based on (7.38) is less appropriate. Note that choosing the "off-diagonal" elements in (7.37) nonzero results in a coupling between the subsystems (7.38), while using the last $m-k$ columns of (7.37) consists of using also these inputs which are superfluous for controllability or pole assignment in the ODE case.

To illustrate the beneficial effect of coupling the subsystems, consider the system (7.36) with

$$A_c = \left[\begin{array}{cc|c} 1 & 1 & 0 \\ 0 & -1 & 0 \\ \hline 0 & 0 & 1 \end{array}\right], \quad B_c = \left[\begin{array}{c|c} 0 & 0 \\ 1 & 0 \\ \hline 0 & 1 \end{array}\right], \quad \tau = 1. \tag{7.39}$$

Since the cyclic index of A is 2, two inputs are necessary. With a decoupling feedback of the form (7.37), we obtain a second subsystem of the form

$$\dot{z}_2(t) = z_2(t) + kz_2(t-\tau), \quad z_2(t) \in \mathbb{R}, \tag{7.40}$$

which was analyzed in Section 7.2.3. Since this resulted in $\inf_k \sup \Re(\lambda) = 0$, asymptotic stabilization is not possible. However, numerical calculations show that the (coupling) feedback gain

$$K_c = \begin{bmatrix} -2.010843017 & 0 \\ -1.064788858 & 0.04112520635 \\ 0.1 & -1.075793522 \end{bmatrix}$$

achieves asymptotic stability, with $\sup \Re(\lambda) = -0.1550803$ (this corresponds to a minimum characterized by a real rightmost characteristic root with multiplicity 5).

To illustrate the beneficial effect of using more inputs than the cyclic index, we refer to the next chapter, where the stabilizability of second-order systems with input delays is studied in detail.

7.5.2 Observer based controllers

Consider the system (7.1) and assume that not the full state x is available for measurement, but only an output $y(t) = C^T x(t) + Du(t-\tau) \in \mathbb{R}$, where (C, A) is observable. In this case the system can be equivalently described by the transfer function

$$\frac{Y(s)}{U(s)} = (C^T(sI-A)^{-1}B + D)e^{-s\tau}. \tag{7.41}$$

Note that *any system* of the form

$$\dot{x}(t) = Ax(t) + Bu(t-\tau_1), \quad y = C^T x(t-\tau_2) + Du(t-\tau), \tag{7.42}$$

with $\tau_1 + \tau_2 = \tau$, is a realization of (7.41). In order to apply the continuous pole placement method, we construct an observer for (7.42),

$$\dot{\hat{x}}(t) = A\hat{x}(t) + Bu(t-\tau_1) + L(C^T\hat{x}(t-\tau_2) + Du(t-\tau) - y), \tag{7.43}$$

with $L \in \mathbb{R}^{n\times 1}$ the observer gain, and apply the feedback law $u = K^T\hat{x}(t)$. With $e(t) = x(t) - \hat{x}(t)$ the observer error, we obtain

$$\begin{cases} \dot{x}(t) = Ax(t) + BK^T x(t-\tau_1) - BK^T e(t-\tau_1), \\ \dot{e}(t) = Ae(t) + LC^T e(t-\tau_2). \end{cases} \tag{7.44}$$

Since in the characteristic equation of the closed loop system (7.44),

$$\det\left(\lambda I-\begin{bmatrix} A & 0 \\ 0 & A \end{bmatrix}-\begin{bmatrix} BK^T & -BK^T \\ 0 & 0 \end{bmatrix}e^{-\lambda\tau_1}-\begin{bmatrix} 0 & 0 \\ 0 & LC^T \end{bmatrix}e^{-\lambda\tau_2}\right)=0, \quad (7.45)$$

all matrices are block triangular, the separation principle is valid, and the closed-loop characteristic roots consist of the solutions of $\det(\lambda I - A - BK^T e^{-\lambda\tau_1}) = 0$, the controller characteristic roots, and of $\det(\lambda I - A - LC^T e^{-\lambda\tau_2}) = \det(\lambda I - A^T - CL^T e^{-\lambda\tau_2}) = 0$, the observer characteristic roots. Hence, to achieve stability the continuous pole placement can be applied twice, to the systems (A, B, τ_1) and (A^T, C, τ_2).

Because in the construction of the realization (7.42) we are free to distribute the delay over input and output, we can use this additional degree of freedom to obtain better performance of plant and observer. It also allows us to extend the class of systems which are stabilizable with (delayed) state feedback. To see this, reconsider the system (7.10) with output $y = x$. Although the full state is available for feedback, it is useful to construct an observer. The transfer function is given by $G_\tau(s) = \frac{e^{-s\tau}}{s-a}$ and we construct an observer based on the realization

$$\dot{x}(t) = ax(t) + u(t-\tau_1), \quad y(t) = x(t-\tau_2), \quad \tau_1 + \tau_2 = \tau.$$

With the observer based controller

$$\dot{\hat{x}}(t) = a\hat{x}(t) + u(t-\tau_1) + l(\hat{x}(t-\tau_2) - y(t)), \quad u(t) = k\tilde{x}(t), \qquad (7.46)$$

the closed-loop characteristic roots consist of the characteristic roots of the systems

$$\dot{x}(t) = ax(t) + kx(t-\tau_1) \text{ and } \dot{e}(t) = ae(t) + le(t-\tau_2). \qquad (7.47)$$

Since these equations are of the form (7.11), the optimal stabilizing controller (i.e., values of k and l which minimize $\sup \Re(\lambda)$) results in

$$\sup \Re(\lambda) = \max\left(a - \frac{1}{\tau_1}, a - \frac{1}{\tau_2}\right),$$

and the best stabilizability results are obtained when distributing the delay equally over input and output. Asymptotic stability can be achieved if and only if

$$a\tau < 2, \qquad (7.48)$$

which is twice as good as (7.12), valid for pure state feedback, and better than (7.13), the "practical" condition in case of finite spectrum assignment.

Remark 7.8. *The idea of the delay distribution over input and output can be generalized. One can theoretically achieve stability for any given delay value by constructing a sufficiently large number of observers, each with a different distribution of the delay over input and output.*

7.5.3 Finite-dimensional dynamic state feedback

Another approach to obtain better stabilizability properties of (7.1) consists of *finite-dimensional* feedback of an *augmented* state; i.e., we may use a controller of the form

$$\begin{aligned} \dot{z}(t) &= Fz(t) + Gx(t), \quad z(t) \in \mathbb{R}^m, \\ u(t) &= Kz(t) + Lx(t). \end{aligned} \tag{7.49}$$

The underlying idea is as follows: in order to cope with a large input delay, the controller should at least implicitly use a prediction of the state variable. When the dimension m of the dynamical system (7.49) becomes larger, it should be able to generate a better prediction of the state variable and, as $m \to \infty$, yield in some sense an approximation of the infinite-dimensional control law (7.7).

Let's now apply the controller (7.49) for $m = 1$ to our theoretical example (7.10). This yields the closed-loop system

$$\left\{ \begin{array}{l} \dot{z}(t) = fz(t) + gx(t), \\ \dot{x}(t) = ax(t) + kz(t-\tau) + lx(t-\tau), \quad f, g, k, l \in \mathbb{R}, \end{array} \right. \tag{7.50}$$

which can be interpreted as a finite-dimensional second-order system with free parameters f, g and controlled with delayed state feedback. In [195] a complete quantitative characterization of the stabilizability of second-order systems is given. Applying these results also leads to the improved stabilizability condition $a\tau < 2$.

7.6 Systems of neutral type

We extend Algorithm 7.1 to systems of neutral type of the form

$$\begin{aligned} \tfrac{d}{dt}\left(x(t) - \textstyle\sum_{k=1}^{m} H_k x(t-\tau_k)\right) = A_0 x(t) + \\ \textstyle\sum_{k=1}^{m} A_k x(t-\tau_k) + B_0 u(t) + \sum_{k=1}^{m} B_k u(t-\tau_k), \end{aligned} \tag{7.51}$$

where

$$x(t) \in \mathbb{R}^n, \ u(t) \in \mathbb{R}, \ 0 < \tau_1 < \cdots < \tau_m,$$

and one aims at finding a stabilizing controller of the form

$$u(t) = K^T x(t). \tag{7.52}$$

In [199] it is explained how various stabilization problems involving systems of neutral type can be rephrased in the form (7.51)–(7.52).

First, we outline the stabilization algorithm. Next, we present a numerical example.

7.6.1 Algorithm

The algorithm of [199] relies on the results presented in Section 1.2 on the sensitivity of stability of the zero solution of (7.51)–(7.52) w.r.t. small delay perturbations. The main steps are as follows (using the notations of Section 1.2).

ALGORITHM 7.2.

Continuous pole placement for neutral systems

A. Determine whether the associated difference equation,

$$x(t) = \sum_{k=1}^{m} H_k x(t - \tau_k),$$

is strongly stable by applying Proposition 1.28. If not, stop. If yes, compute the spectral upper bound $\bar{C}_D(\vec{\tau})$ by applying Theorem 1.30. The properties of $f(c;\ \vec{\tau})$ make a bisection algorithm appropriate to find its zero.

B. Initialize $q = 1$ and choose $\epsilon > 0$ small.

C. Compute the characteristic roots with $\Re(\lambda) \geq \bar{C}_D(\vec{\tau}) + \epsilon$.

D. Compute the sensitivity of the q rightmost characteristic roots w.r.t. changes in the feedback gain K.

E. Move the q rightmost characteristic roots in the direction of the LHP by applying a small change to the feedback gain K, using the computed sensitivities.

F. Monitor the uncontrolled characteristic roots with $\Re(\lambda) > \bar{C}_D(\vec{\tau}) + \epsilon$. If necessary, increase the number of controlled characteristic roots q. Stop when stability is reached, or when the available degrees of freedom in the controller do not allow $\sup \Re(\lambda)$ to be further reduced, or when the leftmost of the controlled characteristic roots reaches the upper bound $\bar{C}_D(\vec{\tau})$. In the other case, go to step C.

The main difference w.r.t. Algorithm 7.1 lies in the preliminary step, which consists of analyzing the stability of the associated difference equation and computing the safe upper bound on the real parts of its characteristic roots, $\bar{C}_D(\vec{\tau})$. This step is necessary because

1. Only neutral systems with $\bar{C}_D(\vec{\tau}) < 0$ (which is equivalent to the strong stability condition) can be stabilized safely (that is, with the achieved stability insensitive to small delay variations).

2. If $\bar{C}_D(\vec{\tau}) > 0$, then instability may be caused by characteristic roots with very large imaginary parts, which easily remain undetected when only computing a *finite* number of characteristic roots (inherent to any numerical scheme). In other words, making stability assertions without the explicit computation of $\bar{C}_D(\vec{\tau})$ is unsafe. See the left frame of Figure 1.6 for an example of a root chain which becomes unstable at very high frequencies—notice the difference in scaling of both axes.

3. The knowledge of $\bar{C}_D(\vec{\tau})$ avoids controlling characteristic roots with smaller real part, which would be useless from a stabilization point of view.

For details regarding the implementation of Steps B–F, being analogous to the retarded case, practical aspects, and convergence properties of the iterative scheme, we refer to [199]. A numerical example is presented in the next section.

7.6.2 Illustrative example

Consider the system

$$\frac{d}{dt}\left(x(t) - H_1 x(t-\tau_1) - H_2 x(t-\tau_2)\right) = A\,x(t) + Bu(t-0.5), \tag{7.53}$$

where

$$H_1 = \begin{bmatrix} 0 & 0.2 & -0.4 \\ -0.5 & 0.3 & 0 \\ 0.2 & 0.7 & 0 \end{bmatrix}, \quad H_2 = \begin{bmatrix} -0.3 & -0.1 & 0 \\ 0 & 0.2 & 0 \\ 0.1 & 0 & 0.4 \end{bmatrix},$$

$$A = \begin{bmatrix} -4.8 & 4.7 & 3 \\ 0.1 & 1.4 & -0.4 \\ 0.7 & 3.1 & -1.5 \end{bmatrix}, \quad B = \begin{bmatrix} 0.3 \\ 0.7 \\ 0.1 \end{bmatrix}, \quad \vec{\tau} = (0.7,\ 1.7).$$

The uncontrolled neutral system is unstable, because it has one real characteristic root $\lambda_1 \approx 0.2180$ and one pair of complex conjugate characteristic roots $\lambda_{2,3} \approx 0.0976 \pm 1.0396i$, located to the right of the stability boundary.

Performing step A of Algorithm 7.2 yields $\gamma_0 = 0.7507$, that is, the difference equation is strongly stable, and $\bar{C}_D = -0.2751$. By applying steps B–F we may attempt to stabilize the neutral system with the control law (7.52). Since the feedback gain has three parameters, whose initial values are $K = [k_1\ k_2\ k_3]^T = [0\,0\,0]^T$, up to three characteristic roots (couples of characteristic roots if they are complex) can be controlled, i.e. continuously shifted to the left. The results of iterations of the continuous pole placement algorithm are shown in Figure 7.13. First, only the real characteristic root λ_1 is being shifted to the left until it gets close to the real part of the pair $\lambda_{2,3}$. Then, from iteration number 10 on, the pair is also being shifted. In order to ensure a numerically stable computation, the distances in real parts of the characteristic roots are kept larger than a default value. At iteration number 51, the couple $\lambda_{4,5}$ is attached to the group of the controlled characteristic roots. Since no more characteristic roots can be controlled by three parameters of the feedback gain, the procedure terminates at iteration 98 when characteristic root λ_6 gets close to $\lambda_{4,5}$. Using the resulting feedback gain $K = [-2.593\ 1.284\ 1.826]^T$, the system is asymptotically stable.

To illustrate the impact of the stabilizing feedback gain on the system dynamics, we show in Figure 7.14 the rightmost characteristic roots of the uncontrolled system and of the system stabilized with the feedback gain resulting from Algorithm 7.2.

7.7 Notes and references

We have shown that the classical pole placement method for systems without delays can be adapted to time-delay systems, where the closed loop system is infinite-dimensional and the number of degrees of freedom of the controller is finite. The method consists of controlling only the rightmost or unstable characteristic roots. Because the controlled characteristic roots are shifted in a quasi-continuous way, the method was called continuous pole placement. It was explained by means of the stabilization of a linear finite-dimensional system in the presence of an input delay, using static state feedback. Various extensions of state feedback were considered. Also the stabilization of systems of neutral type was dealt

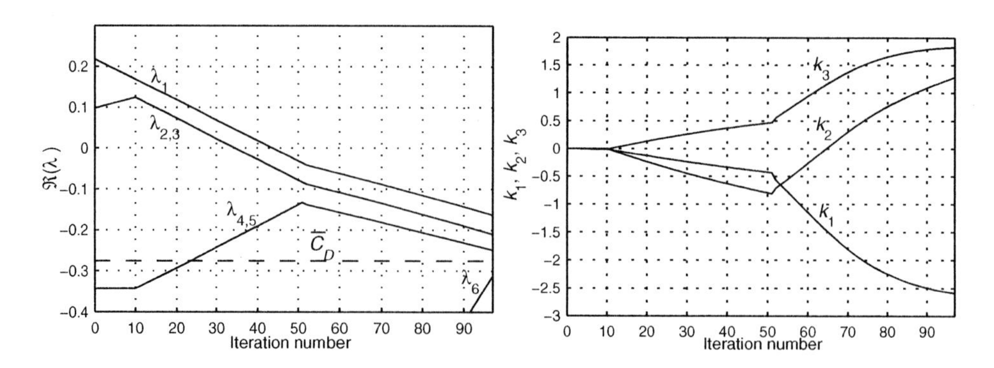

Figure 7.13. *Results of the continuous pole placement procedure applied to the system (7.53). (left) Real parts of the controlled characteristic roots. (right) Coefficients of the feedback gain.*

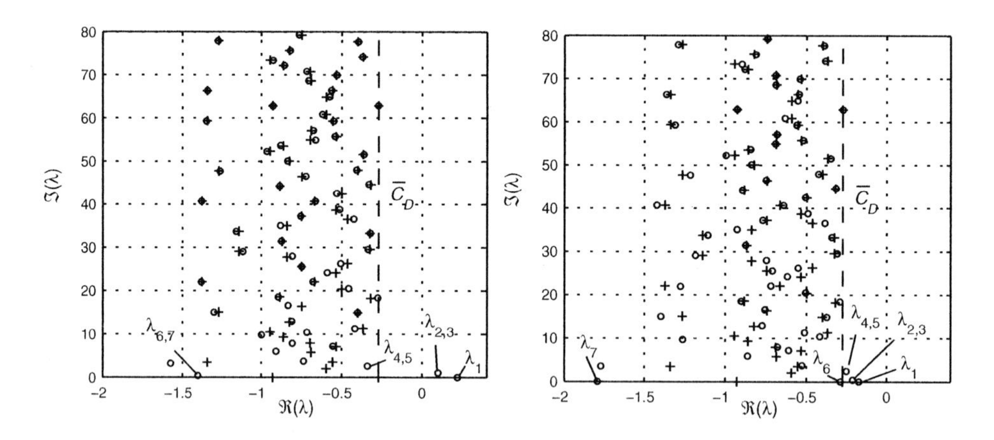

Figure 7.14. *(left) Spectrum of neutral system (7.53) for $u = 0$. (right) Spectrum of stabilized neutral system (7.53). + - spectrum of difference equation associated to (7.53).*

with. The proposed stabilization approach is especially suitable for the (hard) stabilization problem for *fixed* delay values; it is *constructive* and *no conservatism* is introduced, in the sense that the existence of a stabilizing controller implies that the latter can be calculated with the numerical procedure.

As outlined in Section 7.3.3, the stabilization problem considered in this chapter can be interpreted as an optimization problem. A related approach, which is directly based on such optimization point of view toward stabilization, will be presented in Chapter 10.

Note that we have not explicitly discussed controllability and other structural properties. For this, we refer to [267, 82] and the references therein.

The chapter is based on [182, 199, 180, 194] and the references therein.

Chapter 8

Stabilizability with delayed feedback: a numerical case study

8.1 Introduction

In the previous chapter we proposed the continuous pole placement method for the stabilization of linear time-delay systems. It was illustrated by means of the stabilization of the linear controllable single input system (7.1) in the presence of an input delay using static state feedback. This resulted in the closed-loop system

$$\dot{x}(t) = Ax(t) + BK^T x(t-\tau). \tag{8.1}$$

Some restrictions of delayed state feedback were already considered in Sections 7.2.1 and 7.2.3. In this chapter we completely characterize the class of all second-order systems which are stabilizable with delayed state feedback.

Notice that the closed-loop system (8.1) is of the form

$$\dot{x}(t) = Ax(t) + Mx(t-\tau), \tag{8.2}$$

where the "control matrix" $M = BK^T$ is of rank 1. We also consider the case where all the elements of M can be chosen independently. Then (8.2) can be interpreted as the feedback controlled *multiple input* system,

$$\dot{x}(t) = Ax(t) + I\mathbf{u}(t-\tau), \quad \mathbf{u}(t) = Mx(t). \tag{8.3}$$

Even though the controllability assumption of (A, B) in (8.1) implies that A is cyclic and, generically, a rank-1 feedback is sufficient for pole assignment in the ODE case, we will show that the use of two inputs allows us to extend the class of stabilizable systems considerably, as we announced in Section 7.5.1. Furthermore, the availability of two inputs forms no restriction, since it is shown in [194] that it is always possible to construct a stabilizing controller for (7.1), if the stabilization problem (8.3) can be solved. More precisely, the observer-predictor based controller

$$\begin{aligned}\dot{\hat{x}}(t) &= A\hat{x}(t) + Bu(t) + M(\hat{x}(t-\tau) - x(t)),\\ u(t) &= K^T\hat{x}(t)\end{aligned}$$

results in the closed-loop system

$$\begin{cases} \dot{x}(t) = (A + BK^T)x(t) + BK^T e(t), \\ \dot{e}(t) = Ae(t) + Me(t-\tau), \end{cases} \tag{8.4}$$

where $e(t) = \hat{x}(t-\tau) - x(t)$ is the prediction error. Because of the triangular structure, the closed loop system is stable if and only if $(A + BK^T)$ is Hurwitz and the e-subsystem of (8.4) is asymptotically stable. Notice that a similar type of control law is used in the context of predictive synchronization; see [238] and the references therein.

Although the above stabilizability problems are independent of practical numerical methods for calculating the feedback gain, the analysis in this chapter starts from a repeated application of the continuous pole placement procedure, because this reveals some important properties of the systems which form the starting point of the stabilizability analysis. The followed approach further relies on methods and tools from the field of numerical *bifurcation analysis* [270].

In Section 8.2 we characterize all second-order systems which are stabilizable for a fixed, given value of the time delay. In Section 8.3 we discuss the stabilizability with the same controller gain for all delays in a given interval including zero, that is, the stabilizability problem for the corresponding problem without delay and with a guaranteed delay margin.

8.2 Characterization of stabilizable systems

We illustrate the limitations of delayed state feedback and characterize all stabilizable second-order systems. Therefore, consider equation (8.2) with $A, M \in \mathbb{R}^{2\times 2}$. We simultaneously treat both the case where M is of rank 1, for instance $M = BK^T$, and the case where the components of M can be chosen independently. For simplicity we first assume that the time-delay τ is equal to one. The case $\tau \neq 1$ is treated separately. At the end, we also consider the special case where the system matrix A is noncyclic.

8.2.1 System representation

In the second-order case and for $\tau = 1$, system (8.2) can be transformed into the canonical form

$$\dot{x} = \begin{bmatrix} 0 & 1 \\ -a_2 & -a_1 \end{bmatrix} + \begin{bmatrix} -k_4 & -k_3 \\ -k_2 & -k_1 \end{bmatrix} x(t-1). \tag{8.5}$$

Furthermore, when $M = BK^T$, with (A, B) controllable, it is always possible to obtain the form (8.5) with $k_3 = k_4 = 0$, which then reduces to the so-called control canonical form. The characteristic function of (8.5) is given by

$$H(\lambda) = \lambda^2 + (a_1 + \bar{k}_1 e^{-\lambda})\lambda + (a_2 + \bar{k}_2 e^{-\lambda}) + \bar{k}_3 e^{-2\lambda}, \tag{8.6}$$

where

$$\begin{cases} \bar{k}_1 = k_1 + k_4, \\ \bar{k}_2 = k_2 + a_1 k_4 - a_2 k_3, \\ \bar{k}_3 = k_1 k_4 - k_2 k_3. \end{cases} \tag{8.7}$$

In the remainder of this section, we study the stability of the quasipolynomial (8.6) as a function of the three control parameters $\bar{k}_1, \bar{k}_2, \bar{k}_3$. In this way we eliminate the redundancy in (8.5). However, it turns out that $\bar{k}_1, \bar{k}_2, \bar{k}_3$ cannot always be considered as free parameters, since for any given value of these parameters, corresponding values for k_1, k_2, k_3, k_4 in (8.5) should be found (using the relations (8.7)). This is not always possible.

When M is of rank 1, we have $\bar{k}_3 = 0$, and $\bar{k}_1$ and $\bar{k}_2$ can be freely assigned by an appropriate choice of k_1 and k_2.

When the control term is of full rank, i.e., $\bar{k}_3 \neq 0$, we have a constraint on the possible values of the control parameters $\bar{k}_1, \bar{k}_2, \bar{k}_3$. Indeed, eliminating k_1 and k_2 from the first two equations of (8.7) and substituting into the last one leads to the quadratic form

$$k_4^2 + a_2 k_3^2 - a_1 k_3^2 - a_1 k_3 k_4 + \bar{k}_1 k_4 - \bar{k}_2 k_3 - \bar{k}_3 = 0.$$

When $a_2 - a_1^2/4 \neq 0$, this expression can be written as

$$Ax^2 + y^2 - B = 0,$$

where

$$x = k_3 + \frac{\bar{k}_1 a_1/2 - \bar{k}_2}{2(a_2 - a_1^2/4)}, \qquad y = k_4 - a_1 k_3/2 + \bar{k}_1/2,$$
$$A = a_2 - a_1^2/4, \qquad B = \bar{k}_3 + \bar{k}_1^2/4 + \frac{\left(\bar{k}_1 a_1/2 - \bar{k}_2\right)^2}{4(a_2 - a_1^2/4)}.$$

When $A > 0$ this equation can only have real solutions if $B \geq 0$. Hence, the constraint on the control parameters is given by

$$\bar{k}_3 \geq -\left(\bar{k}_1^2/4 + \frac{\left(\bar{k}_1 a_1/2 - \bar{k}_2\right)^2}{4(a_2 - a_1^2/4)}\right), \tag{8.8}$$

whenever $a_2 - \frac{a_1^2}{4} > 0$. When $a_2 = a_1^2/4$, one can show that no constraints on the control parameters are present, except for a degenerate case which does not occur in the procedure followed in this chapter.

8.2.2 Class of stabilizable systems for the unit delay

As follows from the canonical form (8.5) and the characteristic equation (8.6), stabilizability conditions on the system can be expressed as a function of its parameters a_1 and a_2. Note that $\lambda^2 + a_1\lambda + a_2 = 0$ is the characteristic polynomial of the (uncontrolled) system.

Define

$$c(a_1, a_2) = \min_M F(M, a_1, a_2), \tag{8.9}$$

where

$$F(M, a_1, a_2) := \sup\left\{\Re(\lambda) : \ H(\lambda) = \det\left(\lambda I - A - Me^{-\lambda}\right) = 0\right\}. \tag{8.10}$$

Then in the (a_1, a_2)-plane, stabilizable and unstabilizable systems are separated by curves on which $c(a_1, a_2) = 0$. We refer to these curves as the *stabilizability boundary*.

In Figure 8.1, the stabilizability boundary is depicted. Before discussing its properties in Section 8.2.4, we outline its computation.

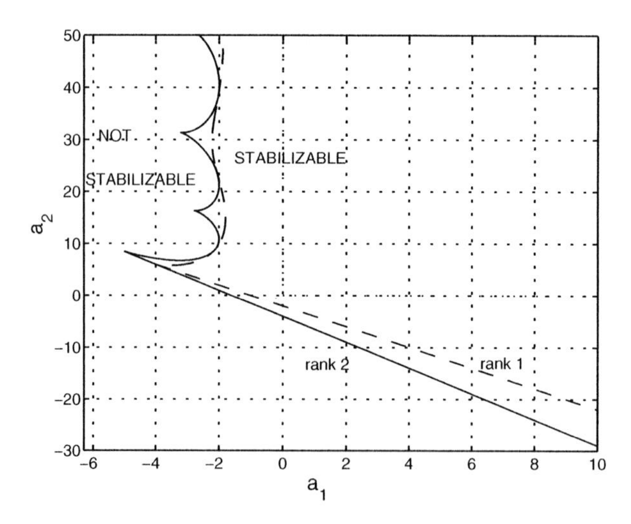

Figure 8.1. *Stabilizability boundary of (8.2)–(8.5) for $\tau = 1$ when M is of rank 1 (dotted line) and M is of rank 2 (full line). The characteristic equation of the uncontrolled system is given by $\lambda^2 + a_1\lambda + a_2 = 0$.*

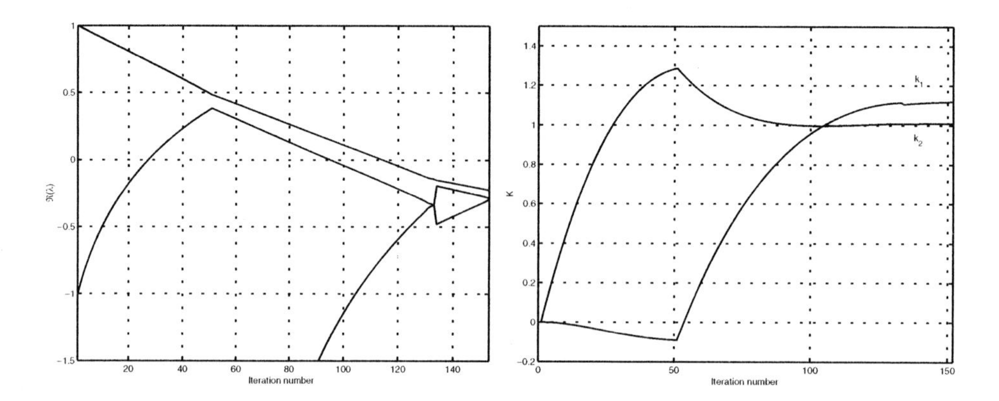

Figure 8.2. *Rightmost characteristic roots (left) and controller parameters, $K = [k_1\ k_2]^T$, (right) for the system (8.5) in case of rank-1 feedback (i.e., k_3=k_4=0), as a function of the iterations of the continuous pole placement algorithm; $(a_1, a_2) = (0, -1)$. The system is stabilizable and the method converges to the optimum, characterized by a real characteristic root with multiplicity 3.*

An exhaustive approach would consist of applying the continuous pole placement method for a large number of values of the plant parameters (a_1, a_2), chosen on a fine grid, to check whether the system is stabilizable. Such a simulation is shown in Figure 8.2. The stabilizability boundary then separates the regions where the system is stabilizable and where it is not. However, a more efficient calculation is possible by taking specific properties of the optimization problem (8.9) into account. This is now explained in detail for the case where M is of rank 1. Similar ideas apply to the case where M is of full rank.

From the analysis in Section 7.3.3, it follows that the optimization problem (8.9) is not differentiable. Moreover, we have observed from our numerical experiments that the derivative of the objective function (8.10) does not always exist in the minimum, due to characteristic root with multiplicity larger than one, and, hence, the optimum cannot be calculated directly using the relations $\frac{\partial F}{\partial m_i} = 0$, where m_i is a component of M. For instance, for the example of Figure 8.2 the minimum is characterized by a rightmost characteristic root with multiplicity one. An application of Theorem 7.6 yields the following general result (which is easily shown to hold also when M has rank 2).

Proposition 8.1. *When the function $M \to F(M, a_1, a_2)$ with $M = BK^T$ is minimal, there are at least three characteristic roots with real part equal to $c(a_1, a_2)$.*

As a consequence, the possible configurations of the rightmost characteristic roots at the global minimum of (8.10) can be reduced to the four basic situations shown in Table 8.1, where we also show the mathematical relations characterizing the position of the rightmost characteristic roots. In our numerical experiments only situations I and II occur. Theoretically situations I′ and II′ are possible, but are less generic than situations I and II.

Indeed, in cases I and II, the mathematical relations allow a direct computation of the minimal value c of (8.10) and the corresponding control parameters $\bar{k}_1$ and $\bar{k}_2$ when good starting values are available. However, in cases I′ and II′ there is one extra unknown variable and therefore, the mathematical relations in Table 8.1 define curves in their unknowns, through the point corresponding to the optimal parameter values. In such situations, a suitable small parameter change generically reduces the value of c, meaning that the situation does not correspond to the global minimum, unless an extra condition is satisfied (e.g., a turning point on the curve, or another characteristic root becoming the rightmost one). Our numerical experiments indicate that this extra condition is characterized by $\omega = 0$ for case I′, which then reduces to case I, and by $\omega_2 = \omega_1$ for case II′, which reduces to case II. These extra conditions are in some sense natural because situation II occurs when the uncontrolled system is highly oscillatory, with ω approximating the natural frequency, whereas situation I occurs when it is highly damped, and in these cases an extra "dominant frequency" in the controlled system is not expected.

Using the information contained in Table 8.1, the calculation of the stabilizability boundary only requires us to apply the continuous pole placement method for plant parameters chosen on a *coarse* grid in the (a_1, a_2)-plane. Indeed, suppose that we have applied the method for the plant parameters $(a_1^{(i)}, a_2^{(i)})$, $i = 1, 2$, leading to $c(a_1^{(1)}, a_2^{(1)}) > 0$ and $c(a_1^{(2)}, a_2^{(2)}) < 0$. Then in the (a_1, a_2)-plane these two pairs are separated by the stabilizability boundary. Moreover, when $c(a_1^{(i)}, a_2^{(i)})$ is sufficiently close to zero, information is available about the type of the minimum (I or II) to be expected on the stabilizability boundary in that neighborhood. This allows us to compute points on the stabilizability boundary directly using the mathematical relations displayed in Table 8.2. Compared to Table 8.1, here the extra condition $c = 0$ is required, while the parameters a_1 and a_2 are freed. Hence, the mathematical relations in Table 8.2 define a branch which can be numerically continued in an efficient way. Good starting values are obtained by the results of the continuous pole placement method for the plant parameters $(a_1^{(i)}, a_2^{(i)})$. The emanating branch forms part of the stabilizability boundary, until it has an intersection with another branch.

Table 8.1. *According to Proposition 8.1, four configurations of the rightmost characteristic roots are possible in the global minimum of (8.10), when M is of rank* 1. *Only situation I and II occur. In the second column the mathematical relations, which characterize each configuration, are displayed.* $H^{(i)}$ *refers to the ith derivative of H w.r.t.* λ.

Poss.	RM eig. (multiplicity)	Equations	Unknowns
I	c (3-f)	$\begin{cases} H(c)=0 \\ H^{(1)}(c)=0 \\ H^{(2)}(c)=0 \end{cases}$	$c, \bar{k}_1, \bar{k}_2$
(I')	c $c \pm j\omega$	$\begin{cases} H(c)=0 \\ \Re(H(c+j\omega))=0 \\ \Im(H(c+j\omega))=0 \end{cases}$	$c, \omega, \bar{k}_1, \bar{k}_2$
II	$c \pm j\omega\ (2-f)$	$\begin{cases} \Re(H(c+j\omega))=0 \\ \Im(H(c+j\omega))=0 \\ \Re\left(H^{(1)}(c+j\omega)\right)=0 \\ \Im\left(H^{(1)}(c+j\omega)\right)=0 \end{cases}$	$c, \omega, \bar{k}_1, \bar{k}_2$
(II')	$c \pm j\omega_1$ $c \pm j\omega_2$	$\begin{cases} \Re(H(c+j\omega_1))=0 \\ \Im(H(c+j\omega_1))=0 \\ \Re(H(c+j\omega_2))=0 \\ \Im(H(c+j\omega_2))=0 \end{cases}$	$c, \omega_1, \omega_2, \bar{k}_1, \bar{k}_2$

According to Table 8.2, the components of the stabilizability boundary are displayed in Figure 8.3. The frequency ω on branch II approximates the natural frequency of the open loop system and tends to zero along the branch when approaching the intersection with branch I.

When the control term is of full rank, there are three control parameters $\bar{k}_1, \bar{k}_2, \bar{k}_3$ in (8.6). The configurations of the rightmost characteristic roots on the stabilizability boundary are analogous to the previous case but, since there is an extra control parameter, one extra condition on the rightmost characteristic roots needs to be fulfilled; see Table 8.2. Four cases can be distinguished and accordingly the stabilizability boundary can be decomposed into four components; see Figure 8.3. The presence of an extra control parameter leads to four coinciding real characteristic roots on branch III, two coinciding pairs of complex conjugate characteristic roots, and one real characteristic root on branch IV. At their intersection we have five characteristic roots equal to zero. The calculation of branches V and VI deserves further attention. Therefore, consider Figure 8.4. Point A lies on branch II, hence $\bar{k}_3 = 0$

Table 8.2. *Positions of the rightmost characteristic roots on the stabilizability boundary of the system (8.1)–(8.2) with $\tau = 1$, and determining systems. $H^{(i)}$ refers to the ith derivative of H w.r.t. λ. Situations I and II occur when M is of rank 1, situations III, IV, V, and VI when M is of rank 2. The corresponding mathematical relations define branches, of which the stabilizability boundary is composed.*

Branch	RM eig. (mult.)	Equations	Unknowns (Analytic sol.)
I	0 (3-f)	$\begin{cases} H(0)=0 \\ H^{(1)}(0)=0 \\ H^{(2)}(0)=0 \end{cases}$	$a_1, a_2, \bar{k}_1, \bar{k}_2$ ($a_2 = -2a_1 - 2$)
II	$\pm j\omega$ (2-f)	$\begin{cases} H(j\omega)=0 \\ H^{(1)}(j\omega)=0 \end{cases}$	$a_1, a_2, \bar{k}_1, \bar{k}_2, \omega$
III	0 (4-f)	$\begin{cases} H(0)=0 \\ H^{(1)}(0)=0 \\ H^{(2)}(0)=0 \\ H^{(3)}(0)=0 \end{cases}$	$a_1, a_2, \bar{k}_1, \bar{k}_2, \bar{k}_3$ ($a2 = -2.5a_1 - 4$)
IV	0 $\pm j\omega$ $(2-f)$	$\begin{cases} H(0)=0 \\ H(j\omega)=0 \\ H^{(1)}(j\omega)=0 \end{cases}$	$a_1, a_2, \bar{k}_1, \bar{k}_2, \bar{k}_3, \omega$
V	$\pm j\omega$ (2-f)	$\begin{cases} H(j\omega)=0 \\ H^{(1)}(j\omega)=0 \\ J_{\bar{k}_1,\bar{k}_2,\bar{k}_3,\omega}X=0 \\ X^TX=1 \end{cases}$	$a_1, a_2, \bar{k}_1, \bar{k}_2, \bar{k}_3, \omega$
VI	$\pm j\omega$ (2-f)	$\begin{cases} H(j\omega)=0 \\ H^{(1)}(j\omega)=0 \\ \bar{k}_3 = -\left(\bar{k}_1^2/4 + \frac{(\bar{k}_1 a_1/2-\bar{k}_2)^2}{4(a_2-a_1^2/4)}\right) \end{cases}$	$a_1, a_2, \bar{k}_1, \bar{k}_2, \bar{k}_3, \omega$

and its parameters $\bar{k}_1, \bar{k}_2, \omega, a_1, a_2$ satisfy

$$\begin{cases} \Re\left(H(j\omega)\right) = 0, \\ \Im\left(H(j\omega)\right) = 0, \\ \Re\left(H^{(1)}(j\omega)\right) = 0, \\ \Im\left(H^{(1)}(j\omega)\right) = 0. \end{cases} \tag{8.11}$$

We now free parameter $\bar{k}_3$, and while keeping a_2 constant, we continue the solution of (8.11) as a function of a_1. The turning point B, where a minimum of a_1 is reached, defines

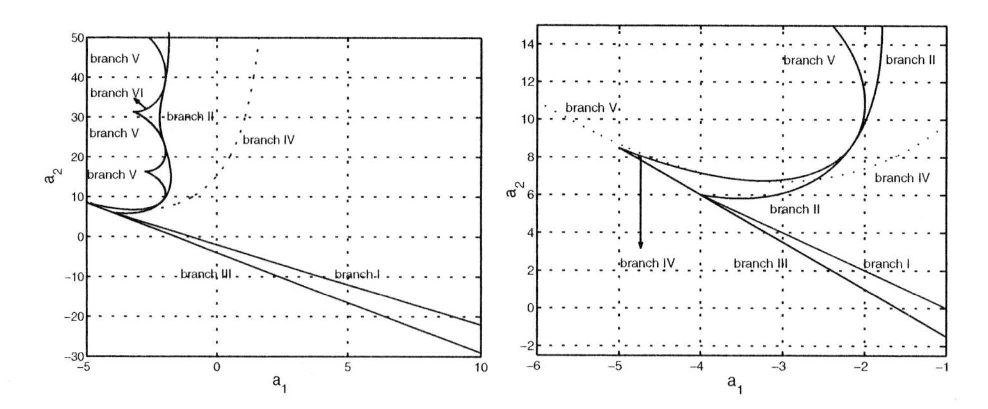

Figure 8.3. *Components of the stabilizability boundary (left) and detail (right) for system (8.1)–(8.2) with $\tau = 1$. The different branches refer to Table 8.2.*

a point of branch V. Let $J_{\bar{k}_1,\bar{k}_2,\bar{k}_3,\omega}(H, H^{(1)})$ be the Jacobian matrix of (8.11) as a function of parameters $\bar{k}_1, \bar{k}_2, \bar{k}_3, \omega$. Since $J_{\bar{k}_1,\bar{k}_2,\bar{k}_3,\omega}(H, H^{(1)})$ is singular in a turning point [270], we can express the extra condition to be fulfilled along branch V as

$$\det\left(J_{\bar{k}_1,\bar{k}_2,\bar{k}_3,\omega}(H, H^{(1)})\right) = 0, \tag{8.12}$$

or as

$$\begin{cases} J_{\bar{k}_1,\bar{k}_2,\bar{k}_3,\omega}(H, H^{(1)})X = 0, \\ X^T X = 1, \end{cases} \tag{8.13}$$

which is better suited for numeral calculations [270]. By computing the branch of turning points, i.e., by continuing the solution of (8.11)–(8.13), branch V is obtained. The sharp edges on Figure 8.3 are caused by the projection on the (a_1, a_2)-plane.

Notice that constraint (8.8) forms a lower bound on $\bar{k}_3$. Hence, it may be that this constraint becomes active before the turning point is reached. In that case the stabilizability boundary is determined by (8.11) and

$$\bar{k}_3 = -\left(\bar{k}_1^2/4 + \frac{\left(\bar{k}_1 a_1/2 - \bar{k}_2\right)^2}{4(a_2 - a_1^2/4)}\right),$$

which expresses that constraint (8.8) is active. This is the case for $32.4 \leq a_2 \leq 34.8$ and here the stabilizability boundary coincides with branch VI. See Figure 8.5 where the relevant part of Figure 8.3 (left) is enlarged.

8.2.3 Class of stabilizable systems for arbitrary delay values

We now consider the computation of the stabilizability boundary of (8.1)–(8.2) when the delay $\tau \neq 1$. Furthermore, we show how contour lines of the function $c(a_1, a_2, \tau)$, defined

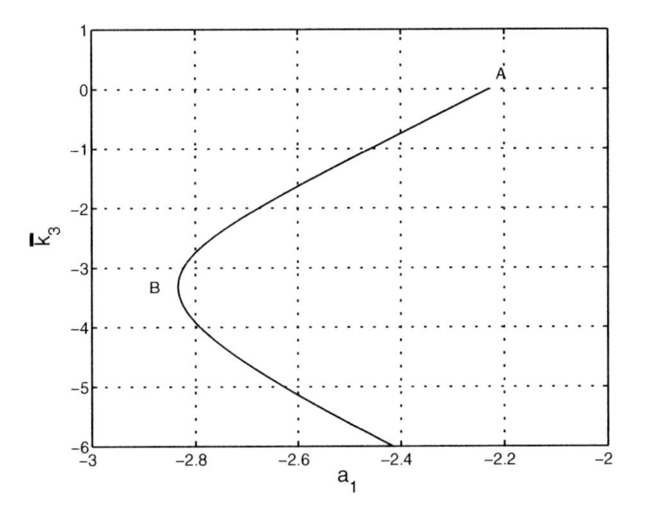

Figure 8.4. *When fixing a_2 and continuing the solution of (8.11) from point A on branch II as a function of parameter a_1, a turning point B occurs. Branch V is composed of such turning points. In point A we have $(a_1, a_2) = (-2.223, 30)$.*

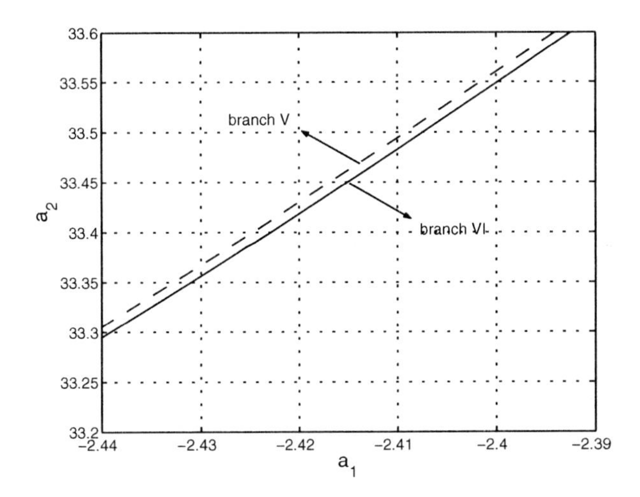

Figure 8.5. *For $32.4 \leq a_2 \leq 34.8$, the stabilizability boundary is not formed by branch V (dashed line), but by branch VI (full line). On branch VI, (8.8) is an equality constraint.*

as

$$c(a_1, a_2, \tau) = \min_M \left\{ \sup \left\{ \Re(\lambda) : \ \det \left(\lambda I - A - M e^{-\lambda \tau} \right) = 0 \right\} \right\},$$

can be calculated. The contour line $c(a_1, a_2, \tau) = 0$ then forms the stabilizability boundary.

With the substitution

$$\bar{\lambda} = \tau(\lambda - \alpha/\tau),$$

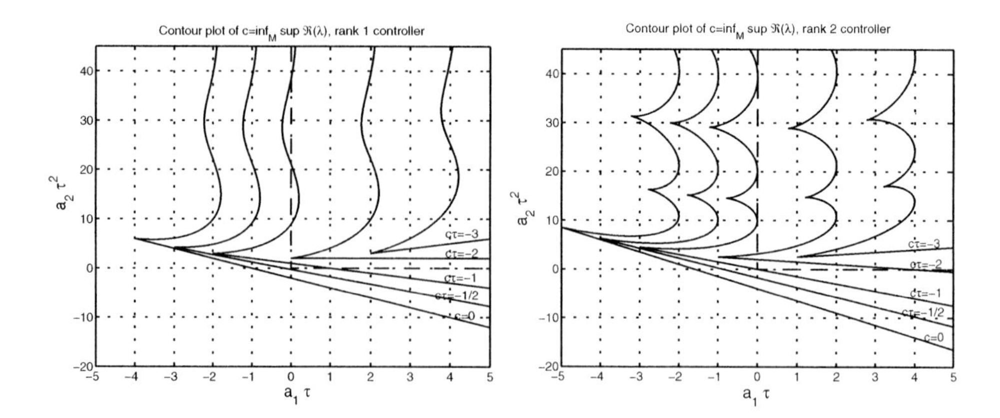

Figure 8.6. *Contour lines of $c(a_1, a_2, \tau)$ as a function of the system parameters a_1, a_2, and τ, where M is of rank 1 (above) and of rank 2 (below). $\lambda^2 + a_1\lambda + a_2 = 0$ is the characteristic equation of the uncontrolled system. These contour lines correspond to the maximal achievable exponential decay rate of the solutions of (8.1) and (8.2) in the second-order case. The stabilizability boundary consists of the contour line $c = 0$.*

the characteristic equation of (8.2),

$$\det\left(\lambda I - A - Me^{-\lambda\tau}\right) = 0, \tag{8.14}$$

becomes

$$\det\left(\bar{\lambda} I - (\tau A - \alpha I) - \tau M e^{\alpha} e^{-\bar{\lambda}}\right) = 0.$$

This can be seen as the characteristic equation of

$$\dot{x} = \bar{A}x + \bar{M}x(t-1), \qquad \bar{A} = \tau A - \alpha I, \;\; \bar{M} = \tau M e^{\alpha}. \tag{8.15}$$

Because of the transformation, the stabilizability boundary of the system (8.15), which corresponds to the contour line $c(a_1, a_2, \tau)\tau = \alpha$ of (8.2), can be computed as outlined in the previous paragraphs and is expressed as a function of the parameters of $\bar{A}$. As follows from the definition of $\bar{A}$, such contour lines can be expressed in the normalized coordinates $a_1\tau$ and $a_2\tau^2$, where $\lambda_2 + a_1\lambda + a_2 = 0$ is the characteristic equation of the uncontrolled system.

In Figure 8.6 contour lines of $c(a_1, a_2, \tau)$ are shown as a function of the parameters a_1, a_2, and τ. These lines correspond to the maximal achievable exponential decay rate of the solutions of (8.1) and (8.2). The contour line $c(a_1, a_2, \tau) = 0$ is the stabilizability boundary.

8.2.4 Discussion

We now summarize some interesting properties of delayed state feedback by means of Figure 8.6.

- The uncontrolled system is only asymptotically stable when $a_1 > 0$ and $a_2 > 0$. With delayed state feedback, the class of stabilizable systems is determined by the stabilizability boundary. This class becomes larger and grows unbounded in all directions in the (a_1, a_2)-plane as the delay τ is reduced. Contrary to the ODE case, where the whole spectrum can be controlled with rank-1 feedback, the class of stabilizable systems is considerably larger with rank-2 feedback in the DDE case, a consequence of the infinite-dimensional nature of DDE.

- From Figure 8.6 one can deduce stabilizability information of a plant with fixed parameters (a_1, a_2) as a function of the delay τ. When the delay changes, the normalized plant parameters $(\hat{a}_1, \hat{a}_2) = (a_1\tau,\ a_2\tau^2)$ move on a (half) parabola. When the plant has an characteristic root in the open RHP, i.e., either $a_1 < 0$ or $a_2 < 0$, this parabola always intersects the stabilizability boundary. Hence, when the uncontrolled system is (exponentially) unstable, it cannot be stabilized for large values of the time-delay, which is expected from Theorems 7.1 and 7.2. On the other hand, when the rightmost characteristic root of the open-loop system lies on the imaginary axis, stabilization is always possible, whatever the value of the time-delay. This result also follows from [230, 178]. Furthermore, the gain can be chosen arbitrarily small [178].

  When the delay value is zero to zero, we have $(\hat{a}_1, \hat{a}_2) \to (0, 0)$ and consequently $c\tau \to -\hat{c}$, with $\hat{c} > 0$ constant. Hence for small τ, the maximal achievable exponential decay rate of the solutions is of order $-1/\tau$. It is easy to verify that this result also holds when the dimension of the plant is larger than two. Note that $c \to -\infty$ as $\tau \to 0+$, which corresponds to the ODE case where the characteristic roots can be assigned arbitrarily.

- When τ is fixed and $a_2 \to \infty$, the stabilizability boundary does not converge to the a_2-axis. Consequently, the set of stabilizable systems includes a class of exponentially unstable systems ($a_1 < 0$) with arbitrarily large natural frequencies (i.e., imaginary parts of the characteristic roots). For such systems with highly oscillatory behavior, stabilization with delayed feedback is intuitively possible because tuning of the matrix M in (8.2) allows us to give the feedback signal Me precisely the necessary phase shift, thereby compensating (only mod 2π) the phase shift introduced by the delay. But this implies that the larger a_2, the more sensitive is the achieved stability w.r.t. changes in the system parameters, and in particular the delay. For instance, the achieved asymptotic stability property will only hold in a small delay interval around the nominal delay.

8.2.5 Noncyclic system matrix

So far we assumed that the system matrix A is cyclic [326, p. 17]. In the other case, it has the form

$$A = diag(a, a),\ a \in \mathbb{R},$$

and for asymptotic stabilization, inherently rank-2 feedback is necessary in (8.2) when $a \geq 0$. Numerical computations show that the optimal stabilizing feedback matrix M is also diagonal, leading to two scalar subsystems like the one treated in Subsection 7.2.3.

Hence, we have

$$\inf_M \left\{\sup \Re(\lambda) : \det(\lambda I - A - Me^{-\lambda\tau})\right\} = a - \frac{1}{\tau}, \tag{8.16}$$

and the stabilizable systems are characterized by $a < 1/\tau$. Note that the best stability results are obtained for a feedback gain which leads to a decoupling into two controlled "single input" subsystems, although the analysis of Section 7.5.1 shows that this is not generally valid.

Using the results of the previous section one can also show that (8.16) remains valid in the case where A is cyclic with a double eigenvalue at a. In that case, a rank-1 feedback results in an optimum at $a - (2 - \sqrt{2})/\tau$.

8.3 Simultaneous stabilization over a delay interval

We reconsider the stabilizability of the second-order systems (8.1). We characterize the class of systems which are *simultaneously* stabilizable for all delay values in an interval $[0,\ \tau_{\max})$. For the sake of conciseness, we restrict ourselves to the case of rank-1 feedback, where the characteristic equation simplifies to

$$\lambda^2 + (a_1 + k_1 e^{-\lambda\tau})\lambda + (a_2 + k_2 e^{-\lambda\tau}) = 0. \tag{8.17}$$

The approach is based on a qualitative study of the evolution of the stability region in the (k_1, k_2)-plane as the delay is increased from zero, and a quantitative characterization of the situations where either stabilizability is lost or the intersection of the stability region with the stability region for $\tau = 0$ becomes empty.

For any pair of system parameters (a_1, a_2) and delay τ, the set S_τ of stabilizing feedback gains in the (k_1, k_2)-plane can be found by searching for characteristic roots on the imaginary axis, i.e., by substituting $\lambda = j\omega$ in (8.17). Therefore, S_τ is determined by the line $k_2 = -a_2$ and by the spiral curve Γ_τ,

$$\Gamma_\tau : \omega \in [0,\ \infty) \rightarrow \left\{ \begin{array}{l} k_1(\omega, \tau), \\ k_2(\omega, \tau), \end{array} \right. \tag{8.18}$$

where

$$\begin{aligned} k_1(\omega,\ \tau) &= \left(\omega^2 \sin(\omega\tau) - a_1\omega\cos(\omega\tau) - a_2 \sin(\omega\tau)\right)/\omega, \\ k_2(\omega, \tau) &= \omega^2 \cos(\omega\tau) + a_1\omega \sin(\omega\tau) - a_2\cos(\omega\tau). \end{aligned}$$

In Figure 8.7 we show the set S_τ for small values of the time-delay. For $\tau = 0$, it is given by

$$k_1 > -a_1,\ k_2 > -a_2,$$

and for small delay values it is enclosed by the line $k_2 = -a_1$ and the curve Γ_τ. Depending on the system parameters (a_1, a_2), there are qualitatively three possible situations as the delay is further increased, which are sketched in Figure 8.8. For these three cases, the values of the feedback gain (k_1, k_2) in point A and the corresponding delay, $\tau = \tau_{\max}$, solve the following optimization problem:

$$\begin{array}{c} \max_{k_1,k_2,\tau}\ \tau, \quad \text{subject to} \\ \max_{\theta\in[0,\ \tau]} \left\{\sup \Re(\lambda) : \lambda^2 + (a_1 + k_1 e^{-\lambda\theta}) + (a_2 + k_2 e^{-\lambda\theta}) = 0\right\} < 0. \end{array} \tag{8.19}$$

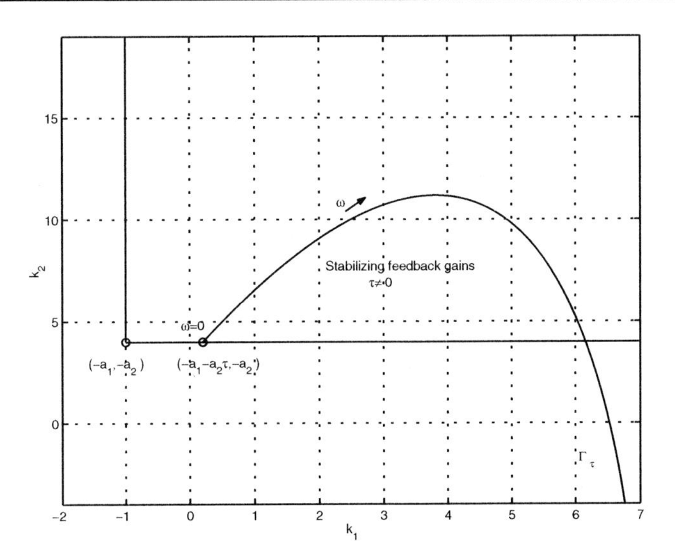

Figure 8.7. *Stabilizing values of* (k_1, k_2) *for equation (8.17) with* $(a_1, a_2) = (1, -4)$, *when* $\tau = 0$ *and* $\tau = 0.3$.

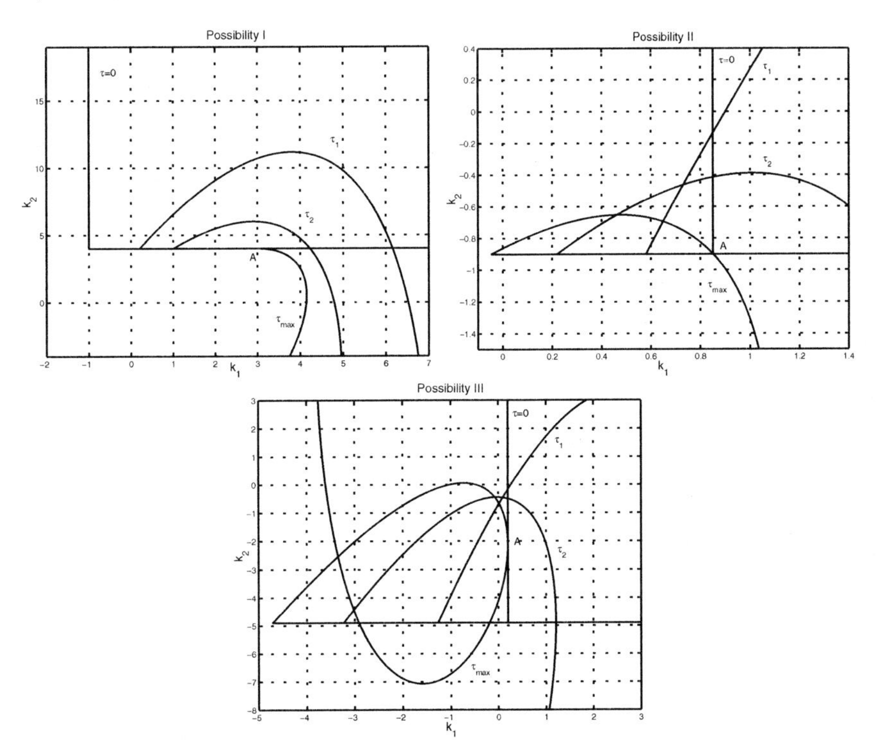

Figure 8.8. *The three possible cases for the evolution of the set* S_τ *as a function of the delay,* $0 < \tau_1 < \tau_2 < \tau_{\max}$. *The feedback gain corresponding to point A and the delay* $\tau_{\max}$ *solve the optimization problem (8.19).*

Table 8.3. *Mathematical relations, which characterize the three possible situations shown in Figure 8.8. When $\tau_{\max}$ is fixed, these relations define branches in the (a_1, a_2)-plane. For $\tau_{\max} = 1$, these branches are shown in Figure 8.9.*

Case	*Equations*	*Unknowns*	*Analytic solution*
I	$\lim_{\omega\to 0+} \frac{\frac{\partial k_2}{\partial\omega}(\omega,\tau_{\max})}{\frac{\partial k_1}{\partial\omega}(\omega,\tau_{\max})} = 0$	$a_1, a_2, \tau_{\max}$	$a_2\tau_{\max}^2 = -2a_1\tau_{\max} - 2$
II	$\begin{cases} k_1(\omega,\tau_{\max}) = -a_1 \\ k_2(\omega,\tau_{\max}) = -a_2 \end{cases}$	$a_1, a_2, \omega, \tau_{\max}$	
III	$\begin{cases} k_1(\omega,\tau_{\max}) = -a_1 \\ \frac{\partial k_1(\omega,\tau_{\max})}{\partial\omega} = 0 \end{cases}$	$a_1, a_2, \omega, \tau_{\max}$	

In other words, the system can be made globally asymptotically stable for all $\tau \in [0,\ \tau_{\max})$, but not in a larger delay interval. Note that in case I, the set S_τ is empty for $\tau = \tau_{max}$. This is not so for cases II and III, which means that stabilization is possible for some fixed delay values $\tau > \tau_{\max}$, while this is not possible over the whole interval $[0,\ \tau]$.

In Table 8.3 we display the mathematical relations describing the three cases depicted in Figure 8.8. When $\tau_{\max}$ is fixed, these relations define branches in the (a_1, a_2)-plane, which determine the class of systems, simultaneously stabilizable in the delay interval $[0,\ \tau_{\max})$. In Figure 8.9 we display these branches for $\tau_{\max} = 1$. Because of the arguments spelled out in Section 8.2.3, the branches corresponding to other values of $\tau_{\max}$ coincide with the branches for $\tau_{\max} = 1$, when normalizing the system parameters to $(a_1\tau_{\max}, a_2\tau_{\max}^2)$.

In Figure 8.10, we compare, for a rank-1 feedback, the stabilizable systems for a fixed delay value τ (characterized in Section 8.2) with the systems, simultaneously stabilizable over the delay interval $[0,\ \tau)$. In the latter case, stabilization is *not* possible for high values of a_2. For example, if the uncontrolled system has characteristic roots on the imaginary axis ($a_1 = 0$), stabilization in the delay interval $[0,\ \tau)$ is only possible if $a_2\tau^2 \le \pi^2$, that is, if $\tau \le T/2$, with T the period of the natural oscillation.

8.4 Notes and references

We explored the limits of stabilizability with delayed state feedback by completely characterizing the class of stabilizable second-order systems. Due to the infinite-dimensional nature of time-delay systems, this is no trivial problem, despite the low dimension, and a combination of both analytical and numerical methods was required. The case study demonstrated a fruitful interaction of methods and tools from control theory and numerical bifurcation analysis.

The chapter is based on [194; 179, Appendix A], and the references therein.

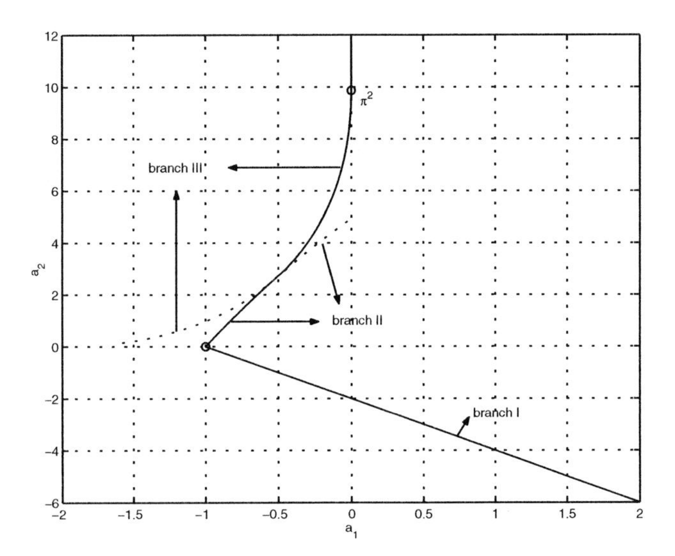

Figure 8.9. *The branches in the (a_1, a_2)-plane, corresponding to the mathematical relations of Table 8.3, for $\tau_{\max} = 1$. The set of systems, simultaneously stabilizable in the delay interval $[0,\ \tau_{\max})$, is determined by the parameter values lying to the right of the solid line, which is composed from these branches.*

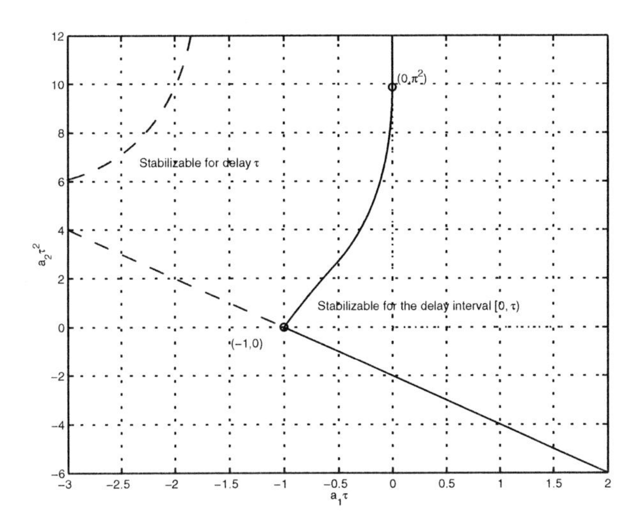

Figure 8.10. *In case of rank-1 feedback, the class of second-order systems, (simultaneously) stabilizable over the delay interval $[0,\ \tau)$, is determined by the solid line, while the dashed line, the stabilizability boundary, determines the class of systems which are stabilizable for the (fixed) delay τ.*

Chapter 9

The robust stabilization problem

9.1 Introduction

In Chapter 7 we proposed a procedure for the stabilization of linear time-delay systems, which was explained by means of the stabilization of a linear finite-dimensional single input system with an input delay using static state feedback,

$$\dot{x}(t) = Ax(t) + Bu(t-\tau), \quad u = K^T x(t). \tag{9.1}$$

The procedure can be applied until the rightmost characteristic roots cannot be moved further to the left using the available controller parameters, that is, when the spectral abscissa function

$$\alpha(K) := \sup\left\{\Re(\lambda) : \det(\lambda I - A - BK^T e^{-\lambda\tau}) = 0\right\} \tag{9.2}$$

is minimal. For stabilizable systems, this means that the exponential decay rate of the closed-loop solutions is maximal.

In this chapter, we assume constant perturbations on the system matrices and the feedback gain in (9.1) and consider the calculation of the (nominal) feedback gain which maximizes some robust stability measures. These measures are expressed by complex stability radii, defined in Chapter 2.

In the literature on time-delay systems, the robust stabilization problem has been widely studied in a Lyapunov context; see, for instance, [152, 151, 65, 41, 221, 170]. This approach allows us to include easily more general types of perturbations (e.g., time-varying). However, practical stability results are typically obtained in the form of sufficient conditions, expressed by the feasibility of linear matrix inequalities or the solvability of AREs. This generally induces conservatism, due to the choice of the form of the functional and/or the estimates involved in the derivation of the stability criteria. The results presented in this chapter can be regarded as a way to tighten the gap between sufficient and necessary conditions by introducing an alternative approach, directly related to the position of the characteristic roots in the complex plane.

The structure of the chapter is as follows. After revisiting the necessary elements from Chapter 2, we illustrate the importance of considering robustness issues in the stabilization procedure with some semianalytical examples. Next, we describe and apply a numerical procedure for the optimization of complex stability radii. Some notes and references end the chapter.

9.2 Stability radii as robustness measures

We assume that the controlled system (9.1) is asymptotically stable and consider the stability of the perturbed system

$$\dot{x}(t) = (A + \delta A)x(t) + (B + \delta B)(K + \delta K)^T x(t - \tau), \tag{9.3}$$

under various classes of perturbations on A, B, and K. The nominal characteristic matrix is given by

$$F(\lambda) := \lambda I - A - BK^T e^{-\lambda\tau}.$$

As a perturbation of A corresponds to an additive perturbation of F,

$$\delta F_A = -\delta A,$$

an application of Theorem 2.15 yields

$$r_{\mathbb{C}}(F;\ \mathbb{C}_-, \Delta_A) = \left(\sup_{\omega \geq 0} \left\| (j\omega I - A - BK^T e^{-j\omega\tau})^{-1} \right\|_2 \right)^{-1}, \tag{9.4}$$

with $\Delta_A = \mathbb{C}^{3\times 3}$. Similarly, if B or K is perturbed, then the additive perturbations are

$$\delta F_B = -\delta B\, K^T, \quad \delta F_K = -B\, \delta K^T,$$

and we get

$$\begin{aligned} r_{\mathbb{C}}(F;\ \mathbb{C}_-, \Delta_B) &= \left(\sup_{\omega \geq 0} \left\| K^T (j\omega I - A - BK^T e^{-j\omega\tau})^{-1} \right\|_2\right)^{-1}, \\ r_{\mathbb{C}}(F;\ \mathbb{C}_-, \Delta_K) &= \left(\sup_{\omega \geq 0} \left\| (j\omega I - A - BK^T e^{-j\omega\tau})^{-1} B \right\|_2\right)^{-1}, \end{aligned} \tag{9.5}$$

with $\Delta_B = \mathbb{C}^{3\times 1}$ and $\Delta_K = \mathbb{C}^{1\times 3}$. The expressions (9.4)–(9.5) have the form

$$\left(\sup_{\omega \geq 0} \| M_K(j\omega) \|_2 \right)^{-1} = \left(\| M_K(j\omega) \|_{\mathcal{H}_\infty} \right)^{-1}, \tag{9.6}$$

where $M_K(j\omega)$ is given by

$$M_K(j\omega) = X \left(j\omega I - A - BK^T e^{-j\omega\tau} \right)^{-1} Y,$$

with the constant matrices X and Y depending on which matrix is perturbed.

In the remainder of this chapter we denote, for the simplicity of the notation,

$$r_{\mathbb{C}}^A := r_{\mathbb{C}}(F;\ \mathbb{C}_-, \Delta_A), \quad r_{\mathbb{C}}^B := r_{\mathbb{C}}(F;\ \mathbb{C}_-, \Delta_B), \quad r_{\mathbb{C}}^K := r_{\mathbb{C}}(F;\ \mathbb{C}_-, \Delta_K). \tag{9.7}$$

By means of the optimization of these complex stability radii for (9.3) as a function of the feedback gain K, the robust stabilization procedure of [196] is described. Applications to the system (9.3) with combined uncertainty on A, B, and K, as well as the optimization of real stability radii, are not explicitly treated but are briefly discussed in the notes and references section. Before presenting the main results, we first illustrate the importance of including robustness aspects in a stabilization procedure with some examples.

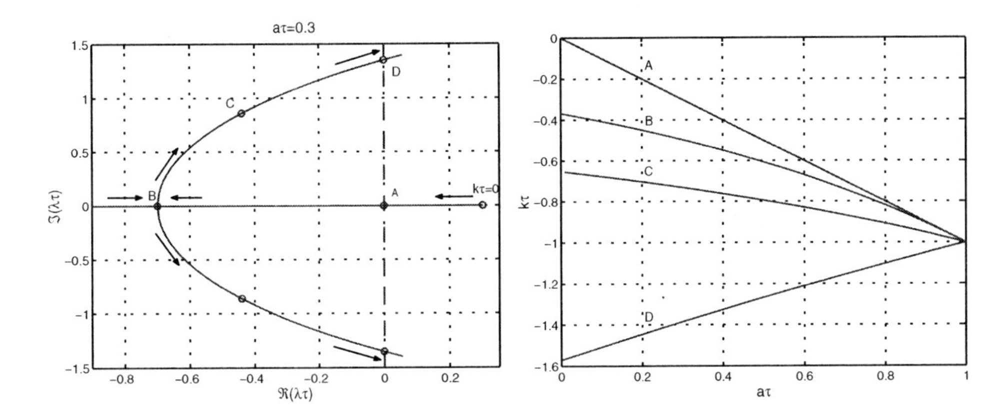

Figure 9.1. *Position of the characteristic roots of (9.8) in the complex plane as a function of k when $a\tau = 0.3$ (left), and values of k corresponding to special cases as a function of a and τ (right). The stabilizing values of k lie between curves A (zero characteristic root) and D (characteristic roots on the imaginary axis). On curve B the exponential decay rate of the solutions is maximal. On curve C, $r_{\mathbb{C}}^k = r_{\mathbb{C}}^a$ is maximal.*

9.3 Stabilization versus robust stabilization

In the examples in [182, 195, 194], the continuous pole placement procedure was applied until a minimum of (9.2) was reached. We now show that both the corresponding feedback gain and the configuration of the rightmost characteristic roots may be considerably different from the cases where certain robustness measures, expressed by stability radii, are maximal and, moreover, a minimization of (9.2) may result in a poor robustness against parameter variations.

First consider the scalar system

$$\dot{x}(t) = ax(t) + kx(t-\tau), \tag{9.8}$$

whose (nonrobust) stability properties are analyzed in Section 7.2.3. We make the assumption that $0 \leq a\tau < 1$, i.e., the uncontrolled system is unstable but stabilizable. In Figure 9.1 (left), we display the position of the rightmost characteristic roots of (9.8) as a function of k when $a\tau = 0.3$. For $k = 0$, there is one unstable characteristic root at $+a$. By decreasing k, this characteristic root is shifted to the left, achieving stability in point A, until it interacts with another real characteristic root in point B, where the spectral abscissa (9.2) is minimal. A further reduction of k leads to a splitting into a complex pair of characteristic roots, which move toward the imaginary axis. In point C the complex stability radii $r_{\mathbb{C}}^a = r_{\mathbb{C}}^k$ are maximal. In point D instability occurs. In Figure 9.1 (right), the values of k corresponding to these special cases are displayed as a function of a and τ.

As a second example, we study the stability of a feedback stabilized double integrator with input delay. Therefore, we consider the system (9.3) with

$$A = \begin{bmatrix} 0 & 1 \\ 0 & 0 \end{bmatrix}, \quad B = \begin{bmatrix} 0 \\ 1 \end{bmatrix}, \quad \tau = 1, \quad K = \begin{bmatrix} -k_2 \\ -k_1 \end{bmatrix}. \tag{9.9}$$

Table 9.1. *Comparison between feedback laws minimizing (9.2) and maximizing the complex stability radius $r_{\mathbb{C}}^K$ for the system (9.3)–(9.9).*

	k_1	k_2	α	$r_{\mathbb{C}}^K$
α minimal	0.461	0.0791	−0.5857	0.0791
$r_{\mathbb{C}}^K$ maximal	0.921	0.228	−0.2572	0.224

Its characteristic equation is given by

$$\lambda^2 + k_1 e^{-\lambda}\lambda + k_2 e^{-\lambda} = 0. \tag{9.10}$$

In Figure 9.2 (top), we plot contour lines of the function (9.2) and (9.9) in the (k_1, k_2)-plane. A contour line $\alpha(K) = -\gamma$, $\gamma > 0$, which determines the values of the feedback gain yielding γ-stability, can be computed by seeking solutions of (9.10) of the form $\lambda = -\gamma + j\omega$. The minimum of the spectral abscissa function is given by $\alpha = -0.5857$ and is obtained for $(k_1, k_2) = (0.461, 0.0791)$. These values can be calculated directly using the approach of [195]. We now take perturbations on K into account and plot in Figure 9.2 (bottom) some contour lines of the complex stability radius $r_{\mathbb{C}}^K$, calculated using formula (9.6). In Table 9.1, a comparison is made between feedback laws minimizing the spectral abscissa function $\alpha(K)$ and maximizing the complex stability radius $r_{\mathbb{C}}^K$, which clearly illustrates the trade-off between performance and robustness which has to be made in a design.

9.4 Robust stabilization procedure

Because of the relation (9.6), the optimization of complex stability radii corresponds to the minimization of $\|M_K(j\omega)\|_{\mathcal{H}_\infty}$ as a function of the feedback gain K. To solve this highly complex optimization problem (a strongly nonquadratic dependence on the variables), we propose a numerical procedure, which applies a local strategy and consists of a quasi-continuous reduction of $\|M_K(j\omega)\|_{\mathcal{H}_\infty}$ by making small changes to the feedback gain. We assume that the initial feedback gain is stabilizing. Such a feasible starting value can be computed with the methods described in Chapters 7 and 10.

Before describing the optimization procedure, we address some continuity properties of $\sigma_1(M_K(j\omega))$ w.r.t. ω and K, on which the numerical procedure relies.

9.4.1 Continuity properties

First we consider a fixed, stabilizing value of K and describe the continuity of the singular values of $M_K(j\omega)$ as a function of ω, based on the results of [21] and the references therein.[15] When a singular value $\sigma_i(M_K(j\omega_0))$ is isolated, the map $\omega \rightarrow \sigma_i(M_K(j\omega))$ is real analytic in a neighborhood of ω_0 and can be expanded in a Taylor series. Moreover, the behavior of all singular values of $M_K(j\omega)$ as a function of ω over the interval $(-\infty, \infty)$ can be

[15] Although the transfer functions in the paper [21] correspond to systems without delay, the results can be extended to the delay case in a trivial way.

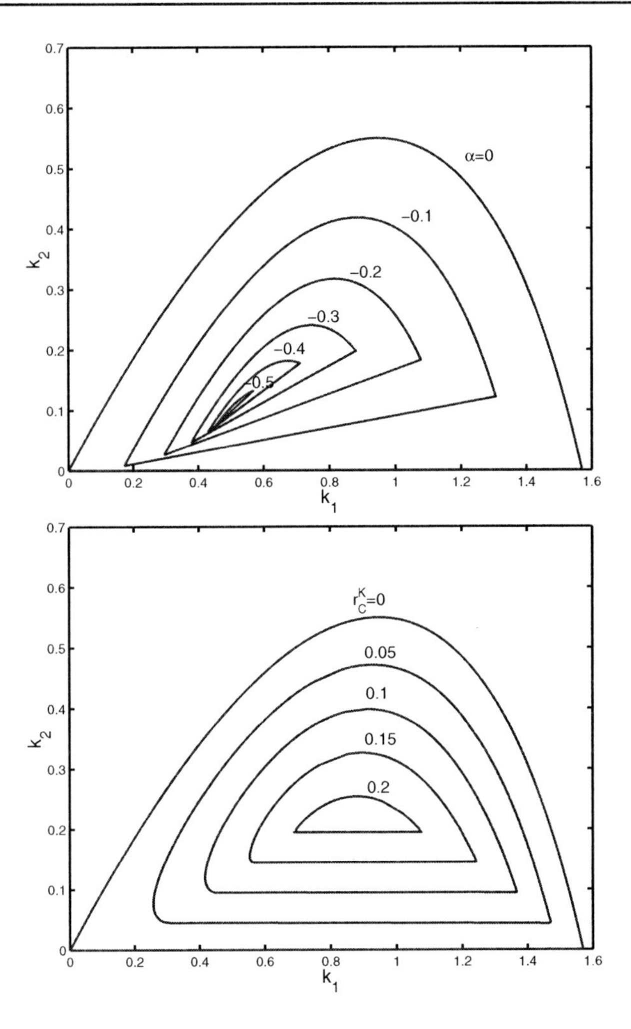

Figure 9.2. *(top) Contour lines of the function (9.2) and (9.9) in the (k_1, k_2)-plane. All stabilizing feedback gains lie inside the contour line $\alpha(K) = 0$. (bottom) Curves corresponding to constant values for the complex stability radius $r_{\mathbb{C}}^K$.*

decomposed as a collection $\{|f_1(\omega)|, \ldots, |f_n(\omega)|\}$, where the n functions f_i, $i = 1, n$, are infinitely smooth. As a consequence, the function $\omega \rightarrow \sigma_1(M_K(j\omega))$ is at least twice-continuously differentiable in the neighborhood of a maximum and infinitely smooth when the maximum is isolated.

We now fix ω and comment on the continuity of $\sigma_1(M_K(j\omega))$ with respect to K. When the singular value is isolated, it is continuously differentiable. For a given component k_j of the feedback gain K and a frequency ω, one can compute

$$\begin{bmatrix} \frac{\partial \sigma_1(M_K(j\omega))}{\partial k_j} \\ \vdots \\ \frac{\partial \sigma_1(M_K(j\omega))}{\partial k_j} \end{bmatrix} = \Re\left(\frac{v_1^* M_K^* X G^{-1} B G^{-1} Y v_1 e^{-j\omega\tau}}{\sigma_1}\right), \tag{9.11}$$

with $G = (j\omega I - A - BK^T e^{-j\omega\tau})$ and v_1 the right singular vector of $M_K(j\omega)$ corresponding to σ_1. In the nongeneric case, where the largest singular value has a multiplicity $l > 1$ for some value $\bar{K}$ of K, there is an l-dimensional space of right singular vectors. Based on the fact that the singular values are the characteristic roots of a normal (hence diagonalizable) matrix, one can show that in a neighborhood of $\bar{K}$ the first l singular values as a function of K can be decomposed into l continuously differentiable functions. For each of these singular value functions, the sensitivity at $\bar{K}$ can be calculated with (9.11), where the vector $v_1(\bar{K})$ must be chosen in such a way that it extends the corresponding map $K \to v(K)$ to a continuous function in a neighborhood of $\bar{K}$.

9.4.2 Algorithm

The basic algorithm is as follows.

ALGORITHM 9.1.

Minimization of $\|M_K(j\omega)\|_{\mathcal{H}_\infty}$ as a function of K

A. Initialize $m = 1$.

B. Compute $\|M_K(j\omega)\|_{\mathcal{H}_\infty}$ and the frequencies ω_i, $i = 1, m$ with $\sigma_1(M_K(j\omega_i)) = \|M_K(j\omega)\|_{\mathcal{H}_\infty}$.

C. Compute the sensitivity of $\sigma_1(M_K(j\omega_i))$, $i = 1, m$ w.r.t. changes in the feedback gain K.

D. Reduce the m peaks $\sigma_1(M_K(j\omega_i))$ by applying small changes to the feedback gain, using the computed sensitivities.

E. Regularly check the presence of other frequencies ω_e with $\sigma_1(M_K(j\omega_e)) \approx \|M_K(j\omega)\|_{\mathcal{H}_\infty}$ and, if necessary, increase the number of controlled peak values in the (ω, σ_1)-plot, m. Stop when the available controller parameters do not allow further reduction of $\|M_K(j\omega)\|_{\mathcal{H}_\infty}$. In the other case, go to step B.

Notice that the structure of the algorithm is similar to that of Algorithm 7.1 for the (nonrobust) stabilization problem. The reduction of the peak values in the singular value plot (step D) corresponds to the shifting of the controlled characteristic roots, while the computationally intensive step E, the determination of the "peaking" frequencies, corresponds to the determination of the rightmost characteristic roots. We now explain some steps of the algorithm in more depth.

Calculation of peak values in the singular value plot

Taking into account Remark 2.8, we propose the classical approach of sketching the singular value plot, based on repeated singular value decompositions for frequencies chosen on a coarse grid, and correcting the approximate maxima with a local optimization routine.

Reduction of peak values

We first calculate the sensitivity of a peak value in the singular value plot w.r.t. the controller gain K. With $\omega^*(K)$ the maximizing frequency as a function of K, the sensitivity of a peak value is defined by

$$\frac{d\sigma_1(M_K(j\omega^*(K)))}{dK}. \tag{9.12}$$

Because of the smoothness properties of $\sigma_1(M_K(j\omega))$ w.r.t. ω and K and the fact that, by the definition of an optimum w.r.t. ω,

$$\left.\frac{\partial\sigma_1(M_K(j\omega))}{\partial\omega}\right|_{\omega^*(K)} = 0,$$

we have

$$\frac{d\sigma_1(M_K(j\omega^*(K)))}{dK} = \left.\frac{\partial\sigma_1(M_K(j\omega))}{\partial K}\right|_{\omega^*(K)},$$

and, hence, expression (9.12) can be calculated using (9.11). Applying this result to the m peaking frequencies ω_i we obtain a sensitivity matrix $S_m \in \mathbb{R}^{m\times n}$, where

$$S_m^{(i,j)} = \frac{\partial\sigma_1(M_K(j\omega_i))}{\partial k_j}.$$

With Δh the desired reduction of $\|M_K(j\omega)\|_{\mathcal{H}_\infty}$ in the iteration step, $m \le n$, and $\text{rank}(S_m) = m$, one can compute the necessary change of the feedback gain by

$$\Delta K = -S_m^{\dagger}\begin{bmatrix} 1 & \cdots & 1 \end{bmatrix}^T \Delta h. \tag{9.13}$$

For the new feedback gain $K + \Delta K$, a correction has to be made on both ω_i and $\sigma_1(M_K(j\omega_i))$ (generally different from the desired value) using the local optimization routine, because (9.13) is based on linearization.

Increasing the number of controlled peak values

When the reduction of the controlled peak values leads to an increase of another peak value, which becomes dominant, the latter also needs to be controlled for a further reduction of $\|M_K(j\omega)\|_{\mathcal{H}_\infty}$.

We now comment on the possible mechanisms of the creation of *additional* local maxima in the singular value plot, which may occur in the optimization procedure. Therefore, consider a trajectory of the feedback gain K in the parameter space, described by a smooth map $\mathbb{R} \ni p \to K(p) \in \mathbb{R}^n$, and define

$$f(\omega;\ p) = \sigma_1(M_{K(p)}(j\omega)),$$

i.e., the first singular value of $M_K(j\omega)$ along the trajectory. Without losing generality, assume that this singular value is isolated for all ω and p. Qualitatively, the mechanisms of the creation (or disappearance) of peak values in the singular value plot can be reduced to the two generic cases depicted in Figure 9.3.

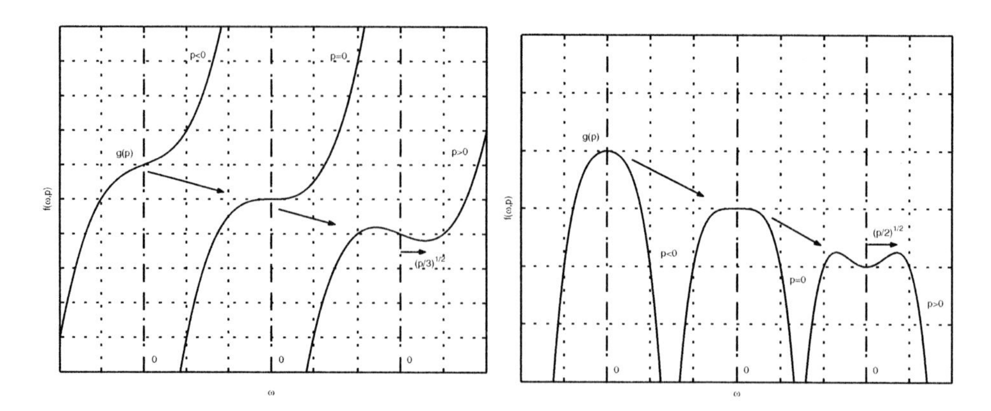

Figure 9.3. *The two mechanisms which create new maxima in the singular value plot, described by the normal form (9.14) (left) and by the normal form (9.15) (right).*

The first case is described by the normal form

$$f(\omega;\ p) = \omega^3 - p\omega + g(p), \tag{9.14}$$

where the continuously differentiable function $g(p)$ is only a drift term. When $p < 0$ there is no extremum, while for $p > 0$, there is a maximum at $\omega^*(p) = -\sqrt{p/3}$; see Figure 9.3 (left). For the critical value $p = 0$ we must have $\frac{\partial f}{\partial \omega} = 0$, the condition for an extremum, and $\frac{\partial^2 f}{\partial \omega^2} = 0$, because otherwise the maximum would be robust against changes of p. This mechanism is not directly of importance in our numerical procedure because it cannot occur at the global maxima in the singular value plot, which are monitored.

The second mechanism is described by the normal form

$$f(\omega;\ p) = -\omega^4 + p\omega^2 + g(p). \tag{9.15}$$

When $p < 0$, there is one maximum at $\omega = 0$, and for $p > 0$ there are two maxima at $\pm\sqrt{p/2}$; see Figure 9.3 (right). With $\omega^*(p)$ the frequency corresponding to a maximum, we have

$$\omega^*(p) = \begin{cases} 0, & p \le 0, \\ \pm\sqrt{p/2}, & p > 0, \end{cases}$$

$$f(\omega^*(p);\ p) = \begin{cases} g(p), & p \le 0, \\ g(p) + p^2/4, & p > 0. \end{cases}$$

This result shows that the function $p \to f(\omega^*(p);\ p)$, that is, the peak value in the singular value plot, is also *continuously differentiable* when the above bifurcation phenomenon occurs (for $p = 0$). As a consequence, the sensitivity of the peak value can always be computed using (9.11). Obviously the function $p \to \omega^*(p)$ is continuous but not differentiable.

Remark 9.1. *The differential equation $\dot{\omega} = \frac{\partial f(\omega;\ p)}{\partial \omega}$, whose stable stationary point is given by $\omega^*(p)$, is the normal form of a turning point in case of (9.14) and of a pitchfork bifurcation in case of (9.15); see [145]. Note that the mechanism related to a transcritical bifurcation does not give rise to the creation or disappearance of extrema.*

Stop criterion

The iterative procedure ends when no further reduction of $\|M_K(j\omega)\|_{\mathcal{H}_\infty}$ is possible, which occurs when $m > n$ or when $\text{rank}(S_m) < m$. In the first case, there are more peak values than available controller parameters. The second case deserves further attention.

First, the condition $\text{rank}(S_m) < m$ is satisfied when a row of S_m is identically zero, i.e., when for some $1 \le i \le n$,

$$\frac{\partial \sigma_1(M_K(j\omega_i))}{\partial k_j} = 0, \quad j = 1, n.$$

This situation corresponds to a minimum of the ith controlled peak value w.r.t. all controller parameters or a saddle-point. Although the latter can theoretically occur, we have not encountered it in our numerical experiments.

When the matrix S_m has no zero rows, the rank deficiency condition implies that, although each controlled peak value can be reduced with a suitable parameter change, a simultaneous reduction of all peak values is not possible (for instance, the reduction of one peak value leads to the increase of another peak value). Here, note that each direction ΔK in the parameter space with

$$\left[\frac{\partial \sigma_1(M_K(j\omega_i))}{\partial k_1} \cdots \frac{\partial \sigma_1(M_K(j\omega_i))}{\partial k_n}\right] \Delta K < 0$$

is a descending direction for the ith peak value and, hence, the rank deficiency of S_m means that no common descending direction exists for all peak values.

9.5 Illustrative example

We reconsider the system (7.27). In Section 7.4.1 the continuous pole placement method was applied to this example, yielding $K = [0.471\ 0.504\ 0.607]^T$ and $\alpha(K) = -0.15$ in the optimum of (9.2). Starting from these values, we optimize $r_{\mathbb{C}}^K$ as a function of K using Algorithm 9.1. In Figure 9.4 we show the singular values of the corresponding matrix $M_K(j\omega) = (j\omega I - A - BK^T e^{-j\omega\tau})^{-1} B$ as a function of ω, at different iterations of the algorithm. In the beginning, $\|M_K(j\omega)\|_{\mathcal{H}_\infty}$ is reduced by controlling one peak value, while from iteration number 55, we simultaneously reduce two peak values. At iteration number 115, the procedure ends because the three available degrees of freedom in the controller do not allow a further reduction. In the optimum we have $\text{rank}(S_2) = 1$. In Figure 9.5 (left), the corresponding values of the feedback gain are depicted as a function of the iterations. The high sensitivity of the three controller parameters in the last iteration steps is caused by the vicinity of the optimum. In Figure 9.5 (right), we depict the rightmost characteristic roots in the first iteration step (corresponding to the optimum of (9.2)) and in the last iteration step. Although the exponential decay of the closed-loop solutions is considerably slower in the latter case, the complex stability radius $r_{\mathbb{C}}^K$ (i.e., $1/\|M_K(j\omega)\|_{\mathcal{H}_\infty}$) has increased from 0.0938 to 0.287. This example once again illustrates the trade-off between performance and robustness.

In Figure 9.6 we plot curves corresponding to the frequencies where $\sigma_1(M_K(j\omega))$ reaches a local maximum or minimum. In the optimization procedure, the new extrema

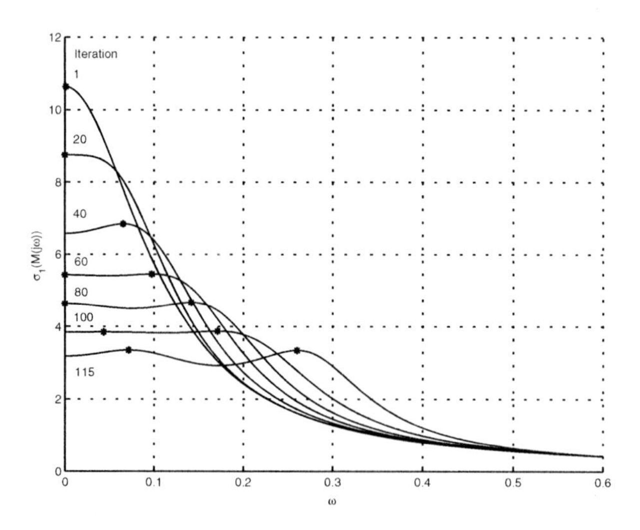

Figure 9.4. *Singular value plots of $M_K(j\omega)$ at different iterations of Algorithm 9.1, applied to the system (9.3) and (7.27) for the optimization of $r_{\mathbb{C}}^K$. The controlled peak values are indicated by $(*)$.*

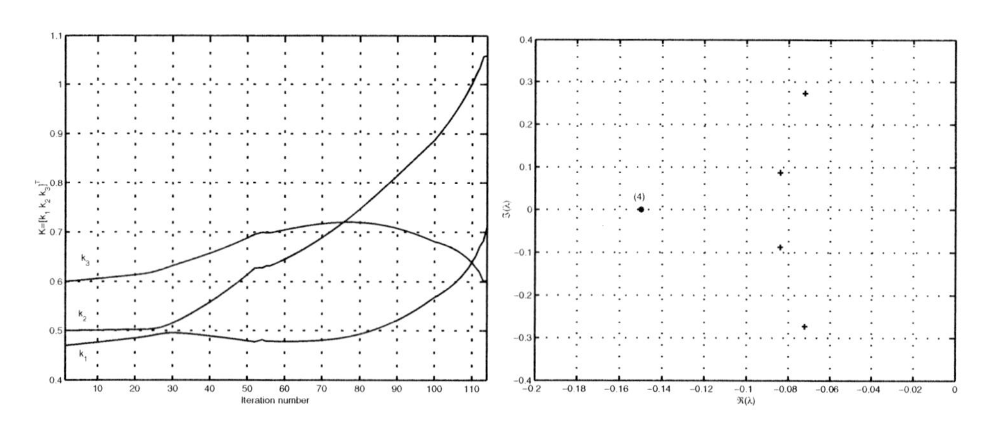

Figure 9.5. *(left) Feedback gain $K = [k_1\ k_2\ k_3]^T$ as a function of the iteration number. (right) Rightmost characteristic roots of the system (9.3) and (7.27) in the optimum of (9.2), indicated by $(*)$, and for $r_{\mathbb{C}}^K$ maximal $(+)$.*

are created with the mechanism described by the normal form (9.15). The nonsmoothness after the first bifurcation is caused by the increase of the number of controlled peak values at iteration number 55, which led to a nonsmooth change of the gain K.

The *delay* sensitivity of the obtained stability can be investigated by means of a numerical continuation of the rightmost characteristic roots. This is illustrated in Figure 9.7. The delay sensitivity in the maximum of $r_{\mathbb{C}}^K$ is better than in the optimum of (9.2), especially around the nominal delay value. This is caused by the characteristic root with multiplicity 4

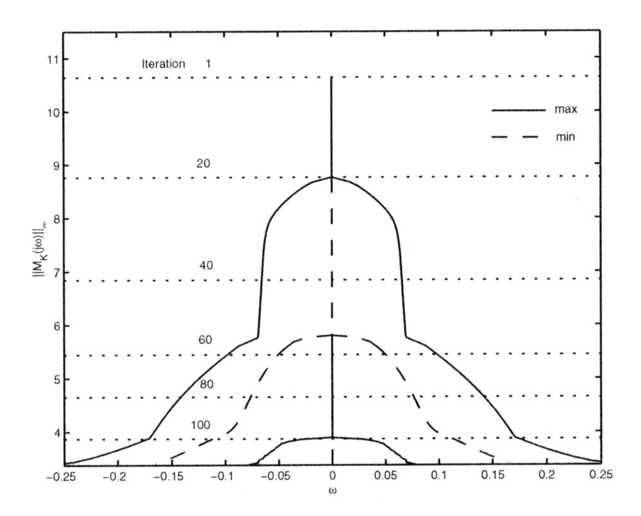

Figure 9.6. *Frequencies where the function $\omega \to \sigma_1(M_K(j\omega))$ reaches a maximum or a minimum for different values of the iteration number (expressed by the corresponding value of $\|M_K(j\omega)\|_{\mathcal{H}_\infty}$). For the iteration numbers indicated with dotted lines, the singular value plots are shown in Figure 9.4.*

in the latter case (see Section 7.4.1) (for more comments on the configuration of the rightmost characteristic roots in the optimum of (9.2), we refer to Chapter 8 and [195, 182]). Although we have only considered the optimization of the robustness against complex perturbations of the system *matrices*, the delay sensitivity in the optima is good, since a small delay mismatch can be interpreted as a (frequency-dependent) complex perturbation on the system matrices.

9.6 Notes and references

We described an iterative numerical procedure to maximize complex stability radii of time-delay systems. The iterative procedure is based on a quasi-continuous increase of the stability radius under consideration by employing a local strategy which consists of reducing peak values in appropriate singular value plots. The procedure assumes that the initial feedback gain is stabilizing. Such a stabilizing value can be calculated with the continuous pole placement procedure of Chapter 7, which has a similar structure, or with the procedure presented in the next chapter. Hence, by adopting a two-step approach, an overall solution for the robust stabilization problem is obtained. As the function $K \to \|M_K(j\omega)\|_{\mathcal{H}_\infty}$ has similar properties as the function $K \to \alpha(K)$, we believe that the gradient sampling algorithm of [36] may be used as an alternative to Algorithm 9.1. This is currently under investigation.

The approach can be extended to systems with structured perturbations which give rise to expressions of stability radii in terms of structured singular values (see Theorem 2.15). It is important to mention that in the cases where only upper bounds on the structured singular value can be computed, an optimization involving such upper bounds eventually yields

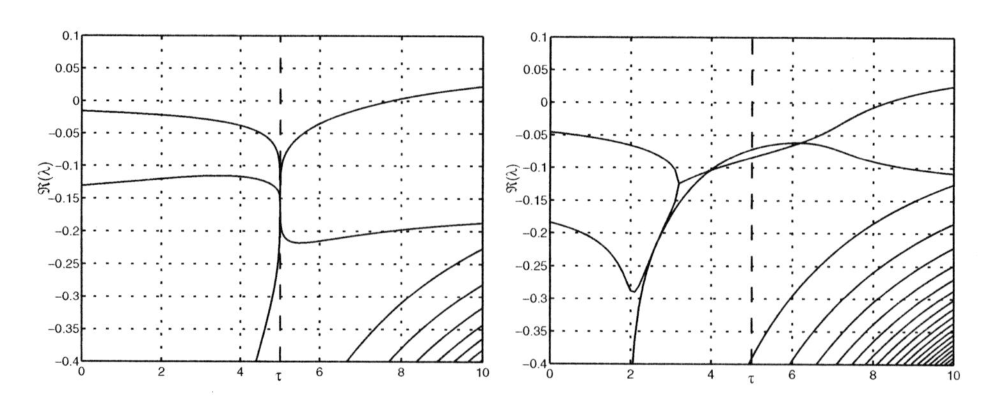

Figure 9.7. *Rightmost characteristic roots as a function of the delay when* $K = [0.471\ 0.504\ 0.607]^T$ *(value in the first iteration step, which corresponds to the optimum of (9.2) for the nominal delay* $\tau = 5$*) (left) and when* $K = [0.71\ 1.06\ 0.6]^T$ *(value in the last iteration step, which corresponds to the maximum of* $r_{\mathbb{C}}^K$ *for the nominal delay* $\tau = 5$*) (right). For the nominal delay, indicated with the dashed line, the rightmost characteristic roots are displayed in Figure 9.5 (right).*

suboptimal values of the stability radius to be optimized. Multiplicative uncertainty, as occurs for instance in the problem (9.3) with combined uncertainty on A, B, and K, can be "linearized" with the descriptor transformation outlined at the end of Section 2.3.2. Finally we notice that an algorithm for the optimization of real stability radii of time-delay systems similar to Algorithm 9.1 is described in [192].

The chapter is based on [196, 192, 183, 321] and the references therein.

Chapter 10

Stabilization using a direct eigenvalue optimization approach

10.1 Introduction

We consider time-delay systems of the form

$$\dot{x}(t) = \sum_{j=0}^{m} A_j(p)\, x(t-\tau_j), \tag{10.1}$$

where $x(t) \in \mathbb{R}^n$ is state, $A_j \in \mathbb{R}^{n\times n}$, $j = 0, \ldots, m$, are the system matrices which smoothly depend on some parameters $p \in \mathbb{R}^{n_p}$, and $\tau_j \geq 0$, $j = 0, \ldots, m$, are fixed delays, with $\tau_0 = 0$. Many stabilization problems for time-delay systems can be rephrased into the form (10.1), where the stabilization problem consists of finding parameter values p for which the spectral abscissa

$$\alpha(p) := \max\left\{\Re(\lambda) : \det\left(\lambda I - A_0(p) - \sum_{j=1}^{m} A_j(p) e^{-\lambda\tau_j}\right) = 0\right\} \tag{10.2}$$

is strictly negative. See, for instance, Section 7.3.3, where such an optimization point of view toward stabilization was briefly discussed for the case of feedback stabilization with static state feedback.

The stabilization approach presented in this chapter consists of directly solving the optimization problem

$$\min_p \alpha(p). \tag{10.3}$$

Recall that the system is stabilizable if either $\alpha(p)$ is unbounded from below, or if a global minimizer p^* exists for which $\alpha(p^*) < 0$. In the latter case, the parameters p^* result in an "optimal spectrum," where the characteristic roots are pushed to the left as far as possible and the trajectories consequently follow a maximal asymptotic decay rate to the steady state. Note that, from a pure stabilization point of view, it is not necessary to go all the way to the minimum, since it suffices to halt the minimization procedure of α once it is strictly negative.

Because the spectral abscissa is a nonsmooth function, standard optimization methods cannot be used to solve (10.3). Instead, we adopt a so-called bundle gradient method, specifically the *gradient sampling algorithm* of [36], an optimization method that is able to find minima of general nonsmooth, nonconvex objective functions. In [34, 36] this algorithm was successfully used for designing stabilizing low-order static output feedback controllers and, recently, it was included in HIFOO [32], a MATLAB package for fixed-order controller design and $\mathcal{H}_\infty$-optimization. In general, these problems can be viewed as finite-dimensional eigenvalue optimization problems where the number of controller parameters is much smaller than the number of eigenvalues to be controlled. The application to the stabilization of time-delay systems can consequently be regarded as an extension to an infinite-dimensional setting. See [150, 236, 172] and the references therein for further results on eigenvalue optimization problems for finite-dimensional systems The structure of the chapter is as follows. In Section 10.2 an approach for the minimization of the objective function (10.2) is proposed. In Section 10.3 two examples illustrate our implementation.

10.2 Eigenvalue optimization approach

We examine some smoothness properties of the function $\alpha(p)$, as defined in (10.2), and explain why it is hard, if not impossible, to optimize with standard optimization tools. Next, we discuss the recently developed gradient sampling algorithm that is able to find local minima for functions satisfying such smoothness properties. Finally, we apply this algorithm to the optimization of the spectral abscissa function (10.2), and briefly discuss the extension toward classes of nonlinear time-delay systems.

10.2.1 Smoothness properties of spectral abscissa function

A first important property of the spectral abscissa $\alpha(p)$ regarding its optimization is that it is nonconvex, and therefore may have many local minima. Consider, as a constructed example, the system

$$\ddot{x}(t) + p^2 x(t) - \epsilon p^2 x(t-1) = 0, \tag{10.4}$$

where p is the variable parameter and ϵ is a small, fixed value.

Figure 10.1 shows the real part of the rightmost characteristic roots versus p for $\epsilon = 0.1$ (left) and $\epsilon = 0.01$ (right). It is clear that for this example the spectral abscissa has several local minima. Moreover, by taking ϵ sufficiently small, one can even make the number of minima arbitrarily large, since the function $\alpha(p)/\epsilon$ uniformly converges on a compact interval to $-\frac{1}{2}p\sin(p)$ if ϵ tends to zero. It is clear that with such a behavior the global minimum is very hard to find, and most optimization algorithms will converge to a local minimum, without being able to give any guarantee about whether or not this is also the global minimum.

But even if we are satisfied with finding a local minimum, standard optimization algorithms will still fail, because the spectral abscissa $\alpha(p)$ is also a *nonsmooth* function. Indeed, despite the fact that isolated characteristic roots behave smoothly with respect to changes of the parameters, the spectral abscissa does not. A first reason for this is the presence of the maximum operator appearing in (10.2). If for certain parameter values there is more

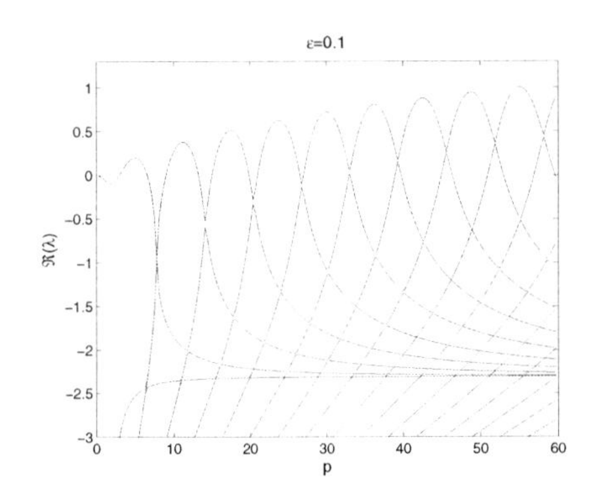

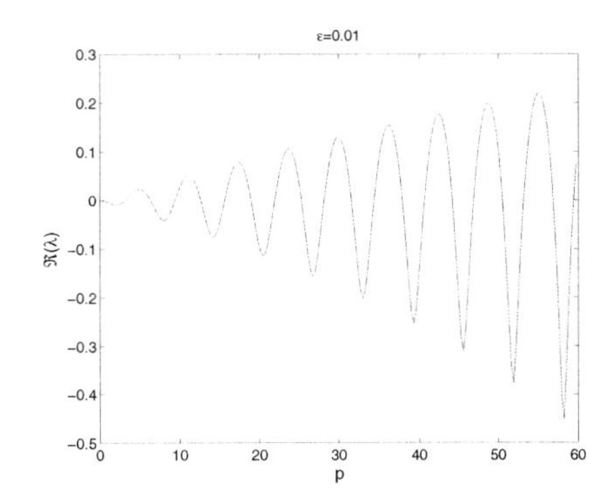

Figure 10.1. *The real part of the rightmost characteristic roots of system (10.4) as a function of p, for* $\epsilon = 0.1$ *(left) and* $\epsilon = 0.01$ *(right)*

than one *active* characteristic root, i.e., several characteristic roots satisfy $\Re(\lambda) = \alpha(p)$, it is easy to see that at this point the function $\alpha(p)$ is most likely not differentiable.

A special situation occurs when an active characteristic root has an algebraic multiplicity larger than the geometric multiplicity (i.e., it is a multiple, nonsemisimple root). Then, $\alpha(p)$ is typically not locally Lipschitz even. This is, for example, the case at parameter values where a complex pair of characteristic roots is about to split up into two real characteristic roots, or vice versa. As another example, consider the one-dimensional system $\frac{d^3x}{dt^3}(t) = px(t)$, which yields the spectral abscissa function

$$\alpha(p) = \begin{cases} \sqrt[3]{p}, & p \geq 0, \\ \sqrt[3]{-p}/2, & p < 0. \end{cases} \tag{10.5}$$

Here, α indeed exhibits a non-Lipschitz point at $p = 0$, corresponding to a triple characteristic root $\lambda^\star = 0$. The parameter value that yields this triple root is also the minimizer of the spectral abscissa function. This is actually not a coincidence, as it is a common observation that the solution of a fixed-order stabilization problem occurs at a characteristic root with higher multiplicity; see, for instance, [33].

For ODE systems, the number of active characteristic roots (multiplicity taken into account) cannot be larger than the dimension of the system. Typical for the delay case is that, since the number of characteristic roots is infinite, the number of active characteristic roots and even their multiplicity *can* be larger than the system's dimension n. This is illustrated by the case study in Chapter 8.

10.2.2 Gradient sampling algorithm

For practical nonsmooth functions, such as the spectral abscissa function, nondifferentiable points will in general only occur on a set with measure zero, meaning that such functions are differentiable *almost everywhere*. Indeed, for randomly selected parameter values p, the gradient of $\alpha(p)$ exists with probability one. This property is exploited by the gradient sampling algorithm, an optimization algorithm recently developed by Burke, Lewis, and Overton that uses a bundle gradient strategy to find a local minimum of a nonsmooth objective function [36]. It consists of the following main steps.

ALGORITHM 10.1.

Gradient sampling algorithm for nonsmooth optimization

Input A function $\phi(p)$, continuous and differentiable almost everywhere, and its gradient, $\vec{\nabla}\phi(p)$ (whenever it exists).

Step 0 Initialize a starting value p_0 arbitrarily.

Step 1 Compute the nonsmooth steepest descent direction at p_k, using a gradient sampling approximation. *If* the norm of this direction is very small, *then* stop (succeed).

Step 2 Perform a line search along the direction computed in the previous step to determine a step size that has a lower function value.

Step 3 *If* Step 2 succeeded, update p_k to p_{k+1} and go back to Step 1, *else* stop (fail).

Essentially, the algorithm has the same outline as the classical steepest descent method, except for using the *nonsmooth steepest descent direction* as the search direction. This direction is defined as

$$\arg\max_{\|d\|\leq 1} \min_{z\in\partial_c\phi(p_k)} \langle -z, d\rangle , \tag{10.6}$$

where $\langle \cdot , \cdot \rangle$ is the standard Euclidian inner product, and $\partial_c\phi(p_k)$ denotes the *Clarke subdifferential* (or generalized gradient) at p_k, given by

$$\partial_c\phi(p_k) := \text{conv}\left\{ \lim_{p\to p_k} \vec{\nabla}\phi(p) \,:\, p \in N \right\}. \tag{10.7}$$

Here, "conv" denotes the convex hull and N is any full-measure subset of a neighborhood around p_k containing differentiable points.

To understand why the nonsmooth descent direction is used in the gradient sampling algorithm, we refer to Figures 10.2–10.3, where some typical contour lines of a nonsmooth function, depending on the configuration of the intersection of smooth manifolds, are plotted. The dashed vectors represent the negative gradients of these different manifolds in a certain nondifferentiable point, and their convex hull can be considered as the "negative" of the Clarke subdifferential. A full vector denotes the nonsmooth steepest descent direction. As seen in the left frame of Figure 10.2, the nonsmooth steepest descent direction lies exactly along the negative gradient when there is no nonsmoothness (the two vectors are drawn slightly shifted for visibility reasons). The other two frames show configurations where two, respectively three, manifolds meet, but without their gradients being in conflict. In both cases, the line of nonsmooth points will simply be traversed and the descent can proceed on the lowest manifold without difficulty.

In the left frame of Figure 10.3, another situation occurs. Here, two manifolds come together forming a kind of ridge. A standard optimization algorithm such as the steepest

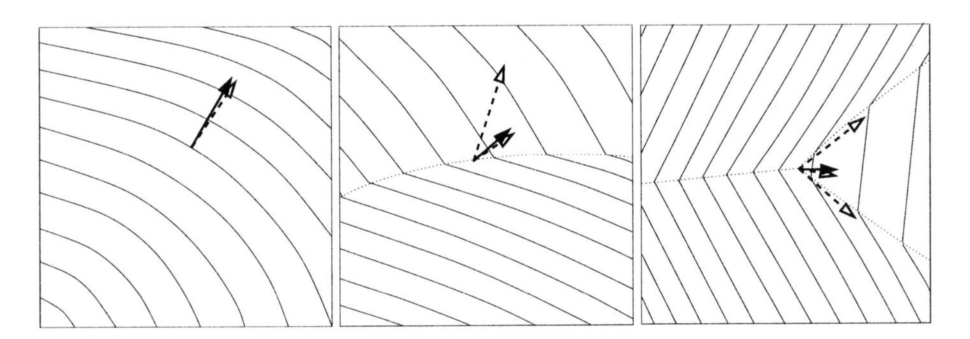

Figure 10.2. *Contours of a nonsmooth function with typical configurations of meeting manifolds. The dotted line consists of nondifferentiable points. The dashed vectors are the negative gradients of the respective manifolds and a full vector shows the nonsmooth steepest descent direction.*

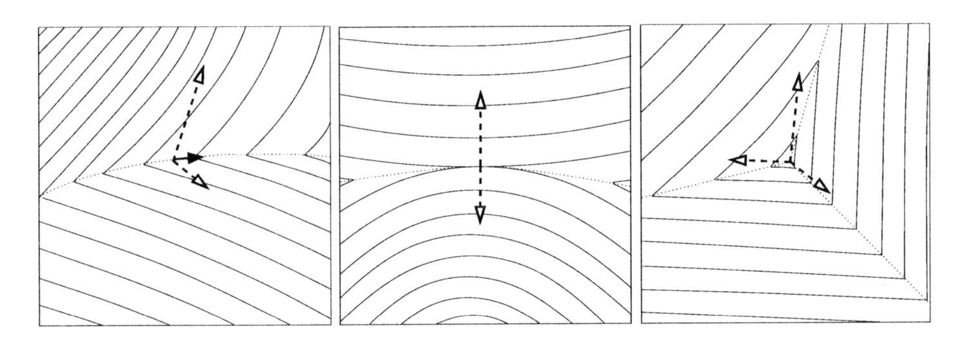

Figure 10.3. *Left, two conflicting manifolds resulting in a nonsmooth steepest direction lying along the intersection. Right, two possible configurations in an optimum.*

descent method will jump back and forth across this ridge, taking smaller and smaller steps, and eventually become jammed. The gradient sampling algorithm, using the nonsmooth steepest descent direction, will behave quite differently. According to (10.6) the nonsmooth steepest descent direction is in the opposite direction of the vector for which the minimum of the inner product with the subgradients is maximal, which corresponds to the vector in the negative Clarke subdifferential with smallest norm. In the case of two intersecting manifolds with conflicting gradients as in Figure 10.3 (left), this vector is tangent to the ridge containing the nonsmooth points. Indeed, in that case the largest decay of the objective function can be achieved by a decrease along the ridge. In the remaining two frames of Figure 10.3, two possible configurations for a nonsmooth local minimum are depicted. It is observed that in such a local minimum the Clarke subdifferential *contains the zero vector* and, hence, (10.6) is equal to zero. Note that all points for which (10.6) is zero are *Clarke stationary* [36].

Figure 10.4 shows a typical evolution of the gradient sampling algorithm, compared with the behavior of the classical steepest descent algorithm. It is seen that the latter, denoted

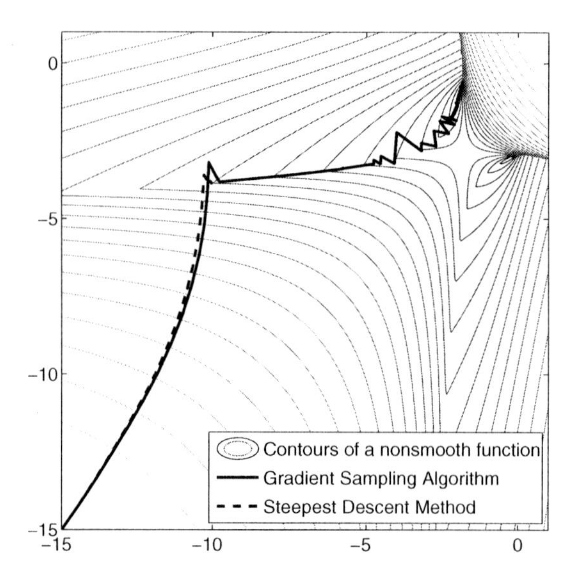

Figure 10.4. *Comparison of the behavior of the classical steepest descent method (dashed line) and the gradient sampling algorithm (full line) on a typical nonsmooth function.*

with a dashed line, indeed strands upon reaching a ridge of nonsmooth points, whereas the full line representing the gradient sampling algorithm, by following the nonsmooth steepest descent direction, is able to proceed until a nonsmooth local minimum is found.

Because the nonsmooth steepest descent direction, to be determined in Step 1 of Algorithm 10.1, is in practice difficult to compute exactly, an approximation is constructed using the following computational procedure.

ALGORITHM 10.2.

Computation of nonsmooth steepest descent direction

Step 1a Sample a number of gradients in a small neighborhood around the current point.

Step 1b Collect these gradients into a bundle, serving as an approximation for the Clarke subdifferential.

Step 1c Compute the vector with smallest norm out of this bundle by solving a quadratic program.

Assume that the current iteration point is p_k. Besides the gradient $\vec{\nabla}\phi(p_k)$, additional gradients are computed in a number of random points sampled in a small neighborhood around p_k. This neighborhood is typically chosen to be the ball with center at p_k and radius ϵ, where ϵ is some small number. If one of the sampled points should happen to be nondifferentiable, it is simply neglected and resampled. Along with the gradient in p_k, these sampled gradients are then regarded as a *bundle* of gradients, and it was shown in [31] that the closure of the convex hull of this bundle can serve as a good approximation to the Clarke subdifferential, as long as a sufficiently large number of samples is taken. The final direction for the line search is then taken as the negative of the vector out of this convex hull that has the smallest norm. This vector can easily be computed by solving a quadratic program [36].

The fact that (10.6) is zero in a (Clarke) stationary point can be used as an indication to decide when to stop the gradient sampling algorithm. In particular, if the norm of the computed search direction is smaller than a threshold value, this point is assumed to be a local minimum (due to the nature of a saddle-point and the sampling step in the algorithm, a breakdown in a saddle-point is not expected). A critical value for the accuracy of this approximate minimum is the sampling radius ϵ that was used in the computation of the nonsmooth steepest descent. Because it is very hard to determine an appropriate value for this ϵ beforehand, the gradient sampling algorithm is usually conducted several times, beginning with a relatively large ϵ, and repeated using a smaller sampling radius, each time starting off at the end point of the previous run. For more practical implementation aspects of the gradient sampling algorithm, and a convergence proof, we refer to [36].

In summary, the gradient sampling procedure can find local minima of a nonsmooth function, under the assumption that it is differentiable almost everywhere. The algorithm relies on an evaluation of the objective function and the corresponding gradient at designated points.

10.2.3 Application to linear time-delay systems

As the spectral abscissa function is continuous, as well as differentiable almost everywhere (see Section 10.2.1), Algorithm 10.1 can be applied for its minimization. The algorithm requires the computation of the objective function and its gradient.

The objective function, $\alpha(p)$, can be evaluated using the procedure for determining the rightmost characteristic roots, described in Section 1.2.6.

The gradient, $\vec{\nabla}\alpha$, can be computed analytically. Indeed, let u and v be the respective left and right null vector of the matrix $\Lambda(\lambda^\star;\ p)$, with $\lambda^\star$ an isolated characteristic root. Differentiating the equation $\Lambda(\lambda^\star;\ p)v = 0$ results, after some calculus, in the following formula for the gradient of α:

$$\vec{\nabla}\alpha(p) = \Re\left(\frac{u^*\left(\frac{\partial A_0}{\partial p} + \sum_{j=1}^{m} e^{-\lambda^\star\tau_j}\frac{\partial A_j}{\partial p}\right)v}{u^*\left(I + \sum_{j=1}^{m}\tau_j e^{-\lambda^\star\tau_j}A_j\right)v}\right). \tag{10.8}$$

However, tests indicate that in general this formula is not always the most robust choice. Alternatively, the gradient $\vec{\nabla}\alpha$ can be approximated by a finite difference formula, as follows. The number of parameters in system (10.1), n_p, equals the dimension of the tangent plane at $\alpha(p)$. Thus, this plane can be constructed—or, equivalently, the gradient $\vec{\nabla}\alpha(p)$—by using values $\alpha(p^i)$ at $n_p + 1$ points p^i close to p. In our implementation, the points p^i are selected as the $n_p + 1$ vertices of a regular polytope in the space of n_p free parameters with center of mass p and radius ε. For our experiments we used $\varepsilon = 10^{-7}$. Thus, in principle, one solves $n_p + 1$ independent eigenvalue problems. However, after computing the active characteristic root $\lambda^\star$ at the center of the polytope p, the active characteristic roots at the vertices p^i can efficiently be computed using Newton's method with starting value $\lambda^\star$.

Finally, we note that the number of optimization parameters n_p affects the number of samples one should take in each iteration to ensure a good approximation of the nonsmooth steepest descent direction, and thus the total number of function evaluations. Practice proved that the double is a safe choice.

10.2.4 Extension to nonlinear time-delay systems

There are two important issues when adapting the approach of the previous section to a nonlinear delay differential system of the form

$$\dot{x}(t) = f\left(x(t), x(t-\tau_1), \ldots, x(t-\tau_m); p\right), \tag{10.9}$$

where $f(\cdot)$ is continuously differentiable. First, the constant $x(t) \equiv 0$ is in general no stationary solution of (10.9). Moreover, the existence and uniqueness of steady states for fixed system parameters no longer hold. For example, if a steady state solution $x^\star \equiv x(t)$ exists for certain system parameters, then a nearby point in the parameter space can have a steady state "close" to $x^\star$ or there can be no steady state at all. Hence, the spectral abscissa α not only depends on the parameters, but also on the considered steady state $x^\star$. This implies that for fixed parameter values, the spectral abscissa is not uniquely defined. The investigation of *branches* of solutions of a nonlinear system for varying parameters is called *bifurcation analysis*; see [270]. The second difference is the meaning of the objective function that is optimized. Consider the linearization of (10.9) about a particular steady state. One can prove that the characteristic roots of this linearization determine the *local stability* of the original nonlinear system in a neighborhood of this steady state. Hence by optimizing the spectral abscissa of the linearization, we optimize the reaction of the system to small perturbations.

The consequences on the computational procedure are the following. Most importantly, the value of the steady state must be corrected after each change in the system's parameters. For this purpose, we employ Newton's method using the previous solution as an initial guess. If for the current point the value of α is negative or zero, then the optimization algorithm will not leave the current branch. Indeed, the values of α along the optimization path decrease monotonically and in a bifurcation point there is at least one characteristic root on the imaginary axis; hence the value of α is positive or zero. Additional branches that exist for the same parameter range should therefore be independently explored if no stabilizing parameters were found on the initial branch. A second consequence is that, since α also depends on the steady state $x^\star$, we have to adapt formula (10.8) by including a term for the partial derivative w.r.t. $x^\star$. When computing the gradient by using finite differences, the steady state $x^\star$ must be corrected by Newton's method for all vertices p^i of the polytope used in the finite difference formula.

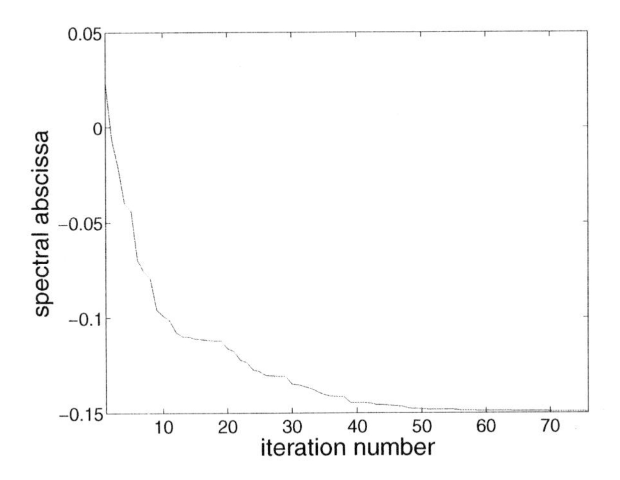

Figure 10.5. *Evolution of the spectral abscissa during the optimization process.*

10.3 Illustrative examples

We present two numerical examples. The software used is based on a MATLAB implementation of the gradient sampling optimization algorithm of [36], combined with the approaches presented in Section 10.2.3 for the computation of the objective function and its gradient.

10.3.1 Model problem: a third-order system

Consider the system

$$\dot{x}(t) = A_0 x(t) + A_1(p) x(t-\tau), \tag{10.10a}$$

with $\tau = 5$ and

$$A_0 = \begin{bmatrix} -0.08 & -0.03 & 0.2 \\ 0.2 & -0.04 & -0.005 \\ -0.06 & -0.2 & -0.07 \end{bmatrix}, \quad A_1 = \begin{bmatrix} -0.1 \\ -0.2 \\ 0.1 \end{bmatrix} \begin{bmatrix} p_1 & p_2 & p_3 \end{bmatrix}. \tag{10.10b}$$

It corresponds to the problem solved in Section 7.4.1 using the continuous pole placement procedure. Applying the gradient sampling algorithm yields (nearly) identical numerical values for the optimal parameters, $p = [\ 0.472 \quad 0.505 \quad 0.603\ ]$. Particularities of the gradient sampling approach are the monotonic decrease of the spectral abscissa function with the iteration number of the algorithm, as seen in Figure 10.5, and the fact that the procedure is fully automatic.

10.3.2 Semiconductor laser

We investigate the stabilization of a system consisting of a DDE coupled to a PDE. The system models a semiconductor laser subject to conventional optical feedback and lateral carrier diffusion [310]. The system variables are the complex scalar variable $E(t)$, representing the electric field, and the real variable $N(x, t)$, representing the carrier density in the interval $x \in [-0.5, 0.5]$, which read as

$$\dot{E}(t) = (1 - i\beta)E(t)\zeta(t) + \eta E(t - \tau)e^{-i\phi} - ibE(t), \tag{10.11}$$

$$T\frac{\partial N(x,t)}{\partial t} = d\frac{\partial^2 N(x,t)}{\partial x^2} - N(x,t) + P(x) \\ -F(x)(1 + 2N(x,t))|E(t)|^2, \tag{10.12}$$

where $\zeta(t)$ is a weighted average of the carrier density $N(x, t)$, specifically,

$$\zeta(t) = \frac{\int_{-0.5}^{0.5} F(x)N(x,t)\,\mathrm{d}x}{\int_{-\infty}^{\infty} F(x)\,\mathrm{d}x}. \tag{10.13}$$

The functions $P(x)$ and $F(x)$ are specified in [310]. We split (10.11) into real and imaginary parts in order to work only with real numbers. The symmetry of the spatial domain about $x = 0$ is exploited by considering only the interval $[0, 0.5]$ and imposing zero Neumann boundary conditions at $x = 0$, i.e., $\partial N(0, t)/\partial x = 0$. We also use zero Neumann boundary conditions at $x = 0.5$.

The partial differential system (10.11)–(10.12) cannot be treated directly by our method. It is first transformed to a system of differential equations of the form (10.9). For this purpose, we use a standard finite-difference discretization in space, in particular a second-order central difference formula with constant stepsize $\Delta x = 0.5/128$. Moreover, the integrals in (10.13) are approximated using the trapezoidal quadrature rule. This transformation gives a time-delay system of size 131 in the unknowns $\Re(E(t))$, $\Im(E(t))$, and $N(x_j, t)$ for $x_j := j\Delta x \in [0, 0.5]$, with $j = 0, \ldots, 128$. Note that this transformation introduces a discretization error. However, since the characteristic roots determining the linear stability after space discretization are computed up to a desired accuracy (due to the Newton corrections; see Section 1.2.6), the computed rightmost characteristic roots correspond to the original system (10.11)–(10.12) if the error of the *spatial* discretization is sufficiently small. That is, if the system's properties are captured sufficiently well by the spatial discretization. The quality of the characteristic roots is confirmed by our experiments using different numbers of mesh intervals in space.

Due to the form of system (10.11)–(10.12), every solution (E, N) belongs to a family of solutions of the form (cE, N) where c is an arbitrary complex number on the unit circle, i.e., $|c| = 1$. It is said that the system is *rotationally symmetric* in E. The tangent at this continuum of solutions (cE, N) is the eigenvector corresponding to a characteristic root at zero. This characteristic root at zero would not occur if one particular solution is selected of the continuum of solutions by imposing a so-called pinning condition or phase condition. It can therefore be safely ignored. For this reason, our algorithm removes the computed characteristic root that is zero, up to rounding error, before the active (pair of) characteristic root(s) is selected.

We fix the parameters to the values $\beta = 3$, $\phi = 0$, $T = 1000$, and delay $\tau = 1000$, and we optimize w.r.t. parameters η and d, representing the feedback strength and diffusion coefficient, respectively. To increase the numerical stability of the computations, the time variable and the time-delay are rescaled by a factor of $1/1000$ and the parameters η and d are rescaled by a factor of 1000. Recall that a nonlinear system can have multiple steady state solutions for a given set of parameters; see Section 10.2.4. In this case, one can show that

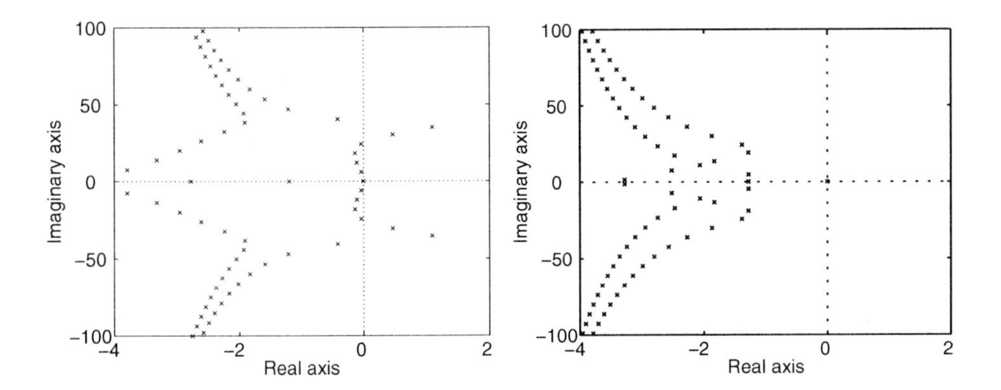

Figure 10.6. *The rightmost characteristic roots corresponding to the starting point of the optimization (left) and the reached minimum of the spectral abscissa function (right). The characteristic root at zero is induced by the rotational symmetry of the system.*

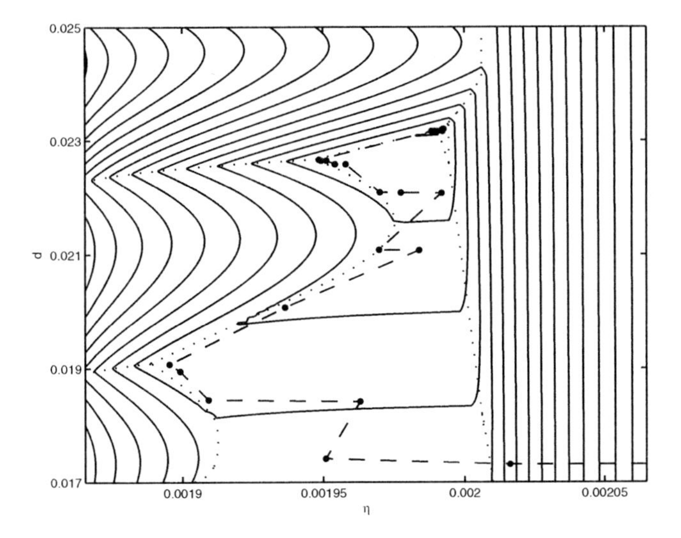

Figure 10.7. *Contour lines of the spectral abscissa function α w.r.t. parameters η and d about the optimum $\approx$ $(1.9926 \times 10^{-3},\ 2.3203 \times 10^{-2})$, the path followed by the optimization algorithm (dashed line and tick markers) and the points of nondifferentiability (dotted lines).*

for η between 0 and 7×10^{-3} and arbitrary d there are four branches that contain *stable* steady state solutions. This is illustrated in [310, Figure 3].

The starting point of the optimization algorithm is an unstable steady state with $\eta = 6.4051 \times 10^{-3}$ and $d = 1.68 \times 10^{-2}$. This initial steady state lies on one of the branches mentioned above. The algorithm returned a local minimum $\alpha^\star = -1.2713$ on this branch at $\eta^\star \approx 1.9926 \times 10^{-3}$ and $d^\star \approx 2.3203 \times 10^{-2}$. Figure 10.6 shows the rightmost characteristic roots for the starting point of the optimization (left) and for the computed local minimum (right).

Figure 10.7 shows the computed local minimum and some contours of the spectral abscissa α (spaced 0.02 apart) in the parameter space η–d. It also shows the path followed by the optimization algorithm (dashed line and tick markers) and the points of nondifferentiability (dotted lines). In this figure one can clearly see how the optimization algorithm tends to follow these lines of nonsmoothness to converge to the local minimum.

10.4 Notes and references

We presented a stabilization method for time-delay systems by tuning a finite number of parameters. Stabilization was achieved by optimizing the spectral abscissa of the system. For this, recently developed nonsmooth, nonconvex optimization algorithms were combined with algorithms for the computation of the rightmost characteristic roots of time-delay systems.

The direct eigenvalue approach has several advantages w.r.t. the other stabilization methods for time-delay systems. First, the optimization procedure induces no conservatism in the sense that stabilizing parameters, can, in principle, always be found when they exist. Second, preliminary model transformations, which may introduce additional dynamics [97, 162], can easily be dealt with. This was illustrated with the semiconductor laser example, where a nonphysical characteristic root at zero, due to the rotational symmetry, was known a priori and explicitly ignored by the optimization procedure. Here, notice that from a mathematical point of view an introduction of an unstable characteristic root renders the system unstable. Hence, sufficient stability conditions for the transformed systems become automatically infeasible. Third, the algorithm is fully automatic and requires no interactions with the user, in contrast with the continuous pole placement procedure of Chapter 7, where the number of controlled characteristic roots needed to be increased manually on stagnation. The algorithm yields a monotonic decrease of the spectral abscissa as a function of the number of iterations until a Clarke stationary point is reached. Finally, the method is very general in the sense that there are no limitations on either the number of time-delays in state, inputs and outputs, or on the type of delays. The extension to periodically varying delays and *distributed* delays is straightforward, since the method only requires that a procedure to compute the stability determining eigenvalues be at hand.

The chapter is based on [308, 36, 182] and the references therein.

Part III

Applications

Chapter 11

Stabilization by delayed output feedback: single delay case

11.1 Introduction

The existence of a time-delay at the actuating input in a feedback control system is usually known to cause *instability* or poor performance for the closed-loop schemes as largely presented and discussed in the literature (see, for instance, [171, 138, 223, 96] and the references therein).

Here we address the opposite problem: characterizing the situations where a delay has a *stabilizing effect*. In other words, we consider the situation where the delay-free feedback system is unstable, but becomes asymptotically stable due to the presence of an appropriate delay in the actuating input.

More precisely, we study the stabilization of the SISO system with the state-space representation ($A \in \mathbb{R}^{n\times n}$, $B, C \in \mathbb{R}^{n\times 1}$)

$$\dot{x}(t) = Ax(t) + Bu(t), \quad y = C^T x(t), \tag{11.1}$$

using the control law

$$u(t) = -ky(t-\tau). \tag{11.2}$$

The *stabilizing delay effect* problem can be defined as follows.

Problem 11.1. *Find explicit conditions on the pair (k, τ) such that the controller (11.2) stabilizes the system (11.1), but the closed-loop system would be unstable if the delay τ is set to zero.*

As we shall see below, the conditions derived will lead to an explicit construction of the controller. Furthermore, for each stabilizing pair, we may define a stabilizing delay interval, which can be seen as a *robustness measure* of the corresponding control law if the delay is subject to parametric uncertainty.

The interest of solving Problem 11.1 is twofold: first, the resulting design procedure is rather simple and the controller is easy to implement; second, it allows us to explore the potential of using such a controller (using the delay as a *design parameter*, defining thus a "wait-and-act" strategy) in situations where it is not easy to design or implement a controller without delay (see, for instance, the congestion controllers in high-speed networks [125, 134, 224], or some discussions in Chapter 13 devoted to the stability analysis of some congestion control algorithms in networks). Some results in this direction have been considered in [1, 227], but without any attempt to treat the problem in the general setting. A Nyquist criterion was used in [1] to prove that a pair (gain, delay) may stabilize some simple second-order linear oscillatory systems. Next, the paper [227] addresses the general static delayed output feedback problem, and some *existence results* (delay-independent, delay-dependent, instability persistence) are derived in terms of generalized eigenvalues distribution of some appropriate matrix pencils, but without any *explicit construction* of the controllers. More specifically, [227] compares the stability of the closed-loop schemes with or without delays in the corresponding control laws.

Finally, it is important to note that the techniques proposed in what follows can also be used for the closed-loop stability analysis of a dead-time plant subject to a proportional controller. We feel that our methodology gives a simpler answer to the corresponding problem (see, for instance, [273], and the references therein, for different frequency-domain approaches).

Although only strictly proper SISO systems are considered above, most of the ideas still work for more general SISO systems, such as a restricted class of (not necessarily strictly) proper systems, or systems with internal delays in addition to the feedback input delay. For the sake of conciseness, we do not present such extensions here.

The structure of the chapter is as follows. In Section 11.2 an illustrative second-order example is discussed. The construction of the corresponding closed-loop stability domain boundaries is completely described, and some useful ideas are explicitly pointed out. Section 11.3 is devoted to necessary stabilizability conditions. Such conditions, easy to check, are expressed in terms of the Hurwitz stability of some parameter-dependent polynomials. In Section 11.4 a procedure is deduced for solving Problem 11.1. The first step consists of analyzing the roots location in the complex plane of two appropriately defined polynomials, which depend on the gain parameter. Next, the sensitivity of the roots in terms of delays is analyzed, and *necessary and sufficient conditions* on the delay values are derived for the asymptotic stability of the closed-loop system. Finally, an algorithm for the explicit construction of the controller is described in the following form: for any gain satisfying some assumptions, a delay interval guaranteeing stability is computed. In Section 11.5 the geometry of the stability regions in the space (controller gain, controller delay) is addressed. Numerical examples are presented in Section 11.6. Some notes and references end the chapter.

11.2 Characterization of all stabilizable second-order systems

In order to illustrate the potential and limitations of the control law (11.2), we completely characterize the output feedback stabilizability of the second-order system,

$$\frac{p(\lambda)}{q(\lambda)} = \frac{c_1\lambda + c_2}{\lambda^2 + a_1\lambda + a_2}, \tag{11.3}$$

as a function of its parameters (a_1, a_2, c_1, c_2).

It is easy to check that with time-invariant output feedback, $u(t) = -ky(t)$, the stabilizability condition is given by

$$\begin{cases} a_2 > \frac{c_2}{c_1} a_1, & \text{if} \quad c_1 c_2 < 0, \\ a_1 > 0, & \text{if} \quad c_1 = 0, c_2 \neq 0, \\ a_2 > 0, & \text{if} \quad c_2 = 0, c_1 \neq 0, \\ a_1, a_2 \in \mathbb{R}, & \text{if} \quad c_1 c_2 > 0. \end{cases}$$

In order to check the stabilizability when the delay is also used as a controller parameter we have to solve an optimization problem for the parameters k and τ, since the stabilizability condition is given by $c(a_1, a_2, c_1, c_2) < 0$, with

$$\begin{aligned} c(a_1, a_2, c_1, c_2) := &\min_{k,\tau} \\ &\max_{\lambda \in \mathbb{C}} \left\{ \Re(\lambda) : \lambda^2 + (a_1 + kc_1 e^{-\lambda\tau})\lambda + (a_2 + kc_2 e^{-\lambda\tau}) = 0 \right\}. \end{aligned} \tag{11.4}$$

We will distinguish between several cases.

Case $c_1 = 0$, $c_2 = 1$. We take a "step-by-step" approach. First, we characterize stabilizability for a *fixed* delay $\tau = 1$ using the approach of Chapter 8. Next, we consider the case where τ is also a controller parameter.

Stabilizability region for a fixed delay. For $\tau = 1$ the optimization problem (11.4) simplifies to the determination of

$$c(a_1, a_2) = \min_k \alpha(k),$$

where

$$\alpha(k) = \max \left\{ \Re(\lambda) : h(\lambda) := \lambda^2 + a_1\lambda + (a_2 + ke^{-\lambda}) = 0 \right\}. \tag{11.5}$$

For different values of (a_1, a_2), chosen on a coarse grid, the rightmost characteristic roots are computed as a function of k, to check whether the system is stabilizable (i.e., $c(a_1, a_2) < 0$) or not ($c(a_1, a_2) \geq 0$). This allows us to sketch the stability domain boundary, which separates stabilizable and nonstabilizable (a_1, a_2)-pairs, and to characterize the configurations of the rightmost characteristic roots, which occur near the minimum of (11.5). It turns out that, close to the stability domain boundary, there are qualitatively the three following possibilities:

(a) a smooth minimum of (11.5), where the real part of a complex conjugate pair of characteristic roots is minimal,

(b) a nonsmooth but Lipschitz minimum where a smooth branch of real characteristic roots and a smooth branch of complex conjugate characteristic roots are involved, and

(c) a nonsmooth, non-Lipschitz minimum where a complex conjugate pair of characteristic roots bifurcates into two real characteristic roots.

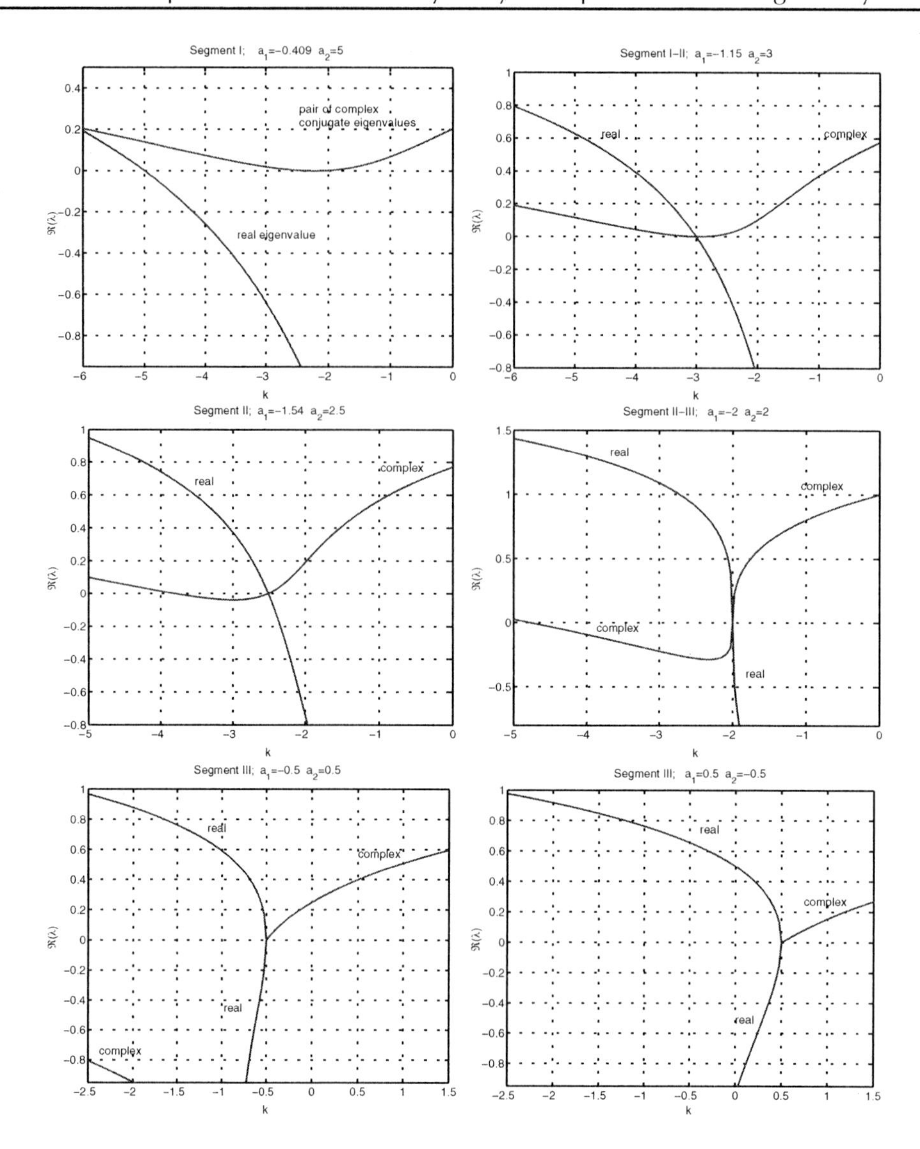

Figure 11.1. *Real parts of the rightmost characteristic roots as a function of k, corresponding to different (a_1, a_2)-pairs lying on the stability domain boundary (for $(c_1, c_2) = (0, 1)$ and $\tau = 1$). The three qualitatively different configurations of the rightmost characteristic roots in the minimum of (11.5) are described in Table 11.1.*

As an illustration, we show in Figure 11.1 such configurations of the characteristic roots for (a_1, a_2)-pairs on the stability domain boundary, together with the transitions. The three possible configurations in the minimum are mathematically characterized in Table 11.1. Since there is one degree of freedom in the mathematical relations after freeing a_1 and a_2,

Table 11.1. *The three possible configurations of the rightmost characteristic roots on the stability domain boundary in the minimum of (11.5) and their mathematical description. These relations define branches in the (a_1, a_2)-plane and allow a fast computation of the stability boundary.*

Segment	Roots configuration	Mathematical description
I	$\begin{cases} h(j\omega) = 0 \\ \frac{\partial\lambda}{\partial k}\Big\vert_{\lambda=j\omega} = 0 \end{cases}$	$\begin{cases} a_2 + k\cos(\omega) - \omega^2 = 0 \\ a_1\omega - k\sin(\omega) = 0 \\ a_1\cos(\omega) - 2\omega\sin(\omega) - k = 0 \end{cases}$
II	$\begin{cases} h(j\omega) = 0 \\ h(0) = 0 \end{cases}$	$\begin{cases} a_2 + k\cos(\omega) - \omega^2 = 0 \\ a_1\omega - k\sin(\omega) = 0 \\ a_2 + k = 0 \end{cases}$
III	$\begin{cases} h(0) = 0 \\ h'(0) = 0 \end{cases}$	$\begin{cases} a_2 + k = 0 \\ a_1 - k = 0 \end{cases}$

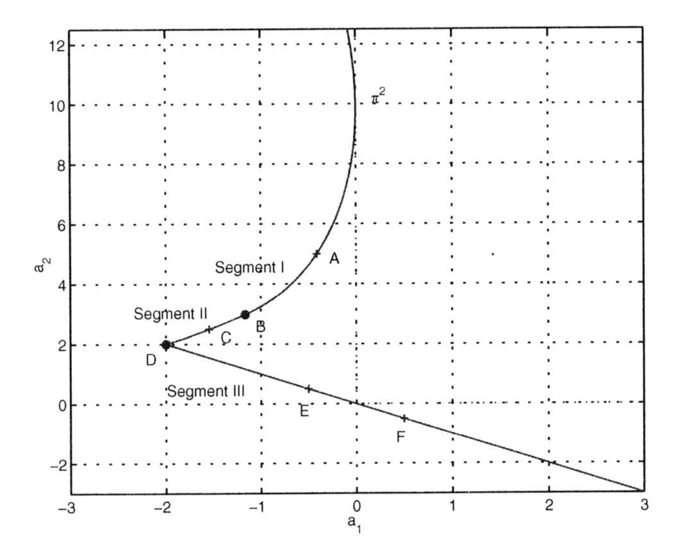

Figure 11.2. *Stability boundary in the (a_1, a_2)-plane for $\tau = 1$ and $(c_1, c_2) = (0, 1)$. The different segments refer to Table 11.1. For parameter values corresponding to the capital letters, the characteristic roots are shown as a function of k in Figure 11.1.*

the latter define branches, which can be continued efficiently in the (a_1, a_2)-space. The stability domain boundary is composed from segments of these branches; see Figure 11.2.

Stabilizability regions with k and τ as parameters. For $\tau \neq 1$, the characteristic equation of the closed-loop system is given by

$$\lambda^2 + a_1\lambda + a_2 + c_2ke^{-\lambda\tau} = 0,$$

which is equivalent to

$$\bar{\lambda}^2 + (a_1\tau)\,\bar{\lambda} + (a_2\tau^2) + c_2 k\tau^2\, e^{-\bar{\lambda}} = 0, \tag{11.6}$$

where $\bar{\lambda} = \lambda\tau$. Therefore, the stability boundary for $\tau \neq 1$ can be directly computed from the one for $\tau = 1$ by normalizing the coefficients (a_1, a_2). In Figure 11.3 the stability domain boundary is shown for different values of τ. Hence, when both k and τ are controller parameters, the stabilizability region extends toward the curves characterized by $\{a_1 = 0,\ a_2 \le 0\}$ and $\{a_1 \le 0,\ a_2 = a_1^2/2\}$. Notice that the former curve coincides with a part of the stability domain boundary for $\tau = 0$ and that the latter curve is characterized by a *zero characteristic root with multiplicity* 3 (Point D, intersection of Segment II and Segment III).

Case $c_1c_2 < 0$. Although the characteristic function

$$h(\lambda) = \lambda^2 + \left(a_1 + c_1 k e^{-\lambda\tau}\right)\lambda + \left(a_2 + c_2 k e^{-\lambda\tau}\right)$$

cannot be rescaled as (11.6), numerical experiments reveal that the stabilizability boundary is also given partially by the stabilizability boundary for $\tau = 0$,

$$a_2 = \frac{c_2}{c_1}a_1 - \left(\frac{c_1}{c_2}\right)^2, \quad a_1 \ge 0,$$

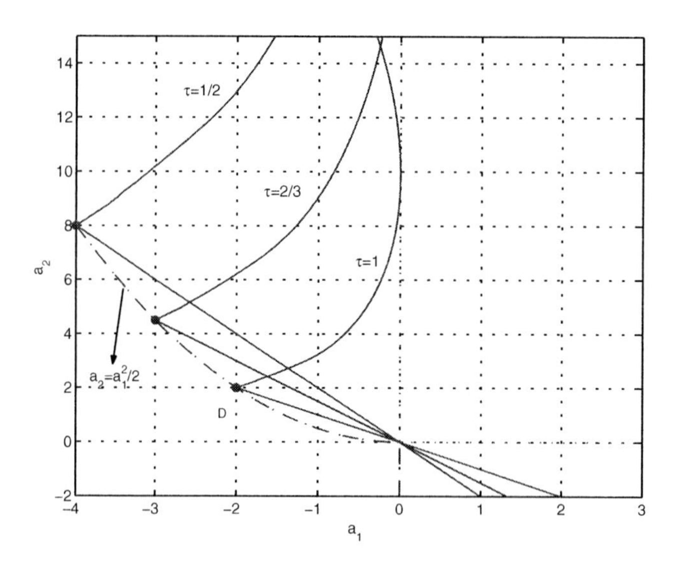

Figure 11.3. *Stability boundary for different values of τ (solid). When τ is also a controller parameter, the stabilizability region consists of the union of the stabilizability regions for all fixed τ. This region is bounded by the dashed line, determined by the conditions $\{a_1 = 0,\ a_2 \le 0\}$ and $\{a_1 \le 0,\ a_2 = a_1^2/2\}$.*

and partially by a curve, characterized by a triple characteristic root at zero (for $a_1 \leq 0$). The conditions

$$h(0) = h'(0) = h''(0) = 0 \Leftrightarrow \begin{cases} a_2 + c_2 k = 0, \\ a_1 + c_1 k - c_2 k\tau = 0, \\ -2c_1 k\tau + c_2 k\tau^2 + 2 = 0 \end{cases}$$

are equivalent to

$$a_2 = -\frac{c_2}{c_1}\left(\frac{c_2}{c_1} + \sqrt{a_1^2 + \left(\frac{c_2}{c_1}\right)^2}\right).$$

Case $c_2 = 0$. A similar approach as in the previous case yields that no improvements with respect to time-invariant output feedback can be achieved. An indication of this result is given by the fact that for $a_2 = 0$, there is characteristic root at zero, independent of the controller parameters.

Finally, combining the three above cases completes the stabilizability study of (11.3) with delayed output feedback. The results are summarized in Table 11.2.

Table 11.2. *Necessary and sufficient conditions for the output feedback stabilizability of the second-order system (11.3).*

$c_1 c_2 < 0$ (relative degree 1, nonminimum phase)	
Time-invariant, $u(t) = -ky(t)$	$a_2 > \frac{c_2}{c_1} a_1$
Delayed, $u(t) = -ky(t-\tau)$	$\left\{a_2 > \frac{c_2}{c_1} a_1\right\} \cup \left\{a_2 > -\frac{c_2}{c_1}\left(\frac{c_2}{c_1} + \sqrt{a_1^2 + \left(\frac{c_2}{c_1}\right)^2}\right)\right\}$
$c_1 = 0,\ c_2 \neq 0$ (relative degree 2)	
Time-invariant	$a_1 > 0$
Delayed	$\{a_1 > 0\} \cup \left\{a_2 > \frac{a_1^2}{2}\right\}$
$c_2 = 0,\ c_1 \neq 0$ (relative degree 1, weakly minimum phase)	
Time-invariant	$a_2 > 0$
Delayed	no improvement
$c_1 c_2 > 0$ (relative degree 1, minimum phase)	
Time-invariant	always stabilizable

11.3 Necessary conditions for stabilizability

Let the transfer function of the system (11.1) be given by

$$\frac{P(\lambda)}{Q(\lambda)} := C^T(\lambda I - A)^{-1}B, \tag{11.7}$$

where P and Q are coprime polynomials satisfying

$$\deg(P(\lambda)) = m, \ \deg(Q(\lambda)) = n.$$

The following necessary stabilizability condition corresponds to [135, Proposition III.3] and is based on an extension of Lucas theorem to classes of entire functions [213].

Proposition 11.2. *A necessary condition for the stabilizability of the system (11.1) with the control law (11.2) is the Hurwitz stability of the polynomial*

$$\gamma(\lambda;\ \tau) := \left(\frac{d}{d\lambda} + \tau\right)^{m+1} Q(\lambda). \tag{11.8}$$

Example 11.3. *We apply Proposition 11.2 to the second-order example (11.3). We distinguish between the following two cases.*

Case $c_1 = 0$, $c_2 \neq 0$. We have $m = 0$ and

$$\gamma(\lambda;\ \tau) = \tau\lambda^2 + (a_1\tau + 2)\lambda + (a_2\tau + a_1).$$

This polynomial is Hurwitz if and only if

$$a_1\tau + 2 > 0, \ \ a_2\tau + a_1 > 0. \tag{11.9}$$

Condition (11.9) is violated for all $\tau \geq 0$ if and only if $a_1 \leq 0$ and $a_2 \leq a_1^2/2$. Hence, a necessary condition for the stabilizability with a control law of the form (11.2) is given by

$$a_2 > \frac{a_1^2}{2} \text{ or } a_1 > 0.$$

In Section 11.2 it was shown that this condition is also sufficient; see Table 11.2.

Case $c_1 \neq 0$. We have $m = 1$ and

$$\gamma(\lambda;\ \tau) = \tau^2\lambda^2 + (a_1\tau^2 + 4\tau)\lambda + (a_2\tau^2 + 2a_1\tau + 2),$$

which is Hurwitz if $\tau = 0$, for all a_1 and a_2. This is expected because the necessary condition from Proposition 11.2 does not depend on the parameters of the nominator polynomial $c_1\lambda + c_2$, while the closed-loop stability can always be achieved by using full state feedback.

A direct corollary of Proposition 11.2 is the following.

Corollary 11.4. *If the polynomial $Q(\lambda)$ has at least one unstable root with multiplicity $\geq m+2$, then the system (11.1) cannot be stabilized by the control law (11.2).*

Notice that, to some extent, Proposition 11.2 concerns the stabilizability for a fixed value of τ with the gain k as control parameter, since the necessary condition explicitly depends on the delay (unlike the condition of Corollary 11.4, which is independent of both controller parameters). In the next section we will address the dual problem of characterizing the stabilizability for a fixed gain with the delay as controller parameter. Combined with a study of the dependence of the resulting stabilizability conditions on the gain and an explicit algorithm for the computation of the stability regions in the delay parameter space, this will, however, lead to a procedure for determining both controller parameters.

11.4 Controller construction

We address Problem 11.1. Before presenting the main results, some prerequisites are needed.

11.4.1 Prerequisites

Using (11.7), the characteristic function of the closed-loop system (11.1)–(11.2) can be written in the form

$$H(\lambda;\ k,\tau) := Q(\lambda) + kP(\lambda)e^{-\lambda\tau}. \tag{11.10}$$

Without any loss of generality we will assume that the polynomials P and Q are coprime, that is, do not have common zeros.

Two quantities will play an important role in the controller design procedure:

1. $\mathrm{card}(\mathcal{U}_+)$, where $\mathcal{U}_+$ is the set of roots of $H(\lambda;\ k,0) = Q(\lambda) + kP(\lambda)$, located in the *closed RHP*, that is the instability degree of the closed-loop system free of delays,

2. $\mathrm{card}(\mathcal{S}_+)$, where $\mathcal{S}_+$ the set of strictly positive roots of the polynomial

$$F(\omega;\ k) = \mid Q(j\omega) \mid^2 - k^2 \mid P(j\omega) \mid^2; \tag{11.11}$$

as we shall see later, $\mathcal{S}_+$ and F characterize the existence of crossing roots on the imaginary axis w.r.t. the delay parameter τ.

Both quantities depend explicitly on the gain parameter only, and we now clarify this dependence in more depth.

The quantity card $(\mathcal{U}_+)$. Its characterization as a function of the gain corresponds to the static (undelayed) output feedback stabilizability problem. The difficulty of this problem is well known (see, for instance, [288] and the references therein). However, in the SISO system case, the problem is reduced to a *one-parameter problem*, which is relatively easy. Indeed, there exist several methods to solve it. These include (standard) graphical tests (root-locus, Nyquist), and computation of the real roots of an appropriate set of polynomials. In addition to these standard methods, we may cite two interesting approaches [44, 112] based on *generalized eigenvalues computation* of some appropriate matrix pencils defined by the corresponding Hurwitz [44], and Hermite [112] matrices. The approach below is inspired by Chen's characterization [44] for systems without delay.

We now introduce the following Hurwitz matrix associated with the denominator polynomial $Q(\lambda) = \sum_{i=0}^{n} q_i \lambda^{n-i}$ of the transfer function:

$$H(Q) = \begin{bmatrix} q_1 & q_3 & q_5 & \dots & q_{2n-1} \\ q_0 & q_2 & q_4 & \dots & q_{2n-2} \\ 0 & q_1 & q_3 & \dots & q_{2n-3} \\ 0 & q_0 & q_2 & \dots & q_{2n-4} \\ \vdots & & & \ddots & \vdots \\ 0 & 0 & 0 & \dots & q_n \end{bmatrix} \in \mathbb{R}^{n \times n}, \tag{11.12}$$

where the coefficients $q_l = 0$, for all $l > n$. Next, corresponding to the numerator polynomial $P(\lambda)$ of the transfer function, we construct $H(P)$ as an $n \times n$ matrix by the same procedure as (11.12) with the understanding that $p_l = 0$ for all $l > m$. The following result is a slight modification and generalization of Theorem 2.1 by Chen [44].

Lemma 11.5. *Let $\lambda_1 < \lambda_2 < \dots \lambda_h$, with $h \leq n$ the real eigenvalues of the matrix pencil*

$$\Lambda(\lambda) = \det(\lambda H(P) + H(Q)).$$

Then the system (11.7) cannot be stabilized by the controller $u(t) = -ky(t)$ for any $k = \lambda_i$, $i = 1, 2, \dots, h$. Furthermore, if there are r unstable closed-loop roots $(0 \leq r \leq n)$ for $k = k^*$, $k^* \in (\lambda_i, \lambda_{i+1})$, then there are r unstable closed-loop roots for any gain $k \in (\lambda_i, \lambda_{i+1})$. In other words, $\operatorname{card}(\mathcal{U}_+)$ *remains constant as k varies within each interval $(\lambda_i, \lambda_{i+1})$. The same holds for the intervals $(-\infty, \lambda_1)$ and (λ_h, ∞).*

Proof. First, we need to show that as k varies, there are closed-loop roots on the imaginary axis if and only if $k = \lambda_i$, $i = 1, 2, \dots, h$. The proof follows the same step as proposed by Chen in [44], and therefore will be omitted.

The above implies that for any gain $k \in (\lambda_i, \lambda_{i+1})$, the corresponding closed-loop system has no roots crossing the imaginary axis. Based on the *continuous dependence* of the roots of the polynomial $Q(s) + kP(s)$ on the parameter k, if there exists a k^* in the interval $(\lambda_i, \lambda_{i+1})$ such that $Q(s) + k^* P(s)$ has exactly r unstable roots, then the property is valid for any $k \in (\lambda_i, \lambda_{i+1})$ since the roots cannot jump from $\mathbb{C}_-$ to $\mathbb{C}_+$, or from $\mathbb{C}_+$ to $\mathbb{C}_-$, without crossing the imaginary axis. ☐

Thus by computing the generalized eigenvalues of the matrix pencil $\Lambda(\lambda)$, yielding the critical gain values, and computing $\mathcal{U}_+$ for intermediate gain values, the function $k \to \operatorname{card}(\mathcal{U}_+)(k)$ is completely determined.

The quantity card $(\mathcal{S}_+)$. Without any loss of generality assume that $P(0) \neq 0$.[16] Its dependence on k is expressed in the following proposition.

[16] If not, then $Q(0) \neq 0$, and the results below still hold by defining the dependence in terms of $\frac{1}{k}$, etc.

Proposition 11.6. *Assume that* $\mathrm{card}(\mathcal{S}_+)$ *changes at a gain value* k^*. *Then there exists a frequency* $\omega^* \geq 0$ *such that for* $\omega = \omega^*$

$$|Q(j\omega)|^2 - k^{*2}|P(j\omega)|^2 = 0 \tag{11.13}$$

and

$$|P(j\omega)|^2 \frac{d}{d\omega}|Q(j\omega)|^2 - |Q(j\omega)|^2 \frac{d}{d\omega}|P(j\omega)|^2 = 0. \tag{11.14}$$

Proof. For any k, F cannot have a root ω where $P(j\omega) = 0$, because this would imply that also $Q(j\omega) = 0$. The root of F therefore coincide with roots of

$$G(\omega;\ k) := \frac{|Q(j\omega)|^2}{|P(j\omega)|^2} - k^2. \tag{11.15}$$

A change of $\mathrm{card}(\mathcal{S}_+)$ at $k = k^*$ implies that $G(\omega;\ k^*)$ has a root with multiplicity larger than one at some frequency ω^*, i.e.,

$$G(\omega^*;\ k^*) = G'(\omega^*;\ k^*) = 0.$$

This leads to (11.13) and (11.14). $\square$

Proposition 11.6 allows computing systematically the behavior of $\mathrm{card}(\mathcal{S}^+)$ as a function of the gain. First, one has to determine the real roots of the polynomial (11.14). Then the critical values of the gain k follow from (11.13). The characterization is complete when computing $\mathcal{S}_+$ for intermediate gain values, which again corresponds to finding the roots of some polynomial.

Remark 11.7. *From the symmetry of the function (11.15), it follows that the pairs* $(\omega, k) = (0, \pm\frac{Q(0)}{P(0)})$ *always satisfy (11.13)–(11.14) and furthermore, at these k-values,* $\mathrm{card}(\mathcal{S}_+)$ *always changes with* ± 1. *It also follows from (11.15) that at other k-values,* $\mathrm{card}(\mathcal{S}_+)$ *can only change with* ± 2.

11.4.2 Stabilization using the delay parameter

For a given value of the gain k we derive conditions for the existence of stabilizing delay values. We make the following technical assumptions.

Assumption 11.8. *Let the gain* $k \in \mathbb{R}$ *be such that*

1. *all the roots of* F *are simple,*
2. $0 \notin \mathcal{U}_+$,
3. $\mathrm{card}(\mathcal{U}_+) \neq 0$.

Notice that the first condition is satisfied for almost all k. The second condition is *necessary* for stabilization because it excludes a characteristic root at zero, the latter being *invariant* w.r.t. delay changes. The third assumption excludes the trivial case where the system is asymptotically stable with the choice $\tau = 0$.

An important result is the following.

Theorem 11.9. *The characteristic equation has a root $j\omega$ for some delay value τ_0 if and only if*

$$\omega \in \mathcal{S}_{+}. \tag{11.16}$$

Furthermore, for any ω satisfying (11.16) the set of corresponding delay values is given by

$$\mathcal{T}_{\omega} = \left\{ \frac{1}{\omega} \left[\angle \left(-\frac{kP(j\omega)}{Q(j\omega)} \right) + 2\pi l \right] \geq 0, \ l \in \mathbb{Z} \right\}. \tag{11.17}$$

When increasing the delay the corresponding crossing direction of characteristic roots is toward instability (stability) when $F'(\omega) > 0$ (< 0).

Proof. The result is a direct consequence of Proposition 4.13 (Chapter 4) applied to the corresponding characteristic function $H(\lambda;\ k, \tau)$. $\square$

Theorem 11.9, combined with the continuous dependence of the characteristic roots w.r.t. the delay, allows us to characterize completely stability/instability regions in the delay parameter. Indeed, the set

$$\mathcal{T} = \bigcup_{\omega \in \mathcal{S}_{+}} \mathcal{T}_{\omega}$$

makes a partition of the delay space ($\mathbb{R}_{+}$) into intervals in which the number of roots in the open RHP is constant. More explicitly, we have the following result.

Remark 11.10. (crossing direction) *Assume that $\mathcal{S}_{+}$ is not empty and denote its elements in descending order with $\omega_1 > \omega_2 > \dots$.*

Because $\lim_{\omega\to\infty} F(\omega) = +\infty$ and the roots $\{\omega_1, \omega_2, \dots\}$ of F are simple, the sign of F' at these roots alternates, with $F'(\omega_1) > 0$.

As a consequence, as the delay is monotonically increased, all root crossings for delay values $\mathcal{T}_{\omega_1}$ are toward instability, all root crossings for delay values $\mathcal{T}_{\omega_2}$ are toward stability, etc.

Taking into account the number of unstable roots for $\tau = 0$ and the crossing direction in each point of $\mathcal{T}$, the number of unstable roots for each delay value can be determined.

Based on Theorem 11.9 and its underlying ideas, we now focus on the delay *stabilization* problem. We have the following results.

Proposition 11.11. *Assume that* card$(\mathcal{U}_{+})$ *is an odd number. Then the delay stabilization problem has no solution.*

Proof. By contradiction. Assume that the closed-loop system is asymptotically stable for some delay value τ_s. Because the number of roots in the closed RHP changes *from odd to even* when increasing the delay from zero (number of closed half-plane roots equal to

$\text{card}(\mathcal{U}_+)$) to τ_s (number of closed half-plane roots equal to zero), a characteristic root at *zero* must occur for some $\tau_0 \in [0,\ \tau_s]$. But $H(0;\ k,\ \tau_0) = 0$ implies $H(0;\ k,\ \tau) = 0,\ \forall \tau \geq 0$, which contradicts the asymptotic stability at $\tau = \tau_s$. $\square$

Remark 11.12. *Such a result is relatively simple, and proves the existence of a strictly real positive root of the characteristic equation for all delay values if such a root exists for the system free of delay. Some discussions have been proposed in Chapter 4 (see the section on Notes and References). Finally, we would like to point out that similar results, with slightly different formulations, and similar (or different) proofs have been already proposed in the literature (see, for instance, [286, 92]).*

Proposition 11.13. *If either* $\text{card}(\mathcal{S}_+) = 0$ *or* $\text{card}(\mathcal{S}_+) = 1$, *then the delay stabilizing problem has no solution.*

Proof. When $\text{card}(\mathcal{S}_+) = 0$, characteristic roots cannot cross the imaginary axis as the delay is varied and the instability for $\tau = 0$ persists for all delay values.

When $\text{card}(\mathcal{S}_+) = 1$, there is one crossing frequency and, from Remark 11.10, the crossing direction is always toward instability as the delay is increased. Combining this fact with the instability for $\tau = 0$ yields the statement of the proposition. $\square$

Notice that the condition $\text{card}(\mathcal{S}_+) = 0$ corresponds to the *delay-independent hyperbolicity* property (fixed number of unstable roots for all positive delay values), as defined in [103] (see also [106]). For the remaining cases, one needs to count the roots crossing the imaginary axis toward stability/instability, and to define the corresponding delay intervals (see also [223], Chapters 4 and 7).

More explicitly, it follows that the first case when the delay has a stabilizing effect may appear if *card*$(\mathcal{U}_+) = 2$, *card*$(\mathcal{S}_+) \in \{2, 3\}$, and the first crossing is toward stability. Furthermore, as proved in what follows, such a condition is also necessary.

Proposition 11.14. *Assume that* $\text{card}(\mathcal{S}_+) = 2$ *or* $\text{card}(\mathcal{S}_+) = 3$. *Then the delay stabilizing problem has a solution if and only if*

1. $\text{card}(\mathcal{U}_+) = 2$,

2. $\tau_- < \tau_+$, *where*

$$\begin{aligned} \tau_- &= \min \bigcup\nolimits_{\omega \in \mathcal{S}_+,\ F'(\omega) < 0} \mathcal{T}_\omega, \\ \tau_+ &= \min \bigcup\nolimits_{\omega \in \mathcal{S}_+,\ F'(\omega) > 0} \mathcal{T}_\omega \setminus \{0\}. \end{aligned}$$

If stabilizable, all delay values $\tau \in (\tau_-,\ \tau_+)$ *are stabilizing.*

Proof. "$\Rightarrow$" Consider first the case where $\text{card}(\mathcal{S}_+) = 2$. Let $\mathcal{S}_+ = \{\omega_1, \omega_2\}$, with $\omega_1 > \omega_2$. Then from Remark 11.10 we have $F'(\omega_1) > 0$ (associated crossings for $\mathcal{T}_{\omega_1}$ toward instability when increasing the delay) and $F'(\omega_2) < 0$ (associated crossings for $\mathcal{T}_{\omega_2}$ toward stability). The set $\mathcal{T}_{\omega_1}$ consists of delay values, equally spaced with $2\pi/\omega_1$, whereas the elements of $\mathcal{T}_{\omega_2}$ are equally spaced with $2\pi/\omega_2 > 2\pi/\omega_1$. As a consequence, between two stability crossings, an instability crossing must occur; i.e., the number of unstable roots in the closed

RHP cannot be reduced by more than two by increasing the delay. Thus the occurrence of a stabilizing delay value necessarily implies that the number of roots in the closed RHP, is two for $\tau = 0$ (i.e., $\mathrm{card}(\mathcal{U}_+) = 2$) and that the first[17] crossing is toward stability, mathematically expressed by $\tau_- < \tau_+$.

If $\mathrm{card}(\mathcal{S}_+) = 3$, then $\mathcal{S}_+ = \{\omega_1, \omega_2, \omega_3\}$ with $\omega_1 > \omega_2 > \omega_3$ and $F'(\omega_1) > 0$, $F'(\omega_2) < 0$, $F'(\omega_3) > 0$. Compared to the previous case, there is one additional crossing frequency ω_3 where additional crossings toward *in*stability occur, and the argument remains the same.

"$\Leftarrow$" The condition $\tau_- < \tau_+$ implies that the first crossing is toward stability when the delay in increased from zero. Since $card(\mathcal{U}_+) = 2$, the closed-loop system is asymptotic stability for any $\tau \in (\tau_-, \tau_+)$. □

Remark 11.15. *The importance of this result lies in the fact that, in order to check stabilizability in the delay parameter, one only has to investigate the* first *root crossing of the imaginary axis as the delay is increased from zero. This is particularly useful when one determines stabilizability by numerically computing the rightmost characteristic roots as a function of the delay. After the first root crossing one can stop the computations.*

In the case when $\mathrm{card}(\mathcal{S}_+) = 2$, the set of *all* stabilizing delay values can be expressed analytically.

Corollary 11.16. *Assume that the conditions of Proposition 11.14 are satisfied, and in addition* $\mathrm{card}(\mathcal{S}_+) = 2$. *Then all the stabilizing delay values are defined by* $\tau \in (\underline{\tau_l}, \overline{\tau_l})$, $l = 0, 1, 2, \ldots, l_m$, *where*

$$\underline{\tau_l} = \tau_- + \frac{2\pi l}{\omega_-}, \overline{\tau_l} = \tau_+ + \frac{2\pi l}{\omega_+},$$

$\mathcal{S}_+ = \{\omega_+, \omega_-\}$ *with* $\omega_+ > \omega_-$, *and* l_m *is the largest integer for which* $\underline{\tau_l} < \overline{\tau_l}$, *which can be expressed as*

$$l_m = \max_{l \in \mathbf{Z}} \left\{ l < \frac{\omega_+ \omega_-}{2\pi} \cdot \frac{\tau_+ - \tau_-}{\omega_+ - \omega_-} \right\}. \tag{11.18}$$

Proposition 11.17. *Assume that* $\mathrm{card}(\mathcal{S}_+) = 2n$ *or* $\mathrm{card}(\mathcal{S}_+) = 2n + 1$, *with* $n \geq 1$. *Assume further that* $\mathrm{card}(\mathcal{U}_+) > 2n$. *Then the delay stabilizing problem has no solution.*

Proof. Let $\mathcal{S}_+ = \{\omega_1, \omega_2, \ldots\}$ with $\omega_1 > \omega_2 > \ldots$. From Theorem 11.9 we have the alternating sequence: $F'(\omega_1) > 0$, $F'(\omega_2) < 0$, $F'(\omega_3) > 0, \ldots$.

Consider the pair (ω_1, ω_2). By the same arguments as used in the proof of Proposition 11.14, there must be an element of $\mathcal{T}_{\omega_1}$ between two elements of $\mathcal{T}_{\omega_2}$. When the delay is increased from zero, the root crossings at $j\omega_1$ and $j\omega_2$ can therefore not contribute to reduction of closed half-plane roots with more than two. When $n > 1$ the same argument can be used for the pairs $(\omega_3, \omega_4), \ldots, (\omega_{2n-1}, \omega_{2n})$. Thus, taking into account the root crossings at $j\omega_1, j\omega_2, \ldots, j\omega_{2n}$ no more than $2n$ unstable roots can be shifted to the LHP

[17] When $0 \in \mathcal{T}_{\omega_1}$ the crossing at $\tau = 0$ is not counted since $\mathrm{card}(\mathcal{U}_+)$ does not change.

and the proof is complete for $\text{card}(\mathcal{S}_+) = 2n$. Finally, in the case $\text{card}(\mathcal{S}_+) = 2n + 1$ the argument remains the same since $F'(\omega_{2n+1}) > 0$. □

Define now the following quantities:

$$n_+(\tau) = \sum_{\omega \in \mathcal{S}_+,\ F'(\omega) > 0} \text{card}\,\{\mathcal{T}_\omega \cap (0, \tau]\}, \tag{11.19}$$

$$n_-(\tau) = \sum_{\omega \in \mathcal{S}_+,\ F'(\omega) < 0} \text{card}\,\{\mathcal{T}_\omega \cap [0, \tau]\}, \tag{11.20}$$

for some positive $\tau > 0$. Furthermore, introduce the sets $\mathcal{T}^+$ and $\mathcal{T}^-$, which represent a partition of $\mathcal{T}$ in function of the sign of the derivative F' evaluated at the corresponding crossing frequency, that is,

$$\mathcal{T}^+ = \bigcup_{\omega \in \mathcal{S}_+,\ F'(\omega) > 0} \mathcal{T}_\omega \setminus \{0\},$$

$$\mathcal{T}^- = \bigcup_{\omega \in \mathcal{S}_+,\ F'(\omega) < 0} \mathcal{T}_\omega.$$

Based on the conditions and the notations above, we conclude with the following result.

Proposition 11.18. *For a given gain k, the stabilizing control problem has a solution of the form* $u(t) = -ky(t - \tau)$ *if and only if the following conditions hold simultaneously:*

(i) $\text{card}(\mathcal{U}_+$ *is a strictly positive even integer, which satisfies the inequality* $\text{card}(\mathcal{U}_+) \leq \text{card}(\mathcal{S}_+)$*, and*

(ii) there exists at least one delay value $\hat{\tau} \in \mathcal{T}$ *such that the following equality is verified:*

$$2n_-(\hat{\tau}) = 2n_+(\hat{\tau}) + \text{card}(\mathcal{U}_+). \tag{11.21}$$

Then all delay values $\tau \in (\hat{\tau}, \hat{\tau}_+)$*, with*

$$\hat{\tau}_+ = \min\left\{\mathcal{T}^+ \cap (\hat{\tau}, +\infty)\right\}, \tag{11.22}$$

guarantee the closed-loop asymptotic stability.

Proof. While condition (i) is clear, the condition (11.21) in (ii) simply characterizes the existence of crossings such that there are no more unstable rightmost roots for delays $\tau = \hat{\tau} + \varepsilon$, for sufficiently small $\varepsilon > 0$, and the definition of the delay interval follows straightforwardly. □

Remark 11.19. *Proposition 11.18 above can be reformulated by defining an appropriate function [18]* $N : \mathbb{R}_+ \mapsto \mathbb{N}$ *defined for some positive delay value* τ*. Such a function computes explicitly the number of strictly unstable roots in* $\mathbb{C}_+$ *of the characteristic function* H *by taking into account all the crossing roots together with their crossing direction for each critical delay inside the interval* $[0, \tau]$*. A similar formula was proposed by [241] but the characterization of the critical delays for which crossing roots exist was derived using a different argument.*

11.4.3 Stabilization using the delay and gain parameter

The main results of the previous section are displayed in Table 11.3. Recall that $\text{card}(\mathcal{U}_+)$ and $\text{card}(\mathcal{S}_+)$ depend *only* on the gain k. The first quantity can be efficiently determined as a function of k by computing the generalized eigenvalues of a matrix pencil (Lemma 11.5), the second quantity by computing the roots of a polynomial (Proposition 11.6).

The procedure to derive a stabilizing pair (k, τ) (if any) can be summarized as follows:

- First, compute $\text{card}(\mathcal{S}_+)$ and $\text{card}(\mathcal{U}_+)$ as functions of the gain parameter k, and next select possible gain intervals $(\underline{k}, \overline{k})$ such that condition (i) of Proposition 11.18 holds.

- Second, for a given k, search for stabilizing delay values. In the special case where $\text{card}(\mathcal{S}_+) = 2$ or $\text{card}(\mathcal{S}_+) = 3$ and $\text{card}(\mathcal{U}_+) = 2$, Proposition 11.14 can be applied; i.e., it is sufficient to investigate only whether the *first* root crossing of the imaginary axis is toward stability as the delay is increased from zero. In general, a more complete characterization of stability/instability regions becomes necessary, as we shall illustrate in Section 11.6. According to condition (ii) of Proposition 11.18 this is, however, a systematic task. Indeed, one has to compute the set $\mathcal{T}$, and next the partition $\mathcal{T}^+$ and $\mathcal{T}^-$, which gives the roots crossings (function of the delay values) toward instability and stability, respectively. Condition (11.21) together with (11.22) will define the corresponding stabilizing delay intervals.

Finally, we note that the set of all stabilizing pairs (k, τ) is bounded if the delay free stabilization problem is not solvable.

Proposition 11.20. *Assume that* $\text{card}(\mathcal{U}_+)(k) > 0$ *for all* $k \in \mathbb{R}$. *Then the set of all stabilizing pairs* (k, τ) *is bounded.*

Proof. For sufficiently large $|k|$ the polynomial $F(\omega;\ k)$ only has one zero. From Theorem 11.9 and, in particular, the crossing direction characterization, it follows that the

Table 11.3. *Output feedback stabilizability conditions when using the delay as controller parameter. Necessary and sufficient conditions are given by Proposition 11.18. In the case of* $\text{card}(\mathcal{U}_+) = 2$ *and* $\text{card}(\mathcal{S}_+) \in \{2, 3\}$, *Proposition 11.14 can be applied.*

	0	1	2	3	4	5	6	7	8	9	$\text{card}(\mathcal{S}_+)$
1	/	/	/	/	/	/	/	/	/	/	
2	/	/	$\tau_- < \tau_+$								
3	/	/	/	/	/	/	/	/	/	/	
4	/	/	/	/							
5	/	/	/	/	/	/	/	/	/	/	
6	/	/	/	/	/	/					
7	/	/	/	/	/	/	/	/	/	/	
8	/	/	/	/	/	/	/	/			
$\text{card}(\mathcal{U}_+)$											

corresponding delay stabilizing problem has no solution. So, there exists an upper bound on the possible gain values of the stabilizing controllers. The same holds for the corresponding delay values, since for any fixed value of k, the function $n_+(\tau) - n_-(\tau)$ tends to infinity as τ tends to infinity. □

11.5 Geometry of stability regions

In Chapter 4 we presented to some extent the geometry of the stability regions in the delay-parameter space. In this section, we adapt the corresponding approach and ideas to the case of the parameter space (k, τ) under consideration, and we compute explicitly the stability crossing curves which give the partition of the parameter space in stable and unstable domains. The difference w.r.t. the approach of the previous section lies in the way in which the stability regions in the (k, τ)-space are computed, or the corresponding stabilization problem is solved. Essentially, the approach of the previous section corresponds to the following three steps: (1) choose and fix the gain k, (2) compute the corresponding crossing frequencies ω, and (3) compute the corresponding delay values τ. On the contrary, the approach presented below corresponds to the following steps: (1) choose possible values of ω, and (2), compute corresponding pairs (k, τ). Hence, ω is explicitly treated as a variable which parameterizes the stability crossing curves in the (k, τ) plane. This allows one to make a geometric characterization of the stability crossing curves. The main steps are as follows:

(a) first, identify the corresponding *crossing points*, that is, the set of frequencies corresponding to all the points on the stability crossing curves, and the associated *frequency crossing set*;

(b) second, classify the corresponding stability crossing curves, including some simple geometric characterizations (tangent, smoothness);

(c) finally, characterize the way in which the roots cross the imaginary axis.

All these steps are detailed in the next sections.

11.5.1 Identification of crossing points

Let $\mathcal{T}_c$ denote the set of all $(k, \tau) \in \mathbb{R} \times \mathbb{R}_+$ such that the characteristic function $H(\lambda;\ k, \tau)$ has at least one zero on imaginary axis. Any $(k, \tau) \in \mathcal{T}_c$ is known as a *crossing point*. The set $\mathcal{T}_c$, which is the collection of all crossing points, forms the so-called *stability crossing curves*. Consider also the set Ω of all real numbers ω such that $j\omega$ is a root of H for at least one pair $(k, \tau) \in \mathbb{R} \times \mathbb{R}_+$. We refer to Ω as the *frequency crossing set*.

Remark 11.21. *If ω is a real number and $(k, \tau) \in \mathbb{R} \times \mathbb{R}_+$, then*

$$Q(-j\omega) + kP(-j\omega)e^{j\omega\tau} = \overline{Q(j\omega) + kP(j\omega)e^{-j\omega\tau}}.$$

Therefore, we only need to consider positive ω. For the sake of simplicity, we make the *additional assumption* that $P'(j\omega) \neq 0$ whenever $P(j\omega) = 0$. We have the following result.

Proposition 11.22. *Given any $\omega > 0$, $\omega \in \Omega$ if and only if it satisfies*

$$|P(j\omega)| \quad > \quad 0, \tag{11.23}$$

and all the corresponding pairs (k, τ) can be calculated as:

$$k(\omega) = \pm \left| \frac{Q(j\omega)}{P(j\omega)} \right|, \tag{11.24}$$

$$\tau_m(\omega) = \frac{1}{\omega}\left(-\angle \left(\frac{Q(j\omega)}{P(j\omega)} \right) + (2m + \epsilon_k + 1)\pi \right), \tag{11.25}$$

where $m = 0, \pm 1, \pm 2, \ldots,$ and

$$\epsilon_k = \begin{cases} 0, & k \geq 0, \\ -1, & k < 0. \end{cases}$$

Proof. For the necessity of (11.23), let ω be a crossing frequency in Ω. Hence, there exists a pair (k, τ) such that

$$Q(j\omega) + kP(j\omega)e^{-j\omega\tau} = 0. \tag{11.26}$$

This implies that

$$|Q(j\omega)| = |k||P(j\omega)| \tag{11.27}$$

is satisfied. It becomes clear that $|P(j\omega)| > 0$ is necessary. Otherwise, $P(j\omega) = 0$, which implies $Q(j\omega) = 0$ for all the gains k, which contradicts the technical assumption that P and Q do not have common zeros.

For the sufficiency of (11.23), we only need to recognize that the pair (k, τ) given by (11.24)–(11.25) makes $\lambda = j\omega$ a root of the corresponding characteristic equation of the closed-loop system. ◻

Remark 11.23. *Assume that Q has no zeros on the imaginary axis. Some simple algebraic manipulations prove that for all the gains k satisfying the inequality*

$$|\, k \,| \quad < \quad \frac{1}{\sup_{\omega > 0}\left\{ \frac{|P(j\omega)|}{|Q(j\omega)|} \right\}}, \tag{11.28}$$

the characteristic function $H(\lambda;\ k, \tau)$ of the closed-loop system is hyperbolic, *that is, characteristic roots crossing the imaginary axis for some positive delays τ do not exist.*

Because of Proposition 11.20 and our prior interest in characterizing *stability* regions, we restrict the gain k in what follows to some *finite interval* $[\alpha, \beta]$ such that either $\alpha \geq 0$ or $\beta \leq 0$, and such that it has an empty intersection with the interval defined by (11.28).

From (11.24), this allows one to restrict also the *crossing set* Ω, to the set of the frequencies $\omega > 0$ satisfying *simultaneously* the inequalities

$$\begin{aligned} \min(|\alpha|, |\beta|)\, |P(j\omega)| \,|\leq |Q(j\omega)| \leq \max(|\alpha|, |\beta|)\, |P(j\omega)|, \\ |P(j\omega)| > 0. \end{aligned} \quad (11.29)$$

Due to the form of (11.29), and from the standard assumptions considered (strictly proper transfer function, P and Q coprime), the corresponding frequency crossing set Ω *consists of a finite number of intervals*. If such an interval contains a frequency where $|Q(j\omega)| = 0$, which is the case if the uncontrolled systems exhibits oscillatory modes, we first subdivide such interval into two intervals with a common end point. This subdivision will be explained later on. Next, we denote the resulting intervals with $\Omega_1, \Omega_2, \ldots, \Omega_N$,

$$\Omega = \bigcup_{k=1}^{N} \Omega_k.$$

In what follows, we let $\Omega_i = [\omega_i^l, \omega_i^r]$, for all $i = 1, 2, \ldots, N$. Without any loss of generality, we can order these intervals from left to right, i.e., for any $\omega_1 \in \Omega_{i_1}$, $\omega_2 \in \Omega_{i_2}$, $i_1 < i_2$, we have $\omega_1 < \omega_2$. Note that ω_1^l can be 0 and in this case Ω_1 is open to the left.

It is clear that $k(\omega_i^l)$, $k(\omega_i^r) \in \{\alpha, \beta\}$ for all $i = 1, \ldots, N$ if $\omega_1^l \neq 0$. In the remainder of this section, we will not restrict $\angle Q(j\omega)/P(j\omega)$ to a 2π range. Rather, we let it vary *continuously* within each interval Ω_i. Note that this is always possible because the continuity of $\angle(Q(j\omega)/P(j\omega))$ can be guaranteed by a suitable choice of the phase angle at every frequency, except at those frequencies where $Q(j\omega) = 0$. But if such a frequency occurs, then it corresponds to an end point of the interval due to the subdivision mentioned above. In summary, for each fixed m, (11.24) and (11.25) give us two continuous curves.

Condition (11.24) and k finite imply that $P(j\omega) \neq 0$ for all $\omega \in \Omega$. We denote the curves defined by (11.24) and (11.25) with $\mathcal{T}_i^{m\pm}$, with the $\pm$ referring to the choice of the sign in (11.24) and i referring to the corresponding interval Ω_i. Therefore,we have an infinite number of continuous stability crossing curves $\mathcal{T}_i^{m\pm}$, $m = 0, \pm1, \pm2, \ldots$, associated to Ω_i.

Finally, it should be noted that, for some m, part or the entire curve may be outside of the range $\mathbb{R} \times \mathbb{R}_+$ and, therefore, may not be physically meaningful. The collection of all the points in $\mathcal{T}_c$ corresponding to Ω_i may be expressed as

$$\mathcal{T}_i := \bigcup_{m=-\infty}^{+\infty} \left[\left(\mathcal{T}_i^{m+} \cap (\mathbb{R} \times \mathbb{R}_+)\right) \cup \left(\mathcal{T}_i^{m-} \cap (\mathbb{R} \times \mathbb{R}_+)\right) \right].$$

Obviously,

$$\mathcal{T}_c = \bigcup_{i=1}^{N} \mathcal{T}_i.$$

Also, it is easy to see that, for each Ω_i, we define two curves, one to the right of the $O\tau$ axis and the other to the left. According to the fixed limits α, β of the interval where k varies we can eliminate some of these curves.

11.5.2 Classification of stability crossing curves, smoothness, and crossing directions

Stability crossing curves. We give a classification of the crossing curves with respect to their shape. In order to do this we first classify the ends of the crossing curves confined to $k \in [\alpha, \beta]$. Each end point ω_i^l or ω_i^r must belong to one of the following five types:

Type 1. It satisfies $k(\omega) = \alpha,\ \omega \neq 0$.

Type 2. It satisfies $k(\omega) = \beta,\ \omega \neq 0$.

Type 3. It equals 0, $Q(0) \neq 0$, and $Q(0)/P(0) \in [\alpha, \beta]$.

Type 4. It equals 0, $Q(0) \neq 0$, and $-Q(0)/P(0) \in [\alpha, \beta]$.

Type 5. It equals 0 and $Q(0) = 0$.

Obviously, only ω_1^l can be of types 3–5. We note that all the crossing curves are situated in the vertical strip $\mathcal{D}$ between the lines $k = \alpha$ and $k = \beta$. Now, let ω_* be an end point of the interval Ω_i. We already said that each $\mathcal{T}_i^{m\pm}$ is an continuous curve, so $(k(\omega_*), \tau_m(\omega_*))$ is an end point of $\mathcal{T}_i^{m\pm}$. In the case $\alpha \geq 0$, it can be characterized as follows:

- If ω_* is of type 1, then $k(\omega_*) = \alpha$ and $\tau(\omega_*)$ are finite. More precisely, $\mathcal{T}_i^{m+},\ m \in \mathbb{Z}$, intersects the vertical line $k = \alpha$, which is the left bound of the strip $\mathcal{D}$.
- If ω_* is of type 2, then $k(\omega_*) = \beta$ and $\tau(\omega_*)$ are finite. Or, we may say that $\mathcal{T}_i^{m+},\ m \in \mathbb{Z}$, intersects the vertical line $k(\omega) = \beta$, which is the right bound of the strip $\mathcal{D}$.
- If ω_* is of type 3, then k approaches $Q(0)/P(0)$ and τ approaches ∞, for all values of m. In other words, $\mathcal{T}_i^{m+},\ m \in \mathbb{Z}$, has a vertical asymptote given by $k = Q(0)/P(0)$.
- If ω_* is of type 4, then k approaches $-Q(0)/P(0)$ and $\mathcal{T}_i^{0+}$ approaches
$$\frac{Q'(0)}{Q(0)} - \frac{P'(0)}{P(0)},$$
whereas $\mathcal{T}_i^{m+},\ m \in \mathbb{Z}_0$, has a vertical asymptote given by $k = -Q(0)/P(0)$.
- If ω_* is of type 5, then k approaches 0 and τ approaches ∞, for all m. In other words, $\mathcal{T}_i^{m+},\ m \in \mathbb{Z}$, has a vertical asymptote given by $k = 0$.

We say that an interval Ω_k is of type lr if its left end is of type l and its right end is of type r. Accordingly, we may divide these intervals into the following nine types:

Type 11. In this case, $\mathcal{T}_i^{m+},\ m \in \mathbb{Z}$, starts at a point on the vertical line $k = \alpha$, and ends at another point on the vertical line $k = \alpha$.

Type 12. In this case, $\mathcal{T}_i^{m+},\ m \in \mathbb{Z}$, starts at a point on the vertical line $k = \alpha$, and ends at a point on the vertical line $k = \beta$.

Type 21. This is the reverse of type 12. $\mathcal{T}_i^{m+}, m \in \mathbb{Z}$, starts at a point on the vertical line $k = \beta$, and ends at a point on the vertical line $k = \alpha$.

Type 22. In this case, $\mathcal{T}_i^{m+}$, $m \in \mathbb{Z}$, starts at a point on the vertical line $k = \beta$, and ends at another point on the vertical line $k = \beta$.

Type 31. In this case, $\mathcal{T}_i^{m+}$, $m \in \mathbb{Z}$, begins at ∞ with a vertical asymptote $k = Q(0)/P(0)$. The other end is on the vertical line $k = \alpha$.

Type 32. In this case, $\mathcal{T}_i^{m+}$, $m \in \mathbb{Z}$, again begins at ∞ with a vertical asymptote $k = Q(0)/P(0)$. The other end is on the vertical line $k = \beta$.

Type 41. In this case, $\mathcal{T}_i^{m+}$, $m \in \mathbb{Z}_0$, begins at ∞ with a vertical asymptote $k = -Q(0)/P(0)$. The other end is on the vertical line $k = \alpha$. The curve $\mathcal{T}_i^{0+}$ begins in $(k, \tau) = (-Q(0)/P(0), Q'(0)/Q(0) - P'(0)/P(0))$ and ends on the line $k = \alpha$. Here it is important to mention that the special case corresponds to $m = 0$ only if the convention of the phase angle is such that

$$\lim_{\omega \to 0+} \angle(Q(j\omega)/P(j\omega)) = \begin{cases} 0, & Q(0)/P(0) \geq 0, \\ \pi, & Q(0)/P(0) < 0. \end{cases} \tag{11.30}$$

This is relevant because, as mentioned before, the classical convention for $\angle(\cdot)$ is sometimes given up because of continuity arguments regarding the curves defined by (11.24)–(11.25).

Type 42. This is the reverse of type 41. Under the assumption (11.30), $\mathcal{T}_i^{m+}$, $m \in \mathbb{Z}_0$, begins at ∞ with a vertical asymptote $k = -Q(0)/P(0)$. The other end is on the vertical line $k = \beta$. The curve $\mathcal{T}_i^{0+}$ begins in
$(k, \tau) = (-Q(0)/P(0), Q'(0)/Q(0) - P'(0)/P(0))$, and ends on the line $k = \beta$.

Type 52. In this case, $\mathcal{T}_i^{m+}$, $m \in \mathbb{Z}$, begins at ∞ with the vertical asymptote $k = 0$. The other end is on the vertical line $k = \beta$.

Remark 11.24. *In the case $\beta \leq 0$ the previous description remains valid provided $\mathcal{T}_i^{m+}$ is replaced with $\mathcal{T}_i^{m-}$, and type 52 is replaced with type 51, which is similarly defined as type 51—the difference lies in the fact that the curves end on the vertical line $k = \alpha$.*

Tangents and smoothness. For a given i, the smoothness properties of the curves in $\mathcal{T}_i^{m\pm}$ and thus in

$$\mathcal{T}_c = \bigcup_{m=-\infty}^{+\infty} \left[\left(\mathcal{T}_i^{m+} \cap (\mathbb{R} \times \mathbb{R}_+)\right) \cup \left(\mathcal{T}_i^{m-} \cap (\mathbb{R} \times \mathbb{R}_+)\right) \right]$$

can be investigated by adopting the approach proposed in Chapter 4 (crossing curves in delay-parameter spaces). We have the following result, which can be proved using the implicit function theorem.

Proposition 11.25. *The curve $\mathcal{T}_i^{m\pm}$ is smooth everywhere except possibly at the point corresponding to $\lambda = j\omega$ in either of the following cases:*

(1) $\lambda = j\omega$ is a multiple root of the characteristic function $H(\lambda;\ k, \tau)$,

(2) $Q(j\omega) = 0$, which is equivalent to $k = 0$.

Directions of crossing. Some appropriate quantities need to be introduced. For a given pair $(k, \tau) \in \mathcal{T}_i$ and corresponding $\omega \in \Omega_i$, let

$$\begin{aligned} R_1 &= \Re\left(\frac{\partial H(\lambda;\ k,\tau)}{\partial k}\right)_{\lambda=j\omega} &&= \Re\left(P(j\omega)e^{-j\omega\tau}\right), \\ I_1 &= \Im\left(\frac{\partial H(\lambda;\ k,\tau)}{\partial k}\right)_{\lambda=j\omega} &&= \Im\left(P(j\omega)e^{-j\omega\tau}\right), \\ R_2 &= \Re\left(\frac{\partial H(\lambda;\ k,\tau)}{\partial \tau}\right)_{\lambda=j\omega} &&= \Im\left(k\omega P(j\omega)e^{-j\omega\tau}\right), \\ I_2 &= \Im\left(\frac{\partial H(\lambda;\ k,\tau)}{\partial \tau}\right)_{\lambda=j\omega} &&= -\Re\left(k\omega P(j\omega)e^{-j\omega\tau}\right). \end{aligned}$$

In order to characterize the direction in which the roots of the characteristic function $H(\lambda;\ k, \tau)$ cross the imaginary axis as (k, τ) deviates from the curve $\mathcal{T}_i^{m\pm}$, we call the direction of the curve that corresponds to increasing ω the *positive direction*. We also call the region on the left-hand side as we head in the positive direction of the curve *the region on the left*.

Similar to the characterization of the crossing direction in the case of delay-parameter spaces discussed in Chapter 4, we have the following result.

Proposition 11.26. *Let $\omega \in (\omega_i^l, \omega_i^r)$ and $(k, \tau) \in \mathcal{T}_i$ such that $j\omega$ is a simple root of the characteristic function $H(\lambda;\ k, \tau)$.*

If (k, τ) moves from the region on the right to the region on the left of a stability crossing curve, then a pair of roots of the characteristic function H crosses the imaginary axis to the right, through $\lambda = \pm j\omega$, whenever $R_2 I_1 - R_1 I_2 > 0$. The crossing is to the left if the inequality is reversed.

Discussions and interpretations. Under standard regularity assumptions, Proposition 11.26 above gives the explicit *crossing direction* toward stability (*reversals*) or instability (*switches*) function of some quantity evaluated at the corresponding crossing point. Such a result, together with Lemma 11.5 and Propositions 11.22 and 11.25, lead to the following simplified procedure:

- First, compute $\text{card}(\mathcal{U}_+)$ (using Lemma 11.5).

- Second, derive the crossing set Ω, and related dependence $(k(\omega), \tau(\omega))$ (using Proposition 11.22, and its corresponding remarks and derivations).

- Next, determine the crossing direction using Proposition 11.26 above.

Finally, in order to find explicitly the corresponding (closed-loop) *stability regions*, one can count the number of crossing roots (toward stability/instability) corresponding to each region whose boundaries are given by the crossing curves, starting from the crossing direction of the roots and the fact that for $\tau = 0$, $\text{card}(\mathcal{U}_+)$ gives the complete (stability/instability) information in terms of k (closed-loop free of delays).

11.6 Illustrative examples

11.6.1 Second-order system

We first consider the system with transfer function

$$\frac{P(\lambda)}{Q(\lambda)} = \frac{1}{\lambda^2 - \gamma\lambda + 2}, \tag{11.31}$$

where $\gamma > 0$ is a real parameter. The polynomial $Q(\lambda) = \lambda^2 - \gamma\lambda + 2$ is unstable and for all $k \in \mathbb{R}$, the polynomial $Q(\lambda) + kP(\lambda)$ has at least one unstable root. Furthermore, if $\gamma = 0$, then (11.31) corresponds to an oscillator (the characteristic equation has two roots on the imaginary axis).

With the controller (11.2), the characteristic equation of the closed-loop system is given by

$$\lambda^2 - \gamma\lambda + 2 + ke^{-\lambda\tau} = 0 \tag{11.32}$$

and polynomial $F(\omega;\ k)$ by

$$\begin{aligned} F(\omega) &= \mid Q(j\omega) \mid^2 - \mid P(j\omega) \mid^2 = (2-\omega^2)^2 + \gamma^2\omega^2 - k^2 \\ &= \omega^4 - (4-\gamma^2)\omega^2 + (4-k^2). \end{aligned} \tag{11.33}$$

For $\gamma > 2$ the quantities $\text{card}(\mathcal{S}_+)$ and $\text{card}(\mathcal{U}_+)$ behave as a function of the gain k as displayed in the following table.

k	< -2	$\in (-2, 2)$	> 2
$\text{card}(\mathcal{S}_+)$	1	0	1
$\text{card}(\mathcal{U}_+)$	1	2	2

For $\gamma < 2$ we have

k	< -2	$\in (-2, -k^*)$	$\in (-k^*, k^*)$	$\in (k^*, 2)$	> 2
$\text{card}(\mathcal{S}_+)$	1	2	0	2	1
$\text{card}(\mathcal{U}_+)$	1	2	2	2	2

where

$$k^* = 2\sqrt{1 - \left(1 - \frac{\gamma^2}{4}\right)^2}.$$

According to the results of the previous section, summarized in Table 11.3, a necessary condition for asymptotic stability of the closed-loop system is therefore given by

$$\gamma < 2, \qquad |k| \in (k^*,\ 2). \tag{11.34}$$

Furthermore, for a gain satisfying (11.34) the existence of a stability region in the delay parameter is determined by the condition $\tau^- < \tau^+$. Summarizing, we have the following.

Proposition 11.27. *The system (11.31) can be stabilized with a controller of the form $u(t) = -ky(t-\tau)$ if and only if the pair (γ, k) satisfies*

$$\gamma \in [0, 2), \quad |k| \in \left(2\sqrt{1 - \left(1 - \frac{\gamma^2}{4}\right)^2}, 2\right), \tag{11.35}$$

and $\tau_- < \tau_+$, *where*

$$\begin{array}{ll} \tau_\pm = \frac{1}{\omega_\pm}\left[\mathrm{Log}\left(\frac{\omega_\pm^2-2}{k} - j\frac{\gamma\omega_\pm}{k}\right) + (1+\operatorname{sign} k)\pi\right], & \gamma \neq 0, \\ \left\{ \begin{array}{l} \tau_+ = (3+\operatorname{sign} k)\frac{\pi}{2\omega_+}, \\ \tau_- = (1+\operatorname{sign} k)\frac{\pi}{2\omega_-}, \end{array} \right. & \gamma = 0, \end{array} \tag{11.36}$$

and

$$\omega_\pm = \sqrt{1-\frac{\gamma^2}{4}} \cdot \sqrt{2\left(1 \pm \sqrt{1 - \frac{1-\frac{k^2}{4}}{\left(1-\frac{\gamma^2}{4}\right)^2}}\right)}. \tag{11.37}$$

A stabilizing controller is then defined by the gain k and $\tau \in (\tau_-, \tau_+)$.

Remark 11.28. *If* $\gamma = 0$ *and* $k \in (-2, 0)$, *we recover the results proposed in [1, 227]:*

$$\tau_- = 0, \ \tau_+ = \frac{\pi}{\sqrt{2+|k|}}.$$

Furthermore, the number of delay intervals is given by

$$\max_{l \in \mathbb{Z}} \left\{ l \leq \frac{1}{2} \cdot \frac{1}{\sqrt{\frac{2+|k|}{2-|k|}} - 1} \right\}.$$

Thus, the smaller the gain is, the larger the number of stabilizing delay intervals is, a property coherent with the graphical representation in [1].

Let us now apply the geometric approach of the previous section to the case $\gamma = 0$. If we consider the interval $[\alpha, \beta] = [-2, 0]$, then we get

$$\Omega_1 = [0, \sqrt{2}], \ \ \Omega_2 = [\sqrt{2}, 2].$$

The curves $\mathcal{T}_1^{m-}$, $m \in \mathbb{Z}$, are of type 42 (notice that $\mathcal{T}_1^{0-}$ is the curve $\tau = 0$), and the curves $\mathcal{T}_2^{m-}$, $m \in \mathbb{Z}$, are of type 21. These stability crossing curves are plotted in Figure 11.4. The stability regions are indicated with the letter S.

11.6.2 Sixth-order system

As a second example we study the stabilization of the system described by

$$\frac{P(\lambda)}{Q(\lambda)} = \frac{1}{\lambda^6 + p_1\lambda^5 + p_2\lambda^4 + p_3\lambda^3 + p_4\lambda^2 + p_5\lambda + p_6}, \tag{11.38}$$

where

$$\begin{array}{lll} p_1 = -6.0000000e-04, & p_2 = 1.4081634e+00, & p_3 = -5.6326533e-04, \\ p_4 = 4.3481891e-01, & p_5 = -8.6963771e-05, & p_6 = 2.6655565e-02, \end{array}$$

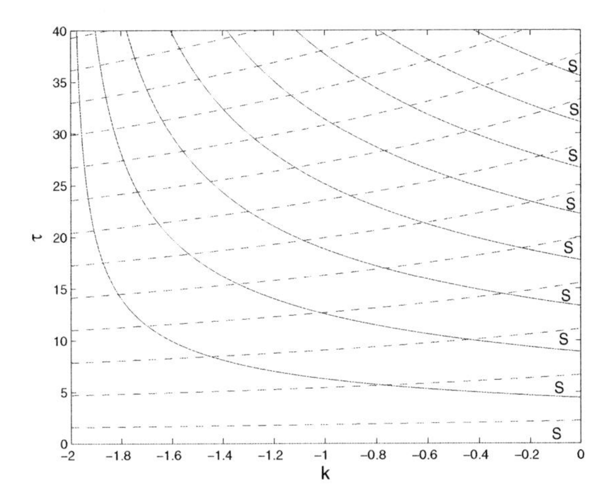

Figure 11.4. *Stability crossing curves for the controlled system (11.31) with $\gamma = 0$. The solid curves correspond to $\mathcal{T}_1^{m-}$, the dashed curves to $\mathcal{T}_2^{m-}$.*

using a controller of the form (11.2). The uncontrolled system has six strictly unstable poles and cannot be stabilized with static, undelayed output feedback.

As a first step in the controller design we compute $\text{card}(\mathcal{U}_+)$ and $\text{card}(\mathcal{S}_+)$ as a function of the gain, resulting in Figure 11.5. The information given in Table 11.3 then helps us to choose possible values of the gain. We take $k = 0.0025$, where $\text{card}(\mathcal{S}_+) = \text{card}(\mathcal{U}_+) = 6$.

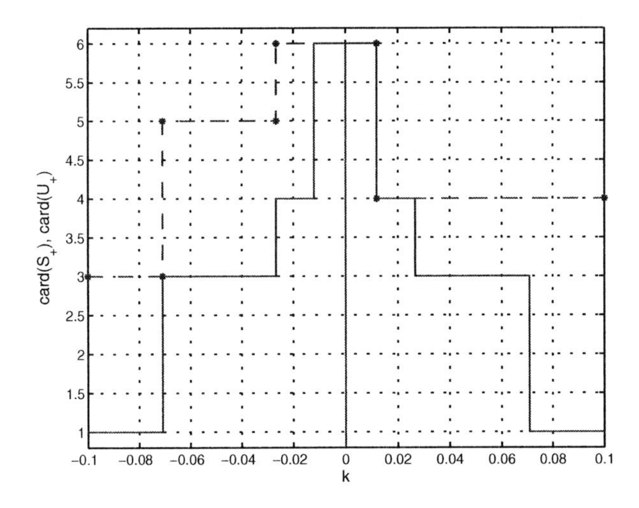

Figure 11.5. $\text{card}(\mathcal{S}_+)$ *(solid line) and* $\text{card}(\mathcal{U}_+)$ *(dashed line) as a function of the gain k.*

Secondly we characterize stability regions in the delay parameter. The set $\mathcal{S}^+$ is given by

$$\omega_1 = 1.0019959e+00, \quad \omega_2 = 9.9795792e-01, \quad \omega_3 = 5.8408171e-01,$$
$$\omega_4 = 5.5740265e-01, \quad \omega_5 = 3.0572050e-01, \quad \omega_6 = 2.6663916e-01.$$

By computing the set $\mathcal{T}_\omega$, using (11.17), and taking into account the crossing direction, we arrive at a complete characterization of stability regions in the delay parameter space, displayed in Table 11.4. Summarizing, the system (11.38) and (11.2) is asymptotically stable for

$$k = 0.0025,$$
$$\tau \in (11.802168, 12.490817) \cup (15.788569, 16.121915) \cup (35.366543, 37.573495).$$

Table 11.4. *Characterization of stability regions in the delay parameter for the system (11.38) and (11.2) with $k = 0.0025$.*

Elements of $\mathcal{T} = \bigcup_{i=1}^{6} \mathcal{T}_{\omega_i}$	Crossing frequency	# unstable roots changes to
		6
1.2048745e-02	ω_4	4
3.1964843e+00	ω_2	2
5.3645410e+00	ω_3	4
6.2201470e+00	ω_1	6
9.4925266e+00	ω_2	4
1.1284305e+01	ω_4	2
1.1802168e+01	ω_6	0
1.2490817e+01	ω_1	2
1.5788569e+01	ω_2	0
1.6121915e+01	ω_3	2
1.8761486e+01	ω_1	4
2.0536234e+01	ω_5	6
2.2084611e+01	ω_2	4
2.2556560e+01	ω_4	2
2.5032156e+01	ω_1	4
2.6879289e+01	ω_3	6
2.8380653e+01	ω_2	4
3.1302825e+01	ω_1	6
3.3828816e+01	ω_4	4
3.4676696e+01	ω_2	2
3.5366543e+01	ω_6	0
3.7573495e+01	ω_1	2
3.7636663e+01	$\omega 3$	4
⋮	⋮	⋮

Notice that stability can be achieved by increasing the delay after having three pairs of unstable roots. Notice also that it is in general not sufficient the investigate only the first root crossings of the imaginary axis.

Using the geometric approach developed in the previous section, the stability crossing curves and the first two stability regions for $k \in (0, 0.16)$ are plotted in Figure 11.6.

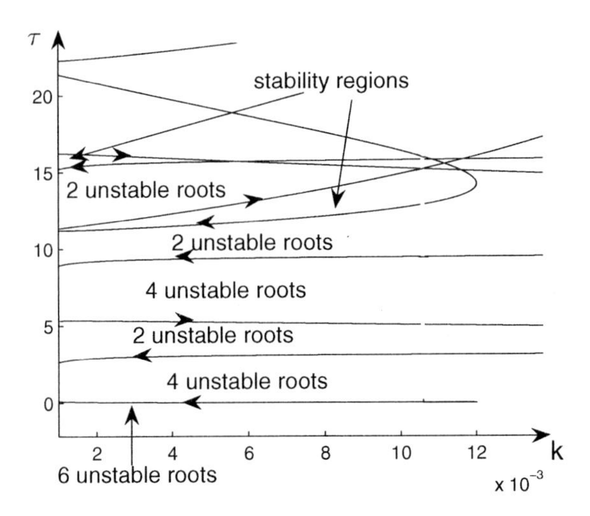

Figure 11.6. *Stability crossing curves for the output feedback controlled system (11.38).*

11.7 Notes and references

This chapter was devoted to the stabilization problem of a class of SISO systems subjected to delayed output feedback. More precisely, we considered the problem where the delay in the control law may induce a *stabilizing* effect, that is, the closed-loop stability is guaranteed due to the presence of the delay existence.

Various stabilization conditions were derived using the eigenvalue approach. The results were based on the characterization of the crossing direction of characteristic roots on the imaginary axis corresponding to some critical delay values. A geometric characterization of stability crossing curves for the closed-loop system was also proposed.

Two examples completed the presentation: a second-order and a sixth-order system, with some emphasis on the stabilization problem for oscillatory systems subjected to small delays. Furthermore, the class of stabilizable second-order systems using delayed output feedback was completely characterized.

The chapter is based on [191, 205, 206, 230, 233] and some of the references therein.

Chapter 12

Stabilization by delayed output feedback: multiple delay case

12.1 Introduction

We reconsider the output feedback stabilizability and stabilization of the SISO system (11.1):

$$\dot{x}(t) = Ax(t) + Bu(t), \quad y = C^T x(t).$$

The analysis of Chapter 11 naturally leads to the question of whether a controller of the form

$$u(t) = -\sum_{i=1}^{l} k_i \, y(t - \tau_i), \tag{12.1}$$

including $l > 1$ delay blocks, may guarantee the *stability* of the closed-loop scheme if one delay block is *not sufficient*. In what follows, by a *delay block i*, we understand the pair (k_i, τ_i). Such a block is reduced to a simple proportional controller if $\tau_i = 0$.

In this chapter, we first discuss the stabilizability problem of the system (11.1) with the control law (12.1), by extending the results presented in Section 11.3 to a multiple delay setting. Second, we consider the stabilization of the particular but important system consisting of a *chain of n integrators*. An application of the obtained stabilizability conditions reveals that at least $l = n$ terms in (12.1) are *necessary* for the stabilization of this system. Then we prove that either n distinct delays in the control law or a proportional+delay compensator with $n - 1$ distinct delays are *sufficient*.

We adopt two different approaches, which are both constructive. The first approach is based on a derivative feedback idea. More precisely, starting from a stabilizing control law without delays but using output derivatives, we show that stability is preserved when the derivatives are closely approximated with (past) measurements of the output. This approach is inspired by the paper of Kokame et al. [137] (see also the references therein) on stabilizing systems with delay difference feedback. The second method is based on an assignment of the n rightmost characteristic roots of the closed-loop system, inspired by the recent work on constrained control of time-delay systems [193, 177, 176, 178]. The chapter is organized as follows. In Section 12.2 necessary stabilizability conditions are presented. In Section 12.3 the stabilization problem for the multiple integrator is completely solved. An illustrative example (stabilization of a triple integrator) is presented in Section 12.4. Some concluding notes end the chapter.

12.2 Necessary conditions for stabilizability

The following result of [135] generalizes Proposition 11.2 to the multiple delay case.

Proposition 12.1. *Let the transfer function of the system (11.1) be given by*

$$H(\lambda) := \frac{P(\lambda)}{Q(\lambda)}, \tag{12.2}$$

where P and Q are coprime polynomials satifying $\deg(P) = n$ *and* $\deg(Q) = m$.

A necessary condition for the stabilizability of (12.2) with the control law (12.1) is the Hurwitz stability of the polynomial

$$\gamma(\lambda;\ \tau_1, \dots, \tau_l) := \left(\frac{d}{d\lambda} + \tau_1\right)^{m+1} \left(\frac{d}{d\lambda} + \tau_2\right)^{m+1} \cdots \left(\frac{d}{d\lambda} + \tau_l\right)^{m+1} Q(\lambda).$$

Example 12.2. *Consider the example (11.3), where $c_1 = 0$ and $c_2 \neq 0$. With $l > 1$, the necessary condition of Proposition 12.1 is satisfied when taking all but one delay equal to zero, and the nonzero delay sufficiently small (whatever the value of a_1 and a_2).*

In fact, two blocks are always *sufficient for stabilization (unlike one block, see Example 11.3), because the control law*

$$\begin{aligned} u(t) &= -\left(k_1 + \tfrac{k_2}{\tau}\right) y(t) + \tfrac{k_2}{\tau} y(t-\tau) \\ &= -k_1 y(t) - k_2 \tfrac{y(t)-y(t-\tau)}{\tau}, \end{aligned}$$

with

$$a_1 + k_2 c_2 > 0,\ a_2 + k_1 c_2 > 0,$$

is stabilizing for sufficiently small values of τ. For this, notice that the control law

$$u(t) = -k_1 y(t) - k_2 \dot{y}(t)$$

is stabilizing. According to [137] asymptotic stability is preserved when replacing the output derivative, $\dot{y}(t)$, with the difference approximation $(y(t) - y(t-\tau))/\tau$, where τ is sufficiently small.

The following corollary of Proposition 12.1 does not depend any more on the delay values.

Corollary 12.3. *Assume that $l(m+1) < n$. If the polynomial $Q(\lambda)$ has at least one unstable root with multiplicity $> l(m+1)$, then the system (12.2) cannot be stabilized by the controller (12.1).*

12.3 Stabilization of multiple integrators

We address the stabilization of a chain including n integrators, whose transfer function is given by

$$H(\lambda) := \frac{1}{\lambda^n}. \tag{12.3}$$

Since (12.3) has a pole at zero with multiplicity n, an application of Corollary 12.3 yields that *at least* n terms in the control law (12.1) are *necessary* for stabilization.

In what follows, we show that n terms are also sufficient by explicitly constructing stabilizing control laws. In the derivation we will employ the following scaling property, which expresses a natural trade-off between "gain" and "delay" in the controller.

Property 12.4. *The control law*

$$u(t) = -\sum_{j=1}^{n} k_j y(t - \tau_j) \tag{12.4}$$

is asymptotically stabilizing if and only if

$$u(t) = -\sum_{j=1}^{n} \frac{k_j}{\rho^n} y(t - \rho\tau_j), \ \rho > 0, \tag{12.5}$$

is asymptotically stabilizing.

Proof. The transformation from (12.4) to (12.5) involves a scaling of the characteristic roots of the closed-loop system by $1/\rho$. □

Note that an analogous scaling property was the basis for the construction of state feedback controllers in the presence of input constraints in [193, 176, 177] and also played a crucial role in the study of the so-called peaking phenomena; see [287, 268] and the references therein.

12.3.1 Control laws based on numerical differentiation with backward differences

This construction is inspired by [137] and consists of approximating output derivatives with (delayed) output measurements, as we have done in Example 12.2. This initially leads to control laws with small delays, but by Property 12.4 control laws with arbitrary delays can be directly derived.

The system (12.3) can be stabilized with the feedback law

$$u(t) = -q_0\, y(t) - q_1\, y'(t) - \cdots - q_{n-1}\, y^{(n-1)}(t), \tag{12.6}$$

where the polynomial $q(\lambda) = \lambda^n + \sum_{k=0}^{n-1} q_k \lambda^k$ is Hurwitz. The latter implies that $q_k > 0$, $k = 1, \ldots, n$. Hence, all the derivatives of the output, up to order $(n-1)$, are needed in the control law.

The key idea in the controller construction consists of approximating the output derivatives in (12.6) with (delayed) output measurements. For instance, we have

$$y'(t) \approx \frac{y(t) - y(t-\epsilon)}{\epsilon} \tag{12.7}$$

for small ϵ, which corresponds to an approximation

$$\lambda \approx \frac{1 - e^{-\lambda\epsilon}}{\epsilon}$$

in the frequency domain. Note that the right-hand side of (12.7) is the derivative of the linear approximation of y through the points $(t, y(t))$ and $(t-\epsilon, y(t-\epsilon))$. We now outline how this idea can be generalized to approximate higher-order derivatives of y also.

Choose a set of n delays satisfying

$$0 \le \tau_1 < \tau_2 < \cdots < \tau_n.$$

We may approximate the output $y(t)$ around any time $t = t_0$ with the polynomial

$$y_p(t) = c_0 + c_1(t - t_0) + c_2(t - t_0)^2 + \cdots + c_{n-1}(t - t_0)^{n-1},$$

which interpolates y(t) at the n past instants $t_0 - \epsilon\tau_1, \ldots, t_0 - \epsilon\tau_n$, i.e.,

$$y_p(t_0 - \epsilon\tau_i) = y(t_0 - \epsilon\tau_i), \; i = 1, \ldots, n. \tag{12.8}$$

Here $\epsilon > 0$ is a small scaling parameter. Since the Vandermonde matrix

$$T(\tau) := \begin{bmatrix} 1 & \tau_1 & \tau_1^2 & \cdots & \tau_1^{n-1} \\ \vdots & & & & \vdots \\ 1 & \tau_n & \tau_n^2 & \cdots & \tau_n^{n-1} \end{bmatrix} \tag{12.9}$$

is invertible when the delays τ_i are different, the conditions (12.8) can be written in matrix form as

$$\begin{bmatrix} c_0 \\ c_1 \\ \vdots \\ c_{n-1} \end{bmatrix} = \begin{bmatrix} 1 & & & \\ & \frac{1}{(-\epsilon)} & & \\ & & \ddots & \\ & & & \frac{1}{(-\epsilon)^{n-1}} \end{bmatrix} T(\tau)^{-1} \begin{bmatrix} y(t_0 - \epsilon\tau_1) \\ y(t_0 - \epsilon\tau_2) \\ \vdots \\ y(t_0 - \epsilon\tau_n) \end{bmatrix} \tag{12.10}$$

and we may approximate

$$y^{(i)}(t_0) \approx y_p^{(i)}(t_0) = i!\, c_i, \;\; i = 1, \ldots, n. \tag{12.11}$$

This way the control law (12.6) at $t = t_0$ can be approximated with

$$u(t_0) = -q_0\, y_p(t_0) - q_1\, y_p'(t_0) - \cdots - q_{n-1}\, y_p^{(n-1)}(t_0).$$

Substituting (12.10) and (12.11) into this expression and applying the same principle for all $t_0 > 0$ leads to the control law

$$u(t) = -\left[q_0 \; \frac{1}{(-\epsilon)}q_1 \; \frac{2!}{(-\epsilon)^2}q_2 \; \cdots \; \frac{(n-1)!}{(-\epsilon)^{n-1}}q_{n-1} \right] T(\tau)^{-1} \begin{bmatrix} y(t-\epsilon\tau_1) \\ \vdots \\ y(t-\epsilon\tau_n) \end{bmatrix}. \quad (12.12)$$

When $\epsilon \to 0+$ the approximation of (12.6) becomes better and we have the following result.

Proposition 12.5. *Assume that the polynomial* $q(\lambda) := \lambda^n + q_{n-1}\lambda^{n-1} + \cdots + q_0$ *is Hurwitz. Assume further that* $0 \le \tau_1 < \tau_2 < \cdots < \tau_n$ *and let* $T(\tau)$ *be defined by (12.9). Then the control law (12.12) achieves asymptotic stability of (12.3) for small values of* ϵ. *Moreover, if* $\epsilon \to 0+$, *then the* n *rightmost characteristic roots of the closed-loop system converge to the* n *zeros of* $q(\lambda)$.

Proof. With the control law (12.6) the characteristic equation of the closed-loop system is given by

$$q(\lambda) = 0,$$

while the control law (12.12) yields

$$q_\epsilon(\lambda) = 0,$$

where

$$q_\epsilon(\lambda) = \lambda^n + \left[q_0 \; \frac{1}{(-\epsilon)}q_1 \; \frac{2!}{(-\epsilon)^2}q_2 \; \cdots \; \frac{(n-1)!}{(-\epsilon)^{n-1}}q_{n-1} \right] T(\tau)^{-1} \begin{bmatrix} e^{-\epsilon\tau_1\lambda} \\ \vdots \\ e^{-\epsilon\tau_n\lambda} \end{bmatrix}. \quad (12.13)$$

We first establish a relation between $q(\lambda)$ and $q_\epsilon(\lambda)$ as $\epsilon \to 0+$. Therefore, consider an arbitrary $\lambda \in \mathbb{C}$. Using a Taylor expansion we have

$$e^{-\epsilon\tau_i\lambda} = 1 + \frac{(-\epsilon\tau_i\lambda)}{1!} + \cdots + \frac{(-\epsilon\tau_i\lambda)^{n-1}}{(n-1)!} + O\left((\epsilon\lambda)^n\right), \quad i = 1, \ldots, n,$$

which can be written as

$$\begin{bmatrix} e^{-\epsilon\tau_1\lambda} \\ e^{-\epsilon\tau_2\lambda} \\ \vdots \\ e^{-\epsilon\tau_n\lambda} \end{bmatrix} = T(\tau) \begin{bmatrix} 1 \\ \frac{(-\epsilon\lambda)}{1!} \\ \vdots \\ \frac{(-\epsilon\lambda)^{n-1}}{(n-1)!} \end{bmatrix} + \begin{bmatrix} O\left((\epsilon\lambda)^n\right) \\ O\left((\epsilon\lambda)^n\right) \\ \vdots \\ O\left((\epsilon\lambda)\right)^n \end{bmatrix}. \quad (12.14)$$

Substituting (12.14) into (12.13) leads to

$$q_\epsilon(\lambda) = q(\lambda) + O(\epsilon\lambda^n). \quad (12.15)$$

Define a *compact* subset S of the complex plane, which contains all the zeros of $q(s)$. From the expression (12.15) it follows that the analytic function $q_\epsilon(\lambda)$ *uniformly* converges to

$q(\lambda)$ on S as $\epsilon \to 0_+$. Therefore, both functions have the same number of zeros in S when ϵ is sufficiently small. Moreover, as $\epsilon \to 0_+$, the n zeros of $q_\epsilon(\lambda)$ in S converge to n corresponding zeros of $q(\lambda)$. These statements follow from Rouché's theorem (see the appendix).

The proof is complete when we also show that in any RHP $q_\epsilon(\lambda)$ has at most n zeros, when ϵ is sufficiently small. This follows from the scaling Property 12.4: the condition $q_\epsilon(\lambda) = 0$ is equivalent to

$$\bar{\lambda}^n + \left[q_0\epsilon^n \ \frac{\epsilon^{n-1}}{(-1)}q_1 \ \frac{2!\epsilon^{n-2}}{(-1)^2}q_2 \ \cdots \ \frac{(n-1)!\epsilon}{(-1)^{n-1}}q_{n-1} \right] T(\tau)^{-1} \begin{bmatrix} e^{-\tau_1\bar{\lambda}} \\ \vdots \\ e^{-\tau_n\bar{\lambda}} \end{bmatrix} = 0,$$

where $\bar{\lambda} = \epsilon\lambda$. This equation can be interpreted as the characteristic equation of a feedback controlled multiple integrator with fixed feedback delays, where the gain can be made arbitrarily small. As proved in [193], n characteristic roots converges to zero as the gain tends to zero, while the real parts of the other characteristic roots move off to minus infinity. This implies that for any $r \in \mathbb{R}$, $q_\epsilon(\lambda)$ has at most n zeros in the half-plane $\Re(\bar{\lambda}) \geq r$ or, equivalently, in $\Re(\lambda) \geq r/\epsilon$, provided ϵ is sufficiently small. $\square$

Remark 12.6. *When the complex variable λ in the characteristic equation of the closed-loop system with control law (12.6) is formally replaced with $(1 - e^{-\lambda\epsilon})/\epsilon$ (except for the term λ^n) and the resulting expression is developed in powers of $e^{-\lambda\epsilon}$, the characteristic equation of a system with proportional+delay compensator with $(n-1)$* commensurate *delays is obtained. This is exactly the controller of Proposition 12.5, when taking one delay equal to zero and the other delays commensurate, i.e., $\tau_i = (i-1)$, $i = 1, \ldots, n$.*

Using Property 12.4 the statements of Proposition 12.5 can be rephrased as follows.

Theorem 12.7. *Assume that $0 \leq \tau_1 < \cdots < \tau_n$ and $q(\lambda)$ is Hurwitz. Then the control law*

$$u(t) = -\left[\epsilon^n q_0 \ \frac{\epsilon^{n-1}}{(-1)}q_1 \ \frac{2!\,\epsilon^{n-2}}{(-1)^2}q_2 \ \cdots \ \frac{(n-1)!\,\epsilon}{(-1)^{n-1}}q_{n-1} \right] T(\tau)^{-1} \begin{bmatrix} y(t-\tau_1) \\ \vdots \\ y(t-\tau_n) \end{bmatrix} \quad (12.16)$$

achieves asymptotic stability of (12.3) for small values of ϵ. As $\epsilon \to 0_+$, the n rightmost characteristic roots converge to $\epsilon\lambda_i$, $i = 1, \ldots, n$, with λ_i the zeros of $q(\lambda)$.

In the next section we outline an alternative approach to design a stabilizing feedback law.

12.3.2 Control laws based on exact pole placement and low-gain design

This approach is inspired by [193, 177, 153, 264] and consists of a placement of the n rightmost characteristic roots of the closed loop by means of low-gain control laws.

In the control law

$$u(t) = -\sum_{j=1}^{n} k_j y(t-\tau_j)$$

there are n degrees of freedom, which allows us to place n characteristic roots at prescribed values. This way stability cannot be insured in general because the number of characteristic roots is infinite and only n of them are controlled. However, this conflict can be solved when using the low-gain approach, developed in [193, 177] in the context of the stabilization of integrators with an input delay and input constraints.

The basic idea is as follows. When the controlled characteristic roots are placed close to zero, it is expected that the gains are low. But when the gains tend to zero, all characteristic roots, excepting n, are shifted far away in the LHP, because the governing DDE behaves as an ODE with a vanishing (delayed) perturbation. We now illustrate this approach with an example, where n characteristic roots are placed at the same position. This will give rise to an explicit formula with a structure analogous to (12.16).

Theorem 12.8. *Assume $0 \le \tau_1 < \tau_2 \ldots < \tau_n$ and let $T(\tau)$ be defined by (12.9). Then control law*

$$u(t) = (-1)^n \left[\epsilon^n \; n\epsilon^{n-1} \ldots n!\,\epsilon\right] T(\tau)^{-1} \begin{bmatrix} e^{-\epsilon\tau_1} & & \\ & \ddots & \\ & & e^{-\epsilon\tau_n} \end{bmatrix} \begin{bmatrix} y(t-\tau_1) \\ \vdots \\ y(t-\tau_n) \end{bmatrix} \tag{12.17}$$

achieves asymptotic stability of (12.3) for small values of ϵ. Moreover, there the closed-loop system has a characteristic root at $\lambda = -\epsilon$, with multiplicity n.

Proof. The characteristic equation of the closed-loop system is given by

$$p(\lambda) := \lambda^n + \sum_{j=1}^{n} k_j e^{-\lambda\tau_j} = 0.$$

Assigning n characteristic roots to $\lambda = \bar{\lambda}$ yields the conditions $p(\bar{\lambda}) = 0, \ldots, p^{n-1}(\bar{\lambda}) = 0$, or

$$\begin{bmatrix} e^{-\bar{\lambda}\tau_1} & \cdots & e^{-\bar{\lambda}\tau_n} \\ -\tau_1 e^{-\bar{\lambda}\tau_1} & \cdots & -\tau_n e^{-\bar{\lambda}\tau_n} \\ \vdots & & \vdots \\ (-\tau_1)^{n-1} e^{-\bar{\lambda}\tau_1} & \cdots & (-\tau_n)^{n-1} e^{-\bar{\lambda}\tau_n} \end{bmatrix} \begin{bmatrix} k_1 \\ k_2 \\ \vdots \\ k_n \end{bmatrix} = - \begin{bmatrix} \bar{\lambda}^n \\ n\bar{\lambda}^{n-1} \\ \vdots \\ n(n-1)\cdots 2\bar{\lambda} \end{bmatrix}.$$

This can be written as

$$\begin{bmatrix} 1 & & & \\ & (-1) & & \\ & & \ddots & \\ & & & (-1)^{n-1} \end{bmatrix} T(\tau)^T \begin{bmatrix} e^{-\bar{\lambda}\tau_1} & & \\ & \ddots & \\ & & e^{-\bar{\lambda}\tau_n} \end{bmatrix} \begin{bmatrix} k_1 \\ k_2 \\ \vdots \\ k_n \end{bmatrix}$$

$$= -\begin{bmatrix} \bar{\lambda}^n \\ n\bar{\lambda}^{n-1} \\ \vdots \\ n(n-1)\cdots 2\bar{\lambda} \end{bmatrix}$$

and, therefore,

$$\begin{bmatrix} k_1 \\ k_2 \\ \vdots \\ k_n \end{bmatrix} = -\begin{bmatrix} e^{\bar{\lambda}\tau_1} & & \\ & \ddots & \\ & & e^{\bar{\lambda}\tau_n} \end{bmatrix} T(\tau)^{-T} \begin{bmatrix} \bar{\lambda}^n \\ (-1)n\bar{\lambda}^{n-1} \\ \vdots \\ (-1)^{n-1}n(n-1)\cdots 2\bar{\lambda} \end{bmatrix}.$$

Choosing $\bar{\lambda} = -\epsilon$ leads to the control law (12.17). When we let $\epsilon \to 0+$, we have

$$K(\epsilon) = [k_1(\epsilon) \ \cdots \ k_n(\epsilon)]^T \to 0.$$

Hence, the n characteristic roots at zero of the uncontrolled system are shifted to $-\epsilon$, while the other characteristic roots cannot cause instability if ϵ (i.e., $K(\epsilon)$), is sufficiently small. □

Remark 12.9. *For $q(\lambda) = (\lambda + 1)^n$ the control law (12.16) reduces to*

$$u(t) = -\begin{bmatrix} \epsilon^n & \frac{n\epsilon^{n-1}}{(-1)} & \frac{n(n-1)\epsilon^{n-2}}{(-1)^2} & \cdots & \frac{n!\,\epsilon}{(-1)^{n-1}} \end{bmatrix} T(\tau)^{-1} \begin{bmatrix} y(t-\tau_1) \\ \vdots \\ y(t-\tau_n) \end{bmatrix}. \qquad (12.18)$$

This control law doesn't coincide with (12.17) because it is based on an asymptotic approximation *of $q(\lambda)$, while (12.17) is based on an* exact *placement of n characteristic roots.*

Remark 12.10. *Both Theorems 12.7 and 12.8 guarantee asymptotic stability for sufficiently small values of ϵ. A threshold can be computed by performing a numerical continuation of the characteristic roots of the closed-loop system s as a function of the parameter ϵ, as illustrated in the next section. Even when the structure of (12.16) or (12.17) is not explicitly used, a stabilizing feedback law may still be synthesized by means of the approaches outlined in Chapter 7 and Section 10.1.*

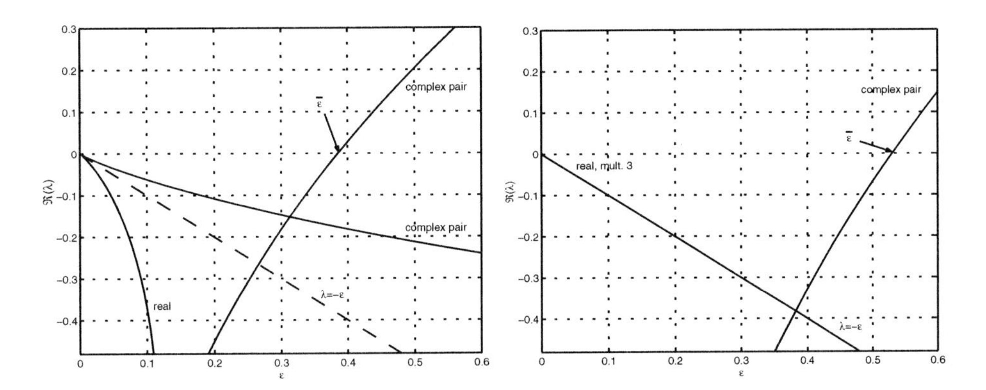

Figure 12.1. *Real parts of the characteristic roots of the triple integrator, controlled with (12.19) (left) and (12.20) (right).*

12.4 Illustrative example

For the *triple integrator* the control law (12.16) with $q(\lambda) = (\lambda + 1)^3$ and $\tau_i = (i-1),\ i = 1, \ldots, 3$, takes the form

$$u(t) = \left(-3\epsilon - \frac{9}{2}\epsilon^2 - \epsilon^3\right)\, y(t) + (6\epsilon + 6\epsilon^2)\, y(t-1) + \left(-3\epsilon - \frac{3}{2}\epsilon^2\right)\, y(t-2). \quad (12.19)$$

In Figure 12.1 (left), the rightmost characteristic roots of the closed-loop system are displayed as a function of the parameter ϵ. For $\epsilon < \bar{\epsilon}$, indicated on the figure, the closed-loop system is asymptotically stable. According to Theorem 12.7, the three rightmost characteristic roots converge to $\lambda = -\epsilon$ as $\epsilon \to 0_+$.

The control law (12.17) with $\tau_i = (i-1),\ i = 1, \ldots, 3$, is given by

$$u(t) = \left(-3\epsilon + \frac{9}{2}\epsilon^2 - \epsilon^3\right)\, y(t) + (6\epsilon - 6\epsilon^2)e^{-\epsilon}\, y(t-1) + \left(-3\epsilon + \frac{3}{2}\epsilon^2\right) e^{-2\epsilon}\, y(t-2) \quad (12.20)$$

and the closed-loop characteristic roots are shown in Figure 12.1 (right). Following Theorem 12.8 three characteristic roots lie at $\lambda = -\epsilon$, for all values of ϵ. Note that when ϵ is small, the spectrum is similar to the previous case, which is not surprising since the dominant terms in (12.19) and (12.20) (the terms $\sim \epsilon$) are equal.

Recall that when ϵ is fixed to a value smaller than $\bar{\epsilon}$, (12.19) and (12.20) actually define a whole one-parameter family of stabilizing feedback laws, by making use of Property 12.4.

12.5 Notes and references

In this chapter, we addressed the output feedback stabilization problem using control laws of the form (12.1), which involve several delay blocks. First, *necessary stabilizability* conditions were given for the general case. Next, it was shown that a chain of n integrators can be stabilized by a proportional+delay controller including $(n-1)$ delays, or by a chain of n delay blocks. Two constructive approaches were presented and applied to a numerical example. The first approach makes use of some numerical differentiation scheme combined with backward differences, and the second approach is based on an exact pole placement method and a low-gain design.

Necessary conditions for the output feedback stabilizability of linear systems with control law of the form (12.1) were given in [135]. The problem of stabilizing a chain of integrators with state feedback and input constraints was treated in [293] (without delay) and in [177, 193] (with a single delay in the control loop).

Concerning the multiple delay case, some particular cases (single integrator, double integrator with one or two delays) were considered [223, 49, 224, 193], while the general case was treated in [232]. A further example where the pole placement idea is used can be found in Chapter 13, which is devoted to the analysis of some control congestion algorithms. Finally, note that the use of delays, in particular delay-difference feedback, is well known in the context of stabilizing unstable orbits in nonlinear systems; see, for instance, [253].

The results presented in this chapter are based on [232, 135, 193] and the references therein.

Chapter 13

Congestion control algorithms in networks

A communication network consists of a collection of "elements" (network users and/or sources) interconnected to transfer information/data from one *node* to another node through some (transmission or communication) *links*. A high-performance network represents a communication network able to support a large variety of applications, which can be transferred at high speed and with low (communication) delay. One of its dominant features is the *large scale* of the system [218].

There exists basically two ways to model communication networks systems using (discrete- or continuous-time) stochastic representations and/or deterministic (continuous-time) approximations.

In what follows, we shall consider only *continuous-time fluid approximations* due to their simplicity of representation as dynamical systems. We do not insist on the way to derive such models or on their explicit interest for congestion control analysis, but we point out the discussions proposed in [272] (see also [19, 43] for some connections between fluid-continuous and discrete-time representations). Further philosophical problems concerning the interactions between networks and feedback control theory can be found in [218].

We will consider in particular the analysis of some congestion control algorithms from the literature. To the best of the authors' knowledge congestion control in high-performance networks was introduced by Van Jacobson [126] by the end of the 1980s. More precisely, we shall analyze the asymptotic stability of two distinct fluid approximation models. In both cases, we will focus on *delay*-induced *instabilities*, that is, the way in which the stability is affected by the delay presence.

The first model was introduced by Izmailov [124, 125], and it represents a second-order linear system that describes the dynamics of a single connection between a source (controlled by an access regulator) and a distant node with a constant transmission capacity. The particularity of the model is the existence of two (independent) delays: round-trip time and the control-time interval, respectively. The stability analysis will be performed in the delay parameter space, and connections with output feedback control by using delay terms will be emphasized.

The second model was introduced by Misra, Gong, and Towsley in [201], and is of transmission control protocol (TCP) congestion-avoidance type. More precisely, this last model is a nonlinear second-order system including only one single delay (round-trip time) with the following features: the first differential equation describes the TCP window dynamics, and the second equation models the bottleneck queue behavior. Some appropriate transformations of the original systems will be considered for deriving the local asymptotic stability w.r.t. the system's parameters. Finally, we shall emphasize also the *chaotic behavior* of the system.

13.1 Algorithms for single connection models with two delays

We present the stability analysis of classes of second-order linear systems including multiple delays in rational dependence, which cover some of the feedback control algorithms proposed by Izmailov [124, 125]. The stability analysis will make use of some algebraic and geometric arguments proposed in the previous chapters. Some connections with the output feedback control problem will also be emphasized.

13.1.1 Model and related remarks

In [124, 125] Izmailov proposed the following deterministic models of a single connection between a source controlled by an access regulator and a distant node with a constant transmission capacity μ:

$$\begin{cases} \dot{x}_1(t) = x_2(t-\tau_1) - \mu, \\ \dot{x}_2(t) = -a(x_1(t-\tau_2) - \bar{X}) - b(x_1(t-\tau_2-r) - \bar{X}), \end{cases} \tag{13.1}$$

and

$$\begin{cases} \dot{x}_1(t) = x_2(t-\tau_1) - \mu, \\ \dot{x}_2(t) = -a(x_1(t-\tau_2) - \bar{X}) - b(x_1(t-\tau_2-r) - \bar{X}) \\ \qquad\qquad -c\left(x_1\left(t-\tau_2-\frac{r}{2}\right) - \bar{X}\right), \end{cases} \tag{13.2}$$

where x_1 represents the buffer contents, x_2 the current input rate, and $\bar{X}$ the target value. Using the new variable $y(t) = x_1(t) - \bar{X}$, systems (13.1) and (13.2) lead to the following second-order delay equations with two discrete and independent delays τ and r,

$$\ddot{y}(t) + ay(t-\tau) + by(t-\tau-r) = 0, \tag{13.3}$$

and,

$$\ddot{y}(t) + ay(t-\tau) + by(t-\tau-r) + cy\left(t-\tau-\frac{r}{2}\right) = 0, \tag{13.4}$$

where the "total" delay $\tau = \tau_1 + \tau_2$ represents the *round-trip time*, and r is the *control time-interval*.

As seen in [224, 226], the delays τ and r have a *stabilizing* effect if the gains a, b, and c satisfy some appropriate assumptions, even if the system free of delays is not asymptotically stable. In this sense consider the case $a > |b|$, $b < 0$ in (13.3), which corresponds to a system that is an *oscillator* if it is free of delays, but which is stable for sufficiently *small* delays $\tau, r \neq 0$.

In what follows, we shall focus on the stability analysis of (13.3) and (13.4). More explicitly, we shall consider a more general model including both systems (13.3) and (13.4) as particular examples. This model takes the form

$$\ddot{y}(t) + \sum_{k=0}^{n} a_k y(t - \tau_k) = 0, \tag{13.5}$$

with appropriate initial condition, and with

$$\tau_k = \frac{k}{n}\tau + r, \qquad k = 1, 2, \ldots, n.$$

In other words, the delays τ_k, $k = 1, 2, \ldots, n$, are *rational dependent* and depend on two rationally independent delays τ and r:

$$\begin{bmatrix} \tau_1 \\ \tau_2 \\ \vdots \\ \tau_n \end{bmatrix} = \begin{bmatrix} \frac{1}{n} & 1 \\ \frac{2}{n} & 1 \\ \vdots & \vdots \\ 1 & 1 \end{bmatrix} \begin{bmatrix} \tau \\ r \end{bmatrix};$$

see Section A.4 for definitions regarding the interdependency of numbers. The system (13.5) can be rewritten in a first-order form as follows:

$$\dot{x}(t) = Ax(t) + \sum_{k=0}^{n} A_k x\left(t - \frac{k}{n}r - \tau\right), \tag{13.6}$$

where

$$A = \begin{bmatrix} 0 & 1 \\ 0 & 0 \end{bmatrix}, \qquad A_k = \begin{bmatrix} 0 & 0 \\ -a_k & 0 \end{bmatrix}, \qquad k = 0, \ldots, n, \tag{13.7}$$

and $x = [y \;\; \dot{y}]^T$.

The analysis in terms of delays can be reduced to the following two steps (one parameter-based analysis at each step):

- First, analyze the stability of (13.5) if $\tau = 0$, but $r \neq 0$, that is, the stability of a system including commensurate delays, multiple of r/n.

- Second, assume also $\tau \neq 0$, and perform the analysis w.r.t. the second delay parameter τ while assuming the first delay r fixed.

In other words, we are searching first for the delay intervals in r for which the corresponding system with $\tau = 0$ is asymptotically stable or the number of unstable roots is as small as possible, and next for any delay value r fixed inside such intervals we are looking for the delay intervals in τ for which the stability is guaranteed, that is, we are looking for the characteristic roots crossing the imaginary axis, and also for their crossing direction.

To summarize, the crossing curves in the parameter space $Or\tau$ are defined by using an explicit definition of the delay τ as a function of the parameter r, $\tau = \tau(r)$. Further remarks on a different way to compute the stability crossing curves and the corresponding stability regions in the delay-parameter space are presented in the next section. In particular, we will exploit an interesting property of this system, namely the fact that *any sufficiently small delay* r can stabilize (13.3) if $\tau = 0$ (oscillator subject to a delay output feedback). Note that the assumptions $a > 0$ and $b < 0$ were already encountered in the work of Izmailov (see [125]), but the argument was completely different.

13.1.2 Linear stability analysis

Let us follow the procedure mentioned above. We start by assuming $\tau = 0$. In this case, the system (13.6) becomes a second-order system with commensurate delays, for which the methodology proposed in Chapter 4 for deriving delay intervals guaranteeing stability/instability works.

In this sense, we now introduce the following matrix pencil $\Lambda_1 \in \mathbb{C}^{2n\times 2n}$:

$$\Lambda_1(z) = z\begin{bmatrix} 1 & & 0 & 0 \\ & \ddots & & \\ 0 & & 1 & 0 \\ 0 & \dots & 0 & a_n \end{bmatrix} + \begin{bmatrix} 0 & -1 & \dots & 0 & 0 & 0 & \dots & 0 \\ & & \ddots & & & & & \\ 0 & & & & & & & -1 \\ -a_n & -a_{n-1} & \dots & -a_1 & 0 & a_1 & \dots & a_{n-1} \end{bmatrix}. \tag{13.8}$$

As mentioned in the previous chapters, the characteristic roots crossing the imaginary axis and the corresponding critical delay values can be obtained from the *generalized eigenvalue distribution* of the matrix pencil Λ_1. Using the formalism and the notations in Chapter 4, the generalized eigenvalue $z_0 \in \sigma(\Lambda_1)$ such that

$$\sigma\left(A_0 + \sum_{k=1}^{n} A_k z_0^k\right) \cap j\mathbb{R}^* \neq \emptyset$$

corresponds to a crossing eigenvalue. Thus, it belongs to the delay crossing generator set $\mathcal{T}_{1,g}$, that will define the corresponding (crossing) frequencies

$$\omega_0 \in \Omega_{1,z_0} := \left\{\omega \in \mathbb{R}_+^* : \quad j\omega \in \sigma\left(A_0 + \sum_{k=1}^{n} A_k z_0^k\right)\right\},$$

and critical delay values

$$r \in \mathcal{T}_{1,\omega_0} := \left\{\frac{\operatorname{Log}(\overline{z}_0)}{j\omega_0} + \frac{2\pi\ell}{\omega_0} > 0 : j\omega_0 \in \sigma\left(A + \sum_{k=1}^{n_d} A_k z_0^k\right) \setminus \{0\}, \quad \ell \in \mathbb{Z}\right\}.$$

As expected, the set of crossing frequencies Ω_1 includes a *finite* number of frequency values, and it is explicitly given by

$$\Omega_1 := \bigcup_{z\in\mathcal{T}_{1,g}} \Omega_{1,z}.$$

By similarity, we will have the set of critical delays $\mathcal{T}_1$ given by

$$\mathcal{T}_1 := \bigcup_{\omega \in \Omega_1} \mathcal{T}_{1,\omega}.$$

Under the assumption of *simple* crossings, the *crossing direction* w.r.t. any critical delay value $r \in \mathcal{T}_1$ corresponding to the crossing frequency ω_0 and to the delay crossing generator z_0 is given by the sign of the following quantity:

$$\Re\left[\sum_{k=1}^{n} k a_k z_0^k\right] > 0 \ (< 0). \tag{13.9}$$

If such a quantity is positive (negative) we will have a crossing toward instability (stability).

In what follows, we reduce our analysis only to the first delay intervals guaranteeing stability with respect to r and τ. In other words, we will compute the *stability region* which is the "closest" to the origin of the parameter space. In the case of *small delays*, we have the following stability result, which simply extends the remarks concerning the stabilization of oscillatory systems by using delays in the output feedback control laws (see Chapter 11), and the remarks above concerning the crossing direction.

Proposition 13.1. *[226] Assume that*

$$\sum_{k=0}^{n} a_k > 0, \qquad \sum_{k=1}^{n} k a_k < 0. \tag{13.10}$$

Then there exists a sufficiently small positive value $\epsilon > 0$ such that (13.6) with $r = \epsilon$, $\tau = 0$ is asymptotically stable.

Sketch of the proof. The first inequality simply says that the characteristic function of the second-order system free of delays has two critical roots on the imaginary axis, and the second inequality ensures that the crossing direction is toward stability for small delays. Indeed, in such a case, the crossing direction is given by the relation (13.9) with $z_0 = 1$. Thus, since the first crossing is toward stability, it follows that the asymptotic stability is guaranteed for small delays.

The next result gives the complete characterization of the *first switch*, that is, the case when the roots of the characteristic equation associated to the original system cross the imaginary axis toward instability when the delay parameter r is varying from 0 to $+\infty$.

Proposition 13.2. *[226] The system (13.6) with $\tau = 0$, satisfying the inequalities (13.10), is asymptotically stable for all delay values r satisfying*

$$0 < r < r_1(a_0, a_1, \ldots, a_n) := \min\{\xi > 0 : \quad \xi \in \mathcal{T}_1\}. \tag{13.11}$$

Furthermore, if $r = 0$ or $r = r_1(a_0, \ldots, a_n)$, the corresponding associated characteristic equation has at least one pair of complex conjugate roots on the imaginary axis.

Proposition 13.3. *[226] The system (13.6) satisfying the constraints (13.10) is asymptotically stable for all delays r and τ and satisfies the following conditions:*

$$\begin{cases} 0 < r < r_1(a_0, \ldots, a_n), \\ 0 \leq \tau < \tau_{1,r} := \min_{\omega_s} \left\{ \dfrac{1}{\omega_s} \cdot \tan^{-1} \left(\dfrac{-\sum_{k=1}^{n} a_k \sin \frac{\omega_s k r}{n}}{\sum_{k=0}^{n} a_k \cos \frac{\omega_s k r}{n}} \right) \right\}, \end{cases} \tag{13.12}$$

where ω_s belongs to the set of positive solutions of the equation

$$\omega^4 = \sum_{k=0}^{n} a_k^2 + 2 \sum_{k=1}^{n} \sum_{h=0}^{k-1} a_k a_h \cos\left(\frac{(k-h)\omega r}{n} \right). \tag{13.13}$$

Furthermore, if the chosen delay r and the solution $\tilde{\omega}_s$ defining the corresponding upper bound $\tau_{1,r}$ in (13.12) satisfy the condition

$$\tilde{\omega}_s^3 > \frac{1}{2} \sum_{k=1}^{n} \sum_{h=0}^{k-1} \frac{a_k a_h (k-h) r}{n} \sin\left(\frac{(k-h)\tilde{\omega}_s r}{n} \right), \tag{13.14}$$

then

(i) The system (13.5) is unstable for $\tau = \tau_{1,r} + \epsilon$, with $\epsilon > 0$ sufficiently small.

(ii) There does not exist any $\tau > \tau_{1,r}$ such that the system (13.5) is asymptotically stable provided the equation (13.13) has only one positive solution.

Condition (i) in Proposition 13.3 above simply says that the first crossing w.r.t. the delay parameter τ is toward instability at $\tau = \tau_{1,r}$ and at the frequency ω_s. Furthermore, an explicit characterization of the corresponding set of crossing frequencies Ω_2 (with respect to the parameter τ for a fixed r) is given in the form of the nonlinear equation (13.13). It is important to point out that, for a fixed r, this equation has a *finite number of roots*. Next, condition (ii) characterizes the situation where, for a fixed delay-parameter r value, $\mathrm{card}(\Omega_2) = 1$.

Remark 13.4. *Equation (13.13) always has at least one positive solution for ω in the interval*

$$\left[\min_{\mathbf{i} \in \mathcal{I}_n} \sqrt{\sum_{k=1}^{n} (-1)^{i_k} a_k^2}, \max_{\mathbf{i} \in \mathcal{I}_n} \sqrt{\sum_{k=1}^{n} (-1)^{i_k} a_k^2} \right], \tag{13.15}$$

where

$$\mathcal{I}_n = \{ i = (i_1, \ldots, i_n) : i_k \in \{1, 2\}, \forall k = 1, \ldots, n \}$$

is an appropriate index family. In [223] one therefore concludes that the upper *bound on τ will* always be finite; *that is, one may expect a sequence of stability/instability delay intervals in terms of τ_0, with* instability persistence *for sufficiently large delays. In other words, there exists a finite value $\bar{\tau}_r$ such that the system is unstable for $\tau > \bar{\tau}_r$.*

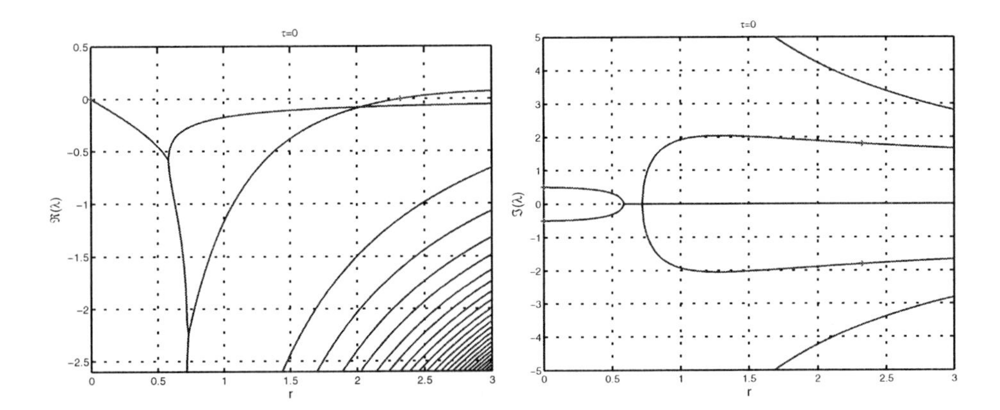

Figure 13.1. *Real and imaginary parts of the rightmost characteristic roots of (13.4) as a function of r when $\tau = 0$. Characteristic roots on the imaginary axis are indicated with "+". The parameter values are $a = 2.25$, $b = c = -1$.*

Remark 13.5. *Based on the results above, it is easy to see that* reducing *the* control-time interval *will increase the* sensitivity *of the control algorithm with respect to the* round-trip time. *In conclusion, a "wait-and-act" strategy in terms of control-time interval will substantially improve the* robustness *of the control algorithm. Such a result was largely discussed in [224, 226].*

If necessary or desirable, more precise stability information can be obtained by directly computing and monitoring the rightmost characteristic roots of the closed-loop system. This is illustrated in Figure 13.1. The characteristic roots of the system (13.4) with $a = 2.25$, $b = c = -1$, $\tau = 0$ are shown as a function of the control interval r, computed with the software package DDE-BIFTOOL [74]. By automatic continuation in the two-parameter space (r, τ) of solutions, for which characteristic roots lie on the imaginary axis, the stability region in this parameter space is obtained; see Figure 13.2 (left). For the optimal value of the control interval (w.r.t. robustness in the round-trip time), the rightmost characteristic roots are shown in Figure 13.2 (right).

13.1.3 Interpretations and discussions

The analysis above was based on analytically computing the bound of τ as a function of r, and taking into account the delay intervals on r, for which (asymptotic) stability exists. In other words, we tried to use r as a *stabilizing* delay-parameter, and next τ as a *destabilizing* delay-parameter; that is, we explicitly computed the delay margin in terms of τ as a function of r. The problem considered in the previous paragraphs concerned the analysis of the effects induced by the delays for some $(n+1)$-tuple of parameters $(a_0, a_1, \ldots, a_n)$.

In what follows, we shall consider two particular ways to reinterpret problem in light of the results proposed in the previous chapters.

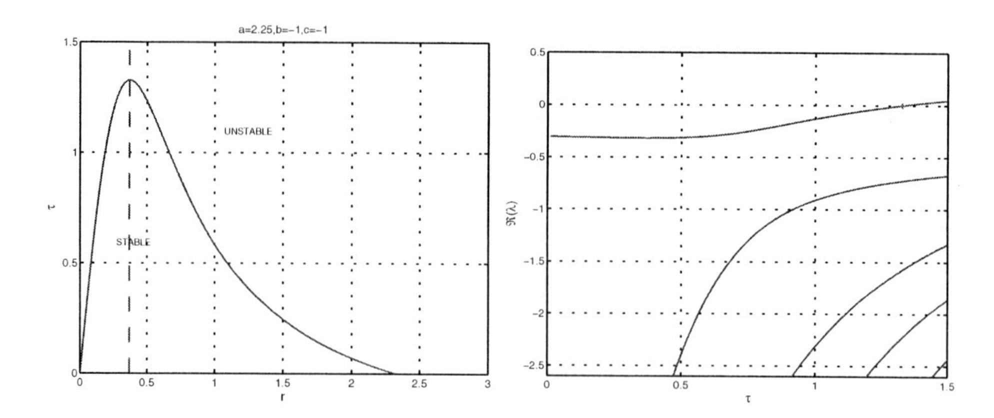

Figure 13.2. *(left) Stability region in the (r, τ)-plane. For the optimal control interval $r \approx 0.37$, indicated with the dashed line, the real parts of the rightmost characteristic roots are shown as a function of the round-trip time τ (right).*

Output feedback stabilization. It is easy to see that the stability of the initial second-order system including two or three (rationally dependent) delays,

$$\ddot{y}(t) + ay(t - \tau) + by(t - \tau - r) = 0$$

and, respectively,

$$\ddot{y}(t) + ay(t - \tau) + by(t - \tau - r) + cy\left(t - \tau - \frac{r}{2}\right) = 0,$$

can be interpreted as an output feedback control problem: *controlling a double integrator by using two or three delay blocks*, that is, the system $H_{yu}(\lambda) = 1/\lambda^2$ with the control laws

$$u(t) = -ay(t - \tau) - by(t - \tau - r)$$

in the first case and

$$u(t) = -ay(t - \tau) - by(t - \tau - r) - cy(t - \tau - r/2)$$

in the second case. It is quite clear that the (two or three) delay blocks above are distinct since $\tau < \tau + r/2 < \tau + r$ for all $(\tau, r) \in \mathbb{R}_+^2$. As discussed in Chapter 12, *two delay blocks* are sufficient for getting asymptotic stability of the corresponding closed-loop scheme.

Let us consider this particular case of two delay blocks controlling the chain including two integrators. One of the possible uses of the algorithm above is to tune the gains a and b by using the delay information available. Indeed, the round-trip time τ is always measured, and the control-time interval r can be selected as discussed by Izmailov in his papers [124, 125]. In this sense, we can apply the pole placement idea mentioned in Chapter 12. The characteristic function of the closed-loop scheme writes as

$$p(\lambda;\ \tau, r) := \lambda^2 + ae^{-\lambda\tau} + be^{-\lambda(\tau+r)}.$$

Consider now that the parameters (a, b) are such that the double root at the origin is "moved" in $\mathbb{C}_-$ at $\lambda = -\varepsilon$ as a double root, for sufficiently small $\varepsilon > 0$. In such a case, we will need the following conditions to hold simultaneously:

$$\begin{cases} p(-\varepsilon;\ \tau, r) = 0, \\ p'(-\varepsilon;\ \tau, r) = 0. \end{cases}$$

This results in the system of linear equations in (a, b)

$$\begin{cases} a + be^{\varepsilon r} = -\varepsilon^2 e^{-\varepsilon\tau}, \\ a\tau + b(\tau + r)e^{\varepsilon r} = 2\varepsilon e^{-\varepsilon\tau}, \end{cases}$$

with solution

$$\begin{cases} a = \varepsilon \frac{2+\tau\varepsilon}{re^{\varepsilon\tau}}, \\ b = -\varepsilon \frac{2-(\tau+r)\varepsilon}{re^{\varepsilon(\tau+r)}}. \end{cases}$$

It is not difficult to prove that for sufficiently small values of $\varepsilon > 0$, the characteristic roots of $p(\lambda;\ \tau, r)$ are located in $\mathbb{C}_-$, and we can use the argument above to "adapt" the gains in function of the parameters (τ, r). Larger values for ε can also be considered, but then we need to be sure that there are no other characteristic roots coming from $\mathbb{C}_-$ to cross the imaginary axis. Reciprocally, the idea above can also be used to find a pair of stabilizing delay values (τ, r) for a given pair of gain values (a, b). Next, some *robustness region* around the corresponding pair $(\tau(a, b), r(a, b))$, for which the closed-loop stability is preserved, can be determined.

On the geometry of the crossing curves in the delay-parameter space. Reconsider the initial problem including only four parameters: (a, b, τ, r). It is easy to see that for fixed parameters (a, b), the problem of characterizing the stability regions in the delay-parameter space is reduced to finding all the stability crossing curves and, next, defining some partition of the delay-parameter space in stability (instability) domains as outlined in Chapter 4. In the particular case under consideration, the corresponding characteristic function is given by

$$\lambda^2 + ae^{-\lambda\tau} + be^{-\lambda(\tau+r)},$$

which is of the form

$$p(\lambda;\ \tau_1, \tau_2) := \lambda^2 + ae^{-\lambda\tau_1} + be^{-\lambda\tau_2}, \tag{13.16}$$

where $\tau_2 \geq \tau_1$. Such a case study enters in the framework of the analysis suggested in Chapter 4, by choosing $p_0(\lambda) = \lambda^2$, and p_1, p_2 as constants, $p_1(\lambda) = a$, $p_2(\lambda) = b$. The corresponding rational transfer functions $a_1(j\omega) := p_1(j\omega)/p_0(j\omega) = a/(j\omega)^2$ and $a_2(j\omega) = p_2(j\omega)/p_0(j\omega) := b/(j\omega)^2$ are properly defined for all positive ω excepting $\omega = 0$. On the other hand, if $a + b \neq 0$, $\lambda = 0$ is not a characteristic root of the quasipolynomial p. As discussed in the previous chapters, such an assumption is standard, and serves to avoid a characteristic root at the origin that is *invariant* w.r.t. delay changes.

Simple computations prove that the corresponding frequency crossing set Ω is reduced to the interval

$$\Omega = \left[\sqrt{\|\,a\,| - |\,b\,\|}, \sqrt{|\,a\,| + |\,b\,|}\right], \tag{13.17}$$

which is of type 13 (or 23), if we assume that $| a |>| b |$ (or $| a |<| b |$). The explicit computation of the stability regions in the delay-parameter space is straightforward. We refer to Chapter 4 for the details.

13.2 TCP/AQM congestion avoidance models with one delay

We study the stability of some nonlinear models for the behavior of congested routers. First we propose and discuss the models. Next, we sketch some important scaling properties of these models. Finally, we study the attractors and their stability properties.

13.2.1 Model and related remarks

Recently, some models describing accurately the behavior of congested routers in TCP/AQM (Transfer Control Protocol/Active Queue Management) networks were presented in [201, 134, 118]. As expected, these kinds of models are described by nonlinear differential equations with time-delay where the delay represents the corresponding round-trip time in the network.

The model of [201, 118, 117] consists of the following coupled nonlinear differential equations with time-varying delay:

$$\dot{W}(t) = \frac{1}{R(t)} - \frac{1}{2}\frac{W(t)W(t-R(t))}{R(t-R(t))}p(t-R(t)), \tag{13.18}$$

$$\dot{Q}(t) = \begin{cases} N(t)\frac{W(t)}{R(t)} - C, & q > 0, \\ \max\left(N(t)\frac{W(t)}{R(t)} - C, 0\right), & q = 0, \end{cases} \tag{13.19}$$

where $W(t)$ denotes the average TCP window size (packets), $Q(t)$ is the average queue length (packets), $R(t)$ is the round-trip time (secs) , C is the queue capacity (packets/sec), $N(t)$ is the number of TCP sessions, and $p(\cdot)$ is the probability function of a packet mark. The queue length $Q(t)$ and window size $W(t)$ are positive. The probability function of a packet mark $p(\cdot)$ takes values only in $[0, 1]$. The round-trip time can be decomposed as

$$R(t) = \frac{Q(t)}{C} + \tau_p, \tag{13.20}$$

where τ_p is the propagation delay (secs).

The first differential equation describes the *TCP window control dynamic*. Indeed, the first term $\frac{1}{R(t)}$ describes the window's additive increase phase, and the second term $\frac{W(t)}{2}$ the multiplicative decreasing phase (including the packet marking probability). Different AIMD (additive-increasing multiplicative-decreasing) continuous-time models can be found in [134, 160]. Note also the excellent overview of existing fluid approximations based approaches proposed in [160]. (13.19) describes the bottleneck queue length as the difference between the packet arrival rate $\frac{NW}{R}$ and the link capacity C, assuming that there are no internal dynamics in the bottleneck (roughly speaking, a simple integrator).

Using fluid flow models like (13.18)–(13.20), AQM can be interpreted as a *feedback control problem*, where the control action consists of marking packets (with probability p) as a function of the measured queue length Q; see [118].

As in [117] we shall assume that the TCP load $N(t)$ and the round-trip time $R(t)$ are time-invariant, i.e., $N(t) \equiv N$ and $R(t) \equiv R$. The latter may be a good approximation when the round-trip time is dominated by the propagation delay. This occurs when the capacity C of the link is large [117]. Furthermore, as it is presented in [117, 184], considering that the probability marking function $p(\cdot)$ is *proportional* to the queue length, i.e., $p(t) = K\, q(t)$, the system under consideration becomes

$$\dot{W}(t) = \frac{1}{R} - \frac{W(t)W(t-R)}{2R} K\, q(t-R), \tag{13.21}$$

$$\dot{Q}(t) = \begin{cases} N(t)\frac{W(t)}{R} - C, & Q > 0, \\ \max\left(N(t)\frac{W(t)}{R} - C, 0\right), & Q = 0. \end{cases} \tag{13.22}$$

The unique equilibrium point of (13.21)–(13.22) is given by

$$W^* = \frac{RC}{N}, \quad Q^* = \frac{2N^2}{R^2C^2K}.$$

In Section 5 of [118] a linearized stability analysis of the equilibrium point was performed in the frequency domain, where the variation of round-trip time was taken into account, yet some of the delay effects were treated as high-frequency uncertainty. The references [117, 184] contain a Lyapunov-based (non)local stability analysis of the equilibrium when making the additional simplification of (13.21) to

$$\dot{W}(t) = \frac{1}{R} - \frac{W(t)^2}{2R} K\, q(t-R). \tag{13.23}$$

In [117] the authors proved that when the delay is equal to zero the equilibrium point of the system (13.23) and (13.22) is asymptotically stable for all $K > 0$. When the delay is different from zero a Lyapunov–Razumikhin approach was used to show the asymptotic stability of the equilibrium point of (13.23) for sufficiently small $\frac{K}{N} > 0$. In [184] this result was refined and sufficient conditions on the parameters for local stability and estimates of the attraction domain were derived using a less conservative Lyapunov–Krasovskii approach.

The structure of this section is as follows: after some brief comments on a transformation of state and time, we completely characterize the linear stability region of the steady state solution of (13.21)–(13.22) as a function of the model parameters. More explicitly, only one delay interval guarantees the asymptotic stability of the linearized model. Then we take the nonlinearities into account and study the global behavior of the solutions.

13.2.2 Transformation

When the round-trip R time is assumed to be constant, one can apply a transformation of state and time to (13.21)–(13.22), yielding:

$$\begin{aligned} \dot{w}(t) &= 1 - \tfrac{w(t)w(t-1)}{2} kq(t-1), \\ \dot{q}(t) &= \begin{cases} w(t) - c, & q > 0, \\ \max(w(t) - c, 0), & q = 0, \end{cases} \end{aligned} \tag{13.24}$$

where

$$w = W, \; q = Q/N, \; t^{(\text{new})} = t^{(\text{old})}/R,$$

and

$$c = \frac{RC}{N}, \; k = KN.$$

The importance of this transformation lies in the fact that the four model parameters (K, N, C, R) are reduced to only two parameters (k, c). This facilitates the study of the dependence of the attractors and their stability properties on the system's parameters. It also allows us to display stability regions w.r.t. *all* parameters in only one figure.

Note that only one of the "new" parameters depends explicitly on the round-trip time R. Furthermore, such a dependence is *linear*, a fact which simplifies the analysis of delay effects on stability.

13.2.3 Stability analysis

Equilibrium

In the normalized coordinates (w, q) the unique equilibrium point is given by

$$(w^*, q^*) = \left(c, \frac{2}{kc^2}\right). \tag{13.25}$$

Linearization around it results in the second-order differential equation in $\tilde{q} := q - q^*$,

$$\ddot{\tilde{q}}(t) + \frac{1}{c}\dot{\tilde{q}}(t) + \frac{1}{c}\dot{\tilde{q}}(t-1) + \frac{kc^2}{2}\tilde{q}(t-1) = 0,$$

whose characteristic equation is given by

$$H(\lambda) := \lambda^2 + \frac{1}{c}\lambda + \frac{1}{c}\lambda e^{-\lambda} + \frac{kc^2}{2}e^{-\lambda} = 0. \tag{13.26}$$

To characterize the stability region in the (k, c)-plane we first fix $c > 0$ and consider the stability region as a function of k. We have the following result.

Proposition 13.6. *For each value of c, there exists exactly* one *stability interval as a function of k; i.e., $k \in (0, \bar{k}(c))$ with $\bar{k}(c) \in \bar{\mathbb{R}}_+$.*

Proof. For $k = 0$ the characteristic equation reduces to $H(\lambda) = \lambda(\lambda + \frac{1}{c} + \frac{1}{c}e^{-\lambda})$. Following from [53], the rightmost eigenvalue is equal to zero and isolated. The continuity of this eigenvalue w.r.t. k implies the existence of a root function $r(k)$, satisfying $r(0) = 0$ and

$$H(r(k)) = 0.$$

Differentiating this identity w.r.t. k at the point $k = 0$ we arrive at

$$r'(0) = -\frac{c^3}{4} < 0. \tag{13.27}$$

Therefore, the linearized system has one unstable real eigenvalue for small $k < 0$. Instability follows for *all* $k < 0$, because a zero eigenvalue cannot occur for $k \neq 0$ and, as a consequence, eigenvalues can only cross the imaginary axis in complex conjugate *pairs* as k is varied.

Equation (13.27) also implies asymptotic stability for small $k > 0$. The stability can only be lost when eigenvalues cross the imaginary axis. When an imaginary eigenvalue $\lambda = j\omega$ would occur for, say, $k = k^*$, one can compute:

$$\left.\frac{d\Re(\lambda)}{dk}\right|_{\lambda=j\omega,k=k^*} = \frac{\omega^2 c^2}{2} \frac{\left(\frac{1}{c}\omega^2 + \frac{k^* c^2}{2}\omega^2 + \frac{k^*}{2} + \frac{k^* c}{2}\right)}{\left(\frac{1}{c}\omega^2 - \frac{k^* c}{2} + \frac{\omega^2}{c^2} + \frac{k^* c^2 \omega^2}{2}\right)^2 + \left(\frac{\omega^3}{c} - \omega k^* c^2 - \frac{k^* c\omega}{2}\right)^2} > 0. \tag{13.28}$$

Therefore, eigenvalues can only cross the imaginary axis toward instability as k is increased, and thus only one stability interval is possible. Notice that (13.28) holds under the implicit assumption that the imaginary eigenvalues are simple. An easy calculation, which is omitted, excludes the nongeneric case of having imaginary eigenvalues with a multiplicity larger than one. □

The critical value $\bar{k}(c)$ in Proposition 13.6 corresponds to a subcritical *Hopf bifurcation* of the original nonlinear system. By numerical continuation of such Hopf bifurcation in the two-parameter space (k, c), the stability region can be computed; the result is shown in Figure 13.3. See [270] for the theory on continuation and bifurcation analysis and [74, 78] for the numerical tool DDE-BIFTOOL.

While the technique of numerical continuation of Hopf bifurcations to separate stability/instability regions of a steady state solution in a two-parameter space is applicable in general, the method of D-subdivision [138] (largely discussed in Chapter 4) only applies to specific problems such as the example above, but allows us to obtain *analytical* expressions for the boundary: when substituting $\lambda = j\omega$ in (13.26), some simple computations yield an implicit expression of the relation $\bar{k}(c)$:

$$\begin{cases} c = \frac{1+\cos\omega}{\omega\sin\omega}, \\ \bar{k} = \frac{2\omega^4(\sin\omega)^2}{(1+\cos\omega)^2}, \quad \omega \in (0, \ \pi). \end{cases} \tag{13.29}$$

The above analysis of the linearized fluid model illustrates how analytical and numerical tools can *complement* each other to obtain a complete solution of an analysis problem. Numerically, we computed a curve separating stable and unstable parameter pairs in the (k, c)-plane. Analytically, we obtained qualitative information which proves that this curve bounds the *whole* stability region.

Other attractors

A first observation is that solutions of (13.24) cannot grow unbounded, even when the steady state solution is (locally) exponentially unstable.

Proposition 13.7. *All the solutions of the system (13.24) are bounded.*

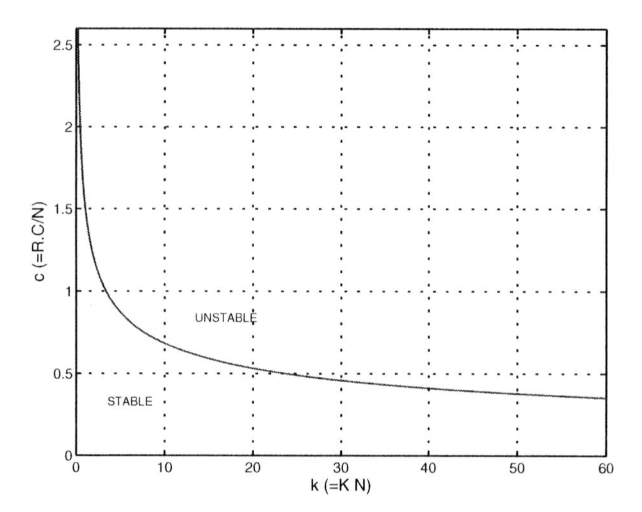

Figure 13.3. *Linear stability region of the steady state solution in the (k, c)-plane. The curve is obtained by computing a branch of Hopf bifurcations.*

Proof. First we show by contradiction that the function $t \to w(t)$ is bounded along a solution. Therefore, assume that w is unbounded and denote by $t = t_m$ the smallest time such that $w(t) = w_m$, where

$$w_m := |w_0| + c + 3 + 2/\sqrt{k}.$$

Since $\dot{w}(t) < 1$ for $t \geq 0$, we have $t_m \geq 2$ and $w(t) \geq w_m - 2 \; \forall t \in [t_m - 2,\ t_m - 1]$. Consequently

$$q(t_m - 1) = q(t_m - 2) + \int_{t_m-2}^{t_m-1} \dot{q}(t)dt \geq \min_{t \in [t_m-2,\ t_m-1]} \dot{q}(t) \geq |w_0| + 1 + 2/\sqrt{k} \geq 1.$$

This implies

$$\dot{w}(t_m) \leq 1 - \frac{w_m(w_m - 2)k}{2} < 0$$

and we have a contradiction; thus w is bounded. Similarly, assume that q is unbounded and denote by $t = t_m$ the first time such that $q(t) = q_m$, where q_m is a sufficiently large number. The boundedness of w implies the boundedness $\dot{q}$, thus t_m must grow unbounded as $q_m \to \infty$. Furthermore, for large t_m there exist a number $M > 0$ such that in the time-interval $[t_m/2,\ t_m]$, the first equation of (13.24) can be written as

$$\dot{w}(t) = 1 - w(t)w(t-1)\alpha(t), \quad \alpha(t) > M. \tag{13.30}$$

By taking q_m sufficiently large, M can also be chosen arbitrarily large. For sufficiently large values, (13.30) and the boundedness of w imply that $w(t_m) < c$, as follows from the method of steps. Thus $\dot{q}(t_m) < 0$ and we have a contradiction. □

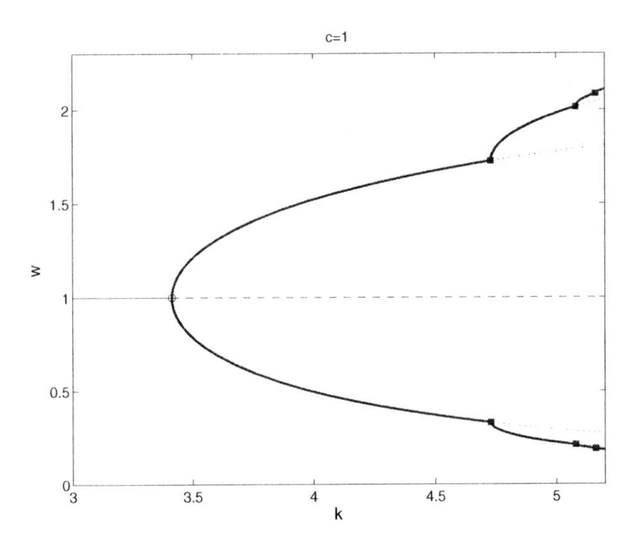

Figure 13.4. *Bifurcation diagram of the system (13.24) with $c = 1$ and free parameter k. Stable (thin line) and unstable (——) steady state solution. Stable (thick lines) and unstable $(\cdots)$ branches of periodic solutions (maximum and minimum values of $w(t)$ are shown). Connections are formed by a Hopf Bifurcation ($\circ$) and by Period Doubling Bifurcations ($\square$). As k is increased a periodic doubling route to chaos occurs. The chaotic attractor is shown in Figure 13.5.*

As a consequence there exist other attractors than the equilibrium point. Now we provide some qualitative and quantitative information on these attractors. Since a complete bifurcation analysis is beyond the scope of this book, we focus on the particularities, due to the delayed damping term in (13.24) (period doubling route to chaos) and the discontinuity in the right-hand side of (13.24) (superstable limit cycles).

Chaotic behavior. In Figure 13.4 we show a bifurcation diagram of (13.24), when k is the free parameter and $c = 1$ is fixed, computed with the help of DDE-BIFTOOL [74]. Recall that for small $k > 0$ the unique steady state solution is locally asymptotically stable and that stability is lost in a subcritical Hopf bifurcation as k is increased. In the Hopf bifurcation a branch of stable *periodic solutions* emanates. The latter become unstable after a period doubling bifurcation, where a new branch of stable, *period doubled* periodic solutions emanates. A sequence of period doubling bifurcations ultimately leads to chaos. In Figure 13.5 we plot the chaotic attractor for $k = 5.3$.

Chaotic behavior is inherent to TCP/IP traffic. The work of Veres and Boda [309], where chaotic behavior was detected and analyzed in simulations with the ns-2 simulator [79], showed that TCP itself can cause or contribute to chaotic behavior as a *deterministic* system (in previous works a large number of ON-OFF sources with *random* periods were rather seen as a source of chaos in TCP). The analysis above shows that already the simple

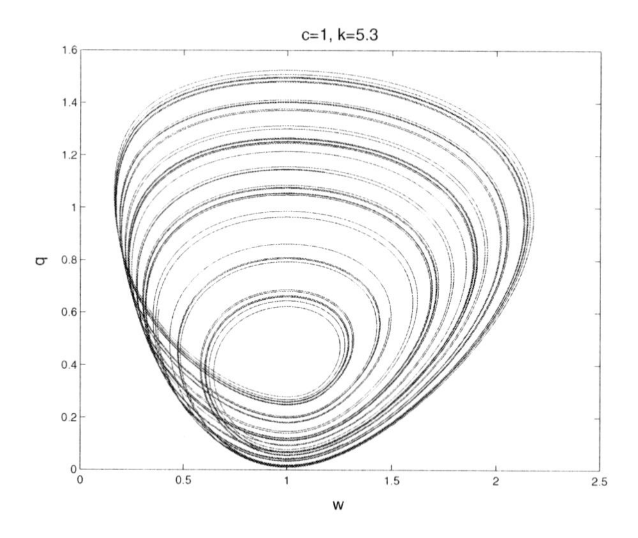

Figure 13.5. *Chaotic attractor of the system (13.24) for $c = 1$ and $k = 5.3$.*

second-order deterministic model (13.21)–(13.22) exhibits chaos, which thus supports this proposition.

The chaotic behavior in (13.24) is clearly caused by the nonlinear delayed damping term in the first equation, which is proportional to $-w(t)w(t-1)$ (notice that q is strictly larger than zero along the attractor shown in Figure 13.5, hence the discontinuity in the right-hand side of (13.24) does not contribute). Therefore, it is expected that the model proposed in [134],

$$\dot{x}(t) = k(w - x(t-\tau)p(x(t-\tau))),$$

which describes the dynamics of a collection of flows all using a single resource and sharing the same gain parameter k, may also exhibit chaotic behavior for particular choices of the function $p(.)$, which can again be interpreted as the fraction of packets indicating congestion. Finally, an analogous instability mechanism leading to chaos occurs in the *delayed logistic equation* [296].

Superstable limit-cycles. When one lets the size of the attractor grow, by changing the system's parameters, it may ultimately simplify to a nonsmooth limit-cycle, which contains a segment where $q \equiv 0$. When $q = 0$ for a sufficiently large time, this limit-cycle is *superstable*, meaning that the effect of perturbations around it disappears in a *finite* time. In Figure 13.6 we plot such a superstable limit-cycle. The discontinuity in the right-hand side of (13.24) creates a mechanism which resets the state to the same value each time before starting a new loop (more precisely to the segment $(w, q) = (c + \theta, 0)$, $\theta \in [-1, 0]$). Notice that such nonsmooth solutions have a physical explanation: when a buffer becomes empty, it remains so until the arrival rate of packages, which increases because the window size is increased, exceeds the maximal capacity of the link.

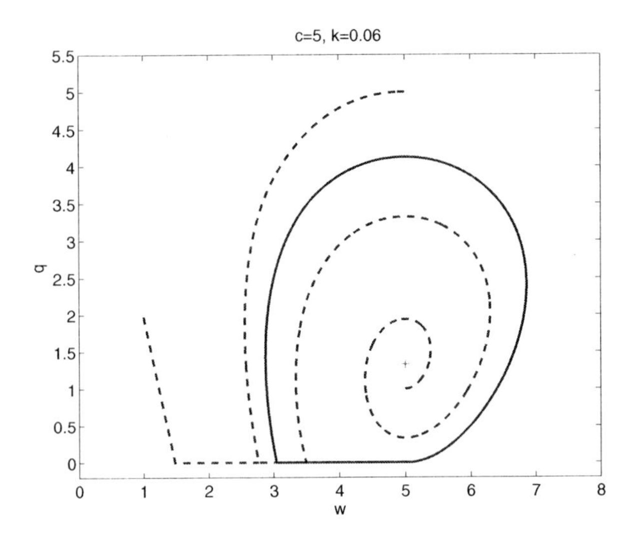

Figure 13.6. *Attractor of (13.24) for* $c = 5$ *and* $k = 0.06$ *(solid line), as well as a few trajectories (dashed lines) and the unstable steady state solution (+). Around the attractor perturbations disappear in a finite time.*

13.3 Notes and references

In this chapter, we presented two particular fluid approximation models encountered in congestion control analysis of high-performance networks. Without discussing the importance and particularities of such models, our interest was in pointing out some specific behaviors induced by the delay presence. Further discussions and a more complete list of references can be found Srikant's overview on existing models and methods for analyzing Internet congestion control algorithms [284]. Other models, some remarks, and an application of Lyapunov techniques for the analysis of some nonlinear fluid approximation models with delays can be found in [184, 234]. An application of the Smith predictor in a congestion control context is analyzed in [174].

The first model under consideration is due to Izmailov [124, 125], and describes the behavior of a single connection between a source controlled by an access regulator and a distant node. One of the main features of this model is the presence of two delays: the round-trip time and some control-time interval. The proposed analysis follows closely the approaches considered in [224, 226]. A different analysis can be found in [328], where the characterization of stability regions in the delay-parameter space was derived by using a dual-locus diagram (an extension of the well-known Nyquist diagram). Some connections with the output feedback stabilization problem in the presence of multiple delays, and some remarks concerning the geometry of the stability regions, have been added and complete the presentation. Further remarks in the case when the round-trip time is time varying can be found in [234] (using the approach based on integral quadratic constraints (IQC)).

Next, we focused on some nonlinear time-delay systems describing the behavior of congested routers in TCP/AQM networks. A stability and bifurcation analysis of a fluid flow–like model was performed. The stability region in the parameter space of the unique equilibrium point was completely characterized by combining analytical and numerical tools. A direct computation of periodic solutions and a continuation procedure revealed a period doubling route to chaos. The presence of a chaotic attractor in the low order, deterministic model supports the assumption that chaotic behavior is inherent to the TCP mechanism. The presentation of the results followed [189] closely.

Chapter 14

Smith predictor for stable systems: delay sensitivity analysis

14.1 Introduction

We consider a SISO stable system with a (discrete) input delay, described by the transfer function

$$H(\lambda) = H_0(\lambda)e^{-\lambda\tau}, \tag{14.1}$$

where the rational function $H_0(\lambda)$ is the transfer function of the system free of delay. One of the simplest methods to control such a system if H_0 is a stable transfer function was proposed by Smith in the 1960s [281].

The idea behind *Smith predictors* is to use a controller structure which takes the delay out of the control loop and allows a feedback design based on $H_0(\lambda)$ only. More precisely, in the classical Smith Predictor [281], a controller of the form

$$C'(\lambda) = \frac{C(\lambda)}{1 + C(\lambda)(H_0(\lambda) - H_0(\lambda)e^{-\lambda\tau})} \tag{14.2}$$

is used, which gives the following closed-loop transfer function from an external reference signal r to the plant output y:

$$H_{y,r}(\lambda) = \frac{C(\lambda)H_0(\lambda)}{1 + C(\lambda)H_0(\lambda)}e^{-\lambda\tau}.$$

Hence, $C(\lambda)$ can be designed based on $H_0(\lambda)$.

The advantage in using such an approach in controlling input delay systems lies in its simplicity. To the best of the authors' knowledge, three main problems of the Smith predictor schemes have been considered in the control literature during the last twenty years: (a) robustness (plant and/or delay) [81, 266, 327], (b) disturbance rejection characteristics [325], and (c) extension of the Smith principle idea to the case of integrative plants [4, 175, 237].

This chapter deals with the first problem. Therefore, we assume that the delay of the plant is given by $\tau + \delta$, with the real constant δ modeling the *delay uncertainty*, and that the controller design is based on the nominal delay value τ. Then the closed-loop transfer function becomes [228]

$$H^{\delta}_{y,r}(\lambda) = \frac{C(\lambda)H_0(\lambda)e^{-\lambda(\tau+\delta)}}{1 + C(\lambda)H_0(\lambda) + C(\lambda)H_0(\lambda)e^{-\lambda\tau}(e^{-\lambda\delta} - 1)}. \tag{14.3}$$

When the delay-free plant and the controller are factorized as

$$H_0(\lambda) = \frac{B_1(\lambda)}{A_1(\lambda)}, \quad C(\lambda) = \frac{B_2(\lambda)}{A_2(\lambda)},$$

we have

$$H^{\delta}_{y,r}(\lambda) = \frac{B(\lambda)e^{-\lambda(\tau+\delta)}}{A(\lambda) + B(\lambda)e^{-\lambda\tau}(e^{-\lambda\delta} - 1)}, \tag{14.4}$$

where $A(\lambda) = A_1(\lambda)A_2(\lambda) + B_1(\lambda)B_2(\lambda)$ and $B(\lambda) = B_1(\lambda)B_2(\lambda)$. Without any loss of generality, we can assume that A and B are coprime, that is, they do not have common zeros.

We are interested in investigating the robustness of stability of the Smith predictor w.r.t. an inaccurate modeling of the delay. This corresponds to the following problem.

Problem 14.1. *Find conditions on Δ such that*

$$A(\lambda) - B(\lambda)e^{-\lambda\tau} + B(\lambda)e^{-\lambda(\tau+\delta)} = 0, \tag{14.5}$$

where $A(\lambda)$, $B(\lambda)$ are polynomials with $\deg(A(\lambda)) \geq \deg(B(\lambda))$ and $A(\lambda)$ Hurwitz, has all its solutions in $\mathbb{C}^-$ when $|\delta| \leq \Delta$.

Remark 14.2. *The transfer function from a plant input disturbance $w(\lambda)$ to the plant output $y(\lambda)$ is given by*

$$\begin{aligned} H^{\delta}_{y,w} &= \frac{H_0(\lambda)\left(1+C(\lambda)H_0(\lambda)-C(\lambda)H_0(\lambda)e^{-\lambda\tau}\right)e^{-\lambda(\tau+\delta)}}{1+C(\lambda)H_0(\lambda)+C(\lambda)H_0(\lambda)e^{-\lambda\tau}(e^{-\lambda\delta}-1)} \\ &= \frac{B_1(\lambda)\left(A(\lambda)-B(\lambda)e^{-\lambda\tau}\right)}{A_1(\lambda)\left(A(\lambda)+B(\lambda)e^{-\lambda\tau}(e^{-\lambda\delta}-1)\right)}; \end{aligned} \tag{14.6}$$

hence, its poles consist of the zeros of (14.5) and the open-loop system poles, i.e., the zeros of $A_1(\lambda)$. Therefore, the Smith predictor can only be applied to stable open-loop systems. For modifications of the scheme, which are applicable to unstable open-loop systems, we refer to Section 15.5.2, the paper [194], and the references therein.

Without being exhaustive, let us cite some of the works in the control literature related to this problem. Some robust stability conditions (plant and delay uncertainty) using frequency-sweeping tests have been proposed for scalar systems in [327]. The corresponding criteria use the growth rate of the plant uncertainty toward high frequencies, and do not include any information on the delay uncertainty upper bound. This result has been extended to handle more general multivariable feedback systems in [81]. A different criterion, including information on the delay uncertainty upper bound, was proposed in [266], where the tuning procedure is based on a proper selection of the closed-loop bandwidth, but the procedure explicitly requires no uncertainty on the (delay-free) plant's parameters.

Notice that the Smith Predictor takes the delay out of the control loop because it involves a dynamic prediction of the output variable. As explained in [194, 248] and the next chapter, it is also possible to use a static prediction, where a predicted output is directly calculated from the current output and an integral over past inputs. It will be shown in the next chapter that the analysis of the delay sensitivity problem of such controllers reduces to a robustness problem of the form (14.5). Therefore, we will analyze the more general Problem 14.1 and only interpret the results in terms of Smith Predictors.

In some previous work [204], the robustness of Smith Predictors w.r.t. delay inaccuracy has been studied in detail and various stability/instability characterizations have been provided, under the condition that the transfer function $C(\lambda)H_0(\lambda)$ is strictly proper or, equivalently, that $\deg(A(\lambda)) > \deg(B(\lambda))$ in equation (14.5).

Here we *also* consider the case where $\deg(A(\lambda)) = \deg(B(\lambda))$, which includes Smith Predictors, where the transfer function $C(\lambda)H_0(\lambda)$ is proper but not strictly proper. As we will see, this can lead to a sensitivity of stability w.r.t. *infinitesimal* delay mismatches; i.e., infinitesimal delay mismatches δ may destroy stability and, hence, equation (14.5) may not be practically stable, the latter defined as in [247, Section 3].

Definition 14.3. *Equation (14.5) is practically stable w.r.t. delay mismatches δ, when there exists a $\bar{\delta} > 0$ such that the solutions of (14.5) are in the open LHP for all $|\delta| < \bar{\delta}$.*

Although the possible lack of practical stability has been observed by other authors in the context of Smith Predictors [248, 247], a complete characterization and interpretation of function of the system parameters will be given in this chapter.

The structure of the chapter is as follows. First we consider the sensitivity of the stability of (14.5) w.r.t. infinitesimal delay mismatches δ. Thereby, we explain the instability mechanism and derive necessary and sufficient conditions for practical stability, based on two different approaches. The first approach, inspired by [5, 181], consists of interpreting (14.5) as the characteristic equation of a neutral differential equation and then studying the sensitivity of the spectrum of the associated difference equation; the other approach consists of relating practical (in)stability of (14.5) with the analysis of infinitesimal delays in feedback loops, studied in [155]. In the second part of the chapter, we assume practical stability and—according to Problem 14.1—derive bounds on the maximal delay mismatch Δ. Then we comment on the geometry of the stability regions, using the results of Chapter 4, and illustrate the obtained results with a numerical example. Finally, we generalize the main results to the multivariable case.

14.2 Sensitivity of stability w.r.t. infinitesimal delay mismatches

We look more closely at the sensitivity w.r.t. small delay mismatches. First, we explain the instability mechanism, and then we present necessary and sufficient conditions for practical stability.

14.2.1 Instability mechanism

Assume that $\deg(A(\lambda)) = n$ and define

$$S = \lim_{|\lambda| \to \infty} \frac{B(\lambda)}{A(\lambda)}. \tag{14.7}$$

Then equation (14.5) can be written as

$$\lambda^n(1 + Se^{-\lambda(\tau+\delta)} - Se^{-\lambda\tau}) + q(\lambda) = 0, \tag{14.8}$$

where $\lim_{|\lambda|\to\infty,\ \Re(\lambda)\geq 0} q(\lambda)/\lambda^n = 0$. This is also the characteristic equation of the DDE

$$\frac{d^n}{dt^n}(x(t) + Sx(t-(\tau+\delta)) - Sx(t-\tau)) + q\left(\frac{d}{dt}\right)x = 0. \tag{14.9}$$

Note that for any $\delta \neq 0$, (14.9) is of *neutral* type when $S \neq 0$, while it is of *retarded* type for $\delta = 0$. In fact, a small perturbation $\delta \neq 0$ forms a noncompact perturbation of the time-integration operator (solution semigroup) associated with equation (14.9), which introduces an essential spectrum.

As shown in Chapter 1 and in [102, 5], precisely this essential spectrum may cause a sensitivity of the stability w.r.t. infinitesimal parameter changes. It corresponds to sequences of roots of (14.5) whose moduli tend to infinity, yet whose real parts have a finite limit. Since the essential spectrum of the solution semigroup of (14.9) coincides with the essential spectrum of the solution semigroup of the delay-difference equation

$$x(t) = -Sx(t-(\tau+\delta)) + Sx(t-\tau), \tag{14.10}$$

see [106], we are led to the study of the stability of (14.10) or, equivalently, of the behavior of the roots of its characteristic equation

$$H(\delta, \lambda) := 1 + Se^{-\lambda(\tau+\delta)} - Se^{-\lambda\tau} = 0. \tag{14.11}$$

We now illustrate that equation (14.10) and, as a consequence, equation (14.5) may be unstable for arbitrarily small values of δ and explain why this is not in conflict with the continuous dependence of the characteristic roots w.r.t. the system parameter δ. Therefore, we plot in Figure 14.1 the characteristic roots of equation (14.10) with $S = 0.6$ and $\delta = \tau/n$, for different integer values of n. Note the spectrum is periodic w.r.t. shifts in the imaginary parts, following from the fact that (14.11) can be written as a polynomial in $e^{-\lambda\tau/n}$. Since the spectrum is empty for the limit case $\delta = 0$, the continuous dependence of the individual characteristic roots w.r.t. δ implies that these have to move off to infinity as $\delta \to 0$. However, since the rightmost characteristic roots grow unbounded without leaving the RHP, instability is preserved for any $\delta \neq 0$.

In Figure 14.2 we show the corresponding (essential) spectrum of the operator $\mathcal{T}(1)$, which performs a time-integration of equation (14.10) over one time-unit. It consists of eigenvalues μ satisfying $\mu = \exp(\lambda)$, where λ is a characteristic root of (14.10). Note that a normalization in the delay is not possible any more. The fact that the essential spectra of the time-integration operators of the equations (14.9) and (14.10) coincide implies that the characteristic roots of (14.10) with a large imaginary part are approximate characteristic roots of (14.9).

14.2.2 Conditions for practical stability

In order to check the stability of (14.10) for small δ, we have to analyze the set

$$Z(\delta) = \{\Re(\lambda) : H(\delta, \lambda) = 0\},$$

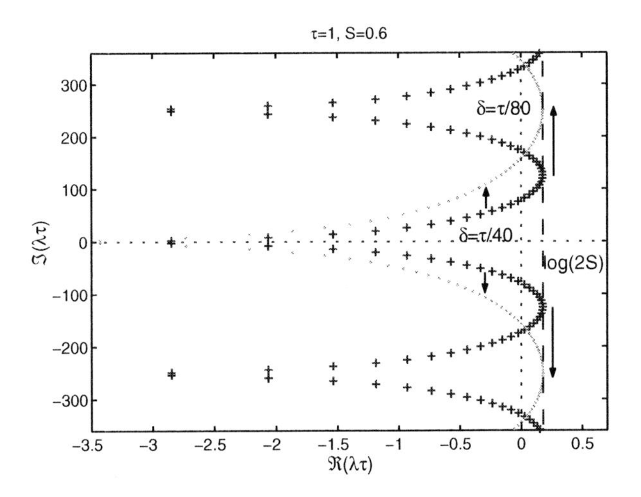

Figure 14.1. *Spectrum of the difference equation (14.10) for* $S = 0.6$ *and* $\delta = \tau/n$ *with* $n = 40$*, indicated with (+), and* $n = 80$*, indicated with (o).*

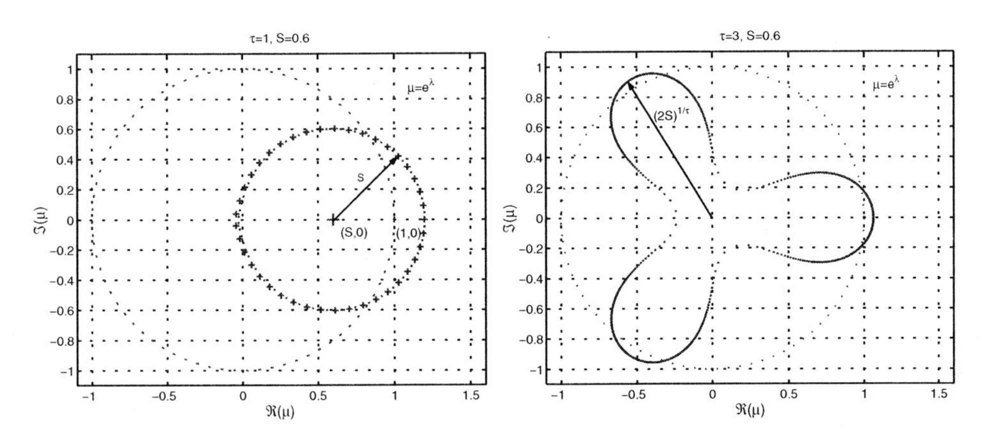

Figure 14.2. *Eigenvalues of the time-integration operator* $\mathcal{T}(1)$ *of the difference equation (14.10) for* $\delta = \tau/n$ *with* $S = -0.6$*,* $n = 40$ *(x), and* $n = 80$ *(o). One can show that for* $\tau = 1$*, the essential spectrum approaches a circle as* $\delta \to 0$ *(left) while for* $\tau = m \in \mathbb{N}$*, one recovers an "m-folded leaf" (right). For all values of* τ *the spectral radius of the essential spectrum is given by* $(2|S|)^{1/\tau}$.

whose properties are described in [5, 181] and now briefly reviewed. In the difference equation (14.10) there are in fact two delays, $\tau_1 = \tau + \delta$ and $\tau_2 = \tau$. When τ_1 and τ_2 are rationally independent (noncommensurate), the spectrum is quasi-periodic and $\bar{Z}(\delta)$ consists of an *interval*, characterized by $\bar{Z}(\delta) = \mathcal{I}(\delta)$, where

$$\begin{aligned}\mathcal{I}(\delta) = \{\alpha \in \mathbb{R} :\ &\exists\theta_1, \theta_2 \in [0,\ 2\pi) \text{ such that} \\ &1 + Se^{-\alpha(\tau+\delta)}e^{-i\theta_1} - Se^{-\alpha\tau}e^{-i\theta_2} = 0\big\}.\end{aligned} \tag{14.12}$$

This result follows from substituting $\lambda = \alpha + i\beta$ in the characteristic equation and the observation that with a suitable choice of β, $(\beta\tau_1,\ \beta\tau_2)\text{mod}2\pi$ is arbitrarily close to any given (θ_1, θ_2) by Kronecker's Theorem [110, Theorem 444].

When the delays τ_1 and τ_2 are rationally dependent (commensurate), the spectrum of (14.10) is periodic. The set $Z(\delta)$ consists of a finite number of *points*, as illustrated with Figure 14.1, and satisfies $Z(\delta) \subset \mathcal{I}(\delta)$. However, when considering a sequence of rationally dependent delays $(\tau + \delta_n, \tau)$ converging to rationally independent delays $(\tau + \delta, \tau)$, the interval $\mathcal{I}(\delta)$, described by (14.12), is arbitrarily well approximated by this set of points as $n \to \infty$, due to the fact that the map $\delta \in \mathbb{R} \to Z(\delta)$ is lower semicontinuous in the Hausdorff metric.

From the above analysis it follows that (practical)[18] stability of the difference equation (14.10) is determined by the maximal value $\alpha_m(\delta)$ of (14.12), defined by

$$1 - |S|e^{-\alpha_m(\delta)(\tau+\delta)} - |S|e^{-\alpha_m(\delta)\tau} = 0, \tag{14.13}$$

which converges[19] to

$$\alpha_M = \frac{\log 2|S|}{\tau} \tag{14.14}$$

as $\delta \to 0$. Taking into account the relation between the spectrum of (14.10) and the roots of (14.5), this leads to the following result.

Proposition 14.4. *Consider equation (14.5) and let S be defined by (14.7). If $|S| < 1/2$, then the asymptotic stability is preserved for small values of δ. If $|S| > 1/2$, then the equation is not practically stable.*

Proof. Instability of the difference equation (14.10), i.e., $\alpha_M > 0$ or $|S| > 1/2$, implies instability of (14.9)–(14.5), as follows from the arguments spelled out before. Hence, we only have to prove that the same holds for asymptotic stability (i.e., $\alpha_M < 0$ or $|S| < 1/2$), when δ is sufficiently small. Here, note that the difference equation only provides information on the essential spectrum of the solution semigroup associated with (14.9), while the latter also has a point spectrum.

Take a number ϵ satisfying $\alpha_M < \epsilon < 0$ and such that $A(\lambda)$ has no roots in the half-plane $\mathbb{C}_\epsilon := \{\lambda \in \mathbb{C} : \Re(\lambda) > \epsilon\}$. Equation (14.8) can be rewritten as

$$1 + Se^{-\lambda(\tau+\delta)} - Se^{-\lambda\tau} = -q(\lambda)/\lambda^n. \tag{14.15}$$

For small values of δ, the modulus of the left-hand side is uniformly bounded below over $\mathbb{C}_\epsilon$ by a strictly positive constant. Since the right-hand size of (14.15) tends to zero for large $|\lambda|$, this implies the existence of numbers $M, \bar{\delta} > 0$ such that all the solutions of (14.15)–(14.5) in $\mathbb{C}_\epsilon$ satisfy $|\lambda| \leq M$ when $|\delta| \leq \bar{\delta}$. Define the compact set $\mathcal{S} := \{\lambda \in \bar{\mathbb{C}}_\epsilon : |\lambda| \leq\}$. Since on $\mathcal{S}$ the function $A(\lambda) - B(\lambda)e^{-\lambda\tau} + B(\lambda)e^{-\lambda(\tau+\delta)}$ uniformly converges to the function $A(\lambda)$ as $\delta \to 0$, we can apply Corollary A.2, which states that the two functions have the

[18] Here we mean that stability should be robust for small deviations around the nominal value δ. Note that rationally dependent delays $(\tau + \delta, \tau)$ can always be perturbed to rationally independent delays by an arbitrarily small perturbation of δ.

[19] Actually α_m is a continuous function of $\delta \geq -\tau$.

same number of zeros in $\mathcal{S}$ when δ is sufficiently small. Therefore, (14.5) has no solutions in $\mathcal{S}$ for small δ and, as a consequence, in $\mathbb{C}_\epsilon$, and is asymptotically stable. □

Remark 14.5. *The condition $|S| < 1/2$ is a necessary and sufficient condition for the strong stability of the difference equation (14.10); see Chapter 1.*

Remark 14.6. *For $|S| = 1/2$ we can only conclude the existence of sequences of roots of (14.5) which approach the imaginary axis. However, the way of approaching the imaginary axis (from the left/from the right/oscillatory) may also depend on other system parameters than those of the difference equation (14.10), which only describes the limit case.*

Remark 14.7. *Stability or instability for small δ does not depend on the value of the nominal delay τ. However, expression (14.14) reveals the softening effect of increasing the delay on the (in)stability of difference equations, observed in [102]. When $\tau \to \infty$, we have $\alpha_M \to 0$ and, hence, unstable equations become less unstable and vice versa.*

Remark 14.8. *In the case where (14.5) represents the (closed-loop) characteristic equation of the Smith Predictor (14.3), we have,*

$$S = \lim_{|\lambda|\to\infty} \frac{B(\lambda)}{A(\lambda)} = \lim_{|\lambda|\to\infty} \frac{C(\lambda)H_0(\lambda)}{1 + C(\lambda)H_0(\lambda)}.$$

The corresponding instability result of Proposition 14.4 has also been derived in [247], based on a Nyquist stability criterion.

We now outline an alternative frequency-domain approach for the analysis of our practical stability problem, based on the results of Logemann et al. [156]. Note that the zeros of (14.5) are the poles of the transfer function

$$\frac{B(\lambda)e^{-\lambda\tau}}{A(\lambda) - B(\lambda)e^{-\lambda\tau} + B(\lambda)e^{-\lambda(\tau+\delta)}} = \frac{M(\lambda)}{1 + M(\lambda)e^{-\lambda\delta}}, \tag{14.16}$$

where the transfer function

$$M(\lambda) = \frac{B(\lambda)e^{-\lambda\tau}}{A(\lambda) - B(\lambda)e^{-\lambda\tau}}$$

is regular. Equation (14.16) can be interpreted as a system represented by $M(\lambda)$ and stabilized with unity feedback, in the presence of a small feedback delay δ. Precisely the robustness of stability of such feedback systems w.r.t. the small feedback delay has been studied in [156]. Denote by $\mathcal{B}_M$ the set of poles of $M(\lambda)$ in $\mathbb{C}_+$. Then a direct application of [156, Theorem 1.1] yields the following.

Proposition 14.9. *Consider the system (14.16) and define*

$$\gamma = \limsup_{\substack{|\lambda| \to \infty \\ \lambda \in \mathbb{C}_+ \setminus \mathcal{B}_M}} |M(\lambda)|.$$

If $\gamma < 1$, then the asymptotic stability is preserved for small feedback delays $\delta > 0$.
If $\gamma > 1$, then the system is not practically stable.

Note that only positive perturbations $\delta > 0$ are considered in the paper [155] (framework of feedback delays), while in our case δ may be negative (recall that δ is a delay *mismatch*). However, based on the continuity properties of characteristic roots one can easily show that the condition $\gamma < 1$ also implies robustness of stability for small $\delta < 0$. Obviously the conditions of Propositions 14.9 and 14.4 are equivalent, since

$$\gamma = \begin{cases} \frac{|S|}{1-|S|}, & |S| < 1, \\ +\infty, & |S| \geq 1. \end{cases}$$

14.3 Stability analysis and critical delay mismatches

Consider (14.5) and assume that $|S| < 1/2$, with S defined by (14.7). By Proposition 14.4, the asymptotic stability is then preserved for a small delay mismatch δ. We now characterize its maximal deviation Δ, i.e., we consider Problem 14.1.

From (14.13) it follows that the difference equation (14.10) remains asymptotically stable for all values of $\delta \geq -\tau$ (although α_m depends on δ, its sign cannot change). Furthermore, from [106, Theorem XII.10.4] it can be deduced that for any $\epsilon > 0$, there are only a finite number of solutions of (14.8) with $\Re(\lambda) > \alpha_m + \epsilon$. Therefore, when increasing the delay mismatch, the essential spectrum of the solution semigroup of (14.9) remains stable and the transition from stability to instability is always of a finite-dimension nature, as in the case of equations of retarded type.

In order to calculate stability regions in δ, we look for solutions of (14.5) on the imaginary axis. Substituting $\lambda = j\omega$ yields

$$A(j\omega) - B(j\omega)e^{-j\omega\tau} = -B(j\omega)e^{-j\omega\tau}e^{-j\omega\delta}. \tag{14.17}$$

Then the critical delay mismatches and the corresponding characteristic roots on the imaginary axis can be computed as follows: the "crossing" frequencies ω_k are the positive solutions of the equation

$$|\, A(j\omega) - B(j\omega)e^{-j\omega\tau} \,| = |\, B(j\omega) \,|, \tag{14.18}$$

and for each ω_k, one can calculate the corresponding delay mismatches $\delta_{k,l}$ using the phase information of equation (14.17):

$$\delta_{k,l} = \frac{\angle\left(\frac{-B(j\omega_k)e^{-j\omega_k\tau}}{A(j\omega_k)-B(j\omega_k)e^{-j\omega_k\tau}}\right) + 2\pi l}{\omega_k}, \quad l \in \mathbb{Z}. \tag{14.19}$$

Remark 14.10. *When rewriting equation (14.18) as*

$$\left| 1 - \frac{B(j\omega)}{A(j\omega)}e^{-j\omega\tau} \right| = \left| \frac{B(j\omega)}{A(j\omega)} \right|,$$

it becomes clear that under the assumed practical stability condition i.e.,

$$|S| = |\lim_{|\lambda|\to\infty} B(\lambda)/A(\lambda)| < 1/2,$$

it only has a finite *number of solutions, which expresses the finite-dimensional nature of the transitions to instability, explained before. On the contrary, when the practical stability condition is not fulfilled (i.e., $|S| > 1/2$), equation (14.18) has an infinite sequence of solutions $\{\omega_k\}_{k\geq 1}$ with $|\omega_k| \to \infty$ and corresponding $|\delta_{k,l}| \to 0$ as $n \to \infty$, which precisely corresponds to the (destabilizing) behavior of the essential spectrum of the associated solution semigroup for small δ, illustrated in Section 14.2.1.*

Define the sets $\Lambda_{+,0}$ and $\Lambda_{-,0}$ as follows [228]: if there does not exist any *positive* $\delta_{k,l}$, then $\Lambda_{+,0} = \{+\infty\}$, and

$$\Lambda_{+,0} = \left\{\delta_{k,l} > 0 \quad : \quad \delta_{k,l} \text{ given by (14.19)}\right\} \tag{14.20}$$

elsewhere. Similarly, if there does not exist any *negative* $\delta_{k,l}$, then $\Lambda_{-,0} = \{-\tau\}$, and

$$\Lambda_{-,0} = \left\{\delta_{k,l} < 0 \quad : \quad \delta_{k,l} \text{ given by (14.19)}\right\} \tag{14.21}$$

elsewhere. The smallest destabilizing delay mismatches, $\Delta_1 < 0$ and $\Delta_2 > 0$, are given by

$$\begin{cases} \Delta_1 = \max \Lambda_{-,0}, \\ \Delta_2 = \min \Lambda_{+,0}, \end{cases} \tag{14.22}$$

and the derived condition is *necessary and sufficient*. The results can be summarized as follows [228].

Proposition 14.11. *Define the real function*

$$\mathcal{F}(\omega) = \mid A(j\omega) + B(j\omega)e^{-j\omega\tau} \mid^2 - \mid B(j\omega) \mid^2 .$$

The stability of the closed-loop system is guaranteed for any inaccurate modeling delay δ, $\mid \delta \mid \leq \Delta$, if

(i) $\mathcal{F}(\omega)$ has no zeros. In such case, the stability property is of delay-independent *type, i.e., it may hold for any $\delta > -\tau$.*

(ii) $\mathcal{F}(\omega)$ has at least one zero. In such case, the stability property is of delay-dependent *type and it holds for any $\delta \in (\Delta_1, \Delta_2)$ $(\Delta_1 < 0 < \Delta_2)$, where Δ_1, Δ_2 are given by*

$$\begin{cases} \Delta_1 = \max \Lambda_{-,0}, \\ \Delta_2 = \min \Lambda_{+,0}, \end{cases}$$

where the sets $\Lambda_{\pm,0}$ are defined in (14.20) and (14.21).

In this case, $\Delta < \min\{\Delta_1, \Delta_2\}$, and is always a finite value.

Remark 14.12. *If the following inequality holds,*

$$2\,|B(j\omega)| < |A(j\omega)| \; \forall \omega \in \mathbb{R},$$

then the closed-loop system stability is guaranteed for all *inaccurate modeling of the delay. The same result can be found in [266], but the stability argument is different.*

14.4 Geometry of stability regions

In Chapter 4 we presented to some extent the geometry of the stability regions in the delay-parameter space of a linear system including two delays. In this section we adapt the corresponding approach and ideas to the (delay, delay uncertainty)-parameter space under consideration.

14.4.1 Identification of crossing points

Define the following auxiliary characteristic function,

$$p(\lambda;\ \tau_1, \tau_2) := A(\lambda) + B(\lambda)e^{-\lambda\tau_1} - B(\lambda)e^{-\lambda\tau_2},$$

which is nothing else than the characteristic function of the closed-loop system of the Smith predictor when taking $\tau_1 = \tau$, and $\tau_2 = \tau + \delta$.

Let $\mathcal{T}$ denote the set of all points of $(\tau_1, \tau_2) \in \mathbb{R}_+^2$ such that $p(\lambda;\ \tau_1, \tau_2)$ has at least one zero on the imaginary axis. Any $(\tau_1, \tau_2) \in \mathcal{T}$ is known as a *crossing point*. The set $\mathcal{T}$, which is the collection of all crossing points, is called the set of *stability crossing curves*. Let $\mathcal{T}_\omega$ denote the set of all $(\tau_1, \tau_2) \in \mathbb{R}_+^2$ such that the auxiliary characteristic function $p(\lambda;\ \tau_1, \tau_2)$ has at least one zero $\lambda = j\omega$ on the imaginary axis. Let Ω be the set of all $\omega > 0$ for which there exists a pair (τ_1, τ_2) such that $p(j\omega;\ \tau_1, \tau_2) = 0$. We will refer to Ω as the *frequency crossing set*. Obviously,

$$\mathcal{T} = \{\mathcal{T}_\omega :\ \omega \in \Omega\}. \tag{14.23}$$

In what follows we consider the *nondegenerate* case satisfying the following assumption:

$$A(j\omega)B(j\omega) \neq 0 \qquad \forall\, \omega \in \Omega. \tag{14.24}$$

Introduce now

$$h(\lambda) := \frac{B(\lambda)}{A(\lambda)}, \tag{14.25}$$

which corresponds to the transfer function from the external reference signal r to the output y for the system free of delay: $C(\lambda)H_0(\lambda)/(1 + C(\lambda)H_0(\lambda))$. Next, introduce

$$D(\lambda;\ \tau_1, \tau_2) := 1 + h(\lambda)e^{-\lambda\tau_1} - h(\lambda)e^{-\lambda\tau_2}. \tag{14.26}$$

For given τ_1 and τ_2, and as long as (14.24) is satisfied, $p(\lambda;\ \tau_1, \tau_2)$ and $D(\lambda;\ \tau_1, \tau_2)$ share all the zeros in a neighborhood of the imaginary axis. As in Chapter 4, we can interpret the three terms in $D(\lambda;\ \tau_1, \tau_2)$ as three vectors in the complex plane, with the magnitudes 1, $|h(\lambda)|$, and $|h(\lambda)|$, respectively. So when we adjust the values of τ_1 and τ_2 we adjust, in fact, the directions of the vectors represented by the second and the third term. Equation (14.26) simply means that if we put the first two vectors head to tail, then we get the third vector. In other words, these vectors form an *isosceles triangle*; see Figure 14.3. These remarks allow us to conclude with the following proposition, whose proof is omitted.

Proposition 14.13. *For some* $(\tau_1, \tau_2) \in \mathbb{R}_+^2$, $D(\lambda;\ \tau_1, \tau_2)$ *has a zero* $\lambda = j\omega$, $\omega \neq 0$, *if and only if*

$$|h(j\omega)| \geq \frac{1}{2}. \tag{14.27}$$

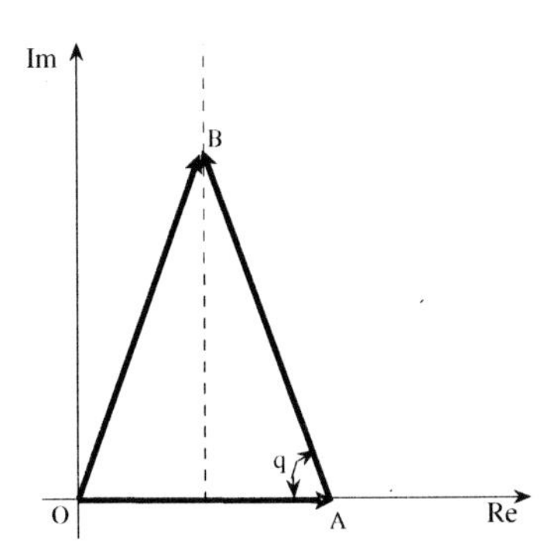

Figure 14.3. *Triangle formed by 1, $h(\lambda)e^{-j\omega\tau_1}$, and $h(\lambda)e^{-j\omega\tau_2}$.*

As in the case of general quasipolynomials including two distinct delays, discussed in Chapter 4, the crossing set Ω consists of a *finite number of intervals of finite length.* In what follows we denote these intervals as Ω_1, Ω_2, ..., Ω_N and without loss of generality we may assume that the intervals are ordered such that $\omega_1 \in \Omega_{k_1}$, $\omega_2 \in \Omega_{k_2}$, $k_1 < k_2$ imply $\omega_1 < \omega_2$. Next, if (14.27) is satisfied for $\omega = 0$ and sufficiently small positive value of ω, then we take 0 as the left end of Ω_1: $\Omega_1 = (0, \omega_1^r]$. In general, an end point of an interval Ω_k must be in one of the following situations:

Type 1. It satisfies the equation $|h(j\omega)| = \frac{1}{2}$.

Type 2. It equals 0.

We say that an interval is of type 11 if both ends are of type 1, and Ω_1 is of type 21 if its left end is 0. Therefore, the crossing set Ω consists of a finite number of intervals of type 11, with the possibility of the first interval Ω_1 of type 21. An application of the results of Chapter 4 then leads to the following.

Proposition 14.14. *The set $\mathcal{T}^k$ corresponding to the crossing set Ω_k consists of a series of curves belonging to one of the following categories:*

(i) *A series of closed curves (Ω_k is of type 11).*

(ii) *A series of open-ended curves with both ends approaching ∞ (Ω_k is of type 21).*

For a discussion on the degenerate cases, where (14.24) is not satisfied, we refer to [208].

14.4.2 Stability crossing curves: smoothness and crossing directions

We briefly discuss some qualitative aspects regarding the crossing curves and the corresponding characteristic roots. The results on smoothness and the direction of crossing are derived straightforwardly from the general theory presented in Chapter 4.

Tangent and smoothness. A direct application of the implicit function theorem allows us to conclude with the following result.

Proposition 14.15. *The curve $\mathcal{T}^k$ is smooth everywhere except possibly at the points corresponding to a multiple root of the auxiliary characteristic function $p(\lambda;\ \tau_1, \tau_2)$.*

Direction of crossing. We discuss the direction in which the characteristic roots of p cross the imaginary axis as (τ_1, τ_2) deviates from the curve $\mathcal{T}^k$. We call the direction of the curve that corresponds to increasing ω the *positive direction*. The region on the left-hand side as we head in the positive direction of the curve is denoted as *the region on the left*. For a given $\omega \in \Omega_k$, let

$$R_l = \Re\left(\frac{1}{\lambda}\frac{\partial D(\lambda;\ \tau_1, \tau_2)}{\partial \tau_l}\right)_{\lambda=j\omega} = (-1)^{l-1}\Re\left(h(j\omega)e^{-j\omega\tau_l}\right),$$

$$I_l = \Im\left(\frac{1}{\lambda}\frac{\partial D(\lambda;\ \tau_1, \tau_2)}{\partial \tau_l}\right)_{\lambda=j\omega} = (-1)^{l-1}\Im\left(h(j\omega)e^{-j\omega\tau_l}\right),\ l = 1, 2.$$

We have the following result.

Proposition 14.16. *Let $\omega \in \Omega_k$ and $(\tau_1, \tau_2) \in \mathcal{T}^k$ such that $j\omega$ is a zero of p with multiplicity one. As (τ_1, τ_2) crosses $\mathcal{T}$ from the region on the right to the region on the left, a pair of zeros of p crosses the imaginary axis to the right, through $\lambda = \pm j\omega$ if $R_2 I_1 - R_1 I_2 > 0$. The crossing is to the left if the inequality is reversed.*

14.5 Illustrative example

We investigate the delay sensitivity of the Smith Predictor (14.2), applied to the system (14.1), where

$$H_0(\lambda) = \frac{1}{\lambda + a},\quad C(\lambda) = k_1(k_2\lambda + 1). \tag{14.28}$$

This corresponds to the analysis of Problem 14.1 with

$$A(\lambda) = (k_1k_2 + 1)\lambda + (a + k_1),\quad B(\lambda) = k_1(k_2\lambda + 1).$$

We assume $a > 0$ and $(a + k_1)/(k_1k_2 + 1) > 0$, which guarantees internal stability of the closed-loop system. In case of a delay mismatch δ in the controller design, the solution semigroup associated with the closed-loop system has an essential spectrum when $k_1k_2 \neq 0$ and, thus, stability may be sensitive to an infinitesimal delay mismatch. According to Proposition 14.4, the practical stability condition is given by

$$|S| = \left|\frac{k_1k_2}{1 + k_1k_2}\right| < \frac{1}{2},$$

which implies $-1/3 < k_1k_2 < 1$.

Now we compute the stability region in δ as a function of the nominal delay τ when

$$a = 1,\ k_1 = 2,\ k_2 = 1/4. \tag{14.29}$$

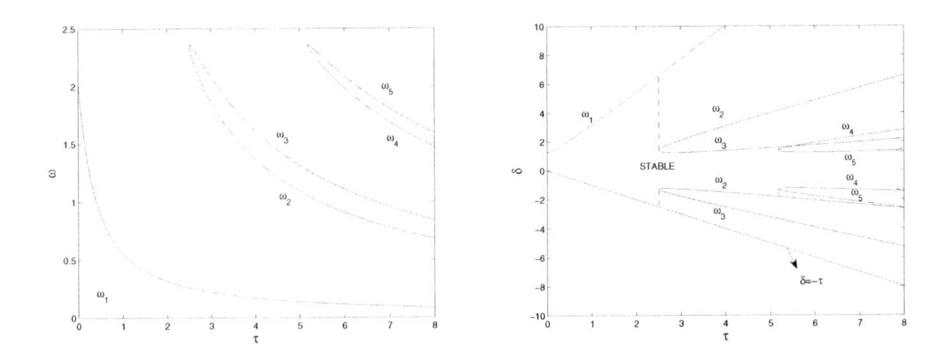

Figure 14.4. *Solutions of equation (14.18) when it represents the characteristic equation of the Smith Predictor scheme, applied to the system (14.28)–(14.29) (left), and corresponding values of the critical delay mismatches δ (right).*

Then the system is practically stable ($S = 1/3$), and for $\delta = 0$, there is one closed-loop characteristic root at $\lambda = -2$. In Figure 14.4 (left) we plot the (positive) solutions of equation (14.18) as a function of τ. As $\tau \to \infty$, the number of solutions tends to infinity.[20] Further, note that when a tuple (ω, τ) satisfies (14.18), this also holds for $(\omega, \tau + m2\pi/\omega)$, $m \in \mathbb{Z}$. In Figure 14.4 (right), we plot for each solution (i.e., possible crossing frequency) ω_k the corresponding smallest (positive and negative) values of the delay mismatch δ, such that the characteristic equation (14.5) has roots at $\pm j\omega_k$. This way the stability region in δ is completely characterized. Notice that the quantities Δ_1 and Δ_2, defined in (14.22), are *no* continuous functions of the nominal delay τ. Furthermore, the maximal allowable delay mismatch Δ is a uniformly bounded function of τ, which implies that the maximal *relative* delay mismatch tends to zero as $\tau \to \infty$.

The geometry of the stability crossing curves presented in the previous sections allows deriving the same condition as above. More precisely, by using the isosceles triangle we obtain that the frequency crossing set Ω is defined by *only one* interval,

$$\Omega = (0, 2.37],$$

which is of type 21. Correspondingly, the set $\mathcal{T}$ in the delay-parameter set (τ_1, τ_2) consists of a series of open-ended curves with both ends approaching infinity, a fact confirmed in Figure 14.4 due to the linear (simple) transformation between (τ, δ) and (τ_1, τ_2). For the sake of brevity, the corresponding figure in $O\tau_1\tau_2$ is omitted.

14.6 Multivariable case

In the multivariable case we have to analyze a characteristic equation of the form

$$\det\{D(\lambda) - N(\lambda)e^{-\tau\lambda}(1 - e^{-\delta\lambda})\} = 0, \tag{14.30}$$

[20] This fact is of importance since it proves that a finite-dimensional characterization of the characteristic roots on the imaginary axis, by, e.g., matrix pencils [228], is not possible.

where $D(\lambda)$, $N(\lambda) \in \mathbb{C}^{n\times n}$ are coprime polynomial matrices [129] such that the Hurwitz polynomial matrix $D(\lambda)$ is column reduced and such that the column degrees of $N(\lambda)$ are less than or equal to the corresponding column degrees of the polynomial matrix $D(\lambda)$; see [204].

14.6.1 Practical stability condition

We provide conditions on the matrix polynomials $D(\lambda)$ and $N(\lambda)$ for the (non)robustness of stability of equation (14.30) w.r.t. infinitesimal delay mismatches δ. Our approach is again based on the interpretation of (14.30) as the characteristic equation of a neutral equation.

Define $d = \deg(D(\lambda))$ and denote by $\{\gamma_i\}_{i=1}^{n}$ the column degrees of $D(\lambda)$. Let

$$P(\lambda) = \begin{bmatrix} \lambda^{d-\gamma_1} & & & \\ & \lambda^{d-\gamma_2} & & \\ & & \ddots & \\ & & & \lambda^{d-\gamma_n} \end{bmatrix}, \tag{14.31}$$

and define the matrices D_0 and N_0 as follows:

$$D_0 = \lim_{|\lambda|\to\infty} \frac{D(\lambda)P(\lambda)}{\lambda^d}, \quad N_0 = \lim_{|\lambda|\to\infty} \frac{N(\lambda)P(\lambda)}{\lambda^d}. \tag{14.32}$$

With $r_\sigma(\cdot)$ denoting the spectral radius, Proposition 14.4 can be generalized to the following.

Proposition 14.17. *Consider the characteristic equation (14.30). Let* $S = r_\sigma(N_0 D_0^{-1})$, *where* N_0 *and* D_0 *are defined in (14.32).*
If $|S| < 1/2$, *then the asymptotic stability of equation (14.30) for* $\delta = 0$ *is preserved for small values of* δ.
If $|S| > 1/2$, *then equation (14.30) is not practically stable.*

Proof. Instead of equation (14.30) we may analyze the spectrum of

$$\det\{\bar{D}(\lambda) - \bar{N}(\lambda)e^{-\tau\lambda}(1 - e^{-\delta\lambda})\} = 0, \tag{14.33}$$

where $\bar{D}(\lambda) = D(\lambda)P(\lambda)$ and $\bar{D}(\lambda) = D(\lambda)P(\lambda)$ (notice that this transformation only introduces additional characteristic roots at zero). Since $D(\lambda)$ is column reduced, $\bar{D}(\lambda)$ is monic. Therefore, (14.33) can be written as

$$\det\{\lambda^d[I - N_0 D_0^{-1} e^{-\tau\lambda}(1 - e^{-\delta\lambda})] + Q(\lambda)\} = 0, \tag{14.34}$$

where $\lim_{|\lambda|\to\infty,\ \Re(\lambda)\geq 0} |Q(\lambda)|/\lambda^d = 0$, D_0 is the leading coefficient matrix of $\bar{D}(\lambda)$ (equivalently the leading column coefficient matrix of $D(\lambda)$), and N_0 is the coefficient matrix of $\bar{D}(\lambda)$ corresponding to the degree d; i.e., these quantities satisfy (14.32).

Since (14.34) can be interpreted as the characteristic equation of a neutral functional differential equation (NFDE), the sensitivity of its spectrum w.r.t. infinitesimal parameter changes is determined by the strong stability of the corresponding difference equation

$$x(t) + N_0 D_0^{-1} x(t - (\tau + \delta)) - N_0 D_0^{-1} x(t - \tau) = 0. \tag{14.35}$$

Recall from Chapter 1 that (14.35) is strongly stable if and only if

$$\sup\left\{r_\sigma(N_0D_0^{-1}e^{i\theta_1} - N_0D_0^{-1}e^{i\theta_2}) : \theta_i \in [0,\ 2\pi],\ i = 1, 2\right\} < 1,$$

which is equivalent to

$$r_\sigma(N_0D_0^{-1}) < 1/2.$$

The statement of the theorem follows. □

Remark 14.18. *When the column degrees of $N(\lambda)$ are strictly smaller than the corresponding column degrees of $D(\lambda)$ or, equivalently, when the transfer matrix $N(\lambda)D(\lambda)^{-1}$ is strictly proper, we have $S = 0$ and thus practical stability, as expected from the high-frequency instability mechanism, outlined in Subsection 14.2.1.*

14.6.2 Stability domain

Under the assumption of *practical stability*, we characterize the stability region in the parameter δ. Substituting $\lambda = j\omega$ in (14.30) and some algebraic manipulations yield

$$\det\left(I - N(j\omega)D(j\omega)^{-1}e^{-j\omega\tau}(1 - e^{-j\omega\delta})\right) = 0. \tag{14.36}$$

Following from the relation

$$(1 - e^{-j\theta})^{-1} = \frac{1}{2} - \frac{j}{2\tan(\theta/2)} \qquad \forall\theta \in (-\pi, \pi),$$

the critical delay mismatches and the corresponding characteristic roots on the imaginary axis can be computed in the following way: the "crossing frequencies" ω_k are the strictly positive solutions of one of the equations

$$\Re\left(\lambda_i\left(N(j\omega)D(j\omega)^{-1}e^{-j\omega\tau}\right)\right) = \frac{1}{2},\ i = 1, \ldots, n, \tag{14.37}$$

with $\lambda_i(\cdot)$ denoting the ith eigenvalue. Next, for each ω_k, one can calculate the corresponding delay mismatches $\delta_{k,l}$ as follows:

$$\delta_{k,l} = \frac{-2\,\mathrm{atan}\left(\left(2\Im\left(\lambda_{\bar{k}}(N(j\omega_k)D(j\omega_k)^{-1}e^{-j\omega_k\tau})\right)\right)^{-1}\right) + 2\pi l}{\omega_k}, \quad l \in \mathbb{Z}, \tag{14.38}$$

where $\bar{k} \in \{1, \ldots, n\}$ is such that

$$\Re\left(\lambda_{\bar{k}}\left(N(j\omega_k)D(j\omega_k)^{-1}e^{-j\omega_k\tau}\right)\right) = \frac{1}{2}.$$

Analogously to the scalar case we define the sets $\Lambda_{+,0}$ and $\Lambda_{-,0}$ as follows: if there does not exist any *positive* $\delta_{k,l}$, then $\Lambda_{+,0} = \{+\infty\}$, and

$$\Lambda_{+,0} = \left\{\delta_{k,l} > 0 \quad : \quad \delta_{k,l} \text{ given by (14.38)}\right\} \tag{14.39}$$

elsewhere. Similarly, if there does not exist any *negative* $\delta_{k,l}$, then $\Lambda_{-,0} = \{-\tau\}$, and

$$\Lambda_{-,0} = \left\{\delta_{k,l} < 0 \quad : \quad \delta_{k,l} \text{ given by (14.38)}\right\} \tag{14.40}$$

elsewhere. The smallest destabilizing delay mismatches, $\Delta_1 < 0$ and $\Delta_2 > 0$, are given by

$$\begin{cases} \Delta_1 &= \max \Lambda_{-,0}, \\ \Delta_2 &= \min \Lambda_{+,0}. \end{cases} \tag{14.41}$$

The results can be summarized as follows.

Proposition 14.19. *Define the real functions*

$$\mathcal{F}_i(\omega) = \Re(\lambda_i(N(j\omega)D(j\omega)^{-1}e^{-j\omega\tau})) - 1/2, \quad i = 1, \ldots, n.$$

The stability of the closed-loop system is guaranteed for any inaccurate modeling delay δ, $|\ \delta\ | \leq \Delta$, *if*

(i) All functions $\mathcal{F}_i(\omega)$, $i = 1, \ldots, n$, *have no zeros. In such case, the stability property is of* delay-independent *type, i.e., it holds for any* $\delta > -\tau$.

(ii) There exists a number $i \in \{1, \ldots, n\}$ *such that* $\mathcal{F}_i(\omega)$ *has at least one zero. In such case, the stability property is of* delay-dependent *type and it holds for any* $\delta \in (\Delta_1, \Delta_2)$ $(\Delta_1 < 0 < \Delta_2)$, *where* Δ_1, Δ_2 *are given by:*

$$\begin{cases} \Delta_1 &= \max \Lambda_{-,0}, \\ \Delta_2 &= \min \Lambda_{+,0}, \end{cases}$$

where the sets $\Lambda_{\pm,0}$ *are defined in (14.39) and (14.40).*

In this case, $\Delta < \min\{\Delta_1, \Delta_2\}$, *and is always a finite value.*

Remark 14.20. *The above analysis is based on isolating the factor* $(1 - e^{-\lambda\delta})$ *in the characteristic equation, whereas the analysis in Section 14.3 is based on isolating* $e^{-\lambda\delta}$, *which is not beneficial in the multivariable case. In the scalar case the two procedures are identical, since (14.37), respectively (14.38), become equivalent to (14.18), respectively (14.19), when taking* $D(\lambda) = A(\lambda)$ *and* $N(\lambda) = B(\lambda)$.

To conclude this section we provide some easy-to-check sufficient stability conditions.

Proposition 14.21. *If for all* $\omega \geq 0$

$$\left\| N(j\omega)D(j\omega)^{-1} \right\|_2 < \frac{1}{2},$$

then the stability of (14.30) is guaranteed for all $\delta \geq -\tau$.

Proof. Since $|e^{-j\omega\tau}(e^{-j\omega\delta} - 1)| \leq 2$ for all ω and δ, the assumption of the proposition implies that

$$\|N(j\omega)D(j\omega)^{-1}e^{-j\omega\tau}(1 - e^{-j\omega\delta})\|_2 < 1 \tag{14.42}$$

for all ω and δ. It follows that (14.30) cannot have roots on the imaginary axis. $\square$

Proposition 14.22. *If there exists a frequency $\omega > 0$ such that*

$$\left\| N(j\omega)D(j\omega)^{-1} \right\|_2 \geq \frac{1}{2}, \tag{14.43}$$

then the stability of (14.30) is guaranteed for

$$|\delta| < \min\left(\inf_{\omega \in S_\omega \setminus \{0\}} \frac{2}{\omega} \arcsin \frac{1}{2\left\| N(j\omega)D(j\omega)^{-1} \right\|_2}, \tau \right), \tag{14.44}$$

where

$$S_\omega = \left\{ \omega \geq 0 : \left\| N(j\omega)D(j\omega)^{-1} \right\|_2 \geq \frac{1}{2} \right\}.$$

Proof. Notice that

$$\|N(j\omega)D(j\omega)^{-1} e^{-j\omega\tau}(e^{-j\omega\delta} - 1)\|_2 = 2\|N(j\omega)D(j\omega)^{-1}\|_2 \left| \sin \frac{\omega\delta}{2} \right|.$$

When δ satisfies (14.44) we have that (14.42) holds for all $\omega \geq 0$ and asymptotic stability follows. □

Remark 14.23. *The above propositions also hold when replacing $\|N(j\omega)D(j\omega)^{-1}\|_2$ with $r_\sigma(N(j\omega)D(j\omega)^{-1})$. Although this significantly reduces conservatism, the obtained conditions may still be far from necessary because no phase information of equation (14.30) is exploited (unlike the conditions of Proposition 14.19).*

14.7 Notes and references

We have analyzed the robustness of Smith Predictors w.r.t. an inaccurate modeling of the delay, which corresponds to Problem 14.1 in the SISO case and to the analysis of equation (14.30) in the MIMO case. When $\deg(A(\lambda)) = \deg(B(\lambda))$ or when not all column degrees of $N(\lambda)$ are strictly smaller than the corresponding column degrees of $D(\lambda)$, the closed-loop system may not be practically stable. We have derived necessary and sufficient conditions for practical stability using two different approaches, one based on an interpretation in terms of neutral equations and the other based on the analysis of small feedback delays in infinite-dimensional control systems.

We have focused on the first approach, since it also provides insight in the instability mechanism. Under the assumption of practical stability, we have characterized the maximal delay mismatch such that stability is maintained.

For the scalar case the geometry of the stability regions in the space (nominal delay, delay mismatch) was briefly discussed by using the methodology presented in Chapter 4.

This chapter is based on the papers [188, 207, 208] and Chapter 3 of [205]. To the best of our knowledge the results of Section 14.6, summarized in Proposition 14.19, have not been published so far in the literature.

Chapter 15

Controlling unstable systems using finite spectrum assignment

15.1 Introduction

We consider the linear finite-dimensional system with input delay

$$\dot{x}(t) = Ax(t) + Bu(t - \tau), \ x \in \mathbb{R}^d, \ u \in \mathbb{R}, \tag{15.1}$$

where we assume that the matrix A is *not* Hurwitz and that the pair (A, B) is stabilizable. An approach for the stabilization and control of (15.1), called *finite spectrum assignment* [173, 323], can be interpreted as follows: a prediction of the state variable over one delay interval is generated first, and then a feedback of the predicted state is applied, thereby compensating the effect of the time-delay. This results in a closed-loop system with a finite number of characteristic roots, which can be freely assigned. Mathematically, with the feedback law

$$\begin{aligned} u(t) &= K^T x_p(t, t+\tau) \\ &= K^T \left(e^{A\tau}x(t) + \int_0^{\tau} e^{A\theta} Bu(t-\theta)d\theta\right), \end{aligned} \tag{15.2}$$

where $x_p(t_1, t_2)$ is the prediction of $x(t)$ at $t = t_2$, based on values of x and u for $t \leq t_1$, the characteristic equation of the closed-loop system is given by

$$\det\left(\lambda I - A - BK^T\right) = 0. \tag{15.3}$$

This elimination of the delay is employed in the so-called process model control techniques [324], such as, the celebrated Smith Predictor discussed in the previous chapter. When applied to (15.1) and (15.2) it can also be interpreted as the effect of a model transformation [3, 146]. For generalizations of the finite spectrum assignment approach to a broader class of time-delay systems than (15.1), we refer to [173, 324, 3].

Note that the application of the control law (15.2) at time-instant t requires the availability of the full state x at time t, unlike the Smith Predictor discussed in Chapter 14. Extensions to the output feedback case and relations with the Smith Predictor are discussed at the end of this chapter.

There are two difficulties related to the implementation of the control law (15.2). First, an exact knowledge of the delay is required for having a finite number of closed-loop characteristic roots, and, hence, an investigation of the effect of a delay mismatch is critical. Second, the application of the control law requires the computation of an integral over past inputs. Obtaining this term as the solution of a differential equation must be discarded because it involves an unstable pole-zero cancellation when the matrix A is unstable; see [173]. It is then suggested that we realize the control law by means of a numerical computation of the integral term *on-line*, which involves some approximation. In this chapter we assume that the distributed delay is approximated with a *sum of pointwise delays* by applying a numerical quadrature rule. To state this more precisely, define $\mathcal{C}([0,\ \tau], \mathbb{C}^d)$ as the space of continuous functions from $[0,\ \tau] \subset \mathbb{R}$ to $\mathbb{C}^d$, equipped with the supremum norm. A quadrature rule on $[0,\ \tau]$ is a sequence of maps $\{\mathcal{I}_n\}_{n\geq 1}$ from $\mathcal{C}([0,\ \tau], \mathbb{C}^d) \rightarrow \mathbb{C}$, defined as

$$\mathcal{I}_n(f) = \sum_{j=1}^{n} h_{j,n} f(\theta_{j,n}), \quad h_{j,n} > 0, \ \theta_{j,n} \in [0,\ \tau], \tag{15.4}$$

where we assume that the following convergence property is satisfied:

$$\forall f \in \mathcal{C}([0,\ \tau], \mathbb{C}^d)\ \forall \epsilon > 0\ \exists \bar{n} \in \mathbb{N} : \ |I_n(f) - \textstyle\int_0^\tau f(\theta)d\theta| < \epsilon\ \forall n \geq \bar{n}. \tag{15.5}$$

When the quadrature formulae (15.4) are used to approximate the integral term in (15.2), we end up with a sequence of control laws

$$u(t) = K^T \left(e^{A\tau} x(t) + \sum_{j=1}^{n} h_{j,n} e^{A\theta_{j,n}} B u(t - \theta_{j,n}) \right). \tag{15.6}$$

The effect of this semidiscretization of the control law on the closed-loop stability will be analyzed in detail.

The structure of the chapter reflects our main goal, that is, giving an overview of the existing stability results on the implementation of distributed delay control laws and on the effects of a delay mismatch. After some preliminaries we discuss the implementation of the integral term in (15.2). We address a possible instability mechanism when using the control law (15.6), as reported in [305, 304, 72, 200]. Then we discuss conditions for a safe implementation [202, 185] and outline modifications of the control law to remove the resulting restrictions [185, 203, 304, 80, 256]. Next we study the effect of a delay mismatch between plant and controller. Finally, we comment on the output feedback case and make connections to the previous chapter. Throughout the chapter plots of characteristic roots are extensively used in order to make the main ideas and results apparent.

15.2 Preliminaries

The initial data for the system (15.1) and (15.2) and the system (15.1) and (15.6) are $x(0) \in \mathbb{R}^d$, $u_0 \in \mathcal{C}([-\tau,\ 0], \mathbb{R})$. For $t \in [0\ \tau]$, the closed-loop system becomes

$$\dot{x}(t) = Ax(t) + B\, u_0(t - \tau).$$

For $t \geq \tau$, we have

$$\begin{aligned} Bu(t-\tau) &= BK^T \left(e^{A\tau}x(t-\tau) + \textstyle\int_0^\tau e^{A\theta} Bu(t-\theta-\tau)d\theta\right) \\ &= BK^T \left(e^{A\tau}x(t-\tau) + \textstyle\int_0^\tau e^{A\theta}(\dot{x}(t-\theta) - Ax(t-\theta))d\theta\right), \end{aligned}$$

where the right-hand derivative of x should be taken at time zero, and we can write the system (15.1) and (15.2) in the form

$$\frac{d}{dt}\mathcal{L}(x_t) = Ax(t) + BK^T e^{A\tau}x(t-\tau) - BK^T A \int_0^\tau e^{A\theta}x(t-\theta)d\theta,$$

where the map $\mathcal{L} : \mathcal{C}([-\tau\ 0], \mathbb{R}^d) \to \mathbb{R}^d$ is defined as

$$\mathcal{L}(\phi) = \phi(0) - BK^T \int_0^\tau e^{A\theta}\phi(-\theta)d\theta.$$

Because (15.1)–(15.2) is a Volterra equation of the second kind, the growth of its solutions is determined by the roots of its characteristic equation (see [173])

$$\begin{aligned} \det\Big\{\lambda\left(I - BK^T \textstyle\int_0^\tau e^{(A-\lambda I)\theta}d\theta\right) - A - BK^T e^{(A-\lambda I)\tau} \\ +BK^T A \textstyle\int_0^\tau e^{(A-\lambda I)\theta}d\theta\Big\} = 0, \end{aligned} \tag{15.7}$$

which can be simplified to (15.3). This makes the finite spectrum assignment property apparent.

Analogously, with the approximated control law (15.6) and for $t \geq \tau$ the closed-loop system can be written as

$$\frac{d}{dt}\mathcal{N}_n(x_t) = Ax(t) + BK^T e^{A\tau}x(t-\tau) - BK^T A \sum_{j=1}^{n} h_{j,n}e^{A\theta_{j,n}}x(t-\theta_{j,n}), \tag{15.8}$$

where the map $\mathcal{N}_n : \mathcal{C}([-\tau_n,\ 0], \mathbb{R}^d) \to \mathbb{R}^d$ is defined as

$$\mathcal{N}_n(\phi) = \phi(0) - BK^T \sum_{j=1}^{n} h_{j,n}e^{A\theta_{j,n}}\phi(-\theta_{j,n}),$$

with $\tau_n = \max_j \theta_{j,n}$.

Note that equation (15.8) is a DDE of neutral type. Under mild assumptions on the integration rule (15.4), the map $\mathcal{N}_n$ is atomic at zero,[21] guaranteeing existence and uniqueness of solutions for initial conditions $\phi \in \mathcal{C}([-\tau_n,\ 0], \mathbb{R}^d)$. The associated delay-difference equation of (15.8) is given by $\mathcal{N}_n(x_t) = 0$, that is,

$$x(t) = BK^T \sum_{j=1}^{n} h_{j,n}e^{A\theta_{j,n}}x(t-\theta_{j,n}). \tag{15.9}$$

Recall from Chapter 1 that the asymptotic behavior of the solutions and, thus, stability of the neutral equation (15.8) is determined by the spectral radius $r(\mathcal{T}_N^n(t))$, satisfying

$$r(\mathcal{T}_N^n(1)) = e^{\alpha},\quad \alpha = \sup\left\{\Re(\lambda) :\ \det\left(\Delta_N^n(\lambda)\right) = 0\right\},$$

[21] This property makes the system (15.1) and (15.6) causal, and allows us to write the control input u at the present time as a function of the present state and *past* inputs.

where the characteristic matrix[22] Δ_N^n is given by

$$\Delta_N^n(\lambda) := \left(\lambda\Delta_D^n(\lambda) - A - BK^T e^{(A-\lambda I)\tau} + BK^T A \sum_{j=1}^{n} h_{j,n} e^{(A-\lambda I)\theta_{j,n}}\right),$$

and $\Delta_D^n(\lambda)$ is defined as follows:

$$\Delta_D^n(\lambda) := \left(I - BK^T \sum_{j=1}^{n} h_{j,n} e^{(A-\lambda I)\theta_{j,n}}\right).$$

In a similar way, stability of the difference equation (15.9) is determined by the spectral radius

$$r_\sigma(\mathcal{T}_D^n(1)) = e^\beta, \quad \beta = \sup\left\{\Re(\lambda) : \det\left(\Delta_D^n(\lambda)\right) = 0\right\}.$$

An important property w.r.t. the stability analysis in this chapter is the relation

$$r_e(\mathcal{T}_N^n(1)) = r_\sigma(\mathcal{T}_D^n(1)), \tag{15.10}$$

where $r_e(.)$ denotes the radius of the essential spectrum.

15.3 Implementation of the integral

We study the effect on stability of the implementation of the control law (15.2) as (15.6).

15.3.1 Instability mechanism

A starting point of the research on the implementation of distributed delay control laws was the paper [305], which illustrated that the closed-loop system (15.1) and (15.6) may be unstable for *arbitrarily large* values of n, even when the ideal closed-loop system (15.1) and (15.2) is exponentially stable, the latter expressed by the Hurwitz stability of matrix $A + BK^T$. This paradox can be explained intuitively with the occurrence of unstable characteristic roots with a large modulus for the approximated closed-loop system. When the approximation becomes better, some characteristic roots tend to the characteristic roots of the limit case, while the others move off to infinity. When some characteristic roots do so without leaving the RHP, instability persists. This is now illustrated with an example.

Example 15.1. *Consider the scalar system*

$$\dot{x}(t) = x(t) + u(t-1), \tag{15.11}$$

and the control law

$$u(t) = -2\,x_p(t, t+1) = -2\left(e\,x(t) + \int_0^1 e^\theta u(t-\theta)d\theta\right), \tag{15.12}$$

[22] The superscript n refers to the number of abscissa of the quadrature rule.

which assigns one closed-loop characteristic root $\lambda = -1$. When the integral term in (15.12) is discretized using the forward rectangular rule, i.e., using (15.4) with

$$\theta_{j,n} = \frac{j-1}{n}, \quad h_{j,n} = \frac{1}{n}, \quad j = 1 \ldots n,$$

the control law becomes

$$u(t) = -2\left(ex(t) + \frac{1}{n}\sum_{j=1}^{n} e^{\frac{j-1}{n}} u\left(t - \frac{j-1}{n}\right)\right). \tag{15.13}$$

In Figure 15.1 (left) the characteristic roots of the closed-loop system (15.11) and (15.13) are shown for $n = 40$ and $n = 60$. As $n \to \infty$, one characteristic root converges to the assigned characteristic root $\lambda = -1$, while all the characteristic roots, introduced by the approximation, move off to infinity. However, stability is not obtained. The sequences of characteristic roots, whose imaginary parts tend to infinity, yet whose real parts have a finite limit, are explained by the neutral type of the closed-loop system. In Figure 15.1 (right) we show the characteristic roots of the associated difference equation

$$u(t) = -2\left(\frac{1}{n}\sum_{j=1}^{n} e^{\frac{j-1}{n}} u\left(t - \frac{j-1}{n}\right)\right). \tag{15.14}$$

As expected from theoretical considerations (related to property (15.10)), the closed-loop characteristic roots with a large modulus, but small real part, are well approximated by characteristic roots of the difference equation [106].

The instability mechanism is due to the *neutral type* of the approximated closed-loop system (15.1) and (15.6), in contrast with the retarded type of the ideal closed-loop system (15.1)–(15.2). Hence, for any n the approximation involves a noncompact perturbation of the solution semigroup $\mathcal{T}_N^n(t)$, associated with (15.1) and (15.2), which introduces an essential spectrum; see [185]. Having the radius of this essential spectrum larger than one results in the sensitivity of stability. Alternatively, a frequency-domain interpretation, including links with (lack of) w-stability [89], is presented in [200]: the integral in (15.2) has a smoothing effect on its input, unlike any finite sum approximation. This is reflected in the property that the sequence

$$\left\{ \sup_{\omega \geq 0} \left| \int_0^{\tau} e^{(A - j\omega I)\theta} d\theta - \sum_{j=1}^{n} h_{j,n} e^{(A - j\omega I)\theta_{j,n}} \right| \right\}_{n \geq 1}$$

does not converge to zero as $n \to \infty$. The high-frequency error is related to the unstable characteristic roots of the closed-loop system with large imaginary parts.

15.3.2 Stability conditions

Because the Hurwitz stability of $A + BK^T$ does not imply the stability of (15.1) and (15.6) for large values of n, additional conditions are needed to guarantee that the characteristic

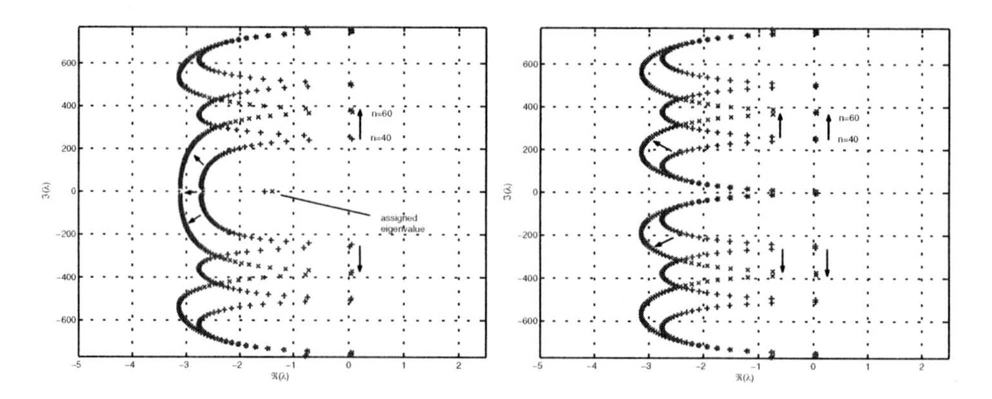

Figure 15.1. *(left) Closed-loop characteristic roots of the system (15.11) and (15.13) for* $n = 40$ *(+) and* $n = 60$ *(x). (right) Characteristic roots of the difference equation (15.14).*

roots, due to the approximation, are also in the open LHP. A comparison of the spectral plots in Figure 15.1 suggests that hyperbolic stability properties of the closed-loop system (15.1) and (15.6) are tightly related to stability properties of the difference equation (15.9). This is indeed the case, leading to very simple and easy-to-check stability criteria, which we now review.

Necessary condition

The well-known results which state that a necessary condition for the stability of the neutral equation (15.8) is given by the stability of the difference equation $\mathcal{N}_n(x_t) = 0$, with

$$\mathcal{N}_n(x_t) = x(t) - \sum_{j=1}^{n} BK^T e^{A\theta_{j,n}} x(t - \theta_{j,n}), \tag{15.15}$$

(a corollary of (15.10), and the relation of (15.15) with the equation $\mathcal{L}(x_t) = 0$, with

$$\mathcal{L}(x_t) = x(t) - \int_0^\tau BK^T e^{A\theta} x(t - \theta) d\theta, \tag{15.16}$$

lead to the following necessary condition for a safe implementation, which slightly generalizes [202, Theorem 1].

Theorem 15.2. *Consider the system (15.1)–(15.2) and a quadrature rule satisfying (15.5). Assume that the closed-loop system (15.1) and (15.6) is asymptotically stable for large values of n. Then the characteristic equation of (15.16),*

$$\det\left\{I - BK^T(\lambda I - A)^{-1}\left(I - e^{-(\lambda I - A)\tau}\right)\right\} = 0, \tag{15.17}$$

has all its roots in the closed LHP.

As an illustration we compare in Figure 15.2 the rightmost roots of equation (15.17), applied to the example (15.11)–(15.12), with characteristic roots of the difference equation (15.14). Notice that the real parts of the roots of (15.17) correspond to the position of chains of characteristic roots of the closed-loop system (15.11) and (15.13) for large n.

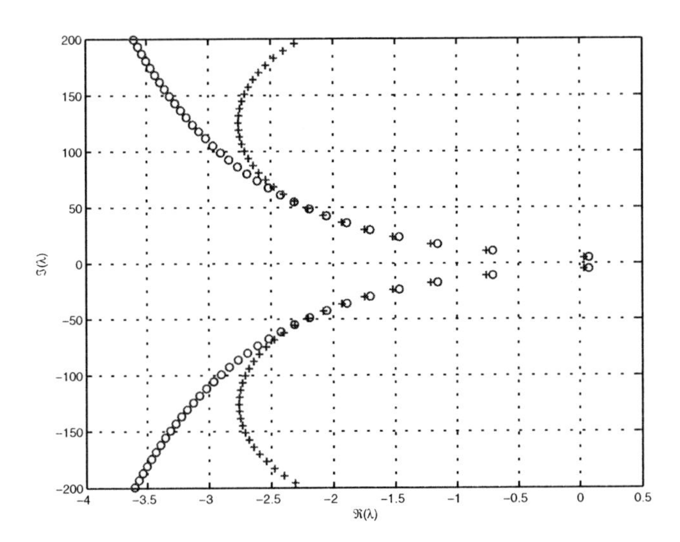

Figure 15.2. *Rightmost roots of Equation (15.17) (o) and characteristic roots of the difference equation (15.14), where $n = 40$ (+). The latter are shown on a different scale in Figure 15.1 (right).*

Necessary and sufficient condition

The numerical experiments with the example could suggest that the necessary condition, following from Theorem 15.2, is close to sufficient. However, in [185] it is proved that this is generally *not* the case, because Theorem 15.2 does not take into account the fact that the radius of the essential spectrum of the solution semigroup, associated with the neutral equation (15.8), i.e., $r_e(\mathcal{T}_N^n(1))$, is *not* continuous in the delays $\theta_{j,n}$. See [5, 102, 107, 181] and the references therein for more information on delay robustness of neutral equations and related questions. A consequence for the system (15.1) and (15.6) is that closed-loop stability may depend on the type of integration rule used and be sensitive to infinitesimal perturbations of the abscissa $\theta_{j,n}$, as we now illustrate.

Example 15.3. *We revisit Example 15.1 and discretize the control law (15.12) as*

$$u(t) = -2\left(e\,x(t) + \sum_{j=1}^{n} h_{j,n} e^{\theta_{j,n}} u\left(t - \theta_{j,n}\right)\right), \qquad (15.18)$$

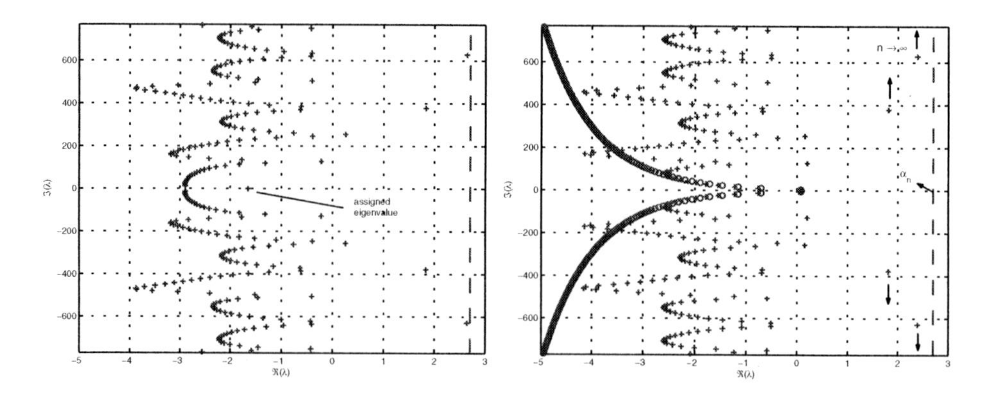

Figure 15.3. *(left) Closed-loop characteristic roots of the system (15.11) and (15.18)–(15.19) for $n = 40$. (right) Characteristic roots of the associated difference equation (15.20) (+) and roots of equation (15.17) (o). Unlike the situation displayed in Figure 15.1, stability of the difference equation for large n is no longer determined by the rightmost roots of equation (15.17).*

where

$$\theta_{j,n} = \begin{cases} \frac{j-1}{n}, & j \text{ even}, \\ \frac{j-4/5}{n}, & j \text{ odd}, \end{cases} \qquad h_{j,n} = \frac{1}{n}, \quad j = 1 \ldots n. \tag{15.19}$$

Notice that the modified integration rule with parameters (15.19) also satisfies the convergence property (15.5). In Figure 15.3 we show the characteristic roots of the closed-loop system (15.11) and (15.18), as well as the characteristic roots of the associated difference equation

$$u(t) = -2 \left(\sum_{j=1}^{n} h_{j,n} e^{\theta_{j,n}} u(t - \theta_{j,n}) \right), \quad h_{j,n}, \theta_{j,n}, \tag{15.20}$$

given by (15.19), and roots of equation (15.17). Although the rightmost roots of (15.17) are also well approximated by characteristic roots of (15.20), they no longer determine the stability of (15.20) and of the closed-loop system (15.11) and (15.18) for large n.

By making infinitesimal perturbations of the abscissa $\theta_{j,n}$ in (15.19), the rightmost characteristic roots of the difference equation can have their real part arbitrarily close to the value α_n, indicated on the figure, but no larger than $\alpha_n + \epsilon$, for any $\epsilon > 0$. In fact, such a value α_n determines stability of the difference equation when subject to small variations in the delays, called strong stability *in [107]. In [185, Section 4] it is described how α_n can be computed analytically. Furthermore, it is shown that $\lim_{n\to\infty} \alpha_n$ exists and is* independent *of the type of quadrature rule.*

Remark 15.4. *The sensitivity of stability w.r.t. infinitesimal perturbations, as well as the high-frequency instability mechanism, described in Section 15.3.1, are phenomena which are related to the sensitivity of stability of some boundary controlled hyperbolic PDEs, feedback controlled descriptor systems, and neutral type systems against small delays in* *the control loop, as reported in [61, 62, 63, 109, 154, 214, 155, 156, 157]. A sensitivity of stability w.r.t. infinitesimal modeling errors may also occur with the Smith predictor control scheme [281]; see Chapter 14 and [248, 188] and the references therein.*

Because arbitrarily small perturbations of the abscissa $\theta_{j,n}$ may destroy stability of the closed-loop system (15.1) and (15.6), yet are inevitable in any practical application, they should be taken into account in a definition of a safe implementation. This leads to the following definition.

Definition 15.5. *Consider the system (15.1) and (15.2), where $A + BK^T$ is Hurwitz, and the quadrature rule (15.4). The implementation (15.6) of the control law (15.2) is safe if the following two conditions are satisfied:*

1. *There exists a number $\bar{n} \in \mathbb{N}$ such that the closed-loop system (15.1) and (15.6) is asymptotically stable for all $n \geq \bar{n}$.*
2. *For each $n \geq \bar{n}$, there exist constants $\Delta\theta_{j,n} > 0$ such that the control law*

$$u(t) = K^T \left(e^{A\tau} x(t) + \sum_{j=1}^{n} h_{j,n} e^{A(\theta_{j,n}+\delta\theta_{j,n})} Bu(t - (\theta_{j,n} + \delta\theta_{j,n})) \right)$$

achieves asymptotic stability for all $|\delta\theta_{j,n}| < \Delta\theta_{j,n}$.

In the sense of this definition, an almost necessary and sufficient stability condition is given in [185, Theorem 1], which does *not* depend on the type of integration rule. Essentially it corresponds to a strong stability requirement for the difference equation (15.9) for large n.

Theorem 15.6. *Consider the system (15.1) and (15.2), and assume that $A+BK^T$ is Hurwitz. Let*

$$S = \int_0^\tau |K^T e^{A\theta} B| d\theta. \tag{15.21}$$

If $S < 1$, then the control law (15.2) can be safely implemented as (15.6), in the sense of Definition 15.5.
If $S > 1$, then the control law (15.2) cannot be safely implemented.

In the multiple input case, a *sufficient* condition for a safe implementation is given by

$$\int_0^\tau \|K^T e^{A\theta} B\| d\theta < 1;$$

see [185, Section 6].

15.3.3 Removing restrictions

The main advantage of the control law (15.2) lies in the fact that all the closed-loop characteristic roots can be freely assigned. A disadvantage is the difficulty of computing the control law on-line, which involves the evaluation of the integral. In particular, for the implementation with a sum of pointwise delays the stability condition of Theorem 15.6, i.e., $S < 1$, puts severe restrictions on stabilizability and performance, which are shown in Section 7.2.3 and [185] to be comparable to the case of a static, nonpredictive, state feedback controller, $u(t) = K^T x(t)$. In this section we briefly comment on possible modifications of the control law (15.6) to remove these restrictions.

Adding a low-pass filter

The instability mechanism, as explained in Section 15.3.1, is a *high-frequency* mechanism, related to the occurrence of unstable characteristic roots with *arbitrarily large* imaginary parts. A closer look at the problem reveals that the latter are caused by the throughput at infinity of past inputs in equation (15.6) and, therefore, can be avoided by including a *low-pass filter* in the control dynamics. This is now illustrated.

Example 15.7. *We reconsider the system (15.11) and (15.13) and modify the control law by adding a first-order low-pass filter to*

$$\begin{cases} \dot{z}(t) = -fz(t) - 2f\left\{ex(t) + \frac{1}{n}\sum_{j=1}^{n} e^{\frac{j-1}{n}} u\left(t - \frac{j-1}{n}\right)\right\}, \\ u = z(t). \end{cases} \tag{15.22}$$

Due to the filter, the closed-loop system is of retarded *type. Its characteristic roots are shown in Figure 15.4 for different values of n. For sufficiently large values the closed-loop system is asymptotically stable.*

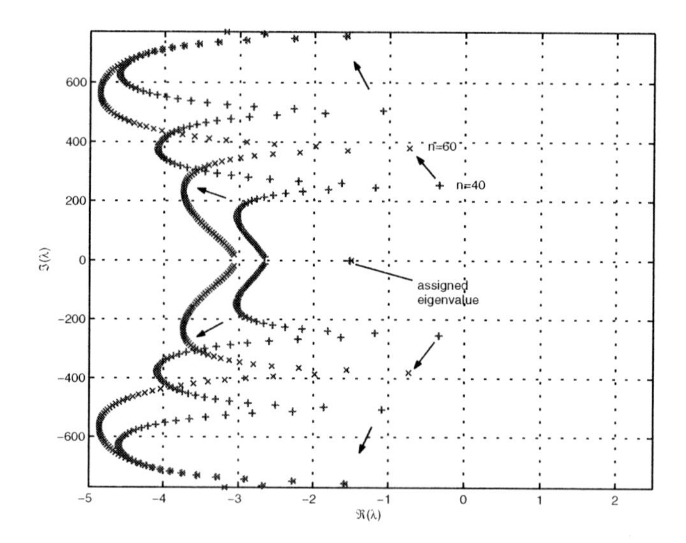

Figure 15.4. *Closed-loop characteristic roots of (15.11) and (15.22) for $f = 100$ and $n = 40$ (+), respectively, $n = 60$ (x).*

This idea is generalized in [203]. Since *any strictly proper linear system* (A_f, B_f, C_f) has a *low-pass* filtering property, the dynamic control law

$$\left\{\begin{array}{ll} \dot{z}(t) & = A_f z(t) + B_f\, x_p(t, t+\tau), \\ u(t) & = C_f z(t), \end{array}\right. \tag{15.23}$$

is suggested. It is shown that the closed-loop characteristic roots are equal to the characteristic roots of the finite-dimensional system

$$\left\{\begin{array}{l} \dot{x}(t) = Ax(t) + BC_f z(t), \\ \dot{z}(t) = A_f z(t) + B_f x(t), \end{array}\right. \tag{15.24}$$

and hence also assignable using standard design methods for systems without delays. Furthermore, a discretization using (15.4) preserves stability for large values of n.

Using piecewise constant inputs

In [304, 256, Section 3] the input $u(t)$ is kept *piecewise constant* in time-intervals of length Δ, inspired by the implementation with a digital controller. Then, at the sampling times, the system (15.1) is completely equivalent to a discrete system. When $\Delta = \tau/p,\ p \in \mathbb{N}$, this discrete system takes the form

$$\bar{x}(k+1) = A_d \bar{x}(k) + B_d \bar{u}(k-p), \tag{15.25}$$

where $\bar{x}(k) = x(k\Delta)$, $\bar{u}(k) = u(k\Delta)$, and

$$A_d = e^{A\Delta}, \quad B_d = \left(\int_0^{\Delta} e^{As}\, ds\right) B.$$

As in the continuous-time case, the delay can be compensated for using a prediction: for the control law

$$\begin{array}{rl} \bar{u}(k) & = K_d^T \bar{x}_p(k, k+p) \\ & = K_d^T \left(A_d^p \bar{x}(k) + \sum_{n=1}^{p} A_d^{n-1} B_d \bar{u}(k-n)\right), \end{array} \tag{15.26}$$

the characteristic equation of the closed-loop system is given by

$$\det\left(zI - (A_d + B_d K_d^T)\right) = 0.$$

Because the system (15.25) and (15.26) is fully discrete, the maximal possible frequency is given by $1/(2\Delta)$ (the celebrated Nyquist–Shannon criterion) and, therefore, sensitivity of stability w.r.t. arbitrarily small perturbations of the parameters of (15.26) is not possible.

Note that the control law (15.26) can be considered as a full discretization of the continuous control law (15.2), using a particular quadrature rule. However, in [304] it is also illustrated that one cannot take any quadrature rule satisfying (15.5) and obtain stability for sufficiently small values of Δ after a full discretization. For instance, applying Simpson's rule may lead to instability. The instability mechanism is related to the one in the continuous-time case, where approximations lead to characteristic roots with arbitrarily large frequencies. In the discrete-time case, unstable modes may occur with the maximal frequency[23] $1/(2\Delta)$, which tends to infinity as $\Delta \to 0$.

[23] Such a mode corresponds to a negative real characteristic root of the fully discretized system.

15.4 Delay mismatch

We investigate the (additional) effect of a delay mismatch between plant and controller. We assume that there is a constant uncertainty δ on the system delay such that (15.1) becomes

$$\dot{x}(t) = Ax(t) + Bu(t - \tau - \delta). \tag{15.27}$$

An application of the control law (15.2) to (15.27) leads to a closed-loop system whose stability properties are determined by the roots of the characteristic equation

$$\begin{aligned} \det \big\{\lambda \left(I - BK^T \textstyle\int_0^\tau e^{(A-\lambda I)\theta} d\theta\right) - A - BK^T e^{(A-\lambda I)\tau} e^{-\lambda\delta} \\ +BK^T A \textstyle\int_0^\tau e^{(A-\lambda I)\theta} d\theta\big\} = 0, \end{aligned} \tag{15.28}$$

which simplifies to

$$\det\left\{D_1(\lambda) - N_1(\lambda;\ \tau)e^{-\lambda\tau}(1 - e^{-\lambda\delta})\right\} = 0, \tag{15.29}$$

where

$$D_1(\lambda) = \lambda I - A - BK^T, \quad N_1(\lambda;\ \tau) = -BK^T e^{A\tau}.$$

Similarly, an application of the control law (15.6) to (15.27) results in the characteristic equation

$$\det\left\{D_2(\lambda;\ \tau) - N_2(\lambda;\ \tau)e^{-\lambda\tau}(1 - e^{-\lambda\delta})\right\} = 0, \tag{15.30}$$

where

$$\begin{aligned} D_2(\lambda;\ \tau) &= \lambda \left(I - BK^T \textstyle\sum_{j=1}^n h_{j,n} e^{(A-\lambda I)\theta_{j,n}}\right) - A - BK^T e^{(A-\lambda I)\tau} \\ &\qquad + BK^T A \textstyle\sum_{j=1}^n h_{j,n} e^{(A-\lambda I)\theta_{j,n}}, \\ N(\lambda;\ \tau) &= -BK^T e^{A\tau}. \end{aligned}$$

Equation (15.29) can be interpreted as the characteristic equation of a system of retarded type with pointwise delays. If the condition $S < 1$ of Theorem 15.6 is satisfied and n is sufficiently large, then (15.30) corresponds to a neutral system with pointwise delays and a corresponding *strongly stable* difference equation. In both cases, a sensitivity of hyperbolic stability properties w.r.t. an infinitesimal perturbation of δ, in particular the transition from $\delta = 0$ to $\delta = 0+$, is not possible. So if $A + BK^T$ is Hurwitz, then the closed-loop system is practically stable in the sense of Chapter 14.

Stability regions with δ as free parameter can be computed using the approach of Section 14.6.2, since (15.29) and (15.30) are of the form (14.30). Note that D and N in (14.30) are assumed to be polynomial matrices that don't depend on τ, yet the analysis and results of Section 14.6.2 do not rely on these assumptions.

15.5 Output feedback

We generalize the results of the previous sections to the static output feedback case. Next we discuss dynamic output feedback and make connections with the Smith predictor scheme, discussed in Chapter 14.

15.5.1 Static output feedback

Assume in what follows that only a scalar output y is available for measurement

$$y(t) = F^T x(t), \ F \in \mathbb{R}^{d\times 1}.$$

In this case the time-delay cannot be directly compensated by using a prediction of the output in the feedback loop, since the computation of the prediction,

$$y_p(t,\ t+\tau) = F^T x_p(t,\ t+\tau) = F^T\left(e^{A\tau}x(t) + \int_0^\tau e^{A\theta}Bu(t-\theta)d\theta\right),$$

requires the availability of the full state $x(t)$. Instead, one may use the *fictitious* output

$$\bar{y}(t) = F^T e^{-A\tau}x(t), \tag{15.31}$$

whose prediction,

$$\begin{aligned} \bar{y}(t,\ t+\tau) &= F^T e^{-A\tau}\left(e^{A\tau}x(t) + \textstyle\int_0^\tau e^{A\theta}Bu(t-\theta)d\theta\right) \\ &= y(t) + \textstyle\int_0^\tau F^T\, e^{A(\theta-\tau)}Bu(t-\theta)d\theta, \end{aligned} \tag{15.32}$$

only depends on the input and the output $y(t)$. The control law

$$\begin{aligned} u(t) &= k\ \bar{y}_p(t,\ t+\tau) \\ &= k\left(y(t) + \textstyle\int_0^\tau F^T\, e^{A(\theta-\tau)}Bu(t-\theta)d\theta\right) \end{aligned} \tag{15.33}$$

results in a closed-loop system whose characteristic equation is given by

$$\det(\lambda I - A - kBF^T e^{-A\tau}) = 0.$$

Hence, the stabilizablity of (15.1) using a control law of the form (15.33) depends on the output feedback stabilizability of the corresponding undelayed system using the fictitious output $\bar{y}$.

To characterize the effects on stability of an implementation of the integral term in (15.33) and of a delay mismatch, the results of Sections 15.3–15.4 can be directly applied, provided that the following substitution is made:

$$K^T \leftarrow kF^T e^{-A\tau}.$$

For instance, the condition for a safe implementation of the integral from Theorem 15.6 becomes

$$|k| < \left(\int_0^\tau \left|F^T e^{A(\theta-\tau)}B\right| d\theta\right)^{-1}.$$

15.5.2 Dynamic output feedback and relations with Smith Predictors

The approach of the previous section can be generalized to dynamic feedback of the prediction of the fictitious output (15.31). Let $(A_c,\ B_c,\ F_c,\ G_c)$ be a minimal realization of the controller and apply the feedback

$$\begin{aligned} \dot{z}(t) &= A_c z(t) + B_c(\bar{y}_p(t,\ t+\tau) - r(t)), \\ u(t) &= F_c z(t) + G_c(\bar{y}_p(t,\ t+\tau) - r(t)), \end{aligned} \tag{15.34}$$

where r is an external reference signal. With

$$\begin{aligned} H_0(\lambda) &= F^T(\lambda I - A)^{-1}B, \\ \bar{H}_0(\lambda) &= F^T e^{-A\tau}(\lambda I - A)^{-1}B, \\ C(\lambda) &= F_c^T(\lambda I - A_c)^{-1}B_c + G_c, \end{aligned}$$

we have (with capital letters denoting the signals in the frequency domain)

$$\begin{aligned} U(\lambda) &= C(\lambda)(R(\lambda) - \bar{Y}_p(\lambda)) \\ &= C(\lambda)\left(R(\lambda) - Y(\lambda) - \int_0^\tau F^T e^{A(\theta-\tau)} B e^{-\lambda\theta} d\theta\, U(\lambda)\right), \end{aligned} \tag{15.35}$$

and we can formally derive

$$\begin{aligned} U(\lambda) &= \frac{C(\lambda)}{1+C(\lambda)\int_0^\tau F^T e^{A(\theta-\tau)} B e^{-\lambda\theta} d\theta}(R(\lambda) - Y(\lambda)) \\ &= \frac{C(\lambda)}{1+C(\lambda)(\bar{H}_0(\lambda) - H_0(\lambda)e^{-\lambda\tau})}(R(\lambda) - Y(\lambda)). \end{aligned}$$

A comparison with the control law (14.2) reveals that the control law (15.34) corresponds to a modification of the Smith predictor. The modification consists of replacing

$$H_0(\lambda) - H_0(\lambda)e^{-\lambda\tau}$$

with

$$\bar{H}_0(\lambda) - H_0(\lambda)e^{-\lambda\tau}, \tag{15.36}$$

where $\bar{H}_0$ is such that the transfer function (15.36) has a *finite impulse response* and is implemented statically by computing an integral on-line. This approach is well known in the context of extending the Smith predictor to open-loop unstable plants; see, for instance, the survey [248] and [329].

The transfer functions of the closed-loop system (15.1) and (15.34) from the reference signal r, as well as an input disturbance w, to the output y are given by

$$\begin{aligned} H_{y,r}(\lambda) &= \frac{C(\lambda)H_0(\lambda)}{1+C(\lambda)\bar{H}_0(\lambda)}e^{-\lambda\tau}, \\ H_{y,w}(\lambda) &= \frac{H_0(\lambda)(1+C(\lambda)(\bar{H}_0(\lambda) - H_0(\lambda)e^{-\lambda\tau}))}{1+C(\lambda)\bar{H}_0(\lambda)}e^{-\lambda\tau}. \end{aligned}$$

Note that due to a static implementation of (15.36), the open-loop poles do not appear any more as poles of the transfer function $H_{y,w}$, in contrast to the classical Smith predictor for which we have expression (14.6).

The implementation of the control law (15.34) requires a discretization of the integral in (15.32). Applying the quadrature rule (15.4) results in:

$$\begin{aligned} \dot{z}(t) &= A_c z(t) + B_c(\bar{y}_p^d(t,\ t+\tau) - r(t)), \\ u(t) &= F_c z(t) + G_c(\bar{y}_p^d(t,\ t+\tau) - r(t)), \\ \bar{y}_p^d(t,\ t+\tau) &= y(t) + \sum_{j=1}^n h_{j,n} F^T e^{A(\theta_{j,n}-\tau)} Bu(t-\theta_{j,n}). \end{aligned} \tag{15.37}$$

From (15.35) and (15.37) it is easy to see that Theorem 15.6 can be generalized to the following.

Theorem 15.8. *Consider the feedback system (15.1) and (15.34), which is assumed to be internally stable. If*

$$|G_c| < \left(\int_0^\tau |F^T e^{A(\theta-\tau)} B| d\theta \right)^{-1},$$

then the control law (15.34) can be safely implemented as (15.37), in the sense of Definition 15.5. If

$$|G_c| > \left(\int_0^\tau |F^T e^{A(\theta-\tau)} B| d\theta \right)^{-1},$$

then the control law (15.34) cannot be safely implemented.

Remark 15.9. *If $G_c = 0$, then the control law can be safely implemented because it behaves as a low-pass filter. The beneficial effect of a low-pass filter in the control loop was illustrated in Section 15.3.3.*

When there is a constant delay mismatch between plant and controller and the plant delay is given by $\tau + \delta$, the transfer function from the reference signal r to the plant output y becomes

$$H_{y,r}^{\delta}(\lambda) = \frac{C(\lambda) H_0(\lambda) e^{-\lambda(\tau+\delta)}}{1 + C(\lambda)\bar{H}_0(\lambda) + C(\lambda) H_0(\lambda) e^{-\lambda\tau}(e^{-\lambda\delta} - 1)},$$

and if, in addition, the integral in the control law is discretized,

$$H_{y,r}^{\delta}(\lambda) = \frac{C(\lambda) H_0(\lambda) e^{-\lambda(\tau+\delta)}}{1 + C(\lambda) \left(\sum_{j=1}^{n} h_{j,n} F^T e^{A(\theta_{j,n}-\tau)} B e^{-\lambda\theta_{j,n}} \right) + C(\lambda) H_0(\lambda) e^{-\lambda(\tau+\delta)}}.$$

In both cases a sensitivity of stability w.r.t. an infinitesimal delay mismatch δ is not possible, since the closed-loop system is of retarded type (due to the fact that H_0 is strictly proper and C proper). The maximal allowable delay mismatch can be computed by applying the techniques of Chapter 14.

15.6 Notes and references

We have studied the problem of the implementation of controllers arising in the context of finite spectrum assignment. Both the effect of an approximation of distributed delays by pointwise delays and the effect of a delay mismatch between plant and controller were studied. In the former case, a high-frequency instability mechanism was outlined, necessary and sufficient conditions for the preservation of stability for a sufficiently fine discretization were provided, and approaches were proposed to overcome the induced limitations. In the latter case, it was shown that a sensitivity of stability w.r.t. infinitesimal delay mismatches does not occur since the proposed controllers have a bounded high-frequency gain. Procedures to compute the maximal achievable delay mismatch were outlined. While the chapter focused on static state feedback, extensions to dynamic output feedback were also discussed, thereby making connections with the results of the previous chapter.

The results presented in this chapter are based on [186, 185, 194, 203, 248] and the references therein.

Chapter 16

Consensus problems in traffic flow applications

16.1 Introduction

It is well known that the traffic dynamics are inherently time-delayed because of the limited sensing and acting capabilities of drivers against velocity and position variations. The undesirable effects of a mismanaged traffic flow in social and economic life has made this interdisciplinary problem interesting and challenging over the years, and it becomes critical in the context of the increasing highway traffic [95, 111].

Without any deep discussions on the modeling of the traffic dynamics, this chapter concerns the stability analysis of some (microscopic) linear system including distributed delays. The idea of using delays in traffic flow dynamics is not new and, to the best of the authors' knowledge, was pointed out in the 1960s (see, for instance, [42]). According to its origin (see [95]), we can classify the delays in the traffic flow dynamics as follows: physiological delays (mainly induced by the human operators), mechanical time-delays (time needed for the vehicle's response after some driver's action), and delays in the vehicle's action. Such a classification is far from complete. For instance, the physiological delay can be further classified as: sensing, perception, response, selection, and programming delay. Finally, it is important to point out that such delays are in order of seconds, although different ranges are stated in the literature, and a function of the correlation of the human reactions with acceleration/deceleration (see, e.g., [277, 95]). Since the presence of delays may drastically change the dynamics of the model free of delays, the analysis of delay effects on the stability properties of the corresponding system is necessary.

One of the simplest models often discussed in the literature is the (microscopic) car following model, describing the behavior of multiple vehicles under the influence of a single constant time-delay [42, 111, 261]. In general, two spatial configurations are dealt with: the linear and the ring configuration. For the sake of brevity, we shall only consider the *ring configuration* when discussing the traffic flow application, but the obtained results can also be applied to the linear configuration. The linear model of [42] can be written conceptually as follows (inspired by some delay-free models of Reuschel in the 1950s):

$$\dot{v}_k(t) = \alpha_k(v_{k-1}(t-\tau) - v_k(t-\tau)),\ k = 1, \dots, p, \tag{16.1}$$

where p is the number of considered vehicles and $v_0 = v_p$. The left-hand side represents the *acceleration* of the *kth* vehicle, and the right-hand side expresses the *velocity difference* of consecutive vehicles. A natural extension of the model above takes into account multiple cars,[24] as discussed in [261]. Such a model is attributed to Kuhne in the survey [278].

However, various issues are not considered in the model (16.1) or its multiple car following version. For instance, humans retain a short-term memory of the past events and this may affect their control decision strategy. Such a behavior cannot be described by using pointwise (or discrete) delays in the model. Furthermore, the drivers' perception and interpretations of the stimuli depend on various parameters, and are different from one driver to another. As pointed out in [278], a more realistic model should include a *delay distribution* over the time that depicts the human behavior in average. Conceptually, defining the delay distribution represents a challenging problem itself, and is far from being solved. In [278], the authors proposed three types of delay distributions: a uniform distribution, a γ-distribution, and a γ-distribution with a gap, where the gap corresponds to the minimum reaction time of the humans with respect to some external signals and/or stimuli. In this chapter we shall assume the third type of distribution. Remarks and discussions on its applications to other problems from engineering and biology can be found in [205].

The above discussions lead us to the stability analysis of the model

$$\dot{v}_k(t) = \sum_{i=1}^{p-1} \alpha_{k,\lfloor k-i \rfloor} \int_0^{\infty} f(\theta)(v_{\lfloor k-i \rfloor}(t-\theta) - v_k(t-\theta))d\theta, \quad k = 1, \dots, p, \tag{16.2}$$

where $f(\cdot)$ denotes the delay kernel, and where the notation $\lfloor \cdot \rfloor$ stands for

$$\lfloor l \rfloor = \begin{cases} l, & l = 1, \dots, p, \\ \lfloor l + p \rfloor & l < 1. \end{cases}$$

We assume that

$$\alpha_{k,l} \geq 0, \qquad k = 1, \dots, p, \ l = 1, \dots, p, \ k \neq l, \tag{16.3}$$

$$\alpha_{k,\lfloor k-1 \rfloor} > 0, \quad k, \dots, p. \tag{16.4}$$

From the application point of view (16.4) is natural, as it expresses that a driver always takes into account the preceding car in his reaction. Since the delay distribution corresponds to a gamma distribution with a gap, the kernel f is given by

$$f(\xi) = \begin{cases} 0, & \xi < \tau, \\ \frac{(\xi-\tau)^{n-1} e^{-\frac{\xi-\tau}{T}}}{T^n (n-1)!}, & \xi \geq \tau, \end{cases} \tag{16.5}$$

where $n \in \mathbb{N}$, $T > 0$, and $\tau \geq 0$. Note that $f(\xi) \geq 0$ for all $\xi \geq 0$ and $\int_0^{\infty} f(\xi) d\xi = 1$. The gap is defined by τ, and the corresponding *average delay* of (16.5) satisfies

$$\tau_m = \int_0^{\infty} \xi f(\xi)\, d\xi = \tau + nT. \tag{16.6}$$

[24]The drivers observe not only the vehicle in front of them, but also other vehicles.

As $T \to 0+$, the kernel (16.5) tends to a Dirac impulse centered at $\xi = \tau$, and (16.2) therefore reduces to a system with a pointwise delay τ. As we shall see, the transition to $T = 0$ is smooth from a stability analysis point of view, as the stability determining eigenvalues are continuous w.r.t. $T \geq 0$.

The aim of this chapter is to perform a stability analysis of (16.2)–(16.5) with respect to the parameters (T, τ) and n, which determine the shape of the delay distribution. More precisely, for a given value of n we will determine regions in the (T, τ) space, such that for all initial conditions a consensus is reached, that is, the cars eventually get to the same speed.

The stability analysis of (16.2)–(16.5) can be interpreted as the study of a consensus protocol of a multiagent system with a fixed, directed network topology and a distributed delay in the communication channels. More precisely, let the directed graph

$$\mathcal{G}(V, E, \mathbb{A}) \tag{16.7}$$

be characterized by the node set $V = \{1, \ldots, p\}$, a set of edges E where $(k, l) \in E$ if and only if $\alpha_{k,l} \neq 0$, and a weighted adjacency matrix $\mathbb{A}$ with zero diagonal and nondiagonal entries $\alpha_{k,l}$. Let each node correspond to an agent whose dynamics are described by

$$\dot{v}_k(t) = u_k(t), \ k = 1, \ldots p. \tag{16.8}$$

Then (16.2) is obtained by applying the following protocol to (16.8):

$$u_k(t) = \sum_{(k,l)\in E} \alpha_{k,l} \int_0^{\infty} f(\theta)(v_l(t-\theta) - v_k(t-\theta))d\theta, \ \ k = 1, \ldots, p.$$

Note that the corresponding problem for an undirected graph (where $\mathbb{A}$ is symmetric) and a constant time-delay was investigated in [240]. Note also that (16.4) implies that the graph $\mathcal{G}$ is strongly connected (see [68] for the definition). It is important to mention that the results derived in this discussion remain valid when this condition is replaced by the weaker condition that the graph is strongly connected.

The structure of the chapter is as follows. Section 16.2 contains the necessary extensions of the stability theory of Chapter 1. Section 16.3 is devoted to the characterization of the region in the delay-parameter space for which the system (16.2)–(16.5) reaches a consensus. Some illustrative examples are presented in Section 16.4. In Section 16.5 the stability analysis of other types of models encountered in the literature is commented on. Some concluding notes and references complete the chapter.

Throughout the chapter, the following functions will be used.

Definition 16.1. *For $n \in \mathbb{N}$, let $g_n : \mathbb{R}_+ \to \mathbb{R}_+$ be such that $y = g_n(x)$ is the positive solution of $|y(1 + jy)^n| = x$.*

16.2 Extension of stability theory to systems with distributed delays

Inspired by the structure of (16.2)–(16.5), we extend the stability theory for systems with pointwise delays, presented in Chapter 1, to systems with distributed delays of the form

$$\dot{x}(t) = A \int_0^{\infty} f(\theta)x(t-\theta)d\theta, \tag{16.9}$$

where $x(t) \in \mathbb{C}^{p\times 1}$, $A \in \mathbb{C}^{p\times p}$, and f is given by (16.5). A second extension consists of characterizing consensus problems in the frequency domain, for systems with both constant and distributed delays.

A solution of (16.9) is uniquely determined for an initial condition ϕ, which belongs to the set $\mathcal{F}((-\infty,\ 0],\ \mathbb{C}^{p\times 1})$, defined as

$$\mathcal{F}((-\infty,\ 0],\ \mathbb{C}^{p\times 1}) := \left\{\phi \in \mathcal{C}((-\infty,\ 0],\ \mathbb{C}^{p\times 1}) :\ \|\phi\|_f := \int_{-\infty}^{0} \|f(-\theta)\phi(\theta)\|_2 d\theta < \infty\right\}$$

and equipped with $\|\cdot\|_f$. Denote by $t \in (-\infty,\ \infty) \to x(\phi)(t)$ the forward solution of (16.9) with initial condition ϕ. In this way, stability definitions can be formulated in a similar way as for systems with constant delays; see [106] for the latter. We say, for instance, that the zero solution of (16.9) is asymptotically stable if and only if

$$\forall\epsilon > 0 \exists\delta > 0\ \forall\phi \in \mathcal{F}((-\infty,\ 0],\ \mathbb{C}^{p\times 1})\ \ \|\phi\|_f < \delta \Rightarrow \forall t \geq 0\ \|x(\phi)(t)\|_2 < \epsilon,$$
$$\forall\phi \in \mathcal{F}((-\infty,\ 0],\ \mathbb{C}^{p\times 1})\ \lim_{t\to\infty} x(\phi)(t) = 0.$$

The substitution of a sample solution of the form $x(t) = e^{\lambda t}X$, with $X \in \mathbb{C}^{p\times 1}$, in (16.9) leads us to the *characteristic equation*

$$\det\left(\lambda I - A\frac{e^{-\lambda\tau}}{(1+\lambda T)^n}\right) = 0, \tag{16.10}$$

which can be factored as

$$\Pi_{k=1}^{p}\left(\lambda - \frac{\mu_k e^{-\lambda\tau}}{(1+\lambda T)^n}\right) = 0, \tag{16.11}$$

with $\mu_k,\ k = 1,\dots,p$, the eigenvalues of A. As we shall see, the roots distribution of (16.10)–(16.11) determines the stability properties of (16.9). However, the commonly used arguments, which are based on a spectral decomposition of the solutions (see, for instance, [67, 106]) cannot be directly applied to a system of the form (16.9). A major obstacle is the fact that functions of the form $e^{\lambda t}X$, $t \leq 0$, do not belong to the space $\mathcal{F}((-\infty,\ 0],\ \mathbb{C}^{p\times 1})$ if $\Re(\lambda) < 1/T$. We shall therefore develop arguments based on a *comparison system*.

Formally, with

$$y(t) = \int_0^{\infty} f(\theta+\tau)x(t-\theta)d\theta = \int_{-\infty}^{t} f(t+\tau-\theta)x(\theta)d\theta,$$

we get

$$\begin{cases} y'(t) &= \int_{-\infty}^{t} f'(t+\tau-\theta)x(\theta)d\theta, \\ &\vdots \\ y^{(n-1)} &= \int_{-\infty}^{t} f^{(n-1)}(t+\tau-\theta)x(\theta)d\theta, \\ y^{(n)}(t) &= f^{(n-1)}(\tau)x(t) + \int_{-\infty}^{t} f^{(n)}(t+\tau-\theta)x(\theta)d\theta, \end{cases}$$

which leads to

$$\begin{aligned} \left(T\tfrac{d}{dt}+I\right)^n y(t) &= T^n f^{(n-1)}(\tau)x(t) + \int_{-\infty}^{t} \left(T\tfrac{d}{dt}+I\right)^n f(t+\tau-\theta)d\theta \\ &= x(t). \end{aligned}$$

We conclude that a solution $x(\phi)(t)$ of (16.9) satisfies

$$\begin{cases} \dot{x}(t) &= Ay(t-\tau), \\ \left(T\tfrac{d}{dt}+I\right)^n y(t) &= x(t) \end{cases} \tag{16.12}$$

for $t \geq 0$, if the initial value problem (16.12) is accordingly initialized with

$$\begin{cases} x(0) = \phi(0), \\ y(\theta) = \int_{-\infty}^{\theta} f(\tau+\theta-\xi)\phi(\xi)d\xi, \quad \theta \in [-\tau,\ 0], \\ y^{(i)}(0) = \int_{-\infty}^{0} f^{(i)}(\tau-\xi)\phi(\xi)d\xi, \quad i = 1,\ldots,n-1. \end{cases} \tag{16.13}$$

Note that the integrals in the right-hand side of (16.13) are defined and bounded because $f^{(i)}(\xi)$, $i = 1,\ldots,n-1$, has the same asymptotic behavior as $f(\xi)$ as $\xi \to \infty$. When letting $z = [x^T\ y^T\ y'^T \cdots (y^{(n-1)})^T]^T$, the *comparison system* (16.12) can be written in a first-order form as

$$\dot{z}(t) = \bar{A}z(t) + \bar{B}z(t-\tau), \tag{16.14}$$

where

$$\bar{A} = \begin{bmatrix} 0 & 0 & 0 & \cdots & 0 \\ 0 & 0 & I & & \\ & & & \ddots & \\ & & & & I \\ \frac{I}{T^n} & -\binom{n}{n}\frac{1}{T^n} & & \cdots & -\binom{n}{1}\frac{1}{T^1} \end{bmatrix} \text{ and } \bar{B} = \begin{bmatrix} 0 & A & 0 & \cdots & 0 \\ 0 & 0 & 0 & & 0 \\ \vdots & & & & \vdots \\ 0 & & \cdots & & 0 \end{bmatrix}.$$

The initial conditions for (16.14) are assumed to belong to the space $\mathcal{C}([-\tau,\ 0],\ \mathbb{C}^{(n+1)p\times 1})$. The next lemma summarizes the established relation between the solutions of (16.9) and (16.14) and addresses a partial converse.

Lemma 16.2. *If $x(t)$, $t \in \mathbb{R}$, is a solution of (16.9), then there exists a solution $z(t)$, $t \geq -\tau$, of (16.14) such that $[I\ 0\cdots 0]z(t) = x(t)$ for all $t \geq -\tau$. If (16.14) has a solution of the form $Ze^{\lambda t}$, $t \geq -\tau$, where $\Re(\lambda) \geq 0$ and $Z \in \mathbb{C}^{(n+1)p\times 1} \setminus \{0\}$, then $[I\ 0\cdots 0]Ze^{\lambda t}$, $t \in \mathbb{R}$, is a nontrivial solution of (16.9).*

Proof. The first assertion follows from the above construction, and an extension of (16.13) on the interval $[-\tau,\ 0]$.

To prove the second assertion, we partition Z according to the structure of $\bar{A}$ and $\bar{B}$ as $Z = [X^T \; Y_0^T \cdots Y_{n-1}^T]^T$. Substituting $Ze^{\lambda t}$ in (16.14) yields

$$\left(\lambda I - A\frac{e^{-\lambda\tau}}{(1+\lambda T)^n}\right) X = 0, \tag{16.15}$$

$$Y_i = \frac{\lambda^i}{(1+\lambda T)^n} X, \quad i = 0, \ldots, n-1. \tag{16.16}$$

It follows that $Z \neq 0$ if and only if $X \neq 0$. Furthermore, (16.15) implies that $Xe^{\lambda t}$ satisfies (16.9) for all $t \in \mathbb{R}$. The function $Xe^{\lambda t}$, $t \in \mathbb{R}$, is a solution since $\Re(\lambda) \geq 0$, and thus $Xe^{\lambda t}$, $t \leq 0$, belongs to $\mathcal{F}((-\infty,\, 0], \mathbb{C}^{p\times 1})$. □

For the system (16.14) a well-established stability theory exists [106, 67]. For instance, its zero solution is asymptotically stable if and only if all the roots of its characteristic equation,

$$\det(\lambda I - \bar{A} - \bar{B}e^{-\lambda\tau}) = 0, \tag{16.17}$$

or, equivalently,

$$\det\left(\begin{bmatrix} \lambda I & -Ae^{-\lambda\tau} \\ -I & (1+\lambda T)^n \end{bmatrix}\right) = 0, \tag{16.18}$$

are in $\mathbb{C}_-$. Note that (16.18) reduces to (16.10) if $\lambda \neq -1/T$. Combining this result with Proposition 16.2 results in the following proposition.

Proposition 16.3. *The zero solution of (16.9) is asymptotically stable if and only if all roots of (16.10) are in $\mathbb{C}_-$.*

Next, we derive conditions on the roots of (16.18) for which the system (16.9) solves a consensus problem. This stability property is defined in the following way.

Definition 16.4. *The system (16.9) solves a consensus problem if and only if*

$$\forall \phi \in \mathcal{F}((-\infty,\, 0],\, \mathbb{C}^{p\times 1}) \;\; \lim_{t\to\infty} x(\phi)(t) = \chi(\phi)\, E_0,$$

where $\chi(\phi) \in \mathbb{C}$ and $E_0 = [1 \cdots 1]^T$. The function $\chi : \mathcal{F}((-\infty,\, 0],\, \mathbb{C}^{p\times 1}) \to \mathbb{C}$ is called the consensus functional. The system (16.9) solves a nontrivial consensus problem if and only if it solves a consensus problem and the consensus functional is not identically zero.

We follow the same methodology as for the aymptotic stability condition: we first address a consensus problem for a system with a constant delay, and next we treat (16.9) using Lemma 16.2.

Lemma 16.5. *The system*

$$\dot{x}(t) = A_0 x(t) + A_1 x(t-\tau), \tag{16.19}$$

with initial condition $\phi \in \mathcal{C}([-\tau,\, 0], \mathbb{C}^{p\times 1})$, solves a nontrivial consensus problem if and only if all its characteristic roots are in the open LHP, excepting a zero root with *multiplicity one, and* $(A_0 + A_1)E_0 = 0$, *with* $E_0 = [1 \ \cdots 1]^T$. *The consensus functional* $\chi : \mathcal{C}([-\tau,\ 0], \mathbb{C}^{p\times 1}) \to \mathbb{C}$ *can be expressed as*

$$\chi(\phi) = \frac{V_0^T \left(x(\phi)(\hat{t}) + A_1 \int_{\hat{t}-\tau}^{\hat{t}} x(\phi)(\theta)\, d\theta\right)}{V_0^T (I + \tau A_1) E_0}, \tag{16.20}$$

where V_0 *is the left null vector of* $(A_0 + A_1)$ *and*

$$\hat{t} \geq p\tau - \limsup_{r\to\infty} \frac{\log F(r)}{r}, \quad F(r) = \max_{|\lambda|=r} \det\left(\lambda I - A - Be^{-\lambda\tau}\right).$$

If, in addition, $A_0 = 0$, *then the consensus functional is given by*

$$\chi(\phi) = \frac{V_0^T \phi(0)}{V_0^T E_0}. \tag{16.21}$$

Proof. The first assertion is a trivial corollary of the spectrum determined growth property of the solutions of (16.19); see, e.g., [106, 67]. The assertions on the form of the consensus functional follow from a spectral decomposition of the solutions, using Theorem 8.4 of [67]. In such a decomposition the stationary term corresponds to the consensus functional, and can be isolated by employing orthogonality properties of left and right eigenfunctions. See [187] for the details. □

Proposition 16.6. *The system (16.9) solves a nontrivial consensus problem if and only if all roots of (16.10) are in the open LHP, excepting a root at zero with multiplicity one, and* $AE_0 = 0$, *with* $E_0 = [1 \ldots 1]^T$. *The corresponding consensus functional* $V : \mathcal{F}((-\infty,\ 0], \mathbb{C}^{p\times 1}) \to \mathbb{C}$ *satisfies*

$$\chi(\phi) = \frac{V_0^T \phi(0)}{V_0^T E_0}, \tag{16.22}$$

where V_0 *is the left eigenvector of* A *corresponding to the zero eigenvalue.*

Proof. The first assertion is based on Lemma 16.2. Expression (16.22) is obtained by simplifying the consensus functional for the corresponding comparison system (16.12), to which Proposition 16.6 is applicable. See [187] for the details. □

Remark 16.7. *Expressions (16.21) and (16.22) also follow from a geometric argument. As in both cases* $V_0^T \dot{x}(t) = 0$ *the solutions* $x(\phi(t))$ *are constrained to the plane* $V_0^T x = V_0^T \phi(0)$ *for all* $t \geq 0$. *Furthermore, a constant stationary solution must be a multiple of* E_0. *Thus* $x^*(\phi) = \lim_{t\to\infty} x(\phi)(t)$ *satisfies the equations*

$$\begin{cases} V_0^T x^*(\phi) = V_0^T \phi(0), \\ x^*(\phi) = \chi(\phi) E_0, \end{cases}$$

which can be interpreted as the intersection of the plane through $\phi(0)$ *and perpendicular to* V_0 *with a line with slope* E_0. *A similar argument was used in Section X of [240].*

16.3 Conditions for the realization of a consensus

We perform a stability analysis of the system (16.2)–(16.5) in the (T, τ) parameter space. In particular, we give necessary and sufficient conditions such that a consensus is reached for all initial conditions. We only present the main results. For the detailed proofs we refer to [187].

The system (16.2)–(16.5) can be written in the form (16.9), yet has some special properties due to the induced structure of A, which we outline first. Next, we present the main results.

16.3.1 Prerequisites

The system (16.2)–(16.5) is of the form (16.9), where $A = [a_{k,l}]$ is defined as

$$a_{k,l} = \begin{cases} \alpha_{k,l}, & k \neq l, \\ -\sum_{i=1,\, i\neq k}^{p} \alpha_{k,i}, & k = l. \end{cases} \tag{16.23}$$

Note that in the context of multiagent systems $-A$ is typically called the graph Laplacian of (16.7). By construction A has the following property.

Property 16.8. *All eigenvalues of A, defined by (16.23), are in $\mathbb{C}_-$, excepting a zero eigenvalue with multiplicity one.*

Note that zero also appears as a root of (16.10) and (16.23), whatever the values of T, τ, and n. If all other roots are in $\mathbb{C}_-$, we have from Proposition 16.6 that the system (16.2)–(16.5) solves a (nontrivial) consensus problem with delay. In the car following application the consensus variables are the speed of the vehicles. This means that, whatever the initial values, the speed of the vehicles will eventually converge to a common value (which depends on the initial values).

In what follows we shall use the following terminology to characterize parameter values in the (T, τ) space for which a consensus is reached.

Definition 16.9. *The consensus region of (16.2)–(16.5) in the (T, τ) parameter space is the set of parameters (T, τ) for which the system (16.2)–(16.5) solves a consensus problem.*

Property 16.8 and the factorization of the characteristic equation (16.10) as (16.11) leads us to study the zeros location of the function

$$\xi(\lambda;\ T, \tau) := \lambda(1 + \lambda T)^n e^{\lambda\tau} - \mu, \quad \mu \in \mathbb{C}_-, \tag{16.24}$$

as a function of the parameters T and τ. We have the following result, which can be proved using the two-step approach of Chapter 4 consisting of first characterizing the crossing frequencies, where zeros can cross the imaginary axis as the parameters are changed, and, second, characterizing the corresponding crossing direction.

Proposition 16.10. *If μ is real and $n = 1$, then the zeros of (16.24) are in $\mathbb{C}_-$ if and only if $T \in [0, \infty)$ and $\tau \in [0, \tau_\mu(T))$. Otherwise, the zeros are in $\mathbb{C}_-$ if and only if $T \in [0, T_\mu)$ and $\tau \in [0, \tau_\mu(T))$.*

16.3.2 Computation of stability regions

Taking into account the factorization of (16.10) as (16.11), Property 16.8, and Proposition 16.10, we obtain the following characterization of the consensus region (cf. Definition 16.9) of the system (16.2)–(16.5) in the (T, τ)-space.

Theorem 16.11. *If $n = 1$ and all eigenvalues of A, defined by (16.23), are real, then the consensus region of (16.2)–(16.5) in the (T, τ)-plane is unbounded and characterized by*

$$T \in [0, \infty), \quad \tau \in [0, \tau^*(T)),$$

where

$$\tau^*(T) = \min_{k=1,\ldots,p,\ \mu_k \neq 0} \frac{|\angle(\mu_k)| - \angle(j\omega_k(T)(1 + j\omega_k(T))^n)}{\frac{\omega_k(T)}{T}} \tag{16.25}$$

and $\omega_k(T) = g_n(T|\mu_k|)$.
Otherwise, the consensus region is bounded and characterized by

$$T \in [0, T^*), \quad \tau \in [0, \tau^*(T)),$$

where

$$T^* = \min_{k=1,\ldots,p,\ \Im(\mu_k)>0} \frac{\tan\left(\frac{\angle(\mu_k) - \frac{\pi}{2}}{n}\right)}{|\mu_k|\left[\cos\left(\frac{\angle(\mu_k) - \frac{\pi}{2}}{n}\right)\right]^n} \tag{16.26}$$

and $\tau^*(T)$ is given by (16.25).

Based on this result the consensus region of (16.2)–(16.5) can be computed fully *automatically*. For large p the overall computational complexity is determined by the computation of the eigenvalues of the p-by-p matrix A.

Theorem 16.11 does not make assumptions on the multiplicity of the eigenvalues of A and is generally applicable. If A has eigenvalues with multiplicity larger than one, then the stability study of (16.2)–(16.5) is even facilitated as not all factors in (16.11) are different. The following proposition clarifies the connection between multiple eigenvalues of A and multiple eigenvalues of the comparison system (16.14) of (16.2)–(16.5).

Proposition 16.12. *Let $\hat{\mu}$ be a nonzero eigenvalue of A with multiplicity m_1 and a corresponding eigenspace of dimension m_2. Then the roots of*

$$\lambda(1 + \lambda T)^n e^{\lambda\tau} - \hat{\mu} = 0 \tag{16.27}$$

with multiplicity m_3 are eigenvalues of the comparison system (16.14) with multiplicity $m_1 m_3$ and an eigenspace of dimension m_2. Furthermore, if $m_3 = 1$, then these roots smoothly depend on the parameters T and τ.

Remark 16.13. *If $m_1 > m_2$, then the roots of (16.27) with multiplicity one (this is, for instance, always the case for roots on the imaginary axis) are multiple, nonsemisimple* *eigenvalues of (16.14), yet they smoothly depend on the parameters T and τ. Small changes of T and τ do not lead to a splitting of these multiple eigenvalues.*

The next proposition reveals a scaling property of the consensus region.

Proposition 16.14. *If the matrix A is scaled with a factor $\epsilon > 0$, then the consensus region of (16.2)–(16.5) in the (T, τ)-plane is scaled with a factor ϵ^{-1} in both directions.*

Remark 16.15. *Proposition 16.14 implies an inherent trade-off between the rate with which the undelayed system ($\tau = T = 0$) reaches a consensus (determined by the rightmost nonzero eigenvalue of A), and the robustness of this stability property w.r.t. delays. Such an observation was already made in [240], where the case of a symmetric matrix A and a pointwise delay was dealt with.*

In the remainder of this section, we refine Theorem 16.11 to two special cases where exploiting the additional structure leads to a *simpler* characterization of the consensus region, and also allows an *analytical* expression for the solutions corresponding to an onset of instability. The latter can be obtained from the eigenfunctions corresponding to the characteristic roots on the imaginary axis (which are the eigenvalues of the infinitesimal generator of the evolution operator associated with the system; see Chapter 1). The following result corresponds to the situation where all cars/drivers have an identical behavior and the reaction of a driver is determined by the preceding car only.

Proposition 16.16. *Consider the system (16.2)–(16.5), where*

$$\alpha_{k,l} = \begin{cases} \alpha, & \lfloor k - l \rfloor = 1, \\ 0, & \text{otherwise}. \end{cases} \tag{16.28}$$

If $n = 1$ and $p = 2$, then the consensus region in the (T, τ)-plane is unbounded and characterized by

$$T \in [0, \infty), \quad \tau \in [0, \tau^*(T)),$$

where

$$\tau^*(T) = \frac{\frac{\pi}{p} - n \arctan(\omega(T))}{\frac{\omega(T)}{T}}, \quad \omega(T) = g_n\left(2\alpha T \sin\left(\frac{\pi}{p}\right)\right). \tag{16.29}$$

Otherwise, the consensus region is bounded and characterized by

$$T \in [0, T^*), \quad \tau \in [0, \tau^*(T)),$$

where

$$T^* = \frac{\tan\left(\frac{\pi}{pn}\right)}{2\alpha \sin\left(\frac{\pi}{p}\right)\left(\cos\left(\frac{\pi}{pn}\right)\right)^n} \tag{16.30}$$

and $\tau^*(T)$ is given by (16.29).

For $\tau = \tau^*(T)$ the stationary solutions are backward travelling waves:

$$\begin{bmatrix} v_1^s(t) \\ v_2^s(t) \\ \vdots \\ v_p^s(t) \end{bmatrix} = C_1 \begin{bmatrix} \cos\left(\frac{\omega(T)}{T}t + \varphi\right) \\ \cos\left(\frac{\omega(T)}{T}t + \varphi - \frac{2\pi}{p}\right) \\ \vdots \\ \cos\left(\frac{\omega(T)}{T}t + \varphi - \frac{2\pi(p-1)}{p}\right) \end{bmatrix} + C_2 \begin{bmatrix} 1 \\ 1 \\ \vdots \\ 1 \end{bmatrix}, \tag{16.31}$$

where $\omega(T)$ is defined in (16.29) and the constants C_1, C_2, and ϕ depend on the initial conditions.

Second, we consider the case where (16.2)–(16.5) is of the form (16.9), with the matrix A symmetric. Although this is not a realistic assumption from the car *following* application point of view, it makes sense in the context of consensus algorithms for multiagent systems. The symmetry of A there corresponds to an *undirected* network topology.

Proposition 16.17. *Consider the system (16.2)–(16.5) with A symmetric. If $n = 1$, then the consensus region of (16.2)–(16.5) in the (T, τ)-plane is unbounded and characterized by*

$$T \in [0, \infty), \quad \tau \in [0, \tau^*(T)),$$

where

$$\tau^*(T) = \frac{\frac{\pi}{2} - n\arctan(\omega(T))}{\frac{\omega(T)}{T}}, \quad \omega(T) = g_n(T|\lambda_{\max}(A)|). \tag{16.32}$$

Otherwise, the consensus region is bounded and characterized by

$$T \in [0, T^*), \quad \tau \in [0, \tau^*(T)),$$

where

$$T^* = \frac{\tan\left(\frac{\pi}{2n}\right)}{|\lambda_{\max}(A)|\left[\cos\left(\frac{\pi}{2n}\right)\right]^n} \tag{16.33}$$

and $\tau^*(T)$ is given by (16.32).

If, in addition,

$$\alpha_{k,l} = \begin{cases} \alpha, & \lfloor k-l \rfloor = 1 \text{ or } \lfloor l-k \rfloor = 1, \\ 0, & \text{otherwise,} \end{cases} \tag{16.34}$$

then

$$\lambda_{\max}(A) = \begin{cases} -4\alpha \ \ (\text{multiplicity } 1), & p \text{ even}, \\ -2\alpha\left(1 + \cos\left(\frac{\pi}{p}\right)\right) \ \ (\text{multiplicity } 2), & p \text{ odd}. \end{cases}$$

The stationary solutions for $\tau = \tau^*(T)$ take the form

$$\begin{bmatrix} v_1^s(t) \\ \vdots \\ v_{p-1}^s(t) \\ v_p^s(t) \end{bmatrix} = C_1 \begin{bmatrix} (-1)^{p-1} \\ \vdots \\ (-1) \\ 1 \end{bmatrix} \cos\left(\frac{\omega(T)}{T}t + \varphi_1\right) + C_2 \begin{bmatrix} 1 \\ \vdots \\ 1 \\ 1 \end{bmatrix} \tag{16.35}$$

if p is even, and

$$\begin{bmatrix} v_1^s(t) \\ \vdots \\ v_{p-1}^s(t) \\ v_p^s(t) \end{bmatrix} = C_3 \begin{bmatrix} (-1)^{p-1}\cos\left(\frac{\pi(p-1)}{p}\right) \\ \vdots \\ (-1)\cos\left(\frac{\pi.1}{p}\right) \\ 1 \end{bmatrix} \cos\left(\frac{\omega(T)}{T}t+\varphi_2\right)$$
$$+C_4 \begin{bmatrix} (-1)^{p-1}\sin\left(\frac{\pi(p-1)}{p}\right) \\ \vdots \\ (-1)\sin\left(\frac{\pi.1}{p}\right) \\ 0 \end{bmatrix} \cos\left(\frac{\omega(T)}{T}t+\varphi_3\right) + C_5 \begin{bmatrix} 1 \\ \vdots \\ 1 \\ 1 \end{bmatrix} \quad (16.36)$$

if p is odd. The constants $C_1, \ldots, C_5$ and $\varphi_1, \ldots, \varphi_3$ depend on the initial conditions.

Remark 16.18. *The consensus functional satisfies*

$$V(\phi) = \frac{1}{p}[1 \cdots 1]\phi(0).$$

This follows from (16.6), taking into account that $V_0 = E_0$ if A is symmetric. Hence, under the conditions of the above proposition an average *consensus problem is solved, in the sense that all components of a solution $x(\phi)(t)$ converge to the average of these components at the starting time, i.e., $\phi(0)$. Note that $\phi(\theta)$, $\theta < 0$, has* no *influence on the limit reached.*

Remark 16.19. *Expression (16.33) reduces to the statement of Theorem 10 in [240] if $T \to 0+$.*

Let us briefly compare the stationary solutions (16.31) with (16.35)–(16.36). In the former case, the directed "network topology" (a driver only reacts—with some delay—on its predecessor, and not the other way around) naturally leads to a backward traveling wave. In the latter case, one would from the symmetry of the coupling intuitively expect a stationary wave, where subsequent agents oscillate in antiphase. This is indeed the case for (16.35) which holds if p is even. However, if p is odd, such a solution is *incompatible* with the ring configuration, and (16.36) holds. If p is large, (16.36) can be seen as an approximation of a stationary wave with subsequent agents oscillating in antiphase that is compatible with the ring configuration.

16.4 Examples

As a first example we compute the consensus regions in the (T, τ)-plane of system (16.2)–(16.5) with $n = 1$ and

$$A = \begin{bmatrix} -5 & 0 & 0 & 5 \\ 1 & -1 & 0 & 0 \\ 0 & 1 & -1 & 0 \\ 0 & 0 & 5 & -5 \end{bmatrix}. \quad (16.37)$$

The eigenvalues of this matrix are given by

$$\mu_1 = -6,\ \mu_2 = \bar{\mu}_3 = -3 + j,\ \mu_4 = 0.$$

An application of Theorem 16.11 yields the consensus region

$$T \in [0,\ 3),\ \tau \in [0,\ \tau^*(T)),$$

where the function $T \to \tau^*(T)$ is displayed in Figure 16.1 as a solid line. The dotted lines bound the "stability" regions of the auxiliary equations

$$\lambda(1 + \lambda T)e^{\lambda\tau} - \mu_{1,2} = 0, \tag{16.38}$$

which are described by Proposition 16.10. The stability region corresponding to μ_1 is unbounded as μ_1 is real and $n = 1$.

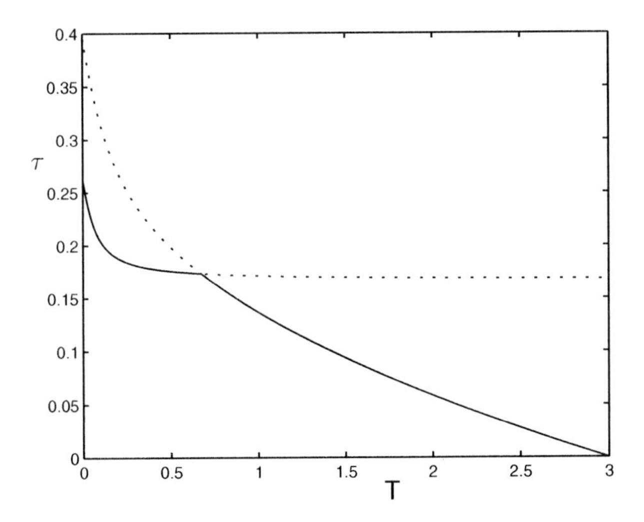

Figure 16.1. *Boundary of the consensus regions of (16.2)–(16.5) with parameters (16.37) (solid curve). Boundaries of stability regions of (16.38) (dotted curves).*

To illustrate the asymptotic behavior when the number of cars is large, we take a system satisfying condition (16.28) of Proposition 16.16. Figure 16.2 shows the consensus region in the (T, τ)-plane for $\alpha = 2$, $n = 1$, and $p = 2^k$, $k = 1, \ldots, 4$. It follows from (16.29) that as $p \to \infty$, the boundary of the consensus region uniformly converges to the function

$$\tau_l^*(T) = \frac{1}{2\alpha} - nT,$$

indicated in Figure 16.2 with a dashed line.

Finally, we consider the system

$$\dot{v}_k(t) = \sum_{l=1}^{3} \alpha_{k,\lfloor k-l \rfloor} \int_0^{\infty} f(\theta)(v_{\lfloor k-l \rfloor}(t-\theta) - v_k(t-\theta))\mathrm{d}\theta,\ \ k = 1, \ldots, 1000, \tag{16.39}$$

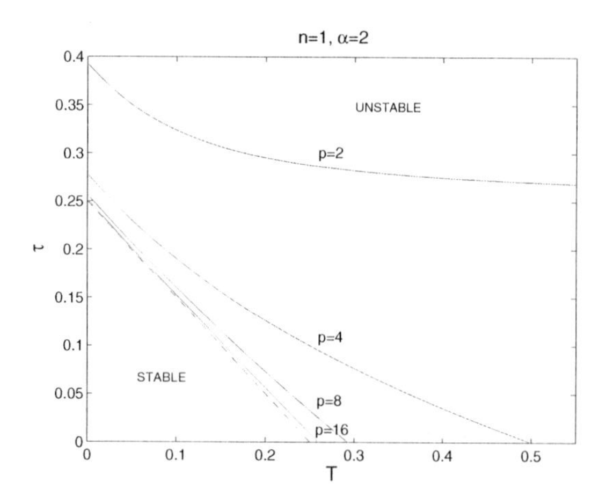

Figure 16.2. *Boundary of the consensus region of a system satisfying (16.28), with parameters $\alpha = 2$ and $n = 1$.*

where f is given by (16.5), with $n = 2$. The parameters

$$\begin{aligned} \alpha_{k,\lfloor k-1 \rfloor} &\in [1,\ 5], \\ \alpha_{k,\lfloor k-2 \rfloor} &\in \left[0,\ \tfrac{3}{4}\alpha_{k,\lfloor k-1 \rfloor}\right], \\ \alpha_{k,\lfloor k-3 \rfloor} &\in \left[0,\ \tfrac{3}{4}\alpha_{k,\lfloor k-2 \rfloor}\right], \qquad k = 1, \dots, 1000 \end{aligned} \tag{16.40}$$

are randomly generated according to a *uniform distribution* over the above intervals. For 30 sets of parameters obtained in this way, the consensus region in the (T, τ)-plane was computed. The results are displayed in Figure 16.3.

16.5 Other models

For general time-delay systems of retarded type with multiple constant delays and distributed delays with gamma-distribution kernels, stability and/or consensus regions of equilibria in two-parameter spaces can be computed semiautomatically by numerical continuation; see, for instance, [162] and the package DDE-BIFTOOL [74]. Such an approach involves the discretization of an infinite-dimensional evolutionary operator, associated with the time-delay system, to compute the rightmost eigenvalues. Roughly speaking the computation of the boundary of a stability or consensus region involves solving r eigenvalue problems of dimension $pq \times pq$, where p is the dimension of the system, q denotes the number of discretization points, and r is the number of points on the stability crossing curves where stability information is checked. When using Theorem 16.11 only one eigenvalue problem of dimension $p \times p$ needs to be solved to determine the complete stability region of (16.2)–(16.5) in the (T, τ)-space. The underlying reason is that the structure of the system allowed a decomposition into small subproblems, which is apparent from the form of the characteristic equation (16.11).

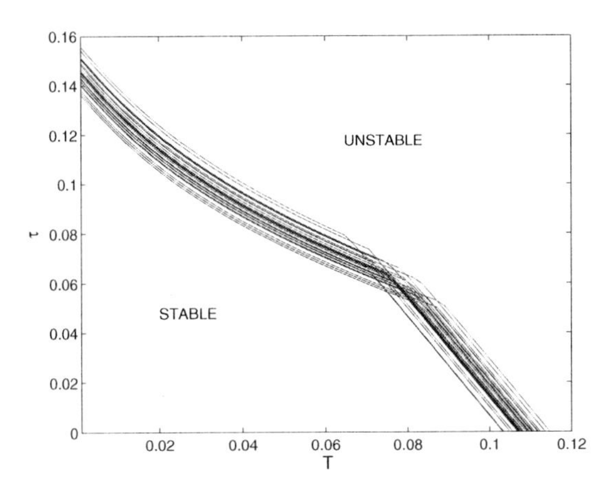

Figure 16.3. *Consensus region of (16.39)–(16.40), for* 30 *different data sets.*

Let us now take a brief look at the so-called optimal velocity models, also frequently encountered in the literature. The linearization around the equilibrium of the models studied in [8], respectively [64, 243, 244] and the references therein, takes the form

$$\tau_k \ddot{x}_k(t) + \dot{x}_k(t-\tau) = \alpha(x_{\lfloor k-1 \rfloor}(t-\tau) - x_k(t-\tau)),\ k = 1, \ldots, p, \tag{16.41}$$

respectively

$$\tau_k \ddot{x}_k(t) + \dot{x}_k(t) = \alpha(x_{\lfloor k-1 \rfloor}(t-\tau) - x_k(t-\tau)),\ k = 1, \ldots, p. \tag{16.42}$$

In both cases, x_k denotes the position of the kth vehicle. The left-hand side models the dynamics of the vehicle and the right-hand side is the reference velocity, which is a function of the distance to the preceding vehicle and models the behavior of the driver. Note that (16.41) and (16.42) can be generalized to

$$\begin{aligned} \ddot{x}_k(t) = \int_0^\infty f(\theta) \bigg(& \Big(\textstyle\sum_{i=1}^{p-1} \alpha_{k,\lfloor k-i \rfloor} (x_{\lfloor k-i \rfloor}(t-\theta) - x_k(t-\theta)) \Big) \\ & -\beta_k \dot{x}_k(t-\theta) \bigg) \mathrm{d}\theta - \gamma_k \dot{x}_k(t),\ k = 1, \ldots, p, \end{aligned} \tag{16.43}$$

with f given by (16.5). General purpose tools for the stability and bifurcation analysis of time-delay systems like DDE-BIFTOOL can be applied directly to (16.43). However, if all vehicles have similar characteristics (but not necessarily drivers), that is, $\alpha_k \equiv \alpha$, $\beta_k \equiv \beta$, then the characteristic equation can again be factorized:

$$0 = \det \left(\lambda(\lambda+\gamma) I - (A - \beta\lambda I) \frac{e^{-\lambda\tau}}{(1+\lambda T)^n} \right) = \Pi_{k=1}^{p} \left(\lambda(\lambda+\gamma) - \frac{(\mu_k - \beta\lambda) e^{-\lambda\tau}}{(1+\lambda T)^n} \right),$$

where A is given by (16.23) and $\mu_1, \ldots \mu_p$ denote its eigenvalues. Also here, it's beneficial to exploit this decomposition into small subproblems, in particular if the number of vehicles is large.

16.6 Notes and references

The stability analysis of a linear system including a γ-distributed delay with a gap for modeling traffic flow dynamics was considered. A complete characterization of the regions in the delay-parameter space, where a consensus is reached for all initial conditions, was obtained. In particular, by exploiting the structure of the system analytical expressions were derived for the bounds on the parameters of the delay distribution. These expressions give rise to a fully automatic computation of the consensus region, whose complexity is determined by the computation of the eigenvalues of one matrix with dimensions equal to the number of vehicles. Some illustrative examples were also presented.

From a theoretical point of view, some stability theory for linear systems with γ-distributed delays was developed. As this type of distributed delay is characterized by kernels with an infinite support, which prohibits a full spectral decomposition of the solutions, the relation between the growth properties of the solutions and the roots of an appropriate characteristic equation was established via a comparison system with constant delays. Necessary and sufficient conditions for the realization of a consensus problem and an explicit construction of the consensus functional were provided, for systems with constant and distributed delays.

The results presented in this chapter are based on [187, 205] and the references therein.

Chapter 17

Stability analysis of delay models in biosciences

17.1 Introduction

Most of the models encountered in biosciences that represent competition between populations, epidemics, or respiration control mechanisms include a particular common element used to describe a reaction chain (distributed character [168]), a transport process (breathing process [10, 11] in the physiological circuit controlling the carbon dioxide level in the blood [316]); storing nutrients or cell cycles (in the case of controlling the supply of nutrients to a growing population of microorganisms in some chemostat [280, 7]); and latency and short intercellular phases (in epidemics, for example, cell-to-cell spread models in a particular compartment, the bloodstream [56]). Such a particular element is nothing else than the *transport* and/or *propagation delay*, which can be discrete or distributed, constant or time-varying.

The list of applications of delay models in biosciences is extensive. For instance, one can cite a density dependent feedback mechanism to respond to changes in population density (that never takes place *instantaneously*, see, e.g., [92, 144]); the spread (*propagation*) of infections in a family, and epidemics with intermediate classes (that is, the presence of "individuals" for a given period such that they are "exposed, but not (*necessarily*) yet infectious," see, e.g., [119]); or *recurrent diseases*, as suggested in various relapse-recovery models (after a given period, an infected individual returns to being fully susceptible again [119]). Further remarks, discussions on delay models in biosciences, and a large list of references can be found in [99, 7, 217].

Without discussing the assumptions needed to derive the biological models considered in what follows, and without considering the problems related to the model representation, this chapter focuses on the stability analysis of the linearization of some of the delay models mentioned above represented by DDEs. The first model is a second-order system including one delay and is encountered in modeling human respiration taken from [316]. The second model, taken from [66], is still a second-order system, but includes four independent delays and is encountered in modeling immune dynamics in chronic leukemia.

Both models have some appealing properties that, to the best of our knowledge, have not been sufficiently exploited in the literature. More precisely, the corresponding delay matrices are of rank one, which allows us to rewrite the characteristic function of the systems as particular quasipolynomials with noninteracting, independent delays. In such a case, the geometrical approach considered in Chapter 4 can be applied, giving some simple characterization of the stability regions with respect to the delay parameters.

The chapter is organized as follows: Section 17.2 is devoted to the stability analysis of the linearization of a two-compartment representation including one transport delay. Next, Section 17.3 is devoted to the stability analysis of some immune dynamics model encountered in chronic leukemia and including four delays in a large range. The geometry of the stability regions is derived in the corresponding delay-parameter space. Various interpretations complete the presentation. Finally, some notes and references end the chapter.

17.2 Delay effects on stability in some human respiration models

The human respiratory system is an extremely complicated mechanism, and a large variety of dynamical models describing its behavior exists. In general, the delays represent the *(circulatory) transport time* between the lung and the peripheral and central chemoreceptors.

17.2.1 Delay model and its linearization

In what follows, we consider a two-compartment representation (lungs and tissues) as an interconnection between some "plant" (in which CO_2 exchange takes place) and some "controller" (which regulates the CO_2 partial pressures in the body), as discussed by Vielle and Chauvet [316].

The model writes as follows:

$$\begin{cases} \dot{P}_T(t) = \dfrac{\bar{Q}}{V_T}(P_L(t) - P_T(t)) + \dfrac{\bar{M}}{\alpha V_T}, \\ \dot{P}_L(t) = \dfrac{\alpha \bar{Q} B}{V_L}(P_T(t) - P_L(t)) - \dfrac{1}{V_L}(P_L(t) - P_1)F(P_L(t-\tau)), \end{cases} \tag{17.1}$$

where the variables P_L, P_T denote the CO_2 partial pressure, and $F(\cdot)$ is the controller function. Here, the subscript $L(T)$ denotes lungs (tissues). Next, the parameters V_T (volume, tissues), V_L (volume, lungs), B (barometric pressure minus the water vapor pressure), $\bar{Q}$ (blood flow), $\bar{M}$ (CO_2 metabolic production rate), α (CO_2 dissociation curve slope), and P_1 are positive.

The transport delay appears in the equation by the controller action F, which is an appropriate nonlinear function of the CO_2 partial pressure P_L. Further discussions on the way such a controller is defined can be found in [316] and the references therein.

It is easy to see that (17.1) has a *unique equilibrium* point $(\bar{P}_T, \bar{P}_L)$. By introducing the new variables

$$x_1(t) = P_T(t) - \bar{P}_T, \qquad x_2(t) = P_L(t) - \bar{P}_L,$$

and neglecting the nonlinear second-order terms, system (17.1) can be linearized as follows:

$$\begin{cases} \dot{x}_1(t) = -ax_1(t) + ax_2(t), \\ \dot{x}_2(t) = bx_1(t) - (b+c)x_2(t) - dx_2(t-\tau), \end{cases} \tag{17.2}$$

where the constants (a, b, c, d) are all *positive*, and given by

$$a = \frac{\bar{Q}}{V_T}, \quad b = \frac{\alpha \bar{Q} B}{V_L}, \quad c = \frac{F(\bar{P}_L)}{V_L}, \quad d = \frac{F'(\bar{P}_L)(\bar{P}_L - P_I)}{V_L}. \tag{17.3}$$

Such a delay system has a particular structure. More precisely, it can be written in the form

$$\dot{x}(t) = A_0 x(t) + b_0 c_0^T x(t-\tau), \tag{17.4}$$

where $x \in \mathbb{R}^2$ is given by $x^T = [x_1 \; x_2]^T$ and b_0 and c_0 are column matrices. As seen in Chapter 1, the characteristic function of such a system writes as follows:

$$p(\lambda;\ \tau) := Q(\lambda) + P(\lambda)e^{-\lambda\tau} = Q(\lambda)\left(1 + h(\lambda)e^{-\lambda\tau}\right), \tag{17.5}$$

where $h(\lambda) = P(\lambda)/Q(\lambda)$, and the polynomials P and Q are given by

$$Q(\lambda) = \lambda^2 + \lambda(a+b+c) + ac, \qquad P(\lambda) = \lambda d + ad. \tag{17.6}$$

Since $a,\ b,\ c,$ and d are strictly positive, it is clear that $p(\lambda;\ 0)$ is Hurwitz stable; that is, all its roots have strictly negative real parts. Furthermore, the polynomial Q is also Hurwitz, and thus the quasipolynomial $p(\lambda;\ \tau)$ shares the same characteristic roots on the imaginary axis with the analytic function

$$q(\lambda;\ \tau) := 1 + h(\lambda)e^{-\lambda\tau}.$$

In what follows we will analyze the stability of the model (17.4) by taking into account the particular form and properties of its characteristic function $p(\lambda;\ \tau)$.

17.2.2 Stability analysis and delay intervals

In Chapter 4, we proposed an algorithm for computing delay intervals guaranteeing asymptotic stability by using the function $q(\lambda;\ \tau)$ instead of the original quasipolynomial $p(\lambda;\ \tau)$. We will explicitly apply such an algorithm to our stability analysis problem.

Since the strictly proper transfer h is stable (the denominator Q is Hurwitz), it follows that it is bounded on the imaginary axis. In this context, it is easy to see that the crossing set Ω, that is, the set of all crossing roots $\omega \in \mathbb{R}_+$ w.r.t. the imaginary axis, is given by the solution of the polynomial equation

$$\mid P(j\omega) \mid = \mid Q(j\omega) \mid .$$

As expected, only two situations can occur:

(i) $\mid P(j\omega) \mid < \mid Q(j\omega) \mid$, for all $\omega \in \mathbb{R}_+^*$, or

(ii) there exists at least one frequency $\omega^* > 0$ such that $\mid P(j\omega^*) \mid = \mid Q(j\omega^*) \mid$.

Let us analyze each case separately.

Delay-independent stability

Condition (i) simply says that the system (17.4) is hyperbolic; that is, there is no crossing w.r.t. the imaginary axis if the delay parameter is increased from 0 to ∞. In such a case, the stability of the system free of delays is preserved for all positive delays. In other words, the system is *delay-independent* asymptotically stable. Under the assumption of stability of the strictly proper transfer function h, the *frequency-sweeping* test

$$| h(j\omega) | < 1 \ \forall \omega \in \mathbb{R}_+$$

is nothing else than the Tsypkin criterion (see [96, 223] and the frequency sweeping tests in the case of matrices of rank one in Chapter 4).

Geometrically speaking, we have delay-independent asymptotic stability if and only if the graph of $-h(j\omega)$, for all $\omega \in \mathbb{R}_+^*$, will stay inside the unit circle of the complex plane, with some eventual tangency at the point $(-1, 0)$ for the frequency $\omega = 0$, under the assumption of a delay-free stable system.

A *necessary condition* for no crossing w.r.t. the imaginary axis is

$$| Q(0) | \geq | P(0) |,$$

that is, $c \geq d$. It is important to point out that we can have the equality above since 0 is not a root of the characteristic function $p(\lambda; \tau)$ for any positive delay $\tau \in \mathbb{R}_+$. Geometrically speaking, the equality simply describes the tangency property mentioned above, since $P(0) = ad$ will be equal to $Q(0) = ac$ if $c = d$.

Now let us check if the condition $c \geq d$ is also *sufficient* for getting delay-independent stability. Simple computations prove that crossing roots do not exist if and only if the second-order equation

$$x^2 + \left[(a+b)^2 + 2bc + c^2 - d^2\right] x + a^2(c^2 - d^2) = 0 \tag{17.7}$$

has *no strictly positive* roots. This last condition holds if and only if the inequalities

$$\begin{cases} a^2(c^2 - d^2) \geq 0, \\ (a+b)^2 + 2bc + c^2 - d^2 \geq 0, \end{cases}$$

are satisfied simultaneously under the constraint of positive parameters a, b, c, and d. In conclusion, we obtain the following simple *delay-independent stability* condition:

$$c \geq d. \tag{17.8}$$

Figure 17.1 depicts such a delay-independent stability condition for some positive c and d such that $c > d$. As explained in Chapter 4, the plot of $-P(j\omega)/Q(j\omega)$ and its intersection with the unit circle are not sufficient to conclude on asymptotic stability of the corresponding system. We can only expect to detect the occurrence of crossing roots for some delay values.

Finally, notice that the same condition was obtained in [316], but using a different argument.

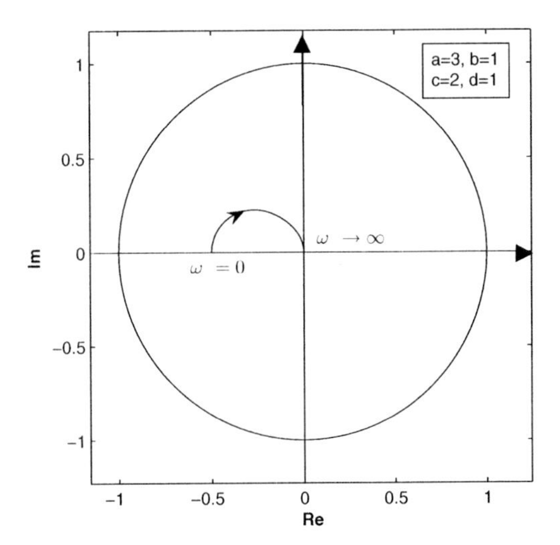

Figure 17.1. *Delay-independent stability since the intersection between the ratio curve $-h(j\omega) = -P(j\omega)/Q(j\omega)$, the unit circle is empty, and the system free of delay is asymptotically stable.*

Delay-dependent stability

As expected, the characteristic function $p(\lambda;\ \tau)$ has roots crossing the imaginary axis if and only if the second-order equation (17.7) has *at least one strictly positive root.* Based on the particular form of this equation, it follows that such a situation appears if and only if

$$d > c. \tag{17.9}$$

More precisely, if $d > c$, (17.7) will always have two real roots of opposite sign. See Figure 17.2 for a graphical interpretation. In other words, *only one root* x_+ will be *positive*, that is, the *frequency crossing set* Ω is given by

$$\Omega = \{\omega_+\},$$

where $\omega_+ = \sqrt{x_+}$, with x_+ the only positive root of (17.7).

In conclusion, since only one crossing root exists, the corresponding crossing direction is always *toward instability*. Thus, the system will be asymptotically stable for *all delays* $\tau \in [0, \tau_m)$, where the delay margin τ_m is given by the formula

$$\tau_m = \min\left\{\tau > 0 : \ \tau = \frac{\angle\left(-\frac{P(j\omega_+)}{Q(j\omega_+)}\right) + 2\pi l, \ \ l \in \mathbb{Z}}{\omega_+}\right\}, \tag{17.10}$$

with ω_+ the unique element of the set Ω.

The corresponding crossing direction is toward instability, and it is independent of the parameter values. Therefore, we can conclude that the linearized second-order system is asymptotically stable for all $\tau \in [0, \tau_m)$. Furthermore, it is *unstable* for all $\tau \geq \tau_m$.

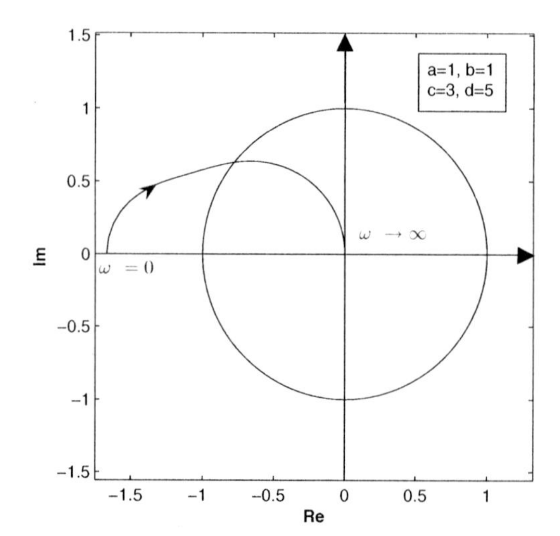

Figure 17.2. *Delay-dependent stability since the intersection between the ratio curve $-h(j\omega) = -P(j\omega)/Q(j\omega)$ and the unit circle consists of only one point, and there exist points outside and inside the unit circle.*

Finally, notice that the same stability conditions were obtained in [316], but using a slightly different argument.

17.2.3 Discussions and interpretations

As pointed out by Vielle and Chauvet [316], the respiration model above includes a two-compartment representation (lungs and tissues) subject to a general class of controllers. The approach considered takes into consideration the particular structure of the linearized two-compartment model. Indeed, the corresponding delay matrix is of rank one, a fact which simplifies the analysis. We believe that such a fact was not sufficiently exploited in the literature. The next section, devoted to the analysis of some models for the human immune system's dynamics in leukemia, provides a further argument to this thesis.

Finally, our sensitivity analysis with respect to the delay-parameter proves that the essential quantities for delay-induced instabilities are the parameters c and d, given by

$$c = \frac{F(\bar{P}_L)}{V_L}, \qquad d = \frac{F'(\bar{P}_L)(\bar{P}_L - P_I)}{V_L},$$

that is, the evaluation of the continuous controller function F and of its derivative F' relating the air flow in lungs with the delayed partial pressure in arterial blood, under the "standard" assumption that the partial pressure in arterial blood is greater than that "outside." Please note that the partial pressures in arterial blood and in lungs are identical (equilibrium) [316].

In conclusion, the delay-induced stability is completely characterized by the value F'/F evaluated at $\bar{P}_L$, and compared to $1/(\bar{P}_L - P_I)$, which are directly defined by the corresponding equilibrium, and by the controller law. Some discussions on the linear and Hill controllers, and appropriate (physiological) assumptions on F, can be found in [316].

17.3 Delays in immune dynamics model of leukemia

We consider a second-order nonlinear model that describes immune dynamics in chronic myelogenous leukemia after a bone marrow transplantation. The particularity of the system is the presence of *four* distinct *delays* in some large range.

17.3.1 Delay model and its linearization

Consider the following nonlinear model proposed by [66, 231] to describe the posttransplantation dynamics of the immune response to chronic myelogenous leukemia:

$$\begin{cases} \dfrac{dT(t)}{dt} = & -d_T T(t) - kC(t)T(t) + p_2 kC(t-\sigma)T(t-\sigma) \\ & +2^N p_1 q_1 kC(t-\rho-N\tau)T(t-\rho-N\tau) \\ & +p_1 q_2 kC(t-\rho-\upsilon)T(t-\rho-\upsilon), \\ \dfrac{dC(t)}{dt} = & rC(t)\left(1-\dfrac{C(t)}{K}\right) - \tilde{p}_1 kC(t-\rho)T(t-\rho), \end{cases} \tag{17.11}$$

where the variable T refers to the anticancer T cell population, and C refers to the cancer cell population, both functions of time t. All the other variables are constant and nonnegative. The constants p_i, q_i, and $\tilde{p}_1$ are probabilities between 0 and 1. Furthermore, $p_1 + p_2 = 1$, and $0 \leq q_1 + q_2 \leq 1$.

As already mentioned, the particularity of the system above is the presence of *four* distinct *delays*, namely σ, $\rho + N\tau$, $\rho + \upsilon$, and ρ. The relevant values considered by [66] are approximately $\sigma = 1$ min, $\rho = 5$ min, $\tau = 1$ day, and $\upsilon = 1$ day. The parameter N is between 1 and 8 and probably close to 3. These constants, respectively, represent the time for unreactive interactions between T cells and cancer cells (σ), the time for reactive interactions (ρ), the time for one round of cell division (τ), the T cell recovery time after killing a cancer cell (υ), and the average number of T cell divisions after stimulation (N). Due to the scale difference, we can define without any loss of generality ρ and σ as *small delays*, and the remaining delays as *large*.

For convenience, let us consider

$$\begin{array}{lll} b_1 = d_T, & b_5 = p_1 q_2 k, & \tilde{\tau} = \rho + N\tau, \\ b_2 = k, & c_1 = r, & \tilde{\upsilon} = \rho + \upsilon, \\ b_3 = p_2 k, & c_2 = r/K, & \\ b_4 = 2^N p_1 q_1 k, & c_3 = \tilde{p}_1 k, & \end{array} \tag{17.12}$$

and rewrite (17.11) as [231]

$$\begin{cases} \frac{dT(t)}{dt} = & -b_1 T(t) - b_2 C(t)T(t) + b_3 C(t-\sigma)T(t-\sigma) \\ & +b_4 C(t-\tilde{\tau})T(t-\tilde{\tau}) + b_5 C(t-\tilde{\upsilon})T(t-\tilde{\upsilon}), \\ \frac{dC(t)}{dt} = & c_1 C(t) - c_2 C(t)^2 - c_3 C(t-\rho)T(t-\rho). \end{cases} \tag{17.13}$$

For future reference, we note that all parameters in (17.12) are positive. Define also $b = -b_2 + b_3 + b_4 + b_5$.

17.3.2 Stability analysis in the delay-parameter space

Fixed points

It is easy to see that the fixed points, (T_0, C_0), of (17.13) are solutions of the following system of equations:

$$\begin{cases} 0 &= -b_1 T_0 + bC_0 T_0 = (-b_1 + bC_0)T_0, \\ 0 &= (c_1 - c_2 C_0 - c_3 T_0)C_0; \end{cases}$$

i.e., the three fixed points are $(T_0, C_0) = (0, 0)$, $(0, \frac{c_1}{c_2})$, and $(\frac{c_1 - c_2 b_1/b}{c_3}, \frac{b_1}{b})$.

The fixed point $(0, 0)$ corresponds to the situation where the cancer population is entirely eliminated and the cancer-reactive T cells become unnecessary and disappear. Unfortunately, this fixed point is a *saddle-point* regardless of the values of the parameters, which means that this fixed point is unattainable [231].

Next, the fixed point $(0, \frac{c_1}{c_2})$ describes the situation where cancer expands to full capacity and the cancer-reactive T cells die off completely. This is the most undesirable state, and it is *unstable* for biologically reasonable parameter choices, as discussed in [231].

The final fixed point, $(\frac{c_1 - c_2 b_1/b}{c_3}, \frac{b_1}{b})$, represents the scenario where the cancer and T cell populations coexist at relatively low populations. This means that cancer is not completely eliminated, but is controlled, in some sense, by the immune response. We will concentrate our attention on this third case.

Linearizing (17.13) around a fixed point (T_0, C_0) yields

$$\begin{cases} \frac{dT(t)}{dt} = & -\tilde{b}_1 T(t) - b_2 T_0 C(t) + b_3 C_0 T(t-\sigma) + b_3 T_0 C(t-\sigma) \\ & + b_4 C_0 T(t-\tilde{\tau}) + b_4 T_0 C(t-\tilde{\tau}) + b_5 C_0 T(t-\tilde{\upsilon}) + b_5 T_0 C(t-\tilde{\upsilon}), \\ \frac{dC(t)}{dt} = & \tilde{c}_1 C(t) - c_3 C_0 T(t-\rho) - c_3 T_0 C(t-\rho), \end{cases} \tag{17.14}$$

where $\tilde{b}_1 = b_1 + b_2 C_0$ and $\tilde{c}_1 = c_1 - 2c_2 C_0$.

The characteristic function of (17.14) can be rewritten as

$$p(\lambda;\ \sigma, \rho, \tilde{\tau}, \tilde{\upsilon}) := \det\Big(\lambda I_2 - A_0 - b_{01} c_{0\rho}^T e^{-\lambda\rho} - b_{02} c_{0\sigma}^T e^{-\lambda\sigma} b_{01} c_{0\rho}^T e^{-\lambda\rho} - b_{02} c_{0\tilde{\upsilon}}^T e^{-\lambda\tilde{\upsilon}} - b_{02} c_{0\tilde{\tau}}^T e^{-\lambda\tilde{\tau}}\Big), \tag{17.15}$$

where $b_{01} = [0\ 1]^T$, $b_{02} = [1\ 0]^T$, and $c_{0\rho}$, $c_{0\sigma}$, $c_{0\tilde{\upsilon}}$, $c_{0\tilde{\tau}}$ are appropriately defined.

In other words, the corresponding "entry" matrices for the delays ρ, $\tilde{\tau}$, $\tilde{\upsilon}$, and σ are *all* of *rank one*, a fact which simplifies the characteristic function to

$$p(\lambda;\ \sigma, \rho, \tilde{\tau}, \tilde{\upsilon}) := p_0(\lambda) + p_1(\lambda)e^{-\rho\lambda} + p_2(\lambda)e^{-\sigma\lambda} + p_3(\lambda)e^{-\tilde{\tau}\lambda} + p_4(\lambda)e^{-\tilde{\upsilon}\lambda},$$

with

$$\begin{aligned} p_0(\lambda) &= -(\tilde{b}_1 + \lambda)(\tilde{c}_1 - \lambda), & p_1(\lambda) &= c_3 T_0(b_1 + \lambda), \\ p_2(\lambda) &= b_3 C_0(\tilde{c}_1 - \lambda), & p_3(\lambda) &= b_4 C_0(\tilde{c}_1 - \lambda), \\ p_4(\lambda) &= b_5 C_0(\tilde{c}_1 - \lambda). \end{aligned}$$

Since $p_0(\lambda)$ has no roots on the imaginary axis, we write

$$p(\lambda;\ \sigma, \rho, \tilde{\tau}, \tilde{\upsilon}) = p_0(\lambda)(1 + a_1(\lambda)e^{-\rho\lambda} + a_2(\lambda)e^{-\sigma\lambda} + a_3(\lambda)e^{-\tilde{\tau}\lambda} + a_4(\lambda)e^{-\tilde{\upsilon}\lambda}), \quad (17.16)$$

where $a_i(\lambda) = p_i(\lambda)/p_0(\lambda)$, for all $i = 1, \ldots, 4$. In conclusion, $p(\lambda;\ \sigma, \rho, \tilde{\tau}, \tilde{\upsilon})$ and the analytic function

$$a(\lambda;\ \sigma, \rho, \tilde{\tau}, \tilde{\upsilon}) := 1 + a_1(\lambda)e^{-\rho\lambda} + a_2(\lambda)e^{-\sigma\lambda} + a_3(\lambda)e^{-\tilde{\tau}\lambda} + a_4(\lambda)e^{-\tilde{\upsilon}\lambda}$$

share the same characteristic roots on the imaginary axis.

Stability crossing curves in the delay-parameter space

In what follows, we apply the methodology proposed in Chapter 4 for characterizing the stability crossing curves. One way to visualize the crossing surface of (17.14) is to fix two delays and determine the crossing curves for the other two delays. Based on the particular form of the characteristic equation and of the delays scales, it seems reasonable to consider the (natural) delays *partition* in *small* and *large* delays.

Consider now an *auxiliary system* associated with the small delays, that is, the system with the characteristic equation

$$p_{\rho,\sigma}(\lambda;\ \rho, \sigma) := p_0(\lambda) + p_1(\lambda)e^{-\rho\lambda} + p_2(s)e^{-\sigma\lambda} = 0. \quad (17.17)$$

Using the *geometric approach* detailed in Chapter 4, we can easily characterize the *stability crossing curves* of $p_{\rho,\sigma}$ given by (17.17) in the delay-parameter space defined by the small delays ρ and σ. Based on such a characterization, and using a standard continuity argument (see, for instance, Chapter 1) of the roots of the characteristic equation (17.16) with respect to the delay parameters, we make the following assumption.

Assumption 17.1. *Let $\mathcal{I}_\rho \subset \mathbb{R}_+$ and $\mathcal{I}_\sigma \subset \mathbb{R}_+$ be some real intervals for which there exists some $\delta > 0$ such that*

$$p_{\rho,\sigma}(\lambda;\ \rho, \sigma) \neq 0 \qquad \forall(\sigma, \rho) \in \mathcal{I}_\sigma \times \mathcal{I}_\rho$$

for all $\lambda \in \mathcal{V}_\delta$, where $\mathcal{V}_\delta$ is defined by

$$\mathcal{V}_\delta = \{\lambda \in \mathbb{C} : \quad -\delta < \Re(\lambda) < \delta\}. \quad (17.18)$$

Such an assumption is not restrictive, and it simply describes some *regularity* condition for the original linearized model. It simply says that there exists some delay intervals such that $p_{\rho,\sigma}$ is *invertible* in some neighborhood $\mathcal{V}_\delta$ of the imaginary axis for all the pairs $(\sigma, \rho) \in \mathcal{I}_\sigma \times \mathcal{I}_\rho$. For the sake of brevity, we do not detail here the situations when the regularity condition above is not satisfied. Further comments can be found in [231].

We will follow the same steps as in Chapter 4 for the characterization of the crossing curves in the delay-parameter space defined by the large delay values. First, we will identify the crossing points, and we will give a complete description of the crossing set characterization. Next, the smoothness, tangency, and crossing direction results will be outlined. All these results are developed in analogy with the characterization for the two delays case.

Identification of the crossing points, and crossing set characterization

We have the following result.

Proposition 17.2. *Assume that the auxiliary system given by (17.17) satisfies Assumption 17.1. Define $a_{\tilde{\tau},\tilde{\upsilon}}$ by*

$$a_{\tilde{\tau},\tilde{\upsilon}}(\lambda;\ \tilde{\tau},\tilde{\upsilon}) = 1 + a_{\tilde{\tau}}(\lambda)e^{-\lambda\tilde{\tau}} + a_{\tilde{\upsilon}}(\lambda)e^{-\lambda\tilde{\upsilon}}, \tag{17.19}$$

where

$$a_{\tilde{\tau}}(\lambda) = \frac{p_3(\lambda)}{p_{\rho,\sigma}(\lambda;\ \rho,\sigma)}, \quad a_{\tilde{\upsilon}}(\lambda) = \frac{p_4(\lambda)}{p_{\rho,\sigma}(\lambda;\ \rho,\sigma)},$$

for all $(\sigma,\rho) \in \mathcal{I}_\sigma \times \mathcal{I}_\rho$.

Then for any $(\sigma,\rho) \in \mathcal{I}_\sigma \times \mathcal{I}_\rho$, the characteristic equation associated to (17.16) and $a_{\tilde{\tau},\tilde{\upsilon}}(\lambda;\ \tilde{\tau},\tilde{\upsilon})$ have the same solutions in a neighborhood $\mathcal{V}_\delta$ of the imaginary axis, where

$$\mathcal{V}_\delta = \{\lambda \in \mathbb{C}:\quad \delta > \Re(\lambda) > -\delta\},$$

for some $\delta > 0$.

Delay-independent type results, and weak T/C interactions

With the notations and the results above, we have the following result (see [231] for more details).

Proposition 17.3. *Assume that the auxiliary system given by the characteristic equation (17.17) satisfies Assumption 17.1, and that $a_{\tilde{\tau},\tilde{\upsilon}}(0;\ 0,0) \neq 0$, where $a_{\tilde{\tau},\tilde{\upsilon}}$ is defined by (17.19).*

Then the following statements are equivalent:

(a) If the auxiliary system (17.17) is stable for some pair $(\rho_0,\sigma_0) \in \mathcal{I}_\rho \times \mathcal{I}_\sigma$, and if the linearized system free of delays ($\sigma = \rho = \tilde{\tau} = \tilde{\upsilon} \equiv 0$) is stable, then the system (17.13) is stable for all pairs $(\tilde{\tau},\tilde{\upsilon}) \in \mathbb{R}_+ \times \mathbb{R}_+$, and there does not exist any root crossing the imaginary axis when the delays $\tilde{\tau}$ and $\tilde{\upsilon}$ are increased in $\mathbb{R}_+$.

(b) The following frequency-sweeping test holds:

$$\frac{\mid C_0 \mid \sqrt{\tilde{c}_1^2 + \omega^2}}{\mid p_{\rho,\sigma}(j\omega;\ \rho_0,\sigma_0) \mid} < \frac{1}{(2^N q_1 + q_2)p_1 k}, \quad \forall\omega > 0, \quad \forall(\rho_0,\sigma_0) \in \mathcal{I}_\rho \times \mathcal{I}_\sigma. \tag{17.20}$$

The same equivalence holds if the stability property is replaced by the instability of the system with a prescribed number of unstable roots.

Such a result simply describes the situation where the stability or instability of the linearized second-order model is *independent* of the delays $\tilde{\upsilon}$ and $\tilde{\tau}$. Since such delays describe the *T/C cell interactions*, such a situation can be simply called *weak T/C cell interaction*, and it defines a reduced probability of reactive interactions between anticancer

T cells and cancer cells. In other words, the weak T/C interaction describes the situations where the anticancer cells will "mostly ignore" the cancer cells. Roughly speaking, such a T/C interaction will be translated into "small" values for the coefficients $b_4 = 2^N p_1 q_1 k$ and $b_5 = p_1 q_2 k$ which may correspond to the case where *no crossing* in the delay-parameter space defined by large delays exists. More discussions can be found in [231].

Remark 17.4. *The frequency-sweeping test (17.20) can be used in defining a* measure *for characterizing the T/C interaction type in the following sense:*

- *The T/C interaction will be called* weak *if the probabilities* (q_1, q_2), *and the average number of cell division N verify the condition*

$$(2^N q_1 + q_2) p_1 k \left(\sup_{\omega \in \mathbb{R}, (\rho_0, \sigma_0) \in \mathcal{I}_\rho \times \mathcal{I}_\sigma} \frac{| C_0 | \sqrt{\tilde{c}_1^2 + \omega^2}}{| p_{\rho,\sigma}(j\omega; \ \rho_0, \sigma_0) |} \right) < 1. \tag{17.21}$$

The left-hand side of (17.21) gives the corresponding T/C interaction measure.

It becomes clear that the average number N of cell divisions plays a central role in defining the T/C interaction character, since the quantity $2^N q_1 + q_2$ *is an increasing function of N.*

Strong T/C interactions, and identification of the crossing points

It is easy to see that the existence of crossing sets in the delay-parameter space defined by $\tilde{\tau}$ and $\tilde{\upsilon}$ is related to the fact that the inequality (17.20) is not satisfied for all $\omega > 0$ or, in other words, that the parameters (q_1, q_2, N) do not satisfy the measure condition (17.21) for the T/C weak interaction. Such a situation is called a *strong T/C cells interaction.*

Based on the remarks above, it follows that we have a relatively simple condition for checking the T/C strong interaction character.[25]

Proposition 17.5. *The T/C interaction is strong if the equilibrium* $(T_0, C_0) = \neq (0, 0)$ *satisfies the following inequality:*

$$\left| (d_T + k(1 - p_2)C_0) \operatorname{sign}(2C_0 - K) + \frac{\tilde{p}_1 k K d_T}{r} \frac{T_0}{| 2C_0 - K |} \right| \\ < | C_0 | (2^N q_1 + q_2) p_1 k. \tag{17.22}$$

Let us characterize now the strong T/C interactions. Based on the discussions in Chapter 4, the condition that $a_{\tilde{\tau},\tilde{\upsilon}}$ defined by (17.19) has at least one root $j\omega_0$ on the imaginary axis is reduced geometrically to the condition that the "lengths" 1, $| a_\upsilon(j\omega_0) |$, and $| a_\mu(j\omega_0) |$ define a triangle. Thus, some simple computations lead to the following criterion for the *identification of the crossing points.*

[25] Such a situation simply parallelizes the situation $c < d$ in $h(0) > 1$ in the respiration second-order model in the previous section.

Proposition 17.6. *Assume that the auxiliary system given by the characteristic equation (17.17) satisfies Assumption 17.1. Then each $\omega \in \mathbb{R}_+$ can be a solution of the characteristic function associated to the original linearized system for some $(\tilde{\tau}, \tilde{\upsilon}) \in \mathbb{R}_+^2$ if and only if*

$$\frac{1}{(2^N q_1 + q_2) p_1 k} \leq \frac{|\, C_0 \,| \sqrt{\tilde{c}_1^2 + \omega^2}}{|\, p_{\rho,\sigma}(j\omega;\ \rho, \sigma) \,|} \leq \frac{1}{|\, 2^N q_1 - q_2 \,|\, p_1 k}. \tag{17.23}$$

Then, the *frequency crossing set* Ω will be defined by all $\omega \in \mathbb{R}_+$, for which the frequency condition (17.23) holds. In conclusion, for a given pair $(\rho_0, \sigma_0) \in \mathcal{I}_\rho \times \mathcal{I}_\sigma$ the algorithm for identifying the crossing points can be summarized as follows:

(i) first, we represent graphically $\dfrac{|\, C_0 \,| \sqrt{\tilde{c}_1^2 + \omega^2}}{|\, p_{\rho,\sigma}(j\omega;\ \rho_0, \sigma_0) \,|}$ against ω, and

(ii) next, we analyze the intersection of this graphic with two parallel lines to ω-axis: $1/((2^N q_1 + q_2) p_1 k)$ and $1/(|\, 2^N q_1 - q_2 \,|\, p_1 k)$, respectively.

Define now $\mathcal{T}_\omega$ as the set of all $(\tilde{\tau}, \tilde{\upsilon})$ such that $a_{\tilde{\tau},\tilde{\upsilon}}$ has one zero on the imaginary axis at $\lambda = j\omega$.

Remark 17.7. *It is easy to see that*

$$\frac{|\, C_0 \,| \sqrt{\tilde{c}_1^2 + \omega^2}}{|\, p_{\rho,\sigma}(j\omega;\ \rho, \sigma) \,|} \to 0$$

as $\omega \to +\infty$, and in conclusion $\infty \notin \Omega$. In other words, Ω is bounded.

In the stability analysis performed in the delay-parameter space in Chapter 4, the characterization of the stability crossing curves for a general system including *two delays* was based on an important property of the corresponding *frequency crossing set*, namely the fact that it consisted of a finite number of intervals of finite length. In our case study, and in order to completely characterize the crossing curves in the parameter space defined by the large delays, it will be interesting to have a similar property. We now introduce the following assumption (see [231]).

Assumption 17.8. *The condition*

$$\frac{d}{d\omega}\left(\frac{|\, C_0 \,| \sqrt{\tilde{c}_1^2 + \omega^2}}{|\, p_{\rho,\sigma}(j\omega;\ \rho, \sigma) \,|} \right) \neq 0$$

holds whenever

$$\frac{|\, C_0 \,| \sqrt{\tilde{c}_1^2 + \omega^2}}{|\, p_{\rho,\sigma}(j\omega;\ \rho, \sigma) \,|} = \frac{1}{|\, 2^N q_1 \pm q_2 \,|\, p_1 k}$$

for some $\omega \in \mathbb{R}_+$.

This assumption simply requires that the corresponding differentiable function satisfies some nonsingularity property at the corresponding upper and lower bounds given by (17.23).

With the remarks and the assumptions above, we have the following proposition.

Proposition 17.9. *Under Assumptions 17.1 and 17.8, the frequency crossing set Ω consists of a finite number of intervals of finite length.*

The proof can be found in [231], and is by contradiction in that it assumes that inside a given frequency interval $[0, \omega_m]$ such that $\Omega \subset [0, \omega_m]$, an appropriate real function cannot have an infinite number of roots.

Characterization of the stability crossing curves

The next step is to characterize the crossing curves of the system (17.13) or, equivalently, all the crossing curves satisfying $a_{\tilde{\tau},\tilde{\upsilon}}(\lambda;\ \tilde{\tau},\tilde{\upsilon}) = 0$ for $\lambda = j\omega$, $\omega \in \Omega$. Using a similar classification of the crossing points as in Chapter 4, define now by $\Omega_k \subset \Omega$ some interval of crossing set Ω, and let $\mathcal{T}^k \subset \mathcal{T}$ be the corresponding stability crossing curves for some positive integer k. We have the following result.

Proposition 17.10. *Under the standard Assumption 17.1, the stability crossing curves $\mathcal{T}^k$ corresponding to Ω_k must be an intersection of $\mathbb{R}^2_+$ with a series of curves belonging to one of the following categories:*

A. a series of closed curves;

B. a series of spiral-like curves with axes oriented either horizontally, vertically, or diagonally;

C. a series of open-ended curves with both ends approaching ∞.

Remark 17.11. *The classification above is given by the way the end points of the corresponding intervals Ω_k are derived.*

Tangent, smoothness, and crossing direction

All these properties follow straightforwardly from the stability crossing curves analysis proposed in Chapter 4.

Tangent, smoothness. We have the following result.

Proposition 17.12. *Under standard assumptions including Assumption 17.1, the curves in $\mathcal{T}^k$ are smooth everywhere except possibly at degenerate points corresponding to a root $\lambda = \omega$ in either of the following cases:*

1. $\lambda = j\omega$ is a multiple solution of $a_{\tilde{\tau},\tilde{\upsilon}}(j\omega) = 0$.

2. *ω is an end point, and*

$$\frac{d}{d\omega}\left(\frac{|C_0|\sqrt{\tilde{c}_1^2+\omega^2}}{|p_{\rho,\sigma}(j\omega)|}\right) = 0.$$

Direction of crossing. Next, for a given $\omega \in \Omega_k$, introduce

$$R_l = -\Re\left(\frac{1}{\lambda}\frac{\partial a_{\tilde{\tau},\tilde{\upsilon}}(\lambda;\ \tilde{\tau},\tilde{\upsilon})}{\partial \tau_k}\right)_{\lambda=j\omega}, \tag{17.24}$$

$$I_l = -\Im\left(\frac{1}{\lambda}\frac{\partial a_{\tilde{\tau},\tilde{\upsilon}}(\lambda;\ \tilde{\tau},\tilde{\upsilon})}{\partial \tau_k}\right)_{\lambda=j\omega}, \tag{17.25}$$

for $l = 1, 2$, and τ_1, τ_2 correspond to $\tilde{\tau}$, and $\tilde{\upsilon}$, respectively. This allows us to arrive to the following proposition.

Proposition 17.13. *Let $\omega \in \Omega_k$, but an end point, and $(\tilde{\tau}_0, \tilde{\upsilon}_0) \in \mathcal{T}^k$ such that $\lambda = j\omega$ is a simple solution of $a_{\tilde{\tau},\tilde{\upsilon}}(\lambda;\ \tilde{\tau}_0, \tilde{\upsilon}_0) = 0$ and*

$$a_{\tilde{\tau},\tilde{\upsilon}}(j\omega';\ \tilde{\tau}_0,\tilde{\upsilon}_0) \neq 0 \text{ for any } \omega' > 0,\ \omega' \neq \omega. \tag{17.26}$$

Then as $(\tilde{\tau}, \tilde{\upsilon})$ moves from the region on the right to the region on the left of the corresponding curve in $\mathcal{T}^k$, a pair of solutions of $a_{\tilde{\tau},\tilde{\upsilon}}(\lambda;\ \tilde{\tau}, \tilde{\upsilon}) = 0$ crosses the imaginary axis to the right if

$$R_2 I_1 - R_1 I_2 > 0. \tag{17.27}$$

The crossing is in the opposite direction if the inequality is reversed.

17.3.3 Illustrative example and discussions

Stability without delays

For our application in [66, 231], we estimated values of the parameters to be approximately

$$\begin{array}{lll} d_T = 0.2, & p_1 = 0.5, & \rho = 0.0035, \\ r = 0.2, & p_2 = 0.5, & \sigma = 0.0007, \\ k = 1, & q_1 = 0.5, & \tilde{\tau} = 2.0035, \\ N = 2, & q_2 = 0.5, & \tilde{\upsilon} = 1.0035. \\ K = 200, & \tilde{p}_1 = 0.5, & \end{array} \tag{17.28}$$

Hence, b_1, c_1, and b are of order 1 or 0.1, while $c_1/c_2 = K$, the carrying capacity of the cancer population, is around 200. As already mentioned in the previous section, we have three *fixed points*. In our example, they have the following properties:

(a) The fixed point I $(T_0, C_0) = (0, 0)$ is a saddle-point.

(b) The fixed point II $(T_0, C_0) = (0, c_1/c_2)$ is unstable.

(c) The fixed point III $(T_0, C_0) = \left(\frac{c_1 - c_2 b_1/b}{c_3}, \frac{b_1}{b}\right)$ is stable.

Stability with delays

We consider the third fixed point. One way to visualize the crossing surface of (17.14) is to fix two delays and determine the crossing curves for the other two delays. This procedure is demonstrated in Figure 17.3.

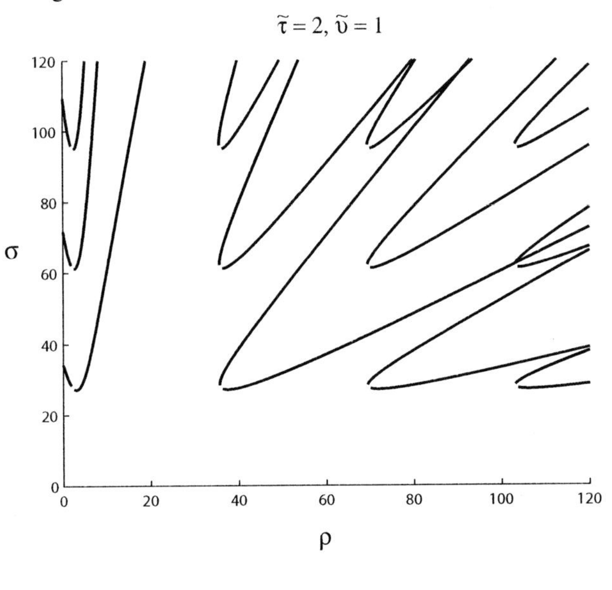

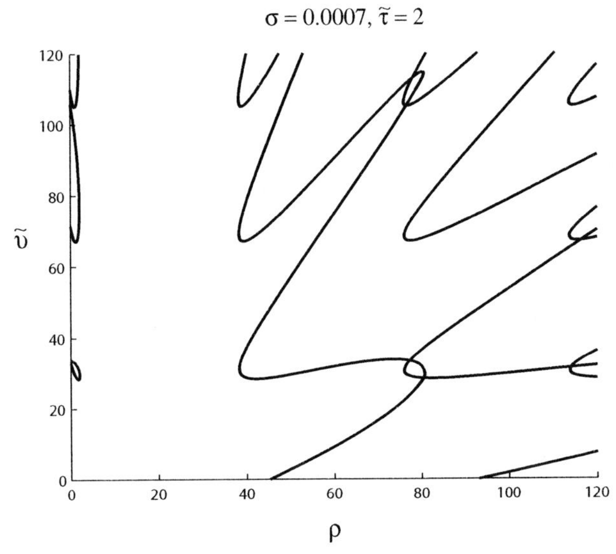

Figure 17.3. *Crossing curves for σ vs. ρ and $\tilde{\upsilon}$ vs. ρ for given values of $(\tilde{\tau}, \tilde{\upsilon})$ and $(\sigma, \tilde{\tau})$, respectively.*

In this case, any pairwise combination of ρ, σ, or τ gives open curves such as the ones shown in Figure 17.3 (top). Any pair containing υ leads to spiral-like curves such as the ones shown in Figure 17.3 (bottom). For the choice of parameters in (17.28), the fixed point

III is stable in the undelayed case, and thus there exists a stability region for sufficiently small delays. However, this region is very small and disappears quickly as the delays are increased. Figure 17.4 shows the crossing curves for σ vs. ρ, when $(\tau, \upsilon) = (0, 0)$ for the linearization of (17.13) around the fixed point III, $(\frac{c_1 - c_2 b_1/b}{c_3}, \frac{b_1}{b})$.

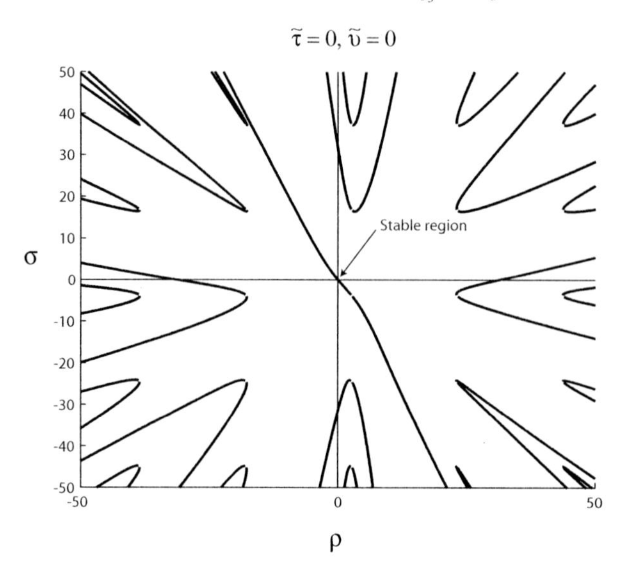

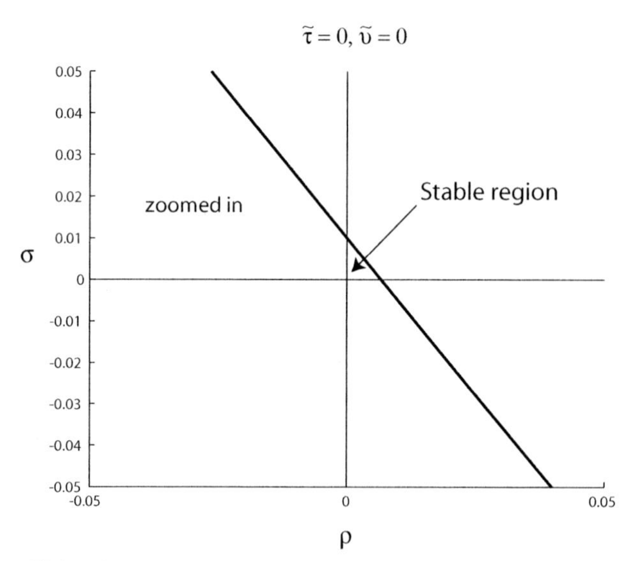

Figure 17.4. *Crossing curves for σ vs. ρ with $(\tau, \upsilon) = (0, 0)$ for the linearization of (17.13) around fixed point III, $\left(\frac{c_1 - c_2 b_1/b}{c_3}, \frac{b_1}{b}\right)$.*

In particular, the delays for T cell division, $\tilde{\tau}$, and recovery from a cytotoxic process, $\tilde{\upsilon}$, are about two and one days, respectively, so fixed point III is unstable. However, for low values of ρ and σ, we find another stable region in $(\tilde{\tau}, \tilde{\upsilon})$-space away from the origin. For

$\rho = 0.0035$ and $\sigma = 0.0007$, the crossing curves for $\tilde{\tau}$ and $\tilde{\upsilon}$ are shown in Figure 17.5(top). A stable solution is shown in Figure 17.5(bottom).

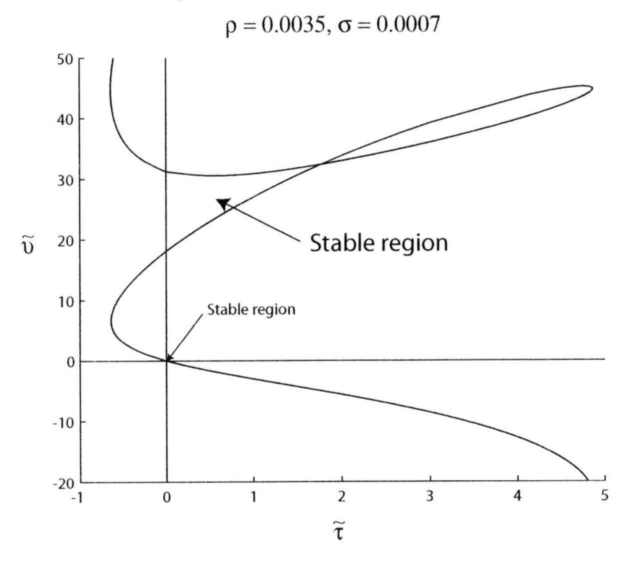

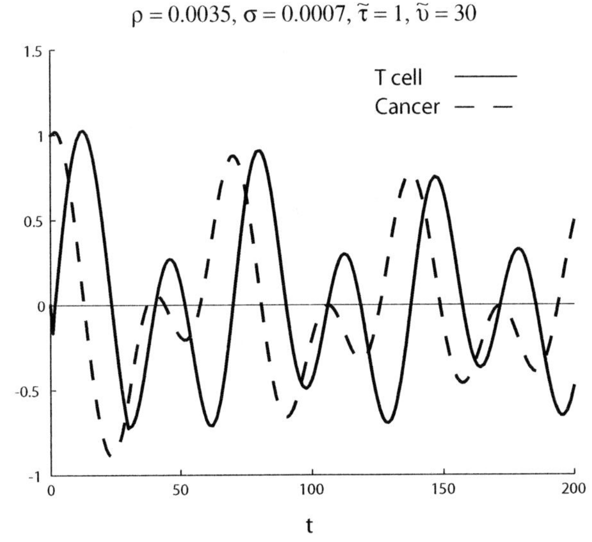

Figure 17.5. *(above) Crossing curves for υ vs. τ with* $(\rho, \sigma) = (0.0035, 0.0007)$. *(below) Solution of (17.14) around fixed point III.*

In this region, the values for ρ and σ are five and one minutes as estimated in (17.28). The delay $\tilde{\tau}$, corresponding to $N = 2$ cell divisions, is about one day, which is a little fast, but still reasonable. On the other hand, the delay $\tilde{\upsilon}$, corresponding to the turnaround time for T cell recovery after cytotoxic responses, is around 20 to 30 days, which is far longer than the expected one-day turnaround time. From the perspective of medical intervention, the larger stable region away from the origin is more interesting, because it is probably easier to slow rates down than to speed them up. For contrast, consider the small stable region around the origin in Figure 17.5. In this region, the delays $\tilde{\tau}$ and $\tilde{\upsilon}$ are constrained to values less than 0.005 (7 min) and 0.02 (30 min), respectively, and it is almost impossible to accelerate T cell division or the T cell recovery time to these rates.

Discussions

Discussion on the stability regions by changing nondelay parameters has been considered by [231]. For the sake of brevity, we will not present such arguments here. However, we point out that for most parameters, the stability region around the origin is very small. The parameters that influence the size of the stability region the most are the kinetic coefficient k and the T cell death rate d_T. Lower kinetic rates k and higher T cell death rates d_T lead to larger stable regions. In either case, the region around the origin is not large enough to allow for biologically meaningful values of $\tilde{\tau}$ and $\tilde{\upsilon}$. However, if we consider the extreme case where we set $k = 0.01$, $d_T = 0.5$, all other parameters according to the estimates in (17.28), and $(\rho, \sigma) = (0.0035, 0.0007)$, we obtain the stable region shown in Figure 17.6. This region covers a large area in $(\tilde{\tau}, \tilde{\upsilon})$-space, including the point $(\tilde{\tau}, \tilde{\upsilon}) = (2, 1)$, which seems "biologically reasonable."

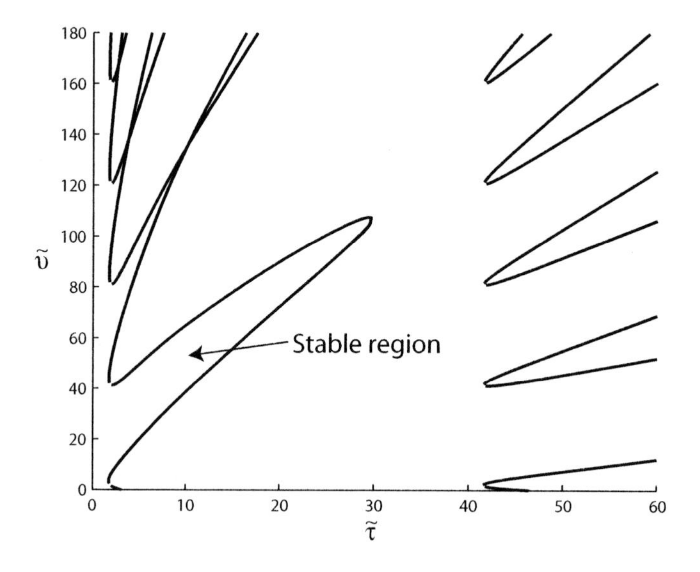

Figure 17.6. *Crossing curves for $\tilde{\upsilon}$ vs. $\tilde{\tau}$ with $k = 0.01$, $d_T = 0.5$, all other parameters according to the estimates in (17.28), and $(\rho, \sigma) = (0.0035, 0.0007)$.*

17.4 Notes and references

This chapter addressed the stability analysis in the delay-parameter space of two time-delay systems encountered in modeling human respiration and modeling immune dynamics in chronic leukemia, respectively. The reason for choosing such different models was dictated by the particular structure of the linearized systems with respect to the delay presence, that helped us in deriving some simple stability conditions and/or characterization by using geometric arguments.

More precisely, in the first part of the chapter, we considered the stability analysis of a simplified system including 1 delay. The analysis was derived by exploiting the rank-1 property of the delayed matrix. The crossing direction characterization was obtained straightforwardly from one of the algorithms proposed in Chapter 4.

Another second-order model including one constant delay can be found in [10], while a third-order model was considered in [11]. The first two equations in both models describe the arterial partial pressures of O_2 and CO_2, and a peripheral controller. The third equation in the second system models the CO_2 level in the brain and a central controller. Both models possess the same particular structure (delayed matrix of rank one) leading to a very simple characteristic function of the form

$$p(\lambda;\ \tau) := Q(\lambda) + P(\lambda)e^{-\lambda\tau}.$$

In this chapter, we considered only a simplified second-order delay model from [316], but the ideas still apply to both models from [10, 11]. It is important to point out that although the geometric argument was mentioned in [10], it was not fully exploited. The characterization of the crossing direction was developed in the light of [51] and following the same procedure as in [54], instead of taking advantage of the particular structure of the system.

Finally, a more general respiration model including five DDEs [136] and three independent delays can be found in [9]. The corresponding delays are as follows: τ_B is the lung to brain delay (equal to lung to tissue transport delay), τ_v is the venous side transport delay from tissue to lung, and τ_a is the lung to carotid artery delay. We believe that the geometric idea proposed in the immune dynamics analysis can also be applied to the stability analysis of the corresponding linearization including three delays.

In the second part we addressed the characterization of stability boundaries in some delay-parameter space for a dynamical system including *four* (independent) *delays*, describing the posttransplantation dynamics of the immune response to chronic myelogenous leukemia. Such a model includes two small delays and two large delays [66]. Similar to the previous case study, the corresponding delayed matrices are of rank one, which simplifies the analysis. More precisely, the stability analysis followed the general lines proposed in [231], and allowed us to briefly discuss two types of T/C cell interactions: weak and strong. Furthermore, a quantitative measure for the weak T/C interaction was introduced, and explicitly computed. We also proved that the large delay values have a small influence on the stability properties in the weak cell interaction case. Next, the strong cell interaction case was analyzed in terms of stability crossing curves, and a classification of such crossing curves was given. Finally, an example was presented to illustrate the derived results. Several discussions completed the presentation.

Appendix

A.1 Rouché's theorem

The proofs of many continuity properties of the spectra of operators associated with linear time-delay systems rely on the celebrated Rouché's theorem (see, e.g., [141, Section 5.3.]).

Theorem A.1. *Let f and g be analytic functions on an (open) domain $\mathcal{D} \subseteq \mathbb{C}$. Let $\mathcal{C} \subset D$ be a closed, simple contour, that is, without self-interactions. If*

$$\forall \lambda \in \mathcal{C} \ |g(\lambda)| < |f(\lambda)|,$$

then the functions f and $f + g$ have the same number of zero inside $\mathcal{C}$, where each zero is counted as many times as its multiplicity.

The following result from [181] is a corollary.

Corollary A.2. *Let f and the sequence $\{f_n\}_{n\geq 1}$ be analytic functions on a domain $\mathcal{D} \subseteq \mathbb{C}$. Assume that $\{f_n(\lambda)\}_{n\geq 1}$ converges uniformly to $f(\lambda)$ on the disk $D = \{\lambda : |\lambda - \lambda_0| \leq R\} \subset \mathcal{D}$ for some $R > 0$. Assume further that on this disk, λ_0 is the only zero of $f(\lambda)$, with multiplicity $k \geq 0$ ($k = 0$ means no zeros in D). Then there exists a number $N \in \mathbb{N}$ such that for all $n \geq N$, $f_n(\lambda)$ has exactly k zeros $\lambda_{n,1}, \ldots, \lambda_{n,k}$ in D and $\lim_{n\to\infty} \lambda_{n,j} = \lambda_0 \ \forall j \in \{1, \ldots, k\}$.*

A.2 Structured singular value (ssv)

We introduce the concept of ssvs of matrices and outline the main principles behind the standard computational schemes. A more elaborate introduction can be found in the review paper [245] and [329, Chap. 11; 115, Chap. 4].

Let $G \in \mathbb{C}^{N\times M}$ and denote its singular values in decreasing order with $\sigma_1(G) \geq \sigma_2(G) \geq \ldots$. A classical result from linear algebra and robust control theory, which lays the basis for the celebrated small gain theorem, relates the largest singular value of G to the solutions of the equation

$$\det(I + G\Delta) = 0 \tag{A.1}$$

in the following way:

$$\sigma_1(G) = \begin{cases} 0, & \text{if } \det(I + G\Delta) \neq 0, \ \forall \Delta \in \mathbb{C}^{M \times N}, \\ \left(\min\left\{\sigma_1(\Delta) : \ \Delta \in \mathbb{C}^{M \times N} \text{ and } \det(I + G\Delta) = 0\right\}\right)^{-1}, & \text{otherwise.} \end{cases} \tag{A.2}$$

We refer to Δ as the uncertainty, as in a robust control framework. (A.1) typically originates from a feedback interconnection of a nominal transfer function and an uncertainty block.

Next we reconsider the solutions of equation (A.1), where Δ is restricted to having a particular structure by imposing $\Delta \in \mathbf{\Delta}$, with $\mathbf{\Delta}$ a closed subset of $\mathbb{C}^{M \times N}$. In analogy with (A.2) one defines the *ssn* of the matrix G with respect to the uncertainty set $\mathbf{\Delta}$ as

$$\mu_{\mathbf{\Delta}}(G) := \begin{cases} 0, & \text{if } \det(I + G\Delta) \neq 0, \ \forall \Delta \in \mathbf{\Delta}, \\ (\min\{\sigma_1(\Delta) : \ \Delta \in \mathbf{\Delta} \text{ and } \det(I + G\Delta) = 0\})^{-1}, & \text{otherwise.} \end{cases} \tag{A.3}$$

It directly follows from the definition that

$$\mu_{\mathbf{\Delta}}(G) \leq \sigma_1(G). \tag{A.4}$$

Furthermore, if $\mathbb{C}\mathbf{\Delta} = \mathbf{\Delta}$, then

$$\mu_{\mathbf{\Delta}}(G) = \max_{\Delta \in \mathbf{\Delta}, \ \sigma_1(\Delta) = 1} r_\sigma(G\Delta), \tag{A.5}$$

with $r_\sigma(\cdot)$ the spectral radius.

In what follows we restrict ourselves for simplicity to an uncertainty set $\mathbf{\Delta}$ of the form

$$\begin{aligned} \mathbf{\Delta} \ := \ & \left\{\operatorname{diag}(\Delta_0, \ldots, \Delta_f, d_0 I_{m_0}, \ldots, d_s I_{m_s}) : \ \Delta_i \in \mathbb{C}^{k_i \times l_i}, \ d_j \in \mathbb{C}, \right. \\ & \left. \qquad 0 \leq i \leq f, \ 0 \leq j \leq s\right\}, \end{aligned} \tag{A.6}$$

where diag$(\cdot)$ represents a block-diagonal matrix, $\sum_{i=0}^{f} k_i + \sum_{i=0}^{s} m_i = M$ and $\sum_{i=0}^{f} l_i + \sum_{i=0}^{s} m_i = N$. Such a set satisfies $\mathbb{C}\mathbf{\Delta} = \mathbf{\Delta}$. Furthermore, based on a slight generalization of [245, Lemma 6.3] to nonsquare block-diagonal perturbations, the search space of the optimization in the right-hand side of (A.5) can be restricted. This results in

$$\mu_{\mathbf{\Delta}}(G) = \max_{U \in \mathcal{U}} r_\sigma(GU), \tag{A.7}$$

where $\mathcal{U} \subseteq \mathbf{\Delta}$ is defined as

$$\begin{aligned} \mathcal{U} := \ & \left\{\operatorname{diag}(U_0, \ldots, U_f, u_0 I_{m_0}, \ldots, u_s I_{m_s}) : \ U_i \in \mathbb{C}^{k_i \times l_i}, \ u_j \in \mathbb{C} \right. \\ & \left. \sigma_k(U_i) = 1, \ 1 \leq k \leq \min(k_i, l_i), \ |u_j| = 1, \ \ 0 \leq i \leq f, \ 0 \leq j \leq s\right\}. \end{aligned}$$

Note that the elements of $\mathcal{U}$ are *unitary* matrices if the uncertainty structure only involves square blocks, that is, $k_i = l_i, \ i = 1, \ldots, f$.

Next, the following invariance property can easily be checked:

$$\mu_{\mathbf{\Delta}}(G) = \mu_{\mathbf{\Delta}}(D_2 G D_1^{-1}), \ \forall (D_1, D_2) \in \mathcal{D}, \tag{A.8}$$

where

$$\begin{aligned} \mathcal{D} := \ & \left\{(D_1, D_2) : \ D_1 = \operatorname{diag}(a_0 I_{k_1}, \ldots, a_f I_{k_f}, D_0, \ldots, D_s), \ \ D_2 = \right. \\ & \left. \operatorname{diag}(a_0 I_{l_1}, \ldots, a_f I_{l_f}, D_0, \ldots, D_s) : \ a_i > 0, D_i \in \mathbb{C}^{m_i \times m_i}, D_i^* = D_i > 0\right\}. \end{aligned}$$

From (A.7) and the combination of (A.8) and (A.4) we finally obtain

$$\max_{U\in\mathcal{U}} r_\sigma(GU) = \mu_\Delta(G) \leq \min_{(D_1,D_2)\in\mathcal{D}} \sigma_1(D_2 G D_1^{-1}). \tag{A.9}$$

Therefore, *optimization* procedures are typically used to compute estimates for $\mu_\Delta(G)$. The function $U \in \mathcal{U} \to r_\sigma(GU)$ may have several local maxima and, for this, a local search for a maximum is not guaranteed to lead to $\mu_\Delta(G)$, but to lower bounds. An appropriate formulation of the optimality condition enables algorithms which resemble power algorithms for computing eigenvalues and singular values; see [246] for an example. Although the convergence of such algorithms to $\mu_\Delta(G)$ is not guaranteed either and they may converge to values corresponding to lower bounds on $\mu_\Delta(G)$, they have proved their effectiveness in practice. The computation of the upper bound in (A.9) can be recast into a standard *convex optimization* problem. However, in general $\mu_\Delta(G)$ is not equal to the upper bound. An exception to this holds if the number of blocks in the matrices belonging to the uncertainty set Δ satisfies $f + 2s \leq 3$ and all the blocks are square, $k_i = l_i$, $i = 0, \ldots, f$.

A.3 Continuity properties

Throughout Chapter 5 we encounter functions from $\mathcal{B}_+^m \subset \mathbb{R}^m$ to $\mathbb{R}$ and from $\mathcal{B}_+^m$ to $\mathcal{P}(\mathbb{C})$. We assume that $\mathbb{R}^m$ is equipped with the Euclidean norm and $\mathcal{P}(\mathbb{C})$ with the Hausdorff metric.

A function $f : \mathcal{B}_+^m \to \mathbb{R}$ is lower semicontinuous, respectively upper semicontinuous, at $\vec{s}$ if and only if (see, for instance, [16])

$$\forall \epsilon > 0\ \exists \delta > 0\ \ \forall \vec{r} \in \mathcal{B}_+^m\ \ \|\vec{r} - \vec{s}\| < \delta \Rightarrow f(\vec{r}) - f(\vec{s}) > -\epsilon,$$

respectively

$$\forall \epsilon > 0\ \exists \delta > 0\ \ \forall \vec{r} \in \mathcal{B}_+^m\ \ \|\vec{r} - \vec{s}\| < \delta \Rightarrow f(\vec{r}) - f(\vec{s}) < \epsilon.$$

It is continuous at $\vec{s}$ when it is both upper and lower semicontinuous at $\vec{s}$.

A function $f : \mathcal{B}_+^m \to \mathcal{P}(\mathbb{C})$ is lower semicontinuous at $\vec{s}$ if and only if

$$\forall \epsilon > 0\ \exists \delta > 0\ \ \forall \vec{r} \in \mathcal{B}_+^m\ \ \|\vec{r} - \vec{s}\| < \delta \Rightarrow D(f(\vec{s}), f(\vec{r})) < \epsilon.$$

When replacing $D(f(\vec{s}), f(\vec{r}))$ with $D(f(\vec{r}), f(\vec{s}))$, respectively $D_h(f(\vec{s}), f(\vec{r}))$, we have upper semicontinuity, respectively continuity, at $\vec{s}$.

A.4 Interdependency of numbers

The real numbers $r_1, r_2, \ldots, r_m$ are *rationally independent* if and only if

$$\sum_{i=1}^{m} z_i r_i = 0, \quad z_i \in \mathbb{Z},$$

implies $z_i = 0,\ i = 1, \ldots, m$. For example, two numbers are rationally independent if and only if their ratio is an irrational number.

If the real numbers $r_1, \ldots, r_m$ are *rationally dependent* (that is, not rationally independent), then there always exists an integer $p < m$ and a matrix $\Gamma \in \mathbb{Z}^{m \times p}$ of full column rank such that

$$\begin{bmatrix} r_1 \\ \vdots \\ r_m \end{bmatrix} = \Gamma \begin{bmatrix} s_1 \\ \vdots \\ s_p \end{bmatrix},$$

with the numbers $s_1, \ldots, s_p$ rationally independent. Thus, rationally dependent numbers depend on a smaller number of rationally independent numbers. In the special case where $p = 1$, the numbers $r_1, \ldots, r_m$ are called *commensurate*, as they are all multiples of the same number.

For example, the numbers $1, \pi$, and $\exp(1)$ are rationally independent; the numbers $1, 2$, and $5/3$ commensurate. The numbers $1,\ \pi$, and $1 + \pi$ are rationally dependent, yet not commensurate, as

$$\begin{bmatrix} 1 \\ \pi \\ 1+\pi \end{bmatrix} = \begin{bmatrix} 1 & 0 \\ 0 & 1 \\ 1 & 1 \end{bmatrix} \begin{bmatrix} 1 \\ \pi \end{bmatrix},$$

with 1 and π rationally independent.

A.5 Software

The following software packages were used for the book:

- DDE-BIFTOOL [74, 78]

  The MATLAB package DDE-BIFTOOL allows a numerical bifurcation and stability analysis of DDEs with several fixed discrete and/or state-dependent delays. It contains routines for the computation, continuation, and stability analysis of steady state solutions, their Hopf and fold bifurcations, periodic solutions, and connecting orbits (but the latter only for the constant delay case). A stability analysis of steady state solutions is achieved through computing approximations and corrections to the rightmost characteristic roots using a linear multistep method. Periodic solutions, their Floquet multipliers, and connecting orbits are computed using piecewise polynomial collocation on adaptively refined meshes. An overview of its application to stabilization problems is presented in [259].

  More information on the package can be found at

  `www.cs.kuleuven.ac.be/cwis/research/twr/research/software/delay/`

  The stabilization approaches presented in Chapters 7–10 rely on algorithms for the computation of characteristic roots of delay equations for which an implementation in the package is available. Furthermore its continuation facility allows an efficient computation of stability regions of linear time-delay systems in a two-parameter space. Unless stated otherwise, the plots of characteristic roots throughout the book were generated by means of the package.

- HIFOO [32]

  The MATLAB package HIFOO aims at solving fixed-order stabilization and local optimization problems for finite-dimensional systems. It depends on a hybrid algorithm for nonsmooth, nonconvex optimization based on quasi-Newton updating, bundling, and gradient sampling.

  More information on the package can be found at

  `www.cs.nyu.edu/overton/software/hifoo/`

  The computations for the numerical examples of Chapter 10, where the stabilization problem for linear time-delay systems is reformulated as an infinite-dimensional eigenvalue optimization problem, were performed by combining the packages HIFOO and DDE-BIFTOOL.

A stability analysis of linear time-delay systems with periodically varying coefficients, as we have encountered in Chapter 6, can be performed using the following package:

- PDDE-CONT [289]

  The package PDDE-CONT allows us to continue periodic solutions, analyze their stability properties, and continue periodic solution bifurcations of autonomous and periodically forced DDEs with several pointwise delays. If one of the three common codimension-one bifurcations (fold, period doubling, Neimark–Sacker) is found along a branch of periodic solutions, the point can be used as a starting point for a continuation of the branch of bifurcation points in a two parameter space.

For an detailed overview of software packages for DDEs we refer to
`www.cs.kuleuven.be/~twr/research/software/delay/software.shtml`

Bibliography

[1] C. T. ABDALLAH, P. DORATO, J. BENITEZ-READ, AND R. BYRNE, *Delayed positive feedback can stabilize oscillatory systems*, in Proceedings of the 1993 American Control Conference, San Francisco, CA, 1993, pp. 3106–3107.

[2] J. ACKERMANN, *Der entwurf linearer regelungssysteme im zustandsraum*, Regelungstechnik und prozessdatenverarbeitung, 7 (1972), pp. 297–300.

[3] Z. ARTSTEIN, *Linear systems with delayed control: A reduction*, IEEE Trans. Automat. Control, 27 (1982), pp. 869–879.

[4] K. J. ASTRÖM, C. C. HANG, AND B. C. LIM, *A new Smith predictor for controlling a process with an integrator and long dead-time*, IEEE Trans. Automat. Control, 39 (1994), pp. 343–345.

[5] C. E. AVELLAR AND J. K. HALE, *On the zeros of exponential polynomials*, Math. Anal. Appl., 73 (1980), pp. 434–452.

[6] Z. BAI, J. DEMMEL, J. DONGARRA, A. RUHE, AND H. VAN DER VORST, eds., *Templates for the Solution of Algebraic Eigenvalue Problems: A Practical Guide*, Software Environ. Tools 11, SIAM, Philadelphia, 2000.

[7] C. T. H. BAKER, G. A. BOCHAROV, AND F. A. RIHAN, *A Report on the Use of Delay Differential Equations in Numerical Modelling in the Biosciences*, Tech. report 343, Manchester Centre for Computational Mathematics, Manchester, UK, 1999.

[8] M. BANDO, K. HASEBE, K. NAKANISHI, AND A. NAKAYAMA, *Analysis of optimal velocity model with explicit delay*, Phys. Rev. E, 58 (1998), pp. 5429–5435.

[9] J. J. BATZEL AND T. H. TRAN, *Modeling instability in the control system for human respiration: Applications to infant non-REM sleep*, Appl. Math. Comput., 110 (2000), pp. 1–51.

[10] ——, *Stability of the human respiratory control system. Part I: Analysis of a two-dimensional delay state-space model*, J. Math. Biol., 41 (2000), pp. 45–79.

[11] ——, *Stability of the human respiratory control system. Part II: Analysis of a three-dimensional delay state-space model*, J. Math. Biol., 41 (2000), pp. 80–102.

[12] A. Bellen and M. Zennaro, *Numerical Methods for Delay Differential Equations*, Clarendon Press, New York, 2003.

[13] R. E. Bellman and K. L. Cooke, *Differential-Difference Equations*, Academic Press, New York, 1963.

[14] A. Ben-Israel and T. N. E. Greville, *Generalized Inverses: Theory and Applications*, Pure Appl. Math., Wiley, New York, 1974.

[15] A. Bensoussan, G. Da Prato, M. C. Defour, and S. K. Mitter, *Representation and Control of Infinite Dimensional Systems*, Systems Control Found. Appl., 2 volumes, Birkhauser, Boston, 1993.

[16] C. Berge, *Espaces topologiques. Fonctions multivoques*, Dunod, Paris, 1966. (in French).

[17] P. A. Bliman, *LMI characterization of strong delay-independent stability of linear delay systems via quadratic Lyapunov-Krasovskii functionals*, Systems Control Lett., 43 (2001), pp. 263–274.

[18] F. G. Boese, *Stability with respect to the delay: On the paper of K. L. Cooke and P. van den Driessche*, J. Math. Anal. Appl., 228 (1998), pp. 293–321.

[19] J. C. Bolot and A. U. Shankar, *Analysis of a fluid approximation to flow control dynamics*, in Proceedings of IEEE INFOCOM, Florence, IT, 1992, pp. 2398–2407.

[20] E.-K. Boukas and Z.-K. Liu, *Deterministic and Stochastic Time Delay Systems*, Birkhäuser, Boston, 2002.

[21] S. Boyd and V. Balakrishnan, *A regularity result for the singular values of a transfer matrix and a quadratically convergent algorithm for computing its $\mathcal{L}_\infty$-norm*, Systems Control Lett., 15 (1990), pp. 1–7.

[22] S. Boyd, V. Balakrishnan, and P. Kabamba, *A bisection method for computing the $\mathcal{H}_\infty$ norm of a transfer matrix and related problems*, Math. Control Signals Systems, 2 (1989), pp. 207–219.

[23] S. Boyd and L. Vandenberghe, *Convex Optimization*, Cambridge University Press, Cambridge, UK, 2004.

[24] F. Brauer, *Characteristic return times for harvested population models with time lag*, Math. Biosci., 45 (1979), pp. 295–311.

[25] D. Breda, S. Maset, and R. Vermiglio, *Computing the characteristic roots for delay differential equations*, IMA J. Numer. Anal., 24 (2004), pp. 1–19.

[26] ———, *Pseudospectral differencing methods for characteristic roots of delay differential equations*, SIAM J. Sci. Comput., 27 (2005), pp. 482–495.

[27] D. Brethé and J. J. Loiseau, *An effective algorithm for finite spectrum assignment of single-input systems with delays*, Math. Comput. Simulation, 45 (1998), pp. 339–348.

[28] S. D. BRIERLEY, J. N. CHIASSON, E. B. LEE, AND S. H. ZAK, *On stability independent of delay for linear systems*, IEEE Trans. Automat. Control, 27 (1982), pp. 252–254.

[29] J. W. BRUCE AND P. J. GIBLIN, *Curves and Singularities*, Cambridge University Press, Cambridge, UK, 1984.

[30] W. E. BRUMLEY, *On the asymptotic behavior of solutions of differential-difference equations of neutral type*, J. Differential Equations, 7 (1970), pp. 175–188.

[31] J. BURKE, A. S. LEWIS, AND M. L. OVERTON, *Approximating subdifferentials by random sampling of gradients*, Math. Oper. Res., 27 (2002), pp. 567–584.

[32] J. V. BURKE, D. HENRION, A. S. LEWIS, AND M. L. OVERTON, *HIFOO: A* MATLAB *package for fixed-order controller design and H-infinity optimization*, in Proceedings of ROCOND, Toulouse, FR, 2006.

[33] ——, *Stabilization via nonsmooth, nonconvex optimization*, IEEE Trans. Automat. Control, 51 (2006), pp. 1760–1769.

[34] J. V. BURKE, A. S. LEWIS, AND M. L. OVERTON, *A nonsmooth, nonconvex optimization approach to robust stabilization by static output feedback and low-order controllers*, in Proceedings of the 4th IFAC Symposium on Robust Control Design, Milan, IT, 2003.

[35] ——, *Optimization and pseudospectra, with applications to robust stability*, SIAM J. Matrix Anal. Appl., 25 (2003), pp. 80–104.

[36] ——, *A robust gradient sampling algorithm for nonsmooth, nonconvex optimization*, SIAM J. Optim., 15 (2005), pp. 751–779.

[37] E. A. BUTCHER, H. MA, E. A. BUELER, V. AVERINA, AND Z. SZABO, *Stability of linear time-periodic delay-differential equations via Chebyshev polynomials*, Internat. J. Numer. Methods Engrg., 59 (2004), pp. 895–922.

[38] R. BYERS, *A bisection method for measuring the distance of a stable matrix to the unstable matrices*, SIAM J. Sci. Statist. Comput., 9 (1988), pp. 875–881.

[39] A. CALLENDER, D. R. HARTREE, AND A. PORTER, *Time lag in a control system*, Philosoph. Transactions of Royal Society London, 235 (1936), pp. 415–444.

[40] A. CALLENDER AND A. G. STEVENSON, *Time lag in a control system*, Proceedings Soc. Chem. Ind., 18 (1936), pp. 108–117.

[41] Y.-Y. CAO, Y.-X. SUN, AND C. CHENG, *Delay-dependent robust stabilization of uncertain systems with multiple state delays*, IEEE Trans. Automat. Control, 43 (1998), pp. 1608–1612.

[42] R. E. CHANDLER, R. HERMAN, AND E. W. MONTROLL, *Traffic dynamics: Studies in car following*, Oper. Res., 6 (1958), pp. 165–183.

[43] F. CHATTÉ, B. DUCOURTHIAL, D. NACE, AND S.-I. NICULESCU, *Results on fluid modelling packet switched networks*, in Proceedings of the 3rd IFAC Workshop on Time-Delay Systems, Santa Fe, NM, 2001.

[44] J. CHEN, *Static output feedback stabilization for SISO systems and related problems: Solutions via generalized eigenvalues*, Control Theory and Advanced Technology, 10 (1995), pp. 2233–2244.

[45] J. CHEN, P. FU, AND S.-I. NICULESCU, *Asymptotic behavior of imaginary zeros of linear systems with commensurate delays*, in Proceedings of the 45th IEEE Conference on Decision and Control, San Diego, CA, 2006.

[46] J. CHEN, G. GU, AND C. A. NETT, *A new method for computing delay margins for stability of linear delay systems*, Systems Control Lett., 26 (1995), pp. 107–117.

[47] J. CHEN AND H. A. LATCHMAN, *Frequency sweeping tests for stability independent of delay*, IEEE Trans. Automat. Control, 40 (1995), pp. 1640–1645.

[48] Y. Q. CHEN AND K. L. MOORE, *Analytical stability bound for delayed second-order systems with repeating poles using Lambert function W*, Automatica, 38 (2002), pp. 891–895.

[49] J. CHIASSON AND C. T. ABDALLAH, *Robust stability of time delay systems: Theory*, in Proceedings of the 3rd IFAC Workshop on Time-Delay Systems, Santa Fe, NM, 2001, pp. 125–130.

[50] J. N. CHIASSON, S. D. BRIERLEY, AND E. B. LEE, *A simplified derivation of the Zeheb-Walach 2-D stability test with applications to time-delay systems*, IEEE Trans. Automat. Control, 30 (1985), pp. 411–414; corrections in IEEE Trans. Automat. Control, 31 (1986) pp. 91–92.

[51] K. COOKE AND J. TURI, *Stability, instability in delay equations modeling human respiration*, J. Math. Biol., 32 (1994), pp. 535–543.

[52] K. L. COOKE, *Delay differential equations*, in Mathematics of Biology, M. Iannelli, ed., Liguori Editore, Napoli, IT, 1981, pp. 5–80.

[53] K. L. COOKE AND Z. GROSSMAN, *Discrete delay, distributed delay and stability switches*, J. Math. Anal. Appl., 86 (1982), pp. 592–627.

[54] K. L. COOKE AND P. VAN DEN DRIESSCHE, *On zeroes of some transcendental equations*, Funkcialaj Ekvacioj, 29 (1986), pp. 77–90.

[55] R. M. CORLESS, G. H. GONNET, D. E. G. HARE, D. J. JEFFREY, AND D. E. KNUTH, *On the Lambert W function*, Adv. Comput. Math., 5 (1996), pp. 329–359.

[56] R. V. CULSHAW, S. RUAN, AND G. WEBB, *A mathematical model of cell-to-cell spread of HIV-1 that includes a time delay*, J. Math. Biol., 46 (2003), pp. 425–444.

[57] R. F. CURTAIN AND A. J. PRITCHARD, *Functional Analysis in Modern Applied Mathematics*, Academic Press, London, 1977.

[58] R. F. CURTAIN AND H. ZWART, *An introduction to Infinite-dimensional Linear Systems Theory*, Texts Appl. Math. 21, Springer, Berlin, 1995.

[59] R. DATKO, *A procedure for determination of the exponential stability of certain differential difference equations*, Quart. Appl. Math., 36 (1978), pp. 279–292.

[60] ——, *Remarks concerning the asymptotic stability and stabilization of linear delay differential equations*, J. Math. Anal. Appl., 111 (1985), pp. 571–584.

[61] ——, *Not all feedback stabilized hyperbolic systems are robust with respect to small time delays in their feedbacks*, SIAM J. Control Optim., 26 (1988), pp. 697–713.

[62] ——, *Two examples of ill-posedness with respect to time delays revisited*, IEEE Trans. Automat. Control, 42 (1997), pp. 434–452.

[63] R. DATKO, J. LAGNESE, AND M. POLIS, *An example on the effect of time delays in boundary feedback stabilization of wave equations*, SIAM J. Control Optim., 24 (1986), pp. 152–156.

[64] L. C. DAVIS, *Modifications of the optimal velocity traffic model to include delay due to driver reaction time*, Phys. A, 319 (2003), pp. 557–567.

[65] C. E. DE SOUZA AND X. LI, *Delay-dependent robust $\mathcal{H}_\infty$ control of uncertain linear state-delayed systems*, Automatica, 35 (1999), pp. 1313–1321.

[66] R. DECONDE, P. S. KIM, D. LEVY, AND P. P. LEE, *Post-transplantation dynamics of the immune response to chronic myelogenous leukemia*, J. Theoret. Biology, 236 (2005), pp. 39–59.

[67] O. DIEKMANN, S. A. VAN GILS, S. M. VERDUYN LUNEL, AND H.-O. WALTHER, *Delay Equations: Functional-, Complex-, and Nonlinear Analysis*, Appl. Math. Sci. 110, Springer, New York, 1995.

[68] R. DIESTEL, *Graph Theory*, 2nd ed., Grad. Texts in Math. 173, Springer-Verlag, Heidelberg, Germany, 2000.

[69] E. J. DOEDEL, A. R. CHAMPNEYS, T. J. FAIRGRIEVE, Y. A. KUZNETSOV, B. SANDSTEDE, AND X.-J. WANG, *AUTO97: Continuation and Bifurcation Software for Ordinary Differential Equations*, Tech. Report, Department of Computer Science, Concordia University, 1998.

[70] L. DUGARD AND E. I. VERRIEST, EDS., *Stability and Control of Time-delay Systems*, Lecture Notes in Control and Inform. Sci. 228, Springer, New York, 1998.

[71] L. E. EL'SGOL'TS AND S. B. NORKIN, *Introduction to the Theory and Applications of Differential Equations with Deviating Arguments*, Math. Sci. Engrg. 105, Academic Press, New York, 1973.

[72] K. ENGELBORGHS, M. DAMBRINE, AND D. ROOSE, *Limitations of a class of stabilization methods for delay equations*, IEEE Trans. Automat. Control, 46 (2001), pp. 336–339.

[73] K. ENGELBORGHS, T. LUZYANINA, K. J. IN 'T HOUT, AND D. ROOSE, *Collocation methods for the computation of periodic solutions of delay differential equations*, SIAM J. Sci. Comput., 22 (2000), pp. 1593–1609.

[74] K. ENGELBORGHS, T. LUZYANINA, AND G. SAMAEY, *DDE-BIFTOOL v. 2.00: A MATLAB Package for Bifurcation Analysis of Delay Differential Equations*, TW Report 330, Department of Computer Science, Katholieke Universiteit Leuven, Belgium, 2001.

[75] K. ENGELBORGHS AND D. ROOSE, *Bifurcation analysis of periodic solutions of neutral functional differential equations: A case study*, Internat. J. Bifur. Chaos Appl. Sci. Engrg., 8 (1998), pp. 1889–1905.

[76] ———, *Numerical computation of stability and detection of Hopf bifurcations of steady state solutions of delay differential equations*, Adv. Comput. Math., 10 (1999), pp. 271–289.

[77] ———, *On the stability of LMS methods and characteristic roots of delay differential equations*, SIAM J. Numer. Anal., 40 (2002), pp. 629–650.

[78] K. ENGELBORGHS, T. LUZYANIA, AND D. ROOSE, *Numerical bifurcation analysis of delay differential equations using DDE-BIFTOOL*, ACM Trans. Math. Software, 28 (2002), pp. 1–21.

[79] K. FALL AND K. VARADHAN, EDS., *The **ns** manual*, http://www-mash.cs.berkeley.edu/ns.

[80] A. FATTOUH, O. SENAME, AND J. M. DION, *Pulse controller design for linear time-delay systems*, in Proceedings of the first IFAC Symposium on Systems, Structure and Control, Prague, Czech Republic, 2001.

[81] W. FENG, *On practical stability of linear multivariable feedback systems with time-delays*, Automatica, 27 (1991), pp. 389–394.

[82] M. FLIESS AND H. MOUNIER, *Quelques propriétés structurelles des systèmes linéaires à retards constants*, Comptes Rendus de L'Académie des Sciences Paris, I-319 (1994), pp. 289–294.

[83] E. FRIDMAN AND U. SHAKED, *A descriptor system approach to $\mathcal{H}_\infty$ control of linear time-delay systems*, IEEE Trans. Automat. Control, 47 (2002), pp. 253–270.

[84] P. FU, J. CHEN, AND S.-I. NICULESCU, *High-order analysis of critical stability properties of linear time-delay systems*, in Proceedings of the American Control Conference, New York, NY, 2007.

[85] E. GALLESTEY, D. HINRICHSEN, AND A. J. PRITCHARD, *Spectral value sets of closed linear operators*, Proc. Roy. Soc. A, 456 (2000), pp. 1397–1418.

[86] F. R. GANTMACHER, *Theory of Matrices*, Chelsea, New York, 1959.

[87] Y. GENIN, R. STEFAN, AND P. VAN DOOREN, *Real and complex stability radii of polynomial matrices*, Linear Algebra Appl., 351–352 (2002), pp. 381–410.

[88] Y. GENIN, P. VAN DOOREN, AND V. VERMAUT, *Convergence of the calculation of $\mathcal{H}_\infty$-norms and related questions*, in Proceedings of the 13th International Symposium on Mathematical Theory of Networks and Systems, Padova, IT, 1998, pp. 429–432.

[89] T. T. GEORGIOU AND M. C. SMITH, *Graphs, causality and stabilizability: Linear, shift-invariant systems on $\mathcal{L}_2([0,\infty))$*, Math. Control Signals Systems, 6 (1993), pp. 195–223.

[90] H. GLUSING-LUERSSEN, *First-order representations of delay-differential systems in a behavioural setting*, Eur. J. Control, 3 (1997), pp. 137–149.

[91] G. H. GOLUB AND C. F. VAN LOAN, *Matrix Computations*, 3rd ed., Johns Hopkins University Press, Baltimore, MD, 1996.

[92] K. GOPALSAMY, *Stability and Oscillations in Delay Differential Equations of Population Dynamics*, Math. Appl. 74, Kluwer Academic Publishers, Dordrecht, NL, 1993.

[93] A. GRAHAM, *Kronecker Products and Matrix Calculus with Applications*, Ellis Horwood, Chichester, 1981.

[94] K. GREEN AND T. WAGENKNECHT, *Pseudospectra and delay differential equations*, J. Comput. Appl. Math., 196 (2006), pp. 567–578.

[95] M. GREEN, *How long does it take to stop?*, Transportation Human Factors, 2 (2000), pp. 195–216.

[96] K. GU, V. L. KHARITONOV, AND J. CHEN, *Stability of Time-Delay Systems*, Birkhäuser, Boston, 2003.

[97] K. GU AND S.-I. NICULESCU, *Additional dynamics in transformed time-delay systems*, IEEE Trans. Automat. Control, 45 (2000), pp. 572–575.

[98] K. GU, S.-I. NICULESCU, AND J. CHEN, *On stability of crossing curves for general systems with two delays*, J. Math. Anal. Appl., 311 (2005), pp. 231–253.

[99] K. P. HADELER, *Delay equations in biology*, in Functional Differential Equations and Approximation of Fixed Points, Lect. Notes Math. 730, Springer-Verlag, Berlin, 1979, pp. 136–157.

[100] A. HALANAY, *Differential Equations: Stability, Oscillations, Time Lags*, Academic Press, New York, 1966.

[101] J. K. HALE, *Theory of Functional Differential Equations*, Appl. Math. Sci. 3, Springer, New York, 1977.

[102] ——, *Effects of delays on dynamics*, in Topological Methods in Differential Equations and Inclusions, A. Granas, M. Frigon, and G. Sabidussi, eds., Kluwer Academic Publishers, Dordrecht, NL, 1995, pp. 191–238.

[103] J. K. HALE, E. F. INFANTE, AND F. S.-P. TSEN, *Stability in linear delay equations*, J. Math. Anal. Appl., 105 (1985), pp. 533–555.

[104] J. K. HALE AND S. M. VERDUYN LUNEL, *Averaging in infinite dimensions*, J. Integral Equations Appl., 2 (1990), pp. 463–494.

[105] ——, *The effect of rapid oscillations in the dynamics of delay equations*, in Mathematical Population Dynamics, O. Arino, D. E. Axelrod, and M. Klimmel, eds., Marcel Dekker, New York, 1991, pp. 211–216.

[106] ——, *Introduction to functional differential equations*, Appl. Math. Sci. 99, Springer, New York, 1993.

[107] ——, *Strong stabilization of neutral functional differential equations*, IMA J. Math. Control Inform., 19 (2002), pp. 5–23.

[108] ——, *Stability and control of feedback systems with time delays*, Internat. J. Systems Sci., 34 (2003), pp. 497–504.

[109] K. B. HANNSGEN, Y. RENARDY, AND R. L. WHEELER, *Effectiveness and robustness with respect to time delays of boundary feedback stabilization in one-dimensional viscoelasticity*, SIAM J. Control Optim., 26 (1988), pp. 1200–1234.

[110] G. H. HARDY AND E. M. WRIGHT, *An Introduction to the Theory of Numbers*, Oxford University Press, New York, 1968.

[111] D. HELBING, *Traffic and related self-driven many-particle systems*, Rev. Modern Phys., 73 (2001), pp. 1067–1141.

[112] U. HELMKE AND B. D. O. ANDERSON, *Hermitian pencils and output feedback stabilization of scalar systems*, Internat. J. Control, 56 (1992), pp. 857–876.

[113] D. HENRY, *Linear autonomous neutral functional differential equations*, J. Differential Equations, 15 (1974), pp. 106–128.

[114] D. HERTZ, E. I. JURY, AND E. ZEHEB, *Simplified analytical stability test for systems with commensurate time delays*, IEE Proc. Control Theory Appl. D, 131 (1984), pp. 52–56.

[115] D. HINRICHSEN AND B. KELB, *Spectral value sets: A graphical tool for robustness analysis*, Systems Control Lett., 21 (1993), pp. 127–136.

[116] D. HINRICHSEN AND A. J. PRITCHARD, *Mathematical Systems Theory I. Modelling, State Space Analysis, Stability and Robustness*, Texts Appl. Math. 48, Springer, Berlin, 2005.

[117] C. V. HOLLOT AND Y. CHAIT, *Nonlinear stability analysis for a class of TCP/AQM networks*, in Proceeding of the 40th IEEE Conference on Decision and Control, Orlando, FL, 2001, pp. 2309–2314.

[118] C. V. HOLLOT, V. MISRA, D. TOWSLEY, AND W. B. GONG, *Analysis and design of controllers for AQM routers supporting TCP flow*, IEEE Trans. Automat. Control, 47 (2002), pp. 945–956.

[119] F. C. HOPPENSTEADT AND C. S. PESKIN, *Mathematics in Medicine and the Life Sciences*, Texts Appl. Math. 10, Springer-Verlag: New York, 1992.

[120] R. A. HORN AND C. R. JOHNSON, *Matrix Analysis*, Cambridge University Press, Cambridge, 1985.

[121] T. INSPERGER AND G. STÉPÁN, *Semi-discretization method for delayed systems*, Internat. J. Numer. Methods Engrg., 55 (2002), pp. 3–18.

[122] ——, *Stability analysis of turning with periodic spindle speed modulation via semi-discretization*, J. Vib. Control, 10 (2004), pp. 1835–1855.

[123] G. IOOSS AND D. D. JOSEPH, *Elementary Stability and Bifurcation Theory*, Undergrad. Texts Math., Springer, New York, 1980.

[124] R. IZMAILOV, *Adaptive feedback control algorithms for large data transfers in high-speed networks*, IEEE Trans. Automat. Control, 40 (1995), pp. 1469–1471.

[125] ——, *Analysis and optimization of feedback control algorithms for data transfers in high-speed networks*, SIAM J. Control Optim., 34 (1996), pp. 1767–1780.

[126] V. JACOBSON, *Congestion avoidance and control*, ACM SIGCOMM Communication Rev., 18 (1988), pp. 314–329.

[127] E. JARLEBRING AND T. DAMM, *The lambert W function and the spectrum of multidimensional time-delay systems*, Automatica, to appear.

[128] S. JAYARAM, S. G. KAPOOR, AND R. E. DEVOR, *Analytical stability analysis of variable spindle speed machines*, J. Manufacturing Sci. Engrg., 122 (2000), pp. 391–397.

[129] T. KAILATH, *Linear Systems*, Prentice-Hall, Englewood Cliffs, NJ, 1980.

[130] E. W. KAMEN, *An operator theory of linear functional differential equations*, J. Differential Equations, 27 (1978), pp. 274–297.

[131] ——, *On the relationship between zero criteria for two-variable polynomials and asymptotic stability of delay differential equations*, IEEE Trans. Automat. Control, 25 (1980), pp. 983–984.

[132] ——, *Linear systems with commensurate time delays: Stability and stabilization independent of delay*, IEEE Trans. Automat. Control, 27 (1982), pp. 367–375.

[133] T. KATO, *Perturbation Theory for Linear Operators*, Springer Verlag, Berlin, 1966.

[134] F. P. KELLY, *Mathematical modelling of the internet*, in Mathematics Unlimited: 2001 and Beyond, B. Engquist and W. Schmid, eds., Springer Verlag, Berlin 2001, pp. 685–702.

[135] V. L. Kharitonov, S.-I. Niculescu, J. Moreno, and W. Michiels, *Static output feedback stabilization: Necessary conditions for multiple delay controllers*, IEEE Trans. Automat. Control, 50 (2005), pp. 82–86.

[136] M. C. Khoo, R. E. Kronauer, K. P. Strohl, and A. S. Slutsky, *Factors inducing periodic breathing in humans: A general model*, J. Appl. Physiology, 53 (1982), pp. 644–659.

[137] H. Kokame, K. Hirata, K. Konishi, and T. Mori, *Difference feedback can stabilize uncertain steady states*, IEEE Trans. Automat. Control, 46 (2001), pp. 1908–1913.

[138] V. B. Kolmanovskii and A. Myshkis, *Introduction to the Theory and Applications of Functional Differential Equations*, Math. Appl. 463, Kluwer Academic Publishers, Dordrecht, NL 1999.

[139] V. B. Kolmanovskii and V. R. Nosov, *Stability of Functional Differential Equations*, Math. Sci. Engrg. 180, Academic Press, London, 1986.

[140] A. M. Krall, *Stability criteria for feedback systems with a time lag*, SIAM J. Control, 3 (1965), pp. 160–170.

[141] S. G. Krantz, *Handbook of Complex Variables*, Birkhäuser, Boston, 1999.

[142] N. N. Krasovskii, *Stability and Motion: Applications of Lyapunov's Second Method to Differential Systems and Equations with Delay*, Stanford University Press, Stanford, CA, 1963.

[143] P. Kravanja and M. Van Barel, *Computing the Zeroes of Analytic Functions*, Lecture Notes in Math. 1727, Springer, New York, 2000.

[144] Y. Kuang, *Delay Differential Equations with Applications in Population Dynamics*, Academic Press, Boston, 1993.

[145] Y. A. Kuznetsov, *Elements of Applied Bifurcation Theory*, Appl. Math. Sci. 112, Springer, New York, 1998.

[146] W. H. Kwon and A. E Pearson, *Feedback stabilization of linear systems with delayed control*, IEEE Trans. Automat. Control, 25 (1980), pp. 266–269.

[147] M. S. Lee and C. S. Hsu, *On the τ-decomposition method of stability analysis for retarded dynamical systems*, SIAM J. Control, 7 (1969), pp. 242–259.

[148] B. Lehman, J. Bentsman, S. M. Verduyn Lunel, and E. I. Verriest, *Vibrational control of nonlinear time lag systems with bounded delay: Averaging theory, stabilizability, and transient behavior*, IEEE Trans. Automat. Control, 39 (1994), pp. 898–912.

[149] B. J. Levin, *A Distribution of Zeros of Entire Functions*, AMS, Providence, RI, 1964.

[150] A. S. Lewis and M. L. Overton, *Eigenvalue optimization*, Acta Numer., 5 (1996), pp. 149–190.

[151] X. Li and C. E. de Souza, *Criteria for robust stability and stabilization of uncertain linear systems with state delay*, Automatica, 33 (1997), pp. 1657–1662.

[152] ——, *Delay-dependent robust stability and stabilization of uncertain linear delay systems: A linear matrix inequality approach*, IEEE Trans. Automat. Control, 42 (1997), pp. 1144–1148.

[153] Z. Lin, *Low Gain Feedback*, Lecture Notes in Control and Inform. Sci. 240, Springer, New York, 1999.

[154] H. Logemann, *Destabilizing effects of small time-delays on feedback-controlled descriptor systems*, Linear Algebra Appl., 272 (1998), pp. 131–153.

[155] H. Logemann and R. Rebarber, *The effect of small time-delays on the closed-loop stability of boundary control systems*, Math. Control Signals Systems, 9 (1996), pp. 123–151.

[156] H. Logemann, R. Rebarber, and G. Weiss, *Conditions for robustness and non-robustness of the stability of feedback systems with respect to small delays in the feedback loop*, SIAM J. Control Optim., 34 (1996), pp. 572–600.

[157] H. Logemann and S. Townley, *The effect of small delays in the feedback loop on the stability of neutral systems*, Systems Control Lett., 27 (1996), pp. 267–274.

[158] J.-J. Loiseau and D. Brethé, *The use of 2D systems theory for the control of time-delay systems*, Journal Européen de Systèmes Automatisés, 31 (1997), pp. 1043–1058.

[159] J. Louisell, *Absolute stability in linear delay-differential systems: Ill-posedness and robustness*, IEEE Trans. Automat. Control, 40 (1995), pp. 1288–1291.

[160] S. Low, F. Paganini, and J. C. Doyle, *Internet congestion control*, IEEE Control Systems Magazine, 22 (2002), pp. 28–43.

[161] T. Luzyanina, K. Engelborghs, and D. Roose, *Computing stability of differential equations with bounded distributed delays*, Numer. Algorithms, 34 (2003), pp. 41–66.

[162] T. Luzyanina and D. Roose, *Equations with distributed delays: Bifurcation analysis using computational tools for discrete delay equations*, Funct. Differ. Equ., 11 (2004), pp. 87–92.

[163] T. Luzyanina, D. Roose, and K. Engelborghs, *Numerical stability analysis of steady state solutions of integral equations with distributed delays*, Appl. Numer. Math., 50 (2004), pp. 75–92.

[164] H. Ma and E. A. Butcher, *Stability of elastic columns with periodic retarded follower forces*, J. Sound Vib., 286 (2005), pp. 849–867.

[165] N. MacDonald, *Comments on a simplified analytical stability test for systems with delay*, IEE Proc. Control Theory Appl. D, 132 (1985), pp. 237–238.

[166] ——, *Two delays may not destabilize although either delay can*, Math. Biosci., 82 (1986), pp. 127–140.

[167] ——, *An interference effect of independent delays*, IEE Proc. Control Theory Appl. D, 134 (1987), pp. 38–42.

[168] ——, *Biological Delay Systems: Linear Stability Theory*, Cambridge University Press, Cambridge, 1989.

[169] F. MAGHAMI ASL AND A. G. ULSOY, *Analytical solution of a system of homogeneous delay differential equations via the Lambert function*, Proc. Amer. Control Conference, 4 (2000), pp. 2496–2500.

[170] M. S. MAHMOUD, *Robust Control and Filtering for Time-delay Systems*, Control Eng. 5, Marcel Dekker, New York, 2000.

[171] M. MALEK-ZAVAREI AND M. JAMSHIDI, *Time Delay Systems: Analysis, Optimization and Applications*, Systems and Control Series 9, North-Holland, Amsterdam, 1987.

[172] M. MAMMADOV AND R. ORSI, *$\mathcal{H}_\infty$ synthesis via a nonsmooth, nonconvex optimization approach*, Pacific J. Optim., 1 (2005), pp. 405–420.

[173] A. MANITIUS AND A. OLBROT, *Finite spectrum assignment problem for systems with delays*, IEEE Trans. Automat. Control, 24 (1979), pp. 541–553.

[174] S. MASCOLO, *Classical control theory for congestion avoidance in high speed internet*, in Proceedings of the 38th IEEE Conference on Decision and Control, Phoenix, AZ, 1999, pp. 2709–2714.

[175] M. R. MATAUŠEK AND A. D. MICIĆ, *On the modified Smith predictor for controlling a process with an integrator and long dead-time*, IEEE Trans. Automat. Control, 44 (1999), pp. 1603–1607.

[176] F. MAZENC, S. MONDIÉ, AND R. FRANCISCO, *Global asymptotic stabilization of feedforward systems with delay in the input*, IEEE Trans. Automat. Control, 49 (2004), pp. 844–850.

[177] F. MAZENC, S. MONDIÉ, AND S. I NICULESCU, *Global asymptotic stabilization for chains of integrators with a delay in the input*, IEEE Trans. Automat. Control, 48 (2003), pp. 57–63.

[178] F. MAZENC, S. MONDIÉ, AND S.-I. NICULESCU, *Global stabilization of oscillators with bounded delayed input*, Systems Control Lett., 53 (2004), pp. 415–422.

[179] V. MEHRMANN AND H. VOSS, *Nonlinear Eigenvalue Problems: A Challenge for Modern Eigenvalue Methods*, Report 83, Arbeitsbereich Mathematik, T.U. Hamburg-Harburg, GAMM Mitteilungen, 27:121–152, 2004.

[180] W. MICHIELS, *Stability and Stabilization of Time-delay Systems*, Ph.D. thesis, Katholieke Universiteit Leuven, Belgium, 2002.

[181] W. MICHIELS, K. ENGELBORGHS, D. ROOSE, AND D. DOCHAIN, *Sensitivity to infinitesimal delays in neutral equations*, SIAM J. Control Optim., 40 (2002), pp. 1134–1158.

[182] W. MICHIELS, K. ENGELBORGHS, P. VANSEVENANT, AND D. ROOSE, *The continuous pole placement method for delay equations*, Automatica, 38 (2002), pp. 747–761.

[183] W. MICHIELS, K. GREEN, T. WAGENKNECHT, AND S.-I. NICULESCU, *Pseudospectra and stability radii for analytic matrix functions with application to time-delay systems*, Linear Algebra Appl., 418 (2006), pp. 315–335.

[184] W. MICHIELS, D. MELCHOR-AGUILAR, AND S.-I. NICULESCU, *Stability analysis of some classes of TCP/AQM networks*, Internat. J. Control, 79 (2005), pp. 1134–1144.

[185] W. MICHIELS, S. MONDIÉ, AND D. ROOSE, *Robust Stabilization of Time-delay Systems with Distributed Delay Control Laws: Necessary and Sufficient Conditions for a Safe Implementation*, TW Report 363, Department of Computer Science, Katholieke Universiteit Leuven, Belgium, 2003.

[186] W. MICHIELS, S. MONDIÉ, D. ROOSE, AND M. DAMBRINE, *The effect of approximating distributed delay control laws on stability*, in Advances in Time-Delay Systems, Lecture Notes in Comput. Sci. Engrg. 38, Springer, New York, 2004, pp. 207–225.

[187] W. MICHIELS, C. I. MORĂRESCU, AND S.-I. NICULESCU, *Consensus Problems with Distributed Delays, with Application to Traffic Flow Models*, TW report, Department of Computer Science, Katholieke Universiteit Leuven, Belgium, 2007.

[188] W. MICHIELS AND S.-I. NICULESCU, *On the delay sensitivity of Smith predictors*, Internat. J. Systems Sci., 34 (2003), pp. 543–552.

[189] ——, *Stability analysis of a fluid flow model for TCP like behavior*, Internat. J. Bifur. Chaos Appl. Sci. Engrg., 15 (2005), pp. 2277–2282.

[190] ——, *Characterization of delay-independent stability and delay-interference phenomena*, SIAM J. Control Optim., 45 (2007), pp. 2138–2155.

[191] W. MICHIELS, S.-I. NICULESCU, AND L. MOREAU, *Using delays and time-varying gains to improve the output feedback stabilizability of linear systems: A comparison*, IMA J. Math. Control Inform., 21 (2004), pp. 393–418.

[192] W. MICHIELS, T. PLOMTEUX, AND D. ROOSE, *Robust stabilization of linear time delay systems via the optimization of real stability radii*, in Proceedings of the 16th IFAC World Congress, Prague, Czech Republic, 2005.

[193] W. MICHIELS AND D. ROOSE, *Global stabilization of multiple integrators with time delay and input constraints*, in Proceedings of the 3rd IFAC Workshop on Time-Delay Systems, Santa Fe, NM, 2001, pp. 266–271.

[194] ——, *Time-delay compensation in unstable plants using delayed state feedback*, in Proceedings of the 40th IEEE Conference on Decision and Control, Orlando, FL, 2001, pp. 1433–1437.

[195] ——, *Limitations of delayed state feedback: A numerical study*, Internat. J. Bifur. Chaos Appl. Sci. Engrg., 12 (2002), pp. 1309–1320.

[196] ——, *An eigenvalue based approach for the robust stabilization of linear time-delay systems*, Internat. J. Control, 76 (2003), pp. 678–686.

[197] W. MICHIELS, V. VAN ASSCHE, AND S.-I. NICULESCU, *Stabilization of time-delay systems with a controlled, time-varying delay and applications*, IEEE Trans. Automat. Control, 50 (2005), pp. 493–504.

[198] W. MICHIELS, K. VERHEYDEN, AND S.-I. NICULESCU, *Mathematical and computational tools for the stability analysis of time-varying delay systems and applications in mechanical engineering*, in Applications of Time-Delay Systems, J. Chiasson and J. J. Loiseau, eds., Lecture Notes in Control and Inform. Sci. 352, Springer, New York, 2007.

[199] W. MICHIELS AND T. VYHLÍDAL, *An eigenvalue based approach for the stabilization of linear time-delay systems of neutral type*, Automatica, 41 (2005), pp. 991–998.

[200] L. MIRKIN AND Q.-C. ZHONG, *Are distributed delay control laws intrinsically unapproximable?*, in Proceedings of the 4th IFAC Workshop on Time-Delay Systems, Rocquencourt, France, 2003.

[201] V. MISRA, W. B. GONG, AND D. TOWSLEY, *Fluid based analysis of a network of AQM routers supporting TCP flows with an application to RED*, Computer Commun. Rev., 30 (2000), pp. 151–162.

[202] S. MONDIÉ, M. DAMBRINE, AND O. SANTOS, *Approximation of control laws with distributed delays: A necessary condition for stability*, Kybernetica, 38 (2002), pp. 541–551.

[203] S. MONDIÉ AND W. MICHIELS, *Finite spectrum assignment of unstable time-delay systems with a safe implementation*, IEEE Trans. Automat. Control, 48 (2003), pp. 2207–2212.

[204] S. MONDIÉ, S.-I. NICULESCU, AND J. J. LOISEAU, *Delay robustness of closed loop finite assignment input delay systems*, in Proceedings of the 3rd IFAC Workshop on Time-Delay Systems, Santa Fe, NM, 2001.

[205] C. I. MORĂRESCU, *Qualitative Analysis of Distributed Delay Systems: Methodology and Algorithms*, Ph.D. thesis, University of Bucharest, Romania / Université de Technologie de Compiègne, France, 2006.

[206] C.-I. MORĂRESCU AND S.-I. NICULESCU, *Stability crossing curves of SISO systems controlled by delayed output feedback*, Dyn. Contin. Discrete Impuls. Syst., to appear.

[207] C. I MORĂRESCU, S.-I. NICULESCU, AND K. GU, *Some remarks on Smith predictors: A geometric point of view*, in Proceedings of the 6th IFAC Workshop on Time-Delay Systems, L'Aquila, IT, 2006.

[208] C.-I. MORĂRESCU, S.-I. NICULESCU, AND K. GU, *On the geometry of stability regions of Smith predictors subject to delay uncertainty*, IMA J. Math. Control Inform., to appear.

[209] C. I. MORĂRESCU, S.-I. NICULESCU, AND W. MICHIELS, *Asymptotic stability of some distributed delay systems: An algebraic approach*, Internat. J. Tomography Statist., 7 (FO7) (2007), pp. 128–133.

[210] L. MOREAU AND D. AEYELS, *Trajectory-based global and semi-global stability results*, in Modern Applied Mathematics Techniques in Curcuits, Systems and Control, N. E. Mastorakis, ed., World Scientific and Engineering Society Press, Athens, Greece, 1999, pp. 71–76.

[211] ——, *Practical stability and stabilization*, IEEE Trans. Automat. Control, 45 (2000), pp. 1554–1558.

[212] L. MOREAU, W. MICHIELS, D. AEYELS, AND D. ROOSE, *Robustness of nonlinear delay differential equations w.r.t. bounded input perturbations: A trajectory based approach*, Math. Control Signals Systems, 15 (2002), pp. 316–335.

[213] J. MORENO, *An extension of Lucas theorem to entire functions*, in Proceedings of the 1st IFAC Workshop on Linear Time-Delay Systems, Grenoble, France, 1998, pp. 159–163.

[214] Ö. MORGÜL, *On the stabilization and stability robustness against small delays of some damped wave equations*, IEEE Trans. Automat. Control, 40 (1995), pp. 1626–1630.

[215] T. MORI AND H. KOKAME, *Stability of* $\dot{x}(t) = Ax(t) + Bx(t-\tau)$, IEEE Trans. Automat. Control, 34 (1989), pp. 460–462.

[216] J. MORO, J. V. BURKE, AND M. L. OVERTON, *On the Lidskii-Vishik-Lyusternik perturbation theory for eigenvalues with arbitrary Jordan structure*, SIAM J. Matrix Anal. Appl., 18 (1997), pp. 793–817.

[217] J. D. MURRAY, *Mathematical Biology*, BioMathematics 18, 2nd. ed., Springer Verlag: Berlin, 1993.

[218] R. M. MURRAY, *Control in an information rich world*, Panel on Future Directions in Control, Dynamics, and Systems, Pasadena, CA, 2002; also available online from http://www.cds.caltech.edu/m̃urray/cdspanel.

[219] A. D. MYSHKIS, *General theory of differential equations with delay*, Uspehi, Mat. Nauk, 4 (1949), pp. 99–141; AMS 55 (1951), pp 1–62 (in English).

[220] J. NEIMARK, *D-subdivisions and spaces of quasi-polynomials*, Prikl. Mat. Meh., 13 (1949), pp. 349–380.

[221] S.-I. NICULESCU, *$\mathcal{H}_\infty$ memoryless control with an α-stability constraint for time-delay systems: A LMI approach*, IEEE Trans. Automat. Control, 43 (1998), pp. 739–743.

[222] ——, *Stability and hyperbolicity of linear systems with delayed state: A matrix pencil approach.*, IMA J. Math. Control Inform., 15 (1998), pp. 331–347.

[223] ——, *Delay Effects on Stability: A Robust Control Approach*, Lecture Notes in Control and Inform. Sci. 269, Springer, New York, 2001.

[224] ——, *On delay robustness analysis of a simple control algorithm in high-speed networks*, Automatica, 38 (2002), pp. 885–889.

[225] ——, *On interference phenomena in stability of linear systems with multiple delays*, in Proceeding of 4th Asian Control Conference, Singapore, 2002.

[226] ——, *On some stability regions of a multiple delays linear systems with applications*, in Proceedings of the 15th IFAC World Congress, Barcelona, Spain, 2002.

[227] S.-I. NICULESCU AND C. T. ABDALLAH, *Delay effects on static output feedback stabilization*, in Proceedings of the 39th IEEE Conference on Decision and Control, Sydney, Australia, 2000.

[228] S.-I. NICULESCU, J.-M. DION, AND L. DUGARD, *Robust stabilization for uncertain time-delay systems containing saturating actuators*, IEEE Trans. Automat. Control, 41 (1996), pp. 742–747.

[229] S.-I NICULESCU, P. FU, AND J. CHEN, *On the stability switches and reversals of linear systems with commensurate delays: A matrix pencil characterization*, in Proceedings of the 16th IFAC World Congress, Prague, Czech Republic, 2005.

[230] S.-I. NICULESCU, K. GU, AND C. T. ABDALLAH, *Some remarks on the delay stabilizing effect in SISO systems*, in Proceedings of the American Control Conference, Denver, CO, 2003.

[231] S.-I. NICULESCU, P. KIM, K. GU, AND D. LEVY, *On the stability crossing boundaries of some delay systems modeling immune dynamics in leukemia*, in Proceedings of the 17th International Symposium on Mathematical Theory of Networks and Systems, Kyoto, Japan, 2006.

[232] S.-I. NICULESCU AND W. MICHIELS, *Stabilizing a chain of integrators using multiple delays*, IEEE Trans. Automat. Control, 49 (2004), pp. 802–807.

[233] S.-I. NICULESCU, W. MICHIELS, K. GU, AND C. T. ABDALLAH, *Delay effects on output feedback control of dynamical systems*, in Complex Time-Delay Systems, F. M. Atay, ed., Lecture Notes in Control and Inform. Sci., Springer, New York, to appear.

[234] S. I. NICULESCU, W. MICHIELS, D. MELCHOR-AGUILAR, T. LUZYANINA, F. MAZENC, K. GU, AND F. CHATTÉ, *Delay effects on the asymptotic stability of various fluid models in high-performance networks*, in Adv. Communication Control Networks 308, Springer-Verlag, Heidelberg, 2004, pp. 87–110.

[235] S.-I. NICULESCU, E. I. VERRIEST, L. DUGARD, AND J.-M. DION, *Stability of linear systems with delayed state: A guided tour*, in Stability and Control of Time-Delay Systems, L. Dugard and E. I. Verriest, eds., Lecture Notes in Control and Inform. Sci. 228, Springer, New York, 1998, pp. 1–71.

[236] D. NOLL AND P. APKARIAN, *Spectral bundle methods for non-convex maximum eigenvalue functions. Part 1: First order methods*, Math. Program. Series B, 104 (2005), pp. 729–747.

[237] J. E. NORMEY-RICO AND E. F. CAMACHO, *Robust tuning of dead-time compensators for processes with an integrator and long dead-time*, IEEE Trans. Automat. Control, 44 (1999), pp. 1597–1603.

[238] T. OGUCHI AND H. NIJMEIJER, *Prediction of chaotic behavior*, IEEE Trans. Circuits Systems I, 52 (2005), pp. 2464–2472.

[239] A. OLBROT, *Stabilizability, detectability and spectrum assignment for linear autonomous systems with general time delays*, IEEE Trans. Automat. Control, 23 (1978), pp. 605–618.

[240] R. OLFATI-SABER AND R. M. MURRAY, *Consensus problems in networks of agents with switching topology and time-delays*, IEEE Trans. Automat. Control, 49 (2004), pp. 1520–1533.

[241] N. OLGAC AND R. SIPAHI, *An exact method for the stability analysis of time-delayed linear time-invariant (LTI) systems*, IEEE Trans. Automat. Control, 47 (2002), pp. 793–797.

[242] A. V. OPPENHEIM AND R. W. SCHAFER, *Digital Signal Processing*, Prentice-Hall, Englewood Cliffs, NJ, 1975.

[243] G. OROSZ, B. KRAUSKOPF, AND R. E. WILSON, *Bifurcations and multiple traffic jams in a car-following model with reaction-time delay*, Phys. D, 211 (2005), pp. 277–293.

[244] G. OROSZ AND G. STÉPÁN, *Subcritical Hopf bifurcations in a car-following model with reaction-time delay*, Proc. Roy. Soc. London A, 462 (2006), pp. 2643–2670.

[245] A. PACKARD AND J. C. DOYLE, *The complex structured singular value*, Automatica, 29 (1993), pp. 71–109.

[246] A. PACKARD, M. FAN, AND J. C. DOYLE, *A power method for the structured singular value*, in Proceedings of the 27th IEEE Conference on Control and Decision, Austin, TX, 1988, pp. 2132–2137.

[247] Z. J. PALMOR, *Stability properties of Smith dead-time compensator controllers*, Internat. J. Control, 32 (1980), pp. 937–949.

[248] ——, *Time-delay compensation: Smith predictor and its modifications*, in The Control Handbook, S. Levine, ed., CRC and IEEE Press, New York, 1996, pp. 224–237.

[249] L. PANDOLFI, *Stabilization of neutral functional differential equations*, J. Optim. Theory Appl., 20 (1976), pp. 191–204.

[250] G. PAPPAS AND D. HINRICHSEN, *Robust stability of linear systems described by higher order dynamic equations*, IEEE Trans. Automat. Control, 38 (1993), pp. 1430–1435.

[251] E. PINNEY, *Ordinary Differential-Difference Equations*, University of California Press, Berkeley, CA, 1958.

[252] E. P. POPOV, *The Dynamics of Automatic Control Systems*, Pergamon Press, New York, 1962.

[253] K. PYRAGAS, *Continuous control of chaos by self-controlling feedback*, Phys. Lett. A, 170 (1992), pp. 421–428.

[254] L. QIU, B. BERNHARDSSON, A. RANTZER, E. DAVISON, P. YOUNG, AND J. DOYLE, *A formula for computation of the real stability radius*, Automatica, 31 (1995), pp. 879–890.

[255] V. RĂSVAN, *Absolute stability of time lag control systems*, Ed. Academiei, Bucharest, 1975 (in Romanian); Russian revised edition, Nauka, Moscow, 1983.

[256] V. RĂSVAN AND D. POPESCU, *Control of systems with input delay: An elementary approach*, in Proceedings of the 1st CNRS-NSF Workshop, Advances in Time-Delay Systems, Paris, France, 2003.

[257] Z. V. REKASIUS, *A stability test for systems with delays*, in Proceedings of 1980 Joint Automatic Control Conference, Vol. TP9-A, San Francisco, CA, 1980.

[258] J.-P. RICHARD, *Time-delay systems: An overview of recent advances and open problems*, Automatica, 39 (2003), pp. 1667–1694.

[259] D. ROOSE, T. LUZYANINA, K. ENGELBORGHS, AND W. MICHIELS, *Software for stability and bifurcation analysis of delay differential equations and application to stabilization*, in Advances in Time-Delay Systems, Lecture Notes in Comput. Sci. Engrg. 38, Springer Verlag, 2004, pp. 167–182.

[260] O. RÖSCH, H. ROTH, AND S.-I. NICULESCU, *Remote control of mechatronic systems over communication networks*, in Proceedings of the IEEE International Conference on Mechatronics & Automation, Niagara Falls, CA, 2005.

[261] R. W. ROTHERY, *Car following models*, in Traffic Flow Theory: A State-of-the-Art Report, H. N. Gartner, C. J. Messner, and A. J. Rathi, eds., 2nd ed., 1998; also available online from http://www.tfhrc.gov/its/tft/tft.htm.

[262] M. R. ROUSSEL, *The use of delay differential equations in chemical kinetics*, J. Phys. Chemistry, 100 (1996), pp. 8323–8330.

[263] W. RUDIN, *Functional Analysis*, McGraw-Hill, New York, 1973.

[264] A. SABERI, A. A. STOORVOGEL, AND P. SANNUTI, *Control of Linear Systems with Regulation and Input Constraints*, Comm. Control Engrg. Ser., Springer, Berlin, 2000.

[265] J. A. SANDERS AND F. VERHULST, *Averaging Methods in Nonlinear Dynamical Systems*, Appl. Math. Sci. 59, Springer, New York, 1985.

[266] C. SANTACESARIA AND R SCATTOLINI, *Easy tuning of Smith predictor in presence of delay uncertainty*, Automatica, 29 (1993), pp. 1595–1597.

[267] O. SENAME, *Sur la commandabilité et le découplage des systèmes linéaires à retards*, Ph.D. thesis, Université de Nantes - Ecole Centrale de Nantes, France, 1994.

[268] R. SEPULCHRE, *Slow peaking and low-gain designs for global stabilization of nonlinear systems*, IEEE Trans. Automat. Control, 45 (2000), pp. 453–461.

[269] J. SEXTON AND B. STONE, *The stability of machining with continuously varying spindle speed*, Ann. CIRP, 27 (1978), pp. 321–326.

[270] R. SEYDEL, *Practical Bifurcation and Stability Analysis: From Equilibrium to Chaos*, Interdiscip. Appl. Math. 5, Springer, New York, 1994.

[271] A. P. SEYRANIAN AND A. A. MAILYBAEV, *Interaction of eigenvalues in multiparameter problems*, J. Sound Vib., 267 (2003), pp. 1047–1064.

[272] S. SHAKKOTTAI AND R. SRIKANT, *How good are deterministic fluid models of internet congestion control?*, in Proceedings of the IEEE INFOCOM 2002, New York, NY, 2002.

[273] G. J. SILVA, A. DATTA, AND S. P. BHATTACHARRYA, *New results on the synthesis of PID controllers*, IEEE Trans. Automat. Control, 47 (2002), pp. 241–252.

[274] G. J. SILVA, A. DATTA, AND S. P. BHATTACHARYYA, *PID Controllers for Time-Delay Systems*, Birkhäuser, Boston, 2005.

[275] J. G. SIMMONDS AND J. E. MANN, JR, *A First Look at Perturbation Theory*, 2nd. ed., Dover, Mineola, NY, 1998.

[276] R. SIPAHI, *Cluster Treatment of Characteristic Roots, CTCR: A Unique Methodology for the Complete Stability Robustness Analysis of Linear Time Invariant Multiple Time Delays against Delay Uncertainties*, Ph.D. thesis, University of Connecticut, Mechanical Engineering Department, 2005.

[277] R. SIPAHI AND S.-I. NICULESCU, *Analytical stability study of a deterministic car following model under multiple delay interactions*, in Proceedings of the 6th IFAC Workshop on Time Delay Systems, L'Aquila, IT, 2006.

[278] ——, *Deterministic models of traffic flow dynamics under time delay influences: A survey*, in Complex Time-Delay Systems, F.M. Atay, ed., Lecture Notes in Control and Inform. Sci., Springer, New York, to appear.

[279] R. Sipahi and N. Olgac, *Complete stability robustness of third-order LTI multiple time-delay systems*, Automatica, 41 (2005), pp. 1413–1422.

[280] H. L. Smith and P. Waltman, *The Theory of the Chemostat*, Cambridge University Press, Cambridge, 1994.

[281] O. J. Smith, *Closer control of loops with dead time*, Chemical Engrg. Progress, 53 (1957), pp. 217–219.

[282] J. Sreedhar, P. Van Dooren, and A. L. Tits, *A fast algorithm to compute the real structured stability radius*, in Stability Theory: Hurwitz Centenary Conference, R. Jeltsch and M. Mansour, eds., Intern. Ser. Numer. Math. 121, Birkhäuser, Basel, 1996, pp. 219–230.

[283] R. Sridhar, R. E. Hohn, and G. W. Long, *A general formulation of the milling process equation*, ASME J. Engrg. Industry, 90 (1968), pp. 317–324.

[284] R. Srikant, *Models and methods for analyzing Internet congestion control algorithms*, in Advances in Communication Control Networks, Lectures Notes in Control and Inform. Sci. 308, Springer, New York, 2005, pp. 65–84.

[285] Editorial Staff, *The damping effect of time lag*, Engineer, 163 (1937), p. 439.

[286] G. Stépán, *Retarded Dynamical Systems: Stability and Characteristic Functions*, Res. Notes Math. 210, Longman Scientific, London, 1989.

[287] H. J. Sussmann and P. V. Kokotovic, *The peaking phenomenon and the global stabilization of nonlinear systems*, IEEE Trans. Automat. Control, 36 (1991), pp. 424–440.

[288] V. L. Syrmos, C. T. Abdallah, P. Dorato, and K. Grigoriadis, *Static output feedback—A survey*, Automatica, 33 (1997), pp. 125–137.

[289] R. Szalai, *PDDE-CONT: A Continuation and Bifurcation Software for Delay Differential Equations*, Technical report, Department of Applied Mechanics, Budapest University of Technology and Economics, 2006.

[290] M. Szydlowski and A. Krawiec, *Differential delay equations in chemical kinetics: Some simple linear model systems*, J. Chemical Phys., 92 (1990), pp. 1702–1712.

[291] ———, *The Kaldor-Kalecki model of business cycles as a two-dimensional dynamical system*, J. Nonlinear Math. Phys., 8 (2001), pp. 266–271.

[292] S. Tarbouriech, C. T. Abdallah, and J. N. Chiasson, eds., *Advances in Communication Control Networks*, Lecture Notes in Control and Inform. Sci. 308, Springer, New York, 2005.

[293] A. R. Teel, *Global stabilization and restricted tracking for multiple integrators with bounded controls*, Systems Control Lett., 18 (1992), pp. 165–171.

[294] A. THOWSEN, *An analytical stability test for a class of time-delay systems*, IEEE Trans. Automat. Control, 25 (1981), pp. 735–736.

[295] ——, *The Routh-Hurwitz method for stability test for a class of time-delay systems*, Internat. J. Control, 33 (1981), pp. 991–995.

[296] Y. TIAN AND F. GAO, *Adaptive control of chaotic continuous-time systems with delay*, Phys. D, 117 (1998), pp. 1–12.

[297] F. TISSEUR AND N. J. HIGHAM, *Structured pseudospectra for polynomial eigenvalue problems with applications*, SIAM J. Matrix Anal. Appl., 23 (2001), pp. 187–208.

[298] F. TISSEUR AND K. MEERBERGEN, *The quadratic eigenvalue problem*, SIAM Rev., 43 (2001), pp. 235–286.

[299] O. TOKER AND H. ÖZBAY, *On the $\mathcal{NP}$-hardness of the purely complex μ computation, analysis/synthesis, and some related problems in multidimensional systems*, in Proceedings of the 1995 American Control Conference, Seattle, WA, 1995, pp. 447–451.

[300] ——, *Complexity issues in robust stability of linear delay-differential systems*, Math. Control Signals Systems, 9 (1996), pp. 386–400.

[301] L. N. TREFETHEN, *Pseudospectra of linear operators*, SIAM Rev., 39 (1997), pp. 383–406.

[302] ——, *Spectral Methods in Matlab*, Software Environ. Tools 10, SIAM, Philadelphia, 2000.

[303] L. N. TREFETHEN AND M. EMBREE, *Spectra and Pseudospectra: The Behavior of Nonnormal Matrices and Operators*, Princeton University Press, Princeton, NJ, 2005.

[304] V. VAN ASSCHE, M. DAMBRINE, J.-F. LAFAY, AND J.-P. RICHARD, *Implementation of a distributed control law for a class of systems with delay*, in Proceedings of the 3nd IFAC Workshop on Time Delay Systems, Santa Fe, NM, 2001, pp. 266–271.

[305] V. VAN ASSCHE, A. GANGULI, AND W. MICHIELS, *Practical stability analysis of systems with delay and periodic coefficients*, in Proceedings of the 5th IFAC Workshop on Time-Delay Systems, Leuven, Belgium, 2004.

[306] P. VAN DOOREN AND V. VERMAUT, *On stability radii of generalized eigenvalue problems*, in Proceedings of the 1997 European Control Conference, Brussels, Belgium, 1997.

[307] G. H. M. VAN TARTWIJK AND D. LENSTRA, *Semiconductor lasers with optical injection and feedback*, J. Opt. B Quantum Semiclass. Opt., 7 (1995), pp. 87–143.

[308] J. VANBIERVLIET, K. VERHEYDEN, W. MICHIELS, AND S. VANDEWALLE, *A nonsmooth optimization approach for the stabilization of linear time-delay systems*, ESAIM Control Optim. Calc. Var., to appear.

[309] A. Veres and M. Boda, *The chaotic nature of TCP-AQM congestion control*, in Proceedings of the IEEE INFOCOM 2000, Tel Aviv, Israel, 2000.

[310] K. Verheyden, K. Green, and D. Roose, *Numerical stability analysis of a large-scale delay system modelling a lateral semiconductor laser subject to optical feedback*, Phys. Rev. E, 69 (2004), 036702.

[311] K. Verheyden and K. Lust, *A Newton-Picard collocation method for periodic solutions of delay differential equations*, BIT, 45 (2005), pp. 605–625.

[312] K. Verheyden, K. Lust, and D. Roose, *Computation and stability analysis of solutions of periodic delay differential algebraic equations*, in Proc. ASME Internat. Design Engrg. Tech. Conferences Computers Infor. Engrg. Conf., Long Beach, CA, 2005.

[313] K. Verheyden, T. Luzyanina, and D. Roose, *Efficient computation of characteristic roots of DDE systems using LMS methods*, TW Report 383, Department of Computer Science, Katholieke Universiteit Leuven, Belgium, 2004.

[314] K. Verheyden and D. Roose, *Efficient numerical stability analysis of delay equations: A spectral method*, in Proceedings of the 5th IFAC Workshop on Time-Delay Systems, Leuven, Belgium, 2004.

[315] E. I. Verriest and W. Michiels, *Inverse Routh table construction and stability of delay equations*, Systems Control Lett., 55 (2006), pp. 711–718.

[316] B. Vielle and G. Chauvet, *Delay equation analysis of human respiratory stability*, Math. Biosci., 152 (1998), pp. 105–122.

[317] M. I. Vishik and L. A. Lyusternik, *The solution of some perturbation problems for matrices and selfadjoint or non-selfadjoint differential equations*, Russian Math. Surveys, 15 (1960), pp. 1–74.

[318] Vyhlídal, T., *Analysis and synthesis of time delay system spectrum*, Ph.D. thesis, Department of Mechanical Engineering, Czech Technical University in Prague, 2003.

[319] T. Wagenknecht, K. Green, S. Adhikari, and W. Michiels, *Structured pseudospectra and random eigenvalue problems in vibrating systems*, AIAA Journal, 44 (2006), pp. 2404–2414.

[320] T. Wagenknecht and J. Agarwal, *Structured pseudospectra in structural engineering*, Internat. J. Numer. Methods Engrg., 64 (2005), pp. 1735–1751.

[321] T. Wagenknecht, W. Michiels, and K. Green, *Structured pseudospectra for nonlinear eigenvalue problems*, J. Comput. Appl. Math., to appear.

[322] K. Walton and J.E. Marshall, *Direct method for TDS stability analysis*, IEE Proceedings, 134, Part D (1987), pp. 101–107.

[323] Q. G. WANG, T. H. LEE, AND K. K. TAN, *Finite Spectrum Assignment for Time-Delay Systems*, Lecture Notes in Control and Inform. Sci. 239, Springer Verlag, New York 1999.

[324] K. WATANABE, *Finite spectrum assignment and observer for multivariable systems with commensurate delays*, IEEE Trans. Automat. Control, 31 (1986), pp. 543–550.

[325] K. WATANABE AND M. ITO, *A process-model control for linear systems with delay*, IEEE Trans. Automat. Control, 26 (1981), pp. 1261–1269.

[326] W. M. WONHAM, *Linear multivariable control: a geometric approach*, Appl. Math., N.Y. 10, Springer Verlag, New York 1979.

[327] K. YAMANAKA AND E. SHIMEMURA, *Effects of mismatched Smith controller on stability in systems with time-delay*, Automatica, 23 (1987), pp. 787–791.

[328] Q.-C. ZHONG, *Robust stability analysis of simple systems controlled over communication networks*, Automatica, 39 (2003), pp. 1309–1312.

[329] ——, *Robust control of time-delay systems*, Springer-Verlag, London, 2006.

[330] K. ZHOU, J. C. DOYLE, AND K. GLOVER, *Robust and Optimal Control*, Prentice-Hall, Upper Saddle River, NJ, 1995.

[331] P. ZÍTEK AND T. VYHLÍDAL, *State feedback control of time delay system: Conformal mapping aided design*, in Proceedings of the 2nd IFAC Workshop on Linear Time Delay Systems, Ancona, Italy, 2000, pp. 146–151.

Index